With Stars in Their Eyes

With Stars in Their Eyes

The Extraordinary Lives and Enduring Genius of Aden and Marjorie Meinel

JAMES B. BRECKINRIDGE
AND ALEC M. PRIDGEON

With an invited chapter by

DONALD E. OSBORN

OXFORD
UNIVERSITY PRESS

OXFORD
UNIVERSITY PRESS

Oxford University Press is a department of the University of Oxford. It furthers
the University's objective of excellence in research, scholarship, and education
by publishing worldwide. Oxford is a registered trade mark of Oxford University
Press in the UK and certain other countries.

Published in the United States of America by Oxford University Press
198 Madison Avenue, New York, NY 10016, United States of America.

Library of Congress Cataloging-in-Publication Data
Names: Breckinridge, Jim B. (Jim Bernard), 1939– author. | Pridgeon, Alec M., author.
Title: With stars in their eyes : the extraordinary lives and enduring
genius of Aden and Marjorie Meinel / James B. Breckinridge and Alec M. Pridgeon.
Description: New York, NY : Oxford University Press, [2022] |
Includes bibliographical references and index.
Identifiers: LCCN 2021047556 (print) | LCCN 2021047557 (ebook) |
ISBN 9780190915674 (hardback) | ISBN 9780190915698 (epub)
Subjects: LCSH: Meinel, Aden B. | Meinel, Marjorie Pettit. |
Astronomers—Biography. | Space astronomy—Instruments.
Classification: LCC QB36.M46 B74 2022 (print) | LCC QB36.M46 (ebook) |
DDC 520.92/2 [B]—dc23/eng/20211108
LC record available at https://lccn.loc.gov/2021047556
LC ebook record available at https://lccn.loc.gov/2021047557

DOI: 10.1093/oso/9780190915674.001.0001

1 3 5 7 9 8 6 4 2

Printed by Sheridan Books, Inc., United States of America

Contents

Foreword

I first met Aden Meinel in 1972 at a meeting of the New England Section of the Optical Society of America (OSA) in Boston, Massachusetts. At that time, Aden was the director of the up-and-coming Optical Sciences Center (OSC) at the UA in Tucson, and he had come to Boston to share his vision for the future of optics and optics education. I was immediately impressed by his knowledge as well as his enthusiasm and foresight. By the end of the meeting, I had "desert fever" and accepted a job at OSC two years later.

Now, almost 50 years later, I am pleased to write this foreword—not only because I have known Jim Breckinridge since the mid-1970s, but because I truly believe there is no one more qualified to write the Meinels' remarkable story. Jim's personal relationship with Aden and Marjorie—first as Aden's student at OSC, then later as Aden and Marjorie's friend and colleague at the Jet Propulsion Laboratory (JPL) in Pasadena, California—gave him the insight to not only write their history but also reflect on their character and the role it played in their success.

Spanning many fields of science and several of the world's most important historical events, *With Stars in Their Eyes* is a biography that will appeal to many different groups of readers. Jim and his coauthor, Alec Pridgeon, weave science, history, and Aden and Marjorie's devotion to each other into a story many will enjoy. The diversity of the authors' backgrounds brings fresh perspective to the breadth of interests and contributions of the Meinels. I believe this book will become a must-read for those wishing to understand more fully the long, storied history of the Optical Sciences Center that started with the incredible vision of one man, Aden Meinel, but also for those with interests including history, astronomy, engineering, and solar energy.

James C. Wyant
Emeritus Dean and Professor of Optical Sciences
College of Optical Sciences, University of Arizona
Tucson, Arizona
January 2021

Preface

This biography traces the professional lives of Aden and Marjorie Meinel from their fascinating ancestry through the Depression, World War II, and advanced academic degrees to careers at Yerkes Observatory, founding of the Kitt Peak National Observatory, Steward Observatory, and the founding of the College of Optical Sciences on the campus of the University of Arizona (UA), and finally to NASA's Jet Propulsion Laboratory in Pasadena. Interspersed with the academic posts were groundbreaking publications on solar energy and also high-level consultancy on aerial surveillance (spy planes and satellites) during the Cold War. Aden's knowledge of rockets and optics placed him in high demand in Washington, DC, and elsewhere around the world, to the extent that he was once picked up in a Colorado pasture by helicopter while on vacation and then flown in a Lockheed JetStar from Durango to an urgent top-secret meeting in Washington.

Aden and Marjorie cultivated several fields of science and technology to produce significant changes to society during the 20th century. Born four years after World War I into middle-class families in Pasadena, California, they used rocket science, astrophysics, optical science, mechanical engineering, and renewable solar energy to change our world. Aden's charismatic leadership, brilliant mind, and amazing ability to see, analyze, and dissect entire complicated systems long before others served him well.

Alec Pridgeon first became interested in the Meinels at a University of Michigan fraternity reunion in 2010. In conversations with his former roommate and mathematics savant, Gerald (Jerry) Newport, Pridgeon learned of the accomplishments and exploits of Jerry's father-in-law, astronomer Aden Meinel, and decided to present a talk on the Meinels to his local astronomy society. In the course of research, he interviewed, among others, Aden's surviving sons Ed and David in Henderson, Nevada, and then drove down to Pasadena, California, to speak with Jim Breckinridge at the Caltech Athenaeum over lunch and persuade him to coauthor a biography of Aden and Marjorie. Years of research into all professional aspects of their lives then followed, fortunately before the Covid-19 pandemic temporarily closed down so many universities and institutions.

This book is based on over 200 hours of oral interviews with more than 30 persons who knew Aden and Marjorie or were familiar with their work. Over the course of more than four years, we conducted research at the National Archives in Washington, DC, and College Park, Maryland, University of California

(UC) Santa Cruz (Lick Observatory), UC Berkeley (Bancroft Library), Caltech, Huntington Library (Mt. Wilson Observatory), University of Chicago, and four Presidential Libraries.

The authors brought diverse backgrounds to the project, which was probably necessary to cover the breadth of interests and contributions of the Meinels. Aden was Breckinridge's master of science advisor at the Optical Sciences Center, UA. Later, both Aden and Marjorie worked in Breckinridge's optics section at NASA/JPL for eight years after Aden retired from UA. Alec's interest in military history, the Cold War, and genealogy made him the ideal person to research and write about the Meinel/Pettit ancestors and Aden's significant roles in the US Air Force and the CIA. Don Osborn was a young engineering student at UA in 1970 looking for a cause when he met Aden and Marjorie. Aden's inspirational and charismatic enthusiasm for solar energy soon drew the young engineer into a career in renewable solar thermal energy. Today, Osborn is president of a major solar thermal energy company, Spectrum Energy Development.

Events in the lives of Aden and Marjorie are generally laid out here in chronological order, but because Aden was involved in so many projects simultaneously in the 1960s and 1970s, for the sake of clarity we have devoted separate chapters to solar work and government consultant work (Chapters 10 and 11, respectively). To respect the privacy of their children, five of whom are still living at this writing, we have focused solely on the ancestry and professional careers of Aden and Marjorie.

The title of the book has a double meaning. One refers to their shared passion for astronomy and optics that dominated their public lives, and the other to the profound love and support for each other in their 63+ years of marriage before Marjorie died in 2008. Theirs was a partnership in every possible way. Marjorie's principal role early in the marriage was raising seven children, but then later she served as Aden's research assistant and coauthored the vast majority of their publications. She assumed a pivotal role in their solar energy research and testimony before congressional energy subcommittees. Growing up she had assisted her father, Mt. Wilson astronomer Edison Pettit, with his solar observations at their home telescope in Pasadena, California, and on Mt. Wilson. In both a figurative and literal sense, she learned astronomy at her father's knee and went on to obtain a master of arts degree in astronomy at Claremont Colleges. With the exception of his travels for classified projects, Aden and Marjorie were inseparable after his return from Europe at the end of World War II.

We owe much to their unpublished autobiographies, principally *Echoes from a Simpler Time* (2002), *The Solar Odyssey: Adventures along the Way* (2003), and *The Golden Age of Astronomy: In the Beginning* (2008). To assemble a more objective account, we have supplemented those reminiscences with interviews of colleagues and some family members, other primary sources (correspondence, published papers), and pertinent secondary sources to provide context for what amounts to an abbreviated history of astronomy, optics, solar energy, and national defense in the 20th century, with emphasis on contributions by the Meinels.

Along with the references, three appendices close this volume: a list of the abbreviations and acronyms used herein, a comprehensive list of Aden and Marjorie's extensive publications, and a list of select publications by Edison and Hannah Pettit and Marjorie's sister, Helen Pettit Knaflich.

In lives as multifaceted and complex as theirs, there are areas and events that are not covered here in as much detail as we would like, including those that are still classified even after more than 60 years. For those who would follow on with our story in the years ahead, we have deposited supporting materials in the Special Collections of UA Libraries and at the James C. Wyant College of Optical Sciences, UA, Tucson.

Acknowledgments

Our extensive research and some illustrations for this volume over the course of five years have been facilitated by knowledgeable and helpful staff members at institutions across the country: Daniel Meyer, Christine Colburn, Barbara Gilbert, and Julia Gardner (Special Collections Research Center, University of Chicago Library); Julie Cooper (NASA/JPL); Kevin M. Bailey and Kathy Struss (Dwight D. Eisenhower Presidential Library & Museum); Aimee Wismar, Michelle DeMartino, Stacey Chandler, and Hailey Philbin (John F. Kennedy Library); John Wilson (Lyndon B. Johnson Presidential Library and Museum); James Neel and Geir Gundersen (Gerald R. Ford Presidential Library); Jessica Hartman, Sarah Waitz, Will Clements, Ian Frederick-Rothwell, Amanda Weimer, and Russell Hill (National Archives and Records Administration, College Park, Maryland, and Washington, DC); Vernica Reyes-Escudero and Molly Strothert-Maurer (UA Libraries, Special Collections); Loma Karklins (Caltech, Archives and Special Collections); Dean Smith (Bancroft Library, UC Berkeley); Pamela Hearn (Director, Society of Fellows of Huntington Library, Art Museum, and Botanical Gardens); Geoff Gentilini (Golden Arrow Research).

We are grateful to the many individuals who sat for long interviews or telephone reminiscences of Aden and Marjorie: Helmut Abt, Roger Angel, Bob Breault, John Glasby, Ronald Hilliard, Joseph Houston, Stephen Jacobs, Buell Jannuzi, Ursula Licon, Bill Livingston, Don Loomis, John Lytle, Dick McEwen, Jim Mayo, Aden Meinel (interview by Robert Shannon), David Drach-Meinel, Diane Drach-Meinel, Ed Meinel, Nancy Meinel, Jim Meinel, Don Osborn, Bob Parks, Robert Shannon, Phil Slater, Peter Strittmatter, Tillman Stuhlinger, Gerard van Belle, Neville Woolf, and James Wyant.

We are pleased to thank others who offered helpful advice, data, or contacts: Ancestry.com; Arnie Bazensky (Schott North America) for information on Hal Bennett; David DeVorkin (Smithsonian Institution) for assistance with photos and a PDF of his biography of Fred Whipple; Gerlinde Doom for English translations from *Die Memoiren des Hu Ningsheng*; Andrea Faling (History Nebraska) for research on George and Edison Pettit in the *Peru Pointer*; Bruce Hevly (Department of History, University of Washington) for a PDF of his dissertation from Johns Hopkins University; David Kohnen (Executive Director, Naval War College Museum) for contact information; Jack Latimer (China Lake Museum Foundation) for his article on the 5-inch HVAR in *The China Laker*; Natalie Luvera (National Atomic Testing Museum); Francis Gary Powers, Jr.;

Phil Pressel for information on HEXAGON; Adam Ratliff (UC Berkeley) for academic records of Aden Meinel; David Thomas; and Al Villaire. Special thanks go to Helmut Abt (National Optical Astronomy Observatory), who as a longtime colleague and friend of Aden and Marjorie gave two interviews and helped with many aspects of this volume, which would be much poorer without his guidance.

Permission to reproduce photos was granted by several individuals, companies, and institutions, and we acknowledge their kindness both here and in the figure captions: Alamy stock images of V-2 rocket diagram and bombed-out Jena; Peter Collopy and Loma Karklins (Caltech Archives) for photos of Caltech luminaries and rocket research; Kate Igoe (Smithsonian Photographic Services); Carol Myers (San Diego History Center) for the photo of the USS *Indianapolis*; Anuja Navare (Pasadena Museum of History); Jessica Rose (NSF NOIRlab); Kathy Struss (Dwight D. Eisenhower Presidential Library and Museum) for the photo of Eisenhower with a CORONA bucket; Christine Wright and Austin Wilson for images of Charles Lindbergh at Romansville, Pennsylvania, dating from 1928.

Jeremy Lewis and Anna Langley of Oxford University Press did a magnificent job of shepherding the original proposal for this book through the review process and then its publication several years later. Dorothy Bauhoff did a superb job of copyediting for OUP. We thank them all for their patience and professionalism.

James C. Wyant, Emeritus Dean and Professor of Optical Sciences, and after whom the College of Optical Sciences is named, was kind enough to write an appreciative foreword on short notice.

Most of all we thank our wives—Ann Breckinridge, Cheryl Pridgeon, and Cindy Osborn—for shouldering many burdens while we were out of town for research, hunched over a computer screen, or otherwise preoccupied with bringing the rich lives and manifold legacy of Aden and Marjorie Meinel to first light.

JBB
AMP
DEO

1

From Tomahawks to Telescopes

Aden Meinel was born in Pasadena, California, on November 25, 1922, to John George Meinel and Gertrude Baker Meinel. From an early age Aden was told never to discuss the family history, and he never did until late in his marriage to Marjorie. Perhaps as a consequence spanning generations, little is known of his ancestry, in marked contrast to a wealth of available information about Marjorie's rich genealogy, which includes Scottish royalty, some of the first members of Plymouth Colony, soldiers in the Revolutionary War, the War of 1812 and the Civil War, a scout for Daniel Boone, state and national legislators, and pioneer settlers in wagon trains. Her heritage is America's.

Aden Baker Meinel

Klingenthal, Saxony, in eastern Germany borders the modern Czech Republic. The town is well known for its long tradition of producing musical instruments, principally strings, wind instruments, harmonicas, and accordions. Violin-making there dates back to the 17th century, when Protestant Bohemian refugees fleeing the persecution and atrocities of the Thirty Years' War brought their skills with them. This musical town of less than 10,000 people also seems to be the epicenter for the surname Meinel and was home to Aden's grandfather, John (Johann) Meinel (1854–1913). He married Rosina Mueller (1861–1920) in 1882, immigrated to the United States that same year, and settled in Pittsburgh, where he was a self-employed house-painter.[1] The first of their six sons, John George (1883–1955), was born the next year. There followed in succession Lawrence (1885–?), George (1888–1985), Harry William (1891–1976), William John (1893–1954), and Albert John (1896–1923).

John George Meinel married Bertha Katherina Schubert (1877–1916) and moved to Pasadena, California, sometime between 1900 and 1910 to escape the industrial pollution in Pittsburgh. There Bertha gave birth to stillborn twin girls in 1907, Mark Perry Meinel (1910–1952), and Harold Schubert Meinel (1912–2007) before she died of tuberculosis at the age of 39. Suddenly a widower, John George moved back to Pittsburgh to live with his brothers, and he likely took his sons with him.

Gertrude Alice Baker (1884–1955) had worked in the silk mills of Harrisburg, Pennsylvania, before moving to Pasadena. In 1907 she married Army veteran

With Stars in Their Eyes. James B. Breckinridge and Alec M. Pridgeon, Oxford University Press. © Oxford University Press 2022. DOI: 10.1093/oso/9780190915674.003.0001

William Guy Manley (1892–1963), who worked as a stock man for the railroad, and they had two boys, Frank C. Manley (1911–2000) and Ralph H. Manley (1912–1913). They eventually divorced, and by 1920 William had remarried.[2] John George Meinel met and courted Gertrude in Pasadena, and soon after they married he built two small cottages on South Greenwood Avenue, selling one to pay for the other. It was there that Aden was born and was named after Gertrude's brother (Aden Sylvester Baker) who had died of pneumonia just before his 30th birthday.[3] Aden grew up with two half-brothers by his father's first marriage (Mark and Harold) and a stepbrother, Frank Manley. Following in his father's footsteps, John George Meinel supported his family of six by painting houses, although he was also commissioned to paint the interior of the Athenaeum at the California Institute of Technology (Caltech) to complement the vaulted ceilings painted by Vatican-trained muralist Giovanni Battista Smeraldi.[4] He died on September 13, 1955, and Gertrude died two days later, circumstances unknown.

Gertrude's heritage was also based in Pennsylvania and was ultimately traceable to Germany. Her great-great-grandfather, Daniel Baker, Sr. (1749–1804), served as a private in the Revolutionary War.[5] His father, George Peter Baker (1706–1788), was born in Brandenburg, Germany, which at that time was part of the Kingdom of Prussia. Several graves in this Baker line are grouped at Bakers Summit in Bedford County, Pennsylvania.[6]

Marjorie Steele Pettit

Marjorie Steele Pettit was born to Edison Pettit (1890–1962) and Hannah Bard Steele Pettit (1886–1961) on May 13, 1922, in Pasadena, California, 25 months after her sister Helen Bard Pettit (1920–1985). Helen married seaman Hanley Knaflich (1921–2000), later divorcing him in 1976 in Seattle, Washington, with no children. Like the rest of her family, Helen became an astronomer and published several papers on the nightglow, ionospheric winds, micropulsations, and solar prominences, first under the name Helen Pettit and then under her married name.[7] She worked out of Claremont Colleges (where she received a Master of Arts degree in astronomy), the US Naval Ordnance Test Station at China Lake, Boeing Scientific Research Laboratories in Seattle, the Smithsonian Astrophysical Observatory, and the High Altitude Observatory in Sunspot, New Mexico.

Edison Pettit

Edison Pettit was born in Peru, Nebraska, on September 22, 1890, the son of George Knox Pettit (1846–1928) and Martha Ann Knox (1854–1939) in an era dominated by the culture-changing inventions of Thomas Edison. According to the Edison Pettit Papers at the Huntington Library in San Marino, California, he originally gave his birth year as 1889, and although he later corrected it to 1890 the 1889 date was perpetuated in his obituaries. He was indeed named after the world-famous Wizard of Menlo Park, a given name so unique that his parents saw no need to add a middle name.[8] George owned and operated a sawmill; using Thomas Edison's cutting-edge technology, he updated its power source from a steam-belt drive to a steam-generated electricity drive. Young Edison helped to finance his education by working in his father's power plant that brought electricity to the town of Peru.[9] He graduated with a Bachelor of Education degree from Nebraska State Normal School in Peru (now known as Peru State College) in 1910.[10] While a student there, he participated in the Boys' Kearney Debating Squad, German Club, Philomathean Literary Society, Ciceronian Debating Society, and Athletic Association, as well as serving as assistant editor of the school yearbook, *The Peruvian*. He decided to graduate again with the first graduating college class in 1911, perhaps because he had married Elizabeth Schmauser, a fellow student from Seward, Nebraska, who was a year behind him.[11]

Following graduation, Edison taught science at Minden High School in Minden, Nebraska, from 1911 through 1914 and then physics and astronomy at Washburn College in Topeka, Kansas, until 1918. He resided in Minden with Elizabeth until she died of tuberculosis,[12] but spent his summers at Yerkes Observatory in Williams Bay, Wisconsin (operated by the Department of Astronomy, University of Chicago), studying solar prominences and photographing the sun with a spectroheliograph.[13] He traveled to Yerkes in 1918 and directed one of three expeditions to photograph the corona of the total solar eclipse on June 8 of that year. Assigned to the station at Matheson, Colorado, he met Hannah Steele there, a bright graduate student at the University of Chicago as well as assistant at Yerkes.[14] They later jointly published their results.[15]

After the sinking of RMS *Lusitania* in 1915 and seven US merchant ships by German submarines and publication of the "Zimmerman Telegram," which asked Mexico to become a German ally against the United States in return for German help in reclaiming Texas, New Mexico, and Arizona, the US Congress declared war on Germany on April 6, 1917. A few weeks later, Congress passed the Selective Service Act, and Edison was drafted by the army in 1918. When the induction staff saw that he had written "instructor in astronomy" on his draft registration form, they assigned him to the US Army Signal Corps, where he worked with physicist Robert Williams Wood (1868–1955), famous for infrared and ultraviolet photography as well as high-powered ultrasound, which had suddenly become important in detecting German submarines. During one experiment with gas-filled balloons as reconnaissance platforms, the two of them reportedly came close to shutting down the municipal gas supply of Baltimore.[16]

What led Robert Wood away from studying for the priesthood and into the field of optics in the first place was his observation of an aurora, which he believed was caused by "invisible rays."[17] It would fall to Edison's son-in-law, Aden Meinel, to explain auroras and their colors conclusively.

When the war ended in November 1918, Edison returned to Yerkes to work toward a PhD, principally under Observatory director Edwin Frost and alongside Hannah Steele, whom he married that year. Hannah, too, was working on her PhD at the University of Chicago, and both she and Edison used the 40-inch telescope at Yerkes. On one occasion, when Edison was rewinding the drive weights, his necktie slipped into the drive mechanism and would have strangled him had not Hannah heard his cries and run to the observatory dome with scissors to free him. After that, Edison wore bow ties exclusively—black bow ties.[18] He received his doctorate in 1920 for his dissertation titled *The Forms and Motions of Solar Prominences* using Hale's spectroheliograph on the 40-inch refractor at Yerkes. Their first daughter Helen was born that year and is pictured in the arms of Edison among the Yerkes staff (Figure 1.1).

Hannah had received her Bachelor of Arts in mathematics at Swarthmore College in 1908 and her master's in 1912. From 1912 to 1915 she was posted as an astronomical observer at the Sproul Observatory at Swarthmore. In 1919 she became the first woman to receive a PhD in astronomy from the University of Chicago for her astrometric dissertation, *Proper Motions and Parallaxes of 359 Stars in the Cluster h Persei.*

The preeminent astronomer of the day was George Ellery Hale (1868–1939), founder of Yerkes Observatory, Mt. Wilson Observatory, and the National Research Council, as well as inventor of the spectrohelioscope (a solar telescope that allows the sun to be viewed in a single wavelength) and the spectroheliograph to take a photographic image of that wavelength. He was the author or coauthor of almost 400 papers and books and longtime editor of the *Astrophysical Journal.* After Hale added a solar observatory at Mt. Wilson, he invited Edison to become a staff member there in 1920 at a salary of $2,000, which did not rise until 1945.[19] Edwin Hubble (1889–1953) had joined the staff the year before. When Edison asked Hale if there was also a position for Hannah, he was told that only men could serve on the staff, an anachronism that persisted for the ensuing five decades; however, she could be a "computer" and grind out the calculations for the male astronomers.[20] Hannah soldiered on despite the overt, institutionalized gender discrimination and published two important papers in 1920: "Parallaxes of fifty-two stars" with Georges van Biesbroeck, and "The parallaxes of fifty stars determined at the Sproul Observatory (Second list)" with John A. Miller and John H. Pitman. Following the births of daughters Helen in 1920 and Marjorie in

Figure 1.1. Yerkes Observatory staff, August 19, 1920. Edison Pettit is second from left in back row holding baby Helen. Hannah Steele Pettit is far left in the middle row.

University of Chicago Photographic Archive, [apf6-00425], Special Collections Research Center, University of Chicago Library.

1922, she participated in expeditions to observe solar eclipses in New Hampshire in 1923, 1930, and 1932 (for this last one joined by Edison).[21]

Edison, Hannah, and baby Helen spent their first Christmas on Mt. Wilson at Kapteyn Cottage in 1920, perhaps by no coincidence where Aden and Marjorie later spent their honeymoon in 1944. Heavy snowfall was forecast, and so the Observatory staff, except for the Pettits, descended the mountain on the primitive toll road to Pasadena before the storm arrived. Several inches fell, rendering the Pettits snowbound and without food. A relief team finally arrived to help Edison transport Hannah and baby Helen across the face of the mountain and down a trail to safety.[22]

Among the many notable achievements of Sir Frederick William Herschel (1738–1822), besides his discovery of Uranus, two of its five largest moons (Oberon and Titania), and two of the 62 moons of Saturn (Mimas and Enceladus), was his discovery of infrared radiation, which he called "calorific rays." William W. Coblentz (1873–1962), the "Father of Infrared Spectroscopy," and Carl O. Lampland (1873–1951) of the Lowell Observatory in Flagstaff, Arizona, had been working on electronic devices called thermocouples, which when arranged in a series or in parallel (known as thermopiles) can measure infrared radiation from stars, the Moon, and planets. In an effort to measure the radiation temperature of sunspots and play catch-up with the Lowell Observatory, Edison began making his own vacuum thermocouples. With his colleague Seth Nicholson (1891–1963) he was able to observe the radiation from stars in different wavelengths and determine their radiometric magnitudes, temperatures, and angular diameters.[23] They also determined the surface temperatures of the Moon, Venus, Mars, Jupiter, and Saturn using the Mt. Wilson 100-inch telescope and also the temperature of the Moon during a lunar eclipse, which revealed the granular composition of its surface.[24]

Perhaps in memory of his first wife, Edison took his family to Tucson in 1925 to the Desert Sanatorium, which was then mainly a tuberculosis treatment center and has been vastly expanded today as the Tucson Medical Center.[25] Just as Herschel had discovered infrared light, Johan Wilhelm Ritter (1776–1810) discovered invisible light at the other end of the spectrum, ultraviolet (UV) light. Arthur Downes and Thomas Blunt later showed in 1877 that UV light has bactericidal action. On the basis of this, Edison invented the UV solar radiometer, which consisted to two quartz lenses, one silvered and one gilded, in airtight cells and mounted on a rotating disk adapted to a telescope mount. One of the lenses admitted only UV light and the other only green light. The radiometer not only measured the intensity of UV light in sunlight, but also enabled precise amounts to be administered to patients as heliotherapy. The treatment was shown to be ineffective against *Mycobacterium tuberculosis*, and a real cure would have to wait until the advent of streptomycin and more powerful antibiotics.

In the 1930s Edison's studies of solar prominences continued in earnest, mostly using Hale's spectroheliograph on the 40-inch refractor at Yerkes but also the spectroheliokinematograph at the McMath-Hulbert Observatory in Lake Angelus, Michigan, in the summers of 1936–1938 to obtain the first motion pictures of the Sun.[26] Building on his dissertation, he was able to categorize prominences by their five forms (active, eruptive, spot, tornado, and quiescent), dimensions, masses, distribution of elements, and height of the chromosphere in H-alpha.[27]

In 1938 Edison and Hannah purchased an Alvan Clark refractor with a 6-inch lens and 8-foot focal length from Mrs. Frank Herte in Lake Orion, Michigan. It had once belonged to her family member named Dr. George Duffield (1794–1868). Eight eyepieces, gears, clock, sector, and small parts for it had been wrapped in a Toronto newspaper dated May 11, 1918.[28] Pettit investigated its origins and found out that Moses Sutton had left $1,000 in his will to the Raisin Valley Seminary in Adrian, Michigan, supported by the Religious Society of

Friends (Quakers), for the construction of an observatory, $1,000 for the purchase of a telescope, and $20,000 as an endowment fund. The observatory was built and furnished with a telescope manufactured in 1875 by the optical firm of Alvan Clark & Sons of West Somerville, Massachusetts.[29] A student at the seminary in 1893–1894, Walter Jones, remembered looking through it to see Jupiter and the four Galilean moons. At the time it was the second-largest telescope in Michigan, surpassed only by the 10-inch refractor at the University of Michigan.[30] Edison cleaned and refurbished it and set it up in his home in Pasadena (Figure 1.2). With it he prepared drawings of Mars and its "canals" and

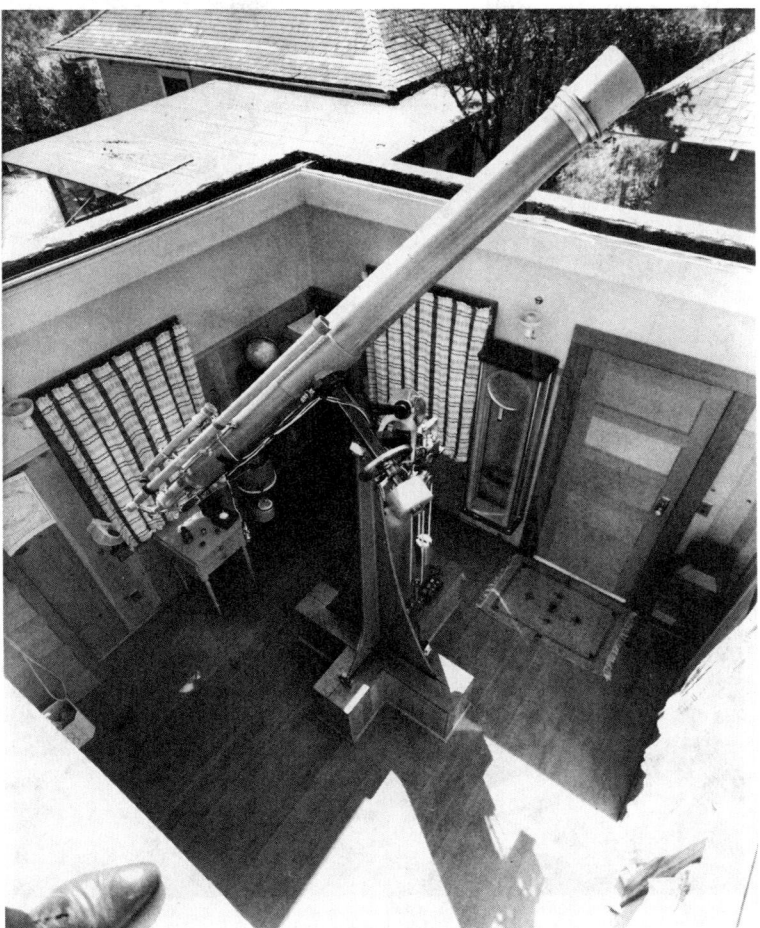

Figure 1.2. Alvan Clark 6-inch refractor at the Pettit home. Undated.
Courtesy of David Drach-Meinel.

made a series of measurements of the brightness of Nova Puppis and magnitudes of over 500 galaxies with a photoelectric photometer that he built.[31] He had also built a quartz-polarizing monochromator at home, which allowed him to observe prominences without an eclipse and to which he attached a time-lapse movie camera.[32] Edison and Hannah were happy to open their home observatory to Girl Scouts, Boy Scouts, and the general public (Figure 1.3).[33]

Edison shared authorship of several papers on solar prominences with Robert R. McMath, who had served as a pilot in World War I and then in the

Figure 1.3. Edison Pettit at home with the Alvan Clark refractor and colleagues. Undated.

Courtesy of David Drach-Meinel.

Signal Corps, like Edison. He was a successful businessman, amateur astronomer, and director of the McMath-Hulbert Observatory at Lake Angelus, Michigan, which was later deeded to the University of Michigan.[34] McMath built a 50-foot solar tower telescope there in 1936 with help from Mt. Wilson Observatory, followed by a 70-foot tower telescope and a 24-inch Cassegrain reflector; Edison spent the summers of 1936–1938 at Lake Angelus using Hale's spectroheliokinematograph to study prominences.[35] In 1938–1939, however, Edison and McMath had a falling-out over authorship of papers. In 1938, McMath "adopted" three years of Edison's work on solar prominences, removed Edison's name from his premier paper, including the first motion pictures of prominences, and professed sole authorship. McMath then presented the paper at the 1938 American Astronomical Society meeting in Ann Arbor, Michigan, as his own work, even though none was his.[36] In a letter to McMath dated May 31, 1939, Edison referred to a manuscript that McMath had sent to Nicholson for review and wrote ". . . you may say 'McMath and Pettit have published [sic] measures of prominence motions, etc.,' but you positively cannot say 'McMath and Pettit have made measures of prominence motions, etc.,' for McMath made none of the measures. Please bear all this in mind in future publications and recognize that I wrote the papers and made all the measurements involved."[37]

In that same letter, Edison vowed to return film negatives and rewind equipment that belonged to the observatory and asked for his own property back. Afterward, Pettit and McMath were not on speaking terms. This is mentioned here because McMath, as professor of astronomy at the University of Michigan, became president of the Board of AURA (Association of Universities for Research in Astronomy) and in that capacity was instrumental in forcing Edison's son-in-law, Aden Meinel, to resign as the first director of Kitt Peak National Observatory southwest of Tucson. This unfortunate episode will be referenced and discussed in more detail in Chapter 7.

In 1944 Pettit was asked to build three thermopiles, and he delivered them in person to Julian E. Mack in Santa Fe, New Mexico, by July 3, 1945, as requested.[38] Edison may have known that the thermopiles were to be used to measure the output from a plutonium-implosion fission device; called "The Gadget," it was detonated on July 16, 1945, near Alamogordo, New Mexico. Mack was in charge of the spectrographic and photographic aspects of the blast. Upon detonation the rise in temperature in a black receiver was translated by the thermopiles into an electrical voltage and recorded on a high-speed oscillograph; the radiant power in calories/cm^2/sec could then be calculated from the heat-content of the receiver, its time-temperature rise, and projected area.[39] The Gadget was virtually identical to the "Fat Man" bomb detonated over Nagasaki, Japan, on August 9, 1945.

To illustrate the deep respect for secrecy of the Manhattan Project, while Edison was making the thermopiles, Aden was part of the Caltech team working on the timing mechanism of The Gadget, and Marjorie was editing Project reports; all were unaware of what the others were doing.[40]

For many years, Hannah had helped Edison, but then became an invalid and then a semi-invalid, first from a goiter (or hypothyroidism) that affected her heart and then a stroke, and so it fell to Marjorie to be her father's assistant at the telescope until she married Aden.[41] Edison visited Hannah in the hospital every day until she died on September 10, 1961.[42] He moved in with Aden and Marjorie in Tucson following a stroke in January 1962 but died about four months later on May 5, 1962, of a second stroke. Both Edison and Hannah are buried in Forest Lawn Memorial Park in Glendale, California. The Alvan Clark refractor was sold in 1975, a century after its manufacture.

Edison received an honorary degree from Carthage College (LLD) in 1935.[43] There is no record of applications for patents for any of the several devices that he created in his machine shop at home, starting a tradition continued by Aden, who felt that he himself should not be enriched by his inventions—they belonged to the world at large. Crater Pettit in the Montes Rook of the Moon and also in the Amazonis quadrangle of Mars are named after Edison. Relatively close to each is crater Nicholson to reflect their significant coauthored works and friendship.

Pettit ancestry

The surname Pettit (also spelled Pettit, Le Petit, Pittitt, Pettitt, Pettye, Pettet, Petty, Petyt, Pettis) is an anglicization of the Norman French name Petit (see Figure 1.4a–c for the Pettit genealogical line). After the Norman Conquest in the 11th century, some of the Petits settled in the English county of Cornwall and later in Kent, Essex, Bedfordshire, Oxfordshire, and Suffolk.[44] For more than three centuries, members of this line lived at Ardevora, Philleigh, Cornwall, beginning with Sir Otes Petit (ca. 1130–), Lord of Ardevora and husband of Elizabeth Fitz Ive (ca. 1155–ca. 1179).[45] A dozen generations of Petits in Cornwall with the given name of John or Michael ensued, most with knighthoods, never straying too far from Ardevora and the village of Philleigh.[46] All were nobles, including Sir John Petit (ca. 1323–1363), Lord of Trenerth, who was a member of Parliament in 1339.[47] Valentine Petit (ca. 1429–1490) may have been the first to venture from Cornwall, as his son John (aka Johannes) Richard Pettit (1456–1532) was born in Kent and served as an alderman of London.[48] John's son Valentine (1486–1545) and grandson Henry (1518–1599) remained in Kent.[49] Henry Pettit II (?–1661) was born and died in Kent, but his son John Pettit I (1608–1662) was born in Essex and immigrated to Roxbury, Massachusetts Colony, most likely as part of

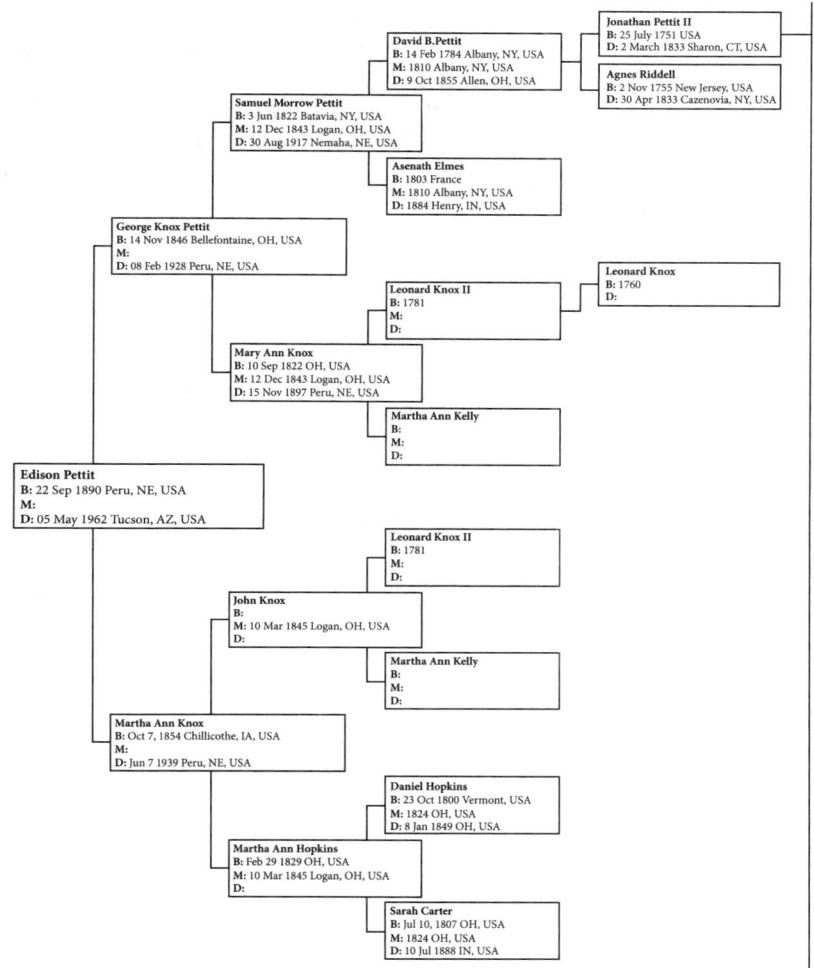

Figure 1.4a–c. Fig. 1.4a-c. Pettit genealogical line (in part) beginning with Marjorie Meinel's father, Edison Pettit, through Henry Pettit (b. 1595) and his grandson, John Pettit I, who immigrated from England to Massachusetts in 1630.

the John Winthrop fleet of 17 ships of Puritans who sailed from Southampton in 1630.[50] John and his first wife Debrow then moved southwest to Stamford, Connecticut (formerly called Rippowam by Native Americans) before 1644, where he was listed among the first pioneers and had children before 1650.[51] No less than 11 clans or tribes of Native Americans (e.g., Mohegans, Mohawks, Delawares) lived in the area and deeded some of their land to the English, but for 30 years following the establishment of Stamford there were skirmishes

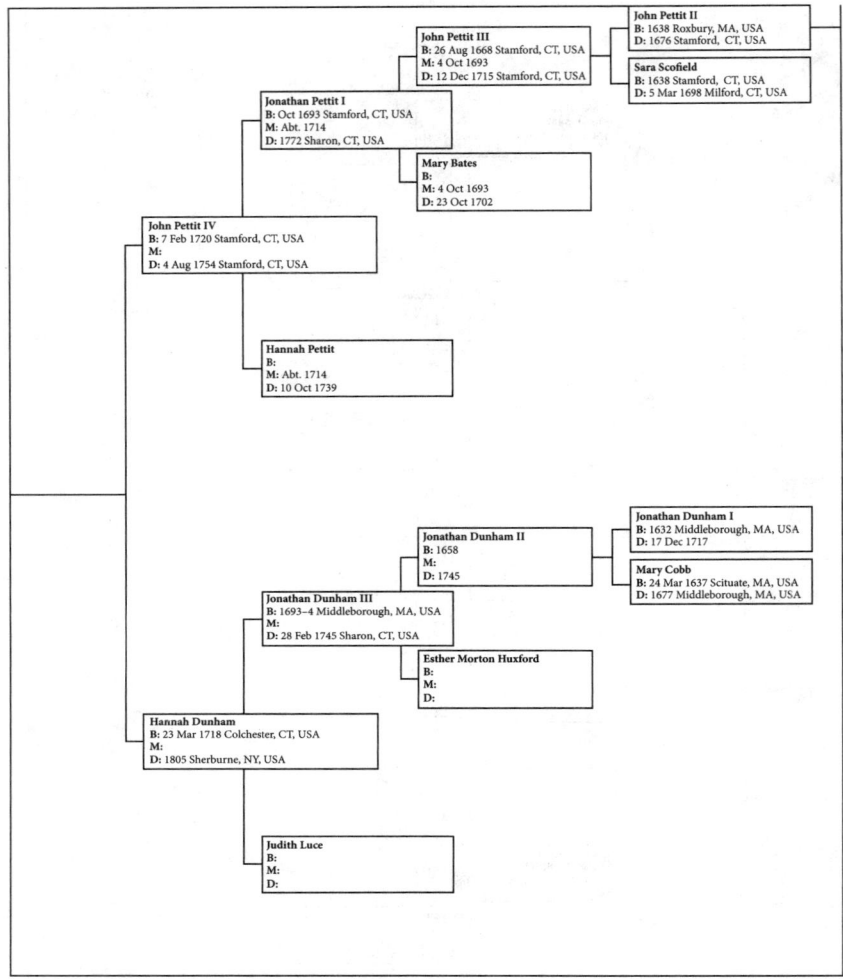

Figure 1.4a–c. Continued.

and horrific massacres, with retaliations by both the settlers and the tribes, as described in gruesome detail by Huntington.[52]

The Pettits owned land in and around Stamford and then in Sharon, Connecticut, for six generations. Jonathan Pettit II (1752–1833) was only two years old when his father, John Pettit IV (1720–1754), died, leaving six fatherless children. Jonathan was raised by his mother, Hannah Dunham Pettit (1718–1805), until the age of five; his grandparents then assumed responsibility for him until he was 14, when he went northwest to Saratoga, New York, and learned shoemaking and tanning.[53] He married Agnes Riddell (1755–1833) in 1775, the

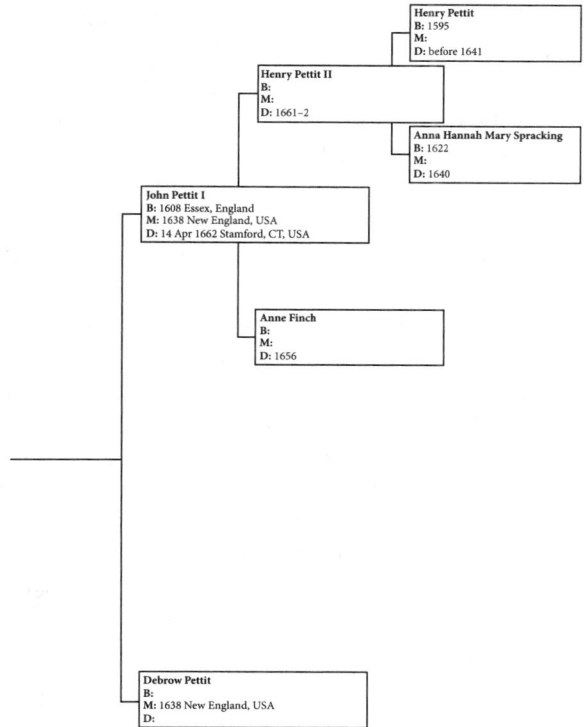

Figure 1.4a–c. Continued.

same year that he enlisted as an ensign under Captain Henry O. Hare in Albany County, New York. Four years later he was made captain of the 16th Regiment militia in Albany County.[54]

Of particular interest here is the genealogical line of Hannah Dunham, mother of Jonathan Pettit II. Her great-great-grandfather, John Dunham I (1589–1668), from Nottinghamshire, England, was a Puritan who fled to Leiden, Holland, where he learned to weave and married Abigail Barlow. He later immigrated with her to Plymouth Colony, where he was listed in the General Court of Free Men in 1633 and received a land grant.[55] Of their 11 children, their daughter Abigail was the first child born in Plymouth (1623).[56]

Phillipe Delano (1603–1681) arrived in Plymouth Colony in 1621 on the ship *Fortune*, the second ship after the *Mayflower*.[57] He married Hester Dewsbury and fathered six children by her.[58] Phillipe was a seventh-generation ancestor of President Ulysses S. Grant on his paternal side[59] and a seventh-generation ancestor of President Franklin D. Roosevelt on his maternal side.[60] Their daughter, Mary Delano (1635–1656), was the first wife of Jonathan Dunham I (son of John

Dunham I) but died childless.[61] Jonathan Dunham I then remarried Mary Cobb (1632–1677) and had five more children, including Jonathan Dunham II (1658–1745), which continued the line to Edison Pettit.[62] Mary Cobb was the daughter of Henry Cobb (1607–1679), who arrived in Plymouth Colony between 1627 and 1631 and married Patience Hurst (1610–1648).[63] Patience was the daughter of James Hurst (ca. 1582–1657) and Garteud (Gertrude) Bennister (1582–before 1670), also members of the Plymouth Colony.[64]

Jonathan and Agnes Pettit II (see earlier mention) had 11 children, although three died in infancy; one of the surviving children was David B. Pettit (1784–1855).[65] David served as a non-commissioned officer in the War of 1812.[66] He later became a teacher in Bellefontaine, Ohio, fathering 10 children with Asenath Elms (1803–1884), one of whom was Jonathan Pettit, killed as a Union soldier at the Civil War siege of Vicksburg in 1863 and buried in an unmarked grave.[67] One of his brothers was Samuel Morrow Pettit (1822–1917), Edison's grandfather.[68]

Samuel Pettit was born near Batavia in Genesee County, New York, on June 3, 1822. As a young boy, probably about four years old, he was said to have witnessed the kidnapping of William Morgan in Batavia.[69] Many of America's founding fathers were Masons (Freemasons) dedicated to liberty, free enterprise, and Deism, and firmly opposed to Catholicism and the British monarchy. When a heavy-drinking, gambling Mason from Canada named William Morgan moved to Batavia, he was welcomed at first, and members passed the hat for him. Later the tide turned against him: "But his intemperate habits, his shabby style of dress, his peccadillos in the way of borrowing and not returning, his vulgar, blasphemous and indecent style of conversation that disgusted the gentler sex and set them against him, all these things growing upon Morgan from month to month, had gradually closed the Lodge-doors against him, and narrowed the circle of his visitations."[70] They later discovered (March 1826) that he in fact was not a Mason. They had entrusted their whole body of secrets to an imposter under no obligations to maintain those secrets.[71] In revenge for being shunned, Morgan threatened to publish a book exposing all their practices and passwords. In response, a group of Masons torched the building where the book was being printed, arrested him on September 11, 1826 (ostensibly on grounds of nonrepayment of a loan and shoplifting a shirt and tie), and jailed him. When Mrs. Morgan arrived at the jail, he was gone and was never seen again.[72] The kidnapping and presumed murder of Morgan provoked outrage and galvanized the formation of America's first third party, the Anti-Masonic Party, which survived until 1840. If four-year-old Samuel was indeed there to witness the kidnapping, it is reasonable to presume that his father David or another responsible adult accompanied him.

Education ended for Samuel by the age of 11, much of it provided by his older sister, Leonora (1818–?).[73] At the age of 21 he married Mary Alice Knox

(1822–1897) of Logan County, Ohio. Her father, Leonard Knox II (1781–?), had been a scout for Daniel Boone, one Benton McCarthy, and General Jacob Dolson Cox.[74] Cox was later recruited for secretary of the interior by President Ulysses S. Grant but resigned when he fell out of favor with Republican Party bosses by establishing a merit system in the civil service to stem corruption.[75]

At about the same time as his marriage, Samuel was apprenticed as a carpenter in Ohio, which provided board and clothing for four years.[76] In 1849 he, Mary, and their three sons moved northwest to Richland Center, Wisconsin, where he worked as a mechanic, contractor, and builder.[77] Spurred by the federal government's offer of cheap purchase prices for land out west to those who would settle it, through the Preemption Act of 1841 and later the Homestead Act of 1862, Samuel and his family formed a wagon train of "prairie schooners" with some neighbors and headed southwest, along with everyone's oxen, mules, and cows. Soon other caravans joined them as they lumbered across the Iowa prairie to farm the fertile land on either side of the Missouri River. The Pettits and the family of Selatheal (also spelled Salathiel) Good and his family crossed the river at Nebraska City on a ferry, leaving the other families behind in Iowa.[78] There was little to the city at that time, only buffalo grass as tall as a man on horseback, a trading post, and Fort Kearney nearby.[79] Samuel then went southwest to the small settlement at Glenrock on the Little Nemaha River and bought the Gill farm there for a yoke of steers and a silver watch.[80]

In 1862 Samuel and his family moved a few miles north along the west bank of the Missouri River and bought 320 acres for $1,000.[81] Although the land was fertile, it was littered with cottonwood trees that were cut down as cord wood, which was used as fuel in the boilers of boats going up and down the river.[82] Tracts more inland were not chosen for homesteading by the settlers for two reasons. First, all supplies needed to be hauled from the river by oxen and mule teams, which had to ford streams in the absence of bridges. And inland Native American tribes were more hostile and dangerous than the peaceful Omahas living near the river; in fact, the Omahas offered protection to the settlers from the Sioux and other tribes.[83]

As the longest river in North America, the Missouri or "Big Muddy" could be both friend and foe to the early pioneers traveling north and south along its banks. Although the river was useful for irrigating crops, watering livestock, transporting people and goods, etc., it also flooded periodically; entire towns have been erased from the map. In 1866, a shipping boat named the *R. M. Bishop*, loaded to the waterline with grain, capsized when the river was running high with swift currents. The crew held on to the upper rigging of the boat, the only portion still above water, crying for help. Samuel and his sons, placing themselves in jeopardy against the current and debris flowing quickly downriver, rescued the men in small skiffs. The crew members were offered hospitality in the

Pettit home until the flood subsided and other boats arrived to pick them up. Ever resourceful, Samuel paid the owner for the sunken boat and salvaged its timbers to build himself a new house with the lumber, used the machinery in a sawmill, and was able to give his wife Mary the other contents—linens, china, silverware, and a walnut-and-marble mantelpiece.[84]

Unfortunately, another flood in 1867 wiped out the new house, so Samuel had to build yet another house, this time safely on a 120-acre tract on bluffs in the town of Peru.[85] He cleared the tract of cottonwood trees and planted a large apple orchard, but lost it in the great flood of 1881 when the river swelled to five miles across;[86] his sawmill and grain elevator on the north side of town survived.[87]

Mary Knox Pettit died on November 15, 1897, at the age of 75 and was buried in Mt. Vernon Cemetery in Peru, a devoted Methodist all her life. Samuel did not belong to a church but was interested in spiritualism and even held séances in his home.[88] After her passing, he built a schoolhouse on his property where the apple orchard had been.[89] Samuel outlived Mary by 20 years, passing away at the age of 95 on August 30, 1917, and was laid to rest next to her.[90]

Their son George (1846–1928), the second of 11 children, inherited the sawmill and grain elevator and operated them the rest of his life. He married Martha Ann Knox in 1876 and had a girl and four boys, among them Edison. Edison surely developed a passion for astronomy from his father George, who built an observatory at the Nebraska State Normal School that Edison attended and softened red elm boards in a steam chamber to fit the wood on the curve of the dome.[91] In 1906 George helped to organize the Light and Power Company of Peru and became vice president; his sons Frank, Leonard, and Wilber and his son-in-law Herbert Hallenbeck had lifelong careers in the company.[92]

Like his father, George was long-lived. Even after his hair and beard had turned white and his shoulders stooped, he still went to work at the grain elevator every day. The family told a story that a traveling salesman was waiting for a train on the platform of the Burlington Railway Depot near the elevator. He saw George and said to a local bystander, "Gosh, that man must be old." The local replied, "Oh, he's not so old. His father [Samuel] is still living here." "In the name of God," exclaimed the salesman, "don't people ever die around here?" "People do die here sometimes," admitted the local, "but anyone who doesn't reach 70 years is said to have died in infancy."[93]

Although George Pettit's interest in astronomy and construction of the observatory at the Normal School clearly influenced Edison's career path, why Edison's wife Hannah turned to astronomy is unknown. She majored in mathematics at Swarthmore College but also belonged to the college's Joseph Leidy Scientific Society, Literary Society, and gymnastics team, according to the college yearbook for 1909. She was born in Coatesville, Chester County, Pennsylvania, on November 6, 1886, four years before Edison, to John Dutton Steele (1850–1886) and Sophia (Sophie) McLaren Bard (1856–1899). Both the Steele and Bard branches illustrate the resourcefulness, self-reliance, and industry necessary for survival in the American frontier, which also characterize the Pettit line.

Steele/Bard ancestry

The Steele family (see Figure 1.5) resided in the parish of Barthomley, Cheshire, England, at an estate known as Taed Hall (later known as Toad Hall and Toad Hole) for five centuries prior to 1795, now located near the present railway station at Crewe.[94] Alan Garner OBE, author of children's fantasy novels and the current owner, said, "We are continually finding artefacts from as far back as 10,000 years, the end of the last Ice Age. We also know that the house is built on a Bronze Age burial mound, one of several, and that beneath the house there are the remains of two further older houses and probably more dwelling places before then. It is a rare site and must be preserved."[95] The house was built like a fortress with a heavy, barred door. The walls of one room, presumably used as a chapel, are inscribed with Scripture verses in Latin, Greek, and Old English. An upper room incorporates a secret passageway leading to a small chamber.[96] It may have been a priest's hole, designed to hide not just the priest himself but also the altar furniture, vestments, paten, and chalice in the event that Royalist troops, intent on imprisoning or killing Roman Catholic priests for treason during the reign of Elizabeth I (1533–1603), stormed the building by surprise.[97]

Apart from census, birth, marriage, and death records, the bulk of what we know about the Steele branch in the 18th and 19th centuries is drawn from two sources: (1) *Descendants of George Steele of Bathomley, Cheshire, England, and Chester County, Pennsylvania*, compiled by Frederick D. Stone, Jr.; and (2) *Recollections of My Life in England and America* by John Dutton Steele (1773–1866) with later introductions by his daughter Hannah Witmer and great-granddaughter Loraine S. McKinstry. George Steele was born Christmas Day 1702, the son of John Dutton and Elizabeth Vernon. He had four children with Esther Broadhead, including George Steele, Jr. (1737–1824). George, Jr., married Hannah Dutton (1735–1811), whose Cheshire family had land deeds dating back to the Norman Conquest, and settled at Toad Hole in Barthomley.[98,99] Their son John Dutton Steele paid off the debts of his spendthrift father and, impressed by the self-evidence of Thomas Paine's *The Rights of Man* and "glowing descriptions" of the transatlantic "land of promise," moved with his parents, three unmarried sisters, and two married sisters (with their families) to America; the group, numbering 16, left Liverpool on August 10, 1795, and stepped ashore in Philadelphia just over two months later on October 12.[100]

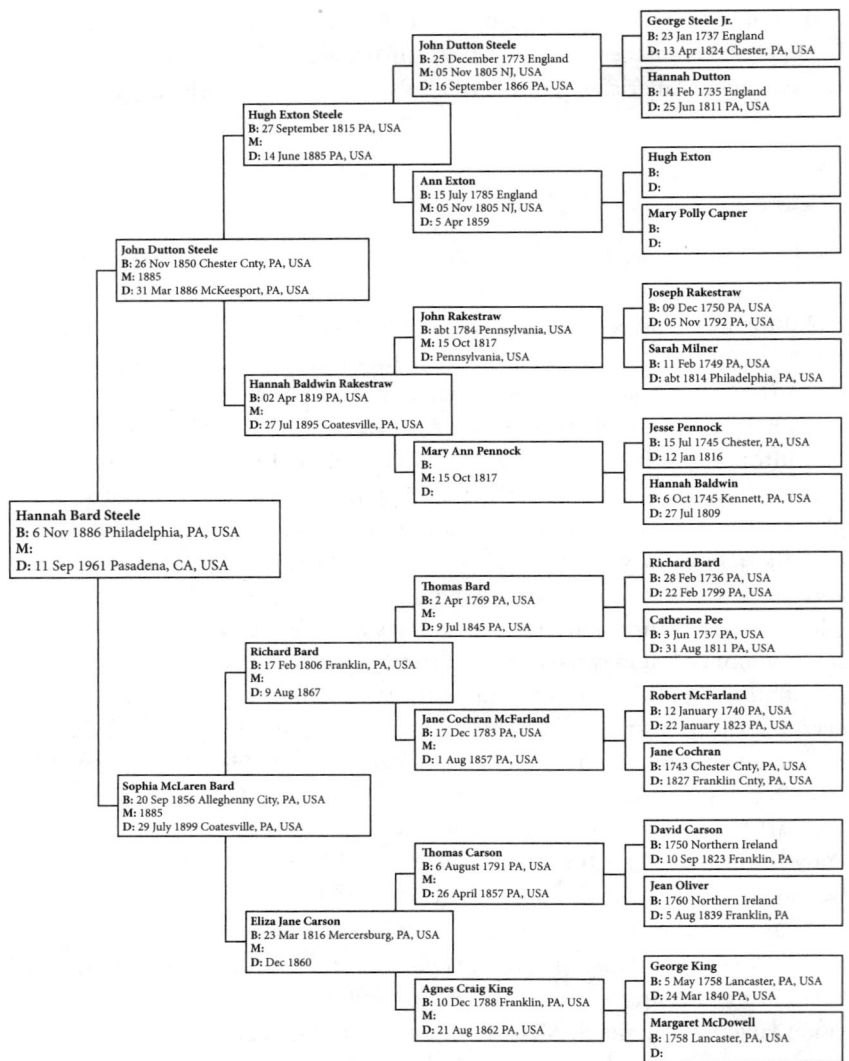

Figure 1.5. Bard/Steele genealogical line (in part) beginning with Marjorie Meinel's mother, Hannah Bard Steele, and ending here with Hannah's great, great-grandparents.

With letters of introduction to the counting house of shipping merchants Messrs. Nicklin and Griffith and a letter of credit, the extended family settled into their lives in Pennsylvania. John heard of a farm of 600 acres for sale in West Bradford Township in Chester County. He traveled by barge, carriage, sailboat, and on foot over two nights and two days to inspect the property. Pleased with what he saw, he and this three unmarried sisters decided to lease it for 10 years at $533/year.[101] In 1805 he purchased the 600 acres for five pounds sterling per acre from General Richard Humpton of the Continental Army and renamed it Stock Grange.[102] John drained the swamps, improved the soil, and built a new house there, known as Wobourn, to improve his chances that Ann Exton (1785–1859) would accept his proposal of marriage that November.[103] They remained happily married until her death.

John managed an extensive dairy business on the property and was well respected throughout Chester County. He was on the Board of the Directors of the Germantown Turnpike Co. and eventually served as a representative in the state legislature.[104] He and Ann had nine children, one of whom was Hugh Exton Steele (1815–1885). Hugh married Hannah Baldwin Rakestraw (1819–1895). One of their children was John Dutton Steele II (1850–1886), who was in the iron business and owned mills in Pottstown.[105] He married Sophia McLaren Bard (1856–1899), whose pioneer line is described later. Their daughter was Hannah Bard Steele, wife of Edison Pettit and mother to Helen Knaflich and Marjorie Pettit Meinel.

Upon John Dutton Steele's death at the age of 93 and the passing of his remaining sisters Esther and Dorothy in January 1910 at the ages of 93 and 89, respectively, Stock Grange was sold after the family had owned it for 105 years.[106] It became even more famous, at least in Chester County, Pennsylvania, for an event 18 years later.

On the foggy morning of March 13, 1928, young Frank Travor was working on the Romansville farm of his grandparents, Emmer and Sarah Way, when he saw a monoplane make some low passes over the field before it landed. Out of the Ryan B-1 Brougham, similar in design to *The Spirit of St. Louis*, climbed Charles Lindbergh and his attorney, Henry Breckinridge. Breckinridge had been President Woodrow Wilson's assistant secretary of war, acted as an intermediary for payment of the ransom for Lindbergh's baby boy who had been kidnapped, represented Lindbergh during the trial of Bruno Hauptmann for kidnapping and killing the baby, and unsuccessfully opposed Franklin D. Roosevelt for the Democratic nomination for US president in 1936. [One of the authors of this book (JBB) is Henry's cousin.] The two had been flying from Roosevelt Field in Long Island, New York, to a dinner in Washington, DC, where they were to be hosted by Nicholas Longworth, speaker of the US House of Representatives.[107] Three hours into the flight they encountered the heavy fog and had to make a forced landing, clipping the top rail of a fence. They walked to the Way Farm, and Mrs. Way offered to let them stay the night, but they ended up going down the hill and staying at Wobourn with Mr. and Mrs. Charles Elkington, the owners of Stock Grange after the Steeles.[108] It was not long before word of Lindbergh's arrival spread throughout the county. After making the first solo transatlantic flight and the first nonstop flight between North America and Europe on May 20–21, 1927, "people behaved as though Lindbergh had walked on water, not flown over it."[109] Major newspapers called the Elkington home, and reporters peered in the windows and repeatedly pounded on the door asking for an interview; Lindbergh briefly stepped outside, smiled and shook hands, but answered no questions. By 5:30 the next morning the crowd numbers had mushroomed at Stock Grange, prompting Lindbergh to remark, "It looks like a big day at the county fair."[110] The swarm followed him and Breckinridge back to the Way Farm, where the plane had been guarded all night (Figure 1.6). When the spectators finally cleared the runway, Lindbergh cranked the engine and took off, circling the field and dipping a wing as if to say goodbye and thank everyone.[111] In June 1945, he would literally be following Aden Meinel's footsteps in postwar Germany (see Chapter 4, section "Nordhausen/Mittelbau-Dora"). From 1941 to 1956, Stock Grange was owned and beloved by the respected actor Claude Rains, who starred in *The Invisible Man, Notorious, Mr. Smith Goes to Washington, The Phantom of the Opera, Casablanca, Lawrence of Arabia,* and many other notable films.[112]

Figure 1.6. Charles Lindbergh preparing to depart in his Ryan B-1 Brougham from a field near Romansville, Pennsylvania, March 14, 1928.

Photographer: Llewellen Jefferson Hoopes. Courtesy of Hoopes's grandson, Austin Wilson, and great-granddaughter, Christine Wright.

Hannah Steele Pettit's mother was Sophia McLaren Bard (1856–1899), born in Allegheny City, Pennsylvania, now part of greater Pittsburgh. She married John Dutton Steele II on April 14, 1884, two and a half years before Hannah was born. A full account of the Bard ancestry was meticulously researched by George Overcash Seilhamer and published in 1908. The surname extends as far back as the 10th century AD, with Baird, Barde, and Beard as orthographic variants; one Seigneur de Barde was among the army of William the Conqueror.[113] The progenitor of the American branch of the Bard family, Archibald Beard (1710–1765), immigrated from County Antrim of present-day Northern Ireland to Delaware. In 1741 he purchased 5,000 acres with three other gentlemen in (what was then) Prince George's County, Maryland, from Charles Carroll of Annapolis (1702–1782), whose son (Charles Carroll of Carrollton) would later sign the Declaration of Independence; they named it "Carroll's Delight" and divided it among themselves.[114] There is no concrete information on the vital statistics of Archibald's wife, but they produced four children, one of whom was Richard Bard (1736–1799).[115] He not only carried through the line to Hannah Bard Steele but was a central figure in a famous pioneer episode that was figuratively—and literally—hair-raising. The graphic and detailed account of what happened was cast in the form of his poetic narrative that came to be known as "The Ballad of Richard Bard." It was that ballad and the papers of Richard's son Archibald (compiled by Archibald Loudon) that formed the basis for G. O. Seilhamer's text, a summary of which follows.[116]

Richard was born near Fairfield, Pennsylvania, not far from Gettysburg. He learned the miller's trade, married Catherine Poe (no relation to Edgar Allan Poe) on December 22, 1756, and built a mill on Mill Creek. Warriors of the Delaware nation burned it down in 1758, but Richard rebuilt it. Richard and Catherine started a family in September 1757 with a boy they named John. On April 13, 1758, 19 Delawares came to the house, occupied then by the Bards, their visiting cousin Thomas Potter, and a little girl and boy. After a skirmish at the front door, the Delawares retreated, but because the roof of the cottage was thatched and ample wood was stacked against the house that could be used to set it afire, the outnumbered Bards surrendered on the condition that their lives be spared. The house was plundered and the new mill burned. Two field workers, Samuel Hunter and Daniel McManimy, and a boy named William White, on his way to the mill, were also captured. Not far from the house, Potter was killed and scalped. Three or four miles from the mill, the Bard baby was tomahawked in the head and chest and then scalped. The prisoners and their captors made their way into the Cumberland Valley and Appalachian Mountains, traveling 30–40 miles before they rested. The Delawares built a fire and placed the scalp of little John Bard near his parents as psychological torture. The next day they climbed to the top of Tuscarora Mountain and without warning tomahawked and scalped Samuel Hunter. On the third day of the march, half of Richard Bard's face was painted red, indicating that his captors were equally divided on whether he should be killed. On the fifth day they all reached Stoney Creek in the Alleghenies, whereupon Richard's hat, taken by one of the Delawares, blew off his head and downstream some distance. When the warrior went to retrieve it, Richard took the opportunity to cross the stream and slip away, but he was recaptured and beaten badly. It was at that point he decided to escape as soon as possible. The opportunity arrived that evening. While Catherine and Richard were plucking a turkey together, Richard covertly told her of his plan.

One of the Delawares' favorite pastimes was to dress themselves in the clothes of their female captives, and that night one of them donned Catherine's gown while Richard was sent to a nearby spring for water. She joined in the fun and kept them occupied while he slipped away into the darkness and was never found after two days of searching. Meanwhile Catherine and the other prisoners were marched from one Delaware encampment to another and were beaten by squaws. At the village of Kaskaskunk, Daniel McManimy was tortured, scalped, and killed. Catherine expected to share his fate but instead was handed over to two Delaware men to replace their deceased sister. The three of them went to the headwaters of the Susquehanna River, and after 500 miles of travel she fell ill from fatigue, cold, and hunger. Recovering after two months, she heard from another captive woman that if she learned their language, one of the Delawares would take her as a wife; if she refused, they would kill her, so she made it a point not to speak in anything other than English for the next two years.

Richard had escaped by hiding in a hollow log as the Delawares passed by, and when they were some distance away he headed in a different direction. His feet were sore and festering, so he bound them with shreds of his breeches and limped along, surviving on buds, berries, and four rattlesnakes eaten raw. On the afternoon of the ninth day he turned a corner on a path and came face to face with three friendly Cherokees, who led him to Fort Lyttleton, where he rested and healed. For the next two years he searched for his wife and inquired as to where she might be.

During the final years of the French and Indian War, British General John Forbes, Colonel George Washington, and a force of up to 8,000 men targeted Fort Duquesne, held by the French, at the confluence of the Allegheny and Monongahela Rivers to form the Ohio River. After the Delaware, Mingo, and Shawnee tribes deserted the French and moved west, the French soldiers then burned the fort and escaped down the Ohio River. Forbes assigned Washington to take it, but when he and his command entered, all they found were charred remains. Fort Pitt, named after British prime minister William Pitt the Elder, would soon be built on the site.[117]

In 1758, Richard Bard went to Fort Pitt at a time when Forbes was attempting to establish a treaty with Delawares across the river, some of whom had been in the party that captured Catherine. They said that they did not know her whereabouts but promised information the next day. Bard heard that if he returned they would kill him for escaping, so he wisely did not.

He later made a second trip toward Fort Pitt with a wagon train and stopped at a British outpost, Fort Bedford, where Richard asked if White Eyes, a leader of the Delawares, would escort him in safety the rest of the way. White Eyes agreed, but a few miles out some of the tribe turned off and brought in horses and a keg of whiskey. All but Richard became intoxicated, and White Eyes raised his gun to shoot Bard for escaping. Bard rounded a tree to stay out of the line of sight, but White Eyes followed him, and round and round they went to the amusement of the others until one young brave took the gun from White Eyes and hid it under a log. White Eyes then hit Richard with a large stick that bruised his arm for weeks, but he managed to mount his horse and speed back to Pittsburgh.

While there, he learned that the two men who adopted Catherine would release her for the sum of 40 pounds sterling and wrote her to that effect. Not receiving an answer, Richard went himself to Cherry Tree on the west branch of the Susquehanna River and met a party of Delawares on the road. He repeated the offer, but the Delawares doubted that they would receive their money if they were among the whites. Richard offered himself as a hostage if Catherine were to leave and return with the money. The offer led the Delawares to trust him, and all went into town, where the exchange was made to everyone's satisfaction.

During the Revolutionary War, Richard was a member of Captain Joseph Culberton's marching company in Philadelphia and later in the ranging company of Captain Walter McKinnie in the western frontier.[118] He was later a member of the Pennsylvania Convention of 1787, to which the Constitution was submitted for ratification. As an anti-Federalist, he refused to lend his signature, which earned him abuse in the county newspaper.[119]

Richard and Catherine Bard left 352 acres to their son Thomas (1769–1845) in their will. He farmed the land and lived in the stone mansion built by his father.[120] In 1814 he volunteered as part of a regiment defending Baltimore in the War of 1812 and was elected to the Pennsylvania legislature in 1832–1833.[121] He married Jane Cochran McFarland (1783–1857). Among their seven children was Richard Bard (1806–1867), who was born in Franklin County, Pennsylvania, but moved to Pittsburgh in 1843, where he worked in the leather business.[122] He was married to Eliza Jane Carson (1816–1860). One of their 11 children (Robert Washington Bard) fought at Chancellorsville in the Civil War. Another was James William Bard, who fought in the battle of Fredericksburg and was wounded in the knee in the battle of Spotsylvania in 1864; he escaped amputation only by threatening the surgeon with a gun.[123] He was fortunate to have suffered only a knee injury because it was one of the bloodiest battles of the Civil War, behind only Gettysburg and Chickamauga in number of casualties. A third child of Richard and Eliza Bard was Sophia McLaren Bard, mother of Hannah Bard Steele and Hugh Exton Steele and grandmother of Marjorie Pettit Meinel.[124]

The nephew of this younger Richard Bard was Thomas Robert Bard (1841–1915), who was left fatherless at the age of 10. Seven years later he studied law at the Chambersburg Academy, but ill health compelled him to go into business.[125] During the Civil War he was a volunteer Union scout during the Confederate invasions of Maryland and Pennsylvania. Captured behind enemy lines, he was about to be hanged as a spy but was saved at the last minute by the Union cavalry.[126] Soon thereafter he was hired by Thomas A. Scott, assistant secretary of war and later president of Pennsylvania Railroad, to oversee 277,000 acres in Ventura and Humboldt Counties, California, at that time sparsely populated and plagued with squatters and desperadoes.[127] Later he entered California politics as a Republican candidate for State Senator from Ventura, Santa Barbara, and San Luis Obispo counties, but lost the last of those by a small margin. As an early commissioner of Ventura County, he was one of its founding fathers and was elected senator from California in 1900.[128] Although he successfully fought against a bill that would have admitted Arizona and New Mexico to the Union as one state and served as chairman of the Committee on Irrigation (reclaiming large arid tracts in the West), he was defeated for re-election when he argued (successfully) against a bill to grant public funds to Catholic and other sectarian Native American schools.[129] While Thomas Bard was senator, the great-grandson of the Delaware chief White Eyes, who had tried to kill the senator's great-grandfather, Richard Bard, appeared in Washington to advance a land claim; Senator Bard agreed to help him.[130] Thomas Robert Bard married Mary Alice Gerberding in April 1876 and with her had eight children.[131] The youngest was Archibald Philip Bard (1898–1977), who was a respected physiologist and later dean of the Johns Hopkins Medical School.[132]

One of us (AMP) has verified the maternal ancestry of Eliza Jane Carson Bard (see earlier mention), wife of the younger Richard Bard, extending from her mother, Agnes Craig King (1788–1862), and her mother, Margaret McDowell (1758–?), through the McDowell line in Northern Ireland up through Alexander McDowell (1595–1652). His father, John McDowell (ca. 1575–1631) emigrated to Northern Ireland from Scotland. The line from his maternal grandmother, Janet Stewart (ca. 1535–1590), to her great-grandfather, James Stewart (1394–1451), lived variously in Fifeshire and Argyllshire, Scotland. The mother of James Stewart, Isobel MacDougall (ca. 1362–1439), was the daughter of Jonet Isaac, who was granddaughter of Robert the Bruce (1274–1329), the first king of Scotland and revered today throughout the nation as one of its founding fathers. He has been the subject of several biographies, notably those by Ronald McNair

Scott (1982) and Michael Penman (2014), tracing a colorful and eventful war of Scottish independence that need only be summarized here.

Robert the Bruce was the sixth generation bearing the same name, all descended directly from Adam de Brus (1051–1081), who had accompanied William the Conqueror to England in 1066 and who later, because he had succeeded in breaking Anglo-Saxon resistance in Yorkshire, was given manors in and around that county.[133] Robert was born July 11, 1274, at Turnberry Castle above the Firth of Clyde on the west coast. He was the son of Robert de Bruce, the sixth Lord of Annandale, and Marjorie, Countess of Carrick. Following the death of Alexander III, king of the Scots, in 1286, King Edward I of England (1239–1307) filled the vacant Scottish throne with his relative, John Balliol. This move impelled Robert, sixth Lord of Annandale, to bequeath the Earldom of Carrick to his son and thereby advance Robert's claim to the throne of Scotland. Robert the future king was more nationalistic than his father and joined in the rebellion against Edward I, setting the stage for guerrilla warfare and the final battle for independence with Edward II (1284–1327) at Bannockburn on June 23–24, 1314. Although the Scots were outnumbered 4–1, their pre-established hedgerow of spears stopped the advance of the English cavalry; the resulting melee of men and horses prevented the English archers in the rear from shooting their arrows because to do so would injure or kill their own men and horses.[134] Edward II and his forces turned around in full retreat, leaving heavy casualties in their wake, and few battles have left the English more humiliated. Later, Pope John XXII (1244–1334) recognized the earlier crowning of Robert the Bruce as king of Scots. In 1328 the Treaty of Edinburgh-Northampton was signed with Edward III, recognizing Robert as king and Scotland as an independent country.

Robert the Bruce died of unknown causes on June 7, 1329. Signifying the respect and worship that the Scots had for him, his heart was removed and ultimately ended up at Melrose Abbey in Roxburghshire on the border with England. His body was buried in the choir at Dunfermline Abbey in county Fife, next to his wife Elizabeth de Burgh and other kings and queens of Scotland.[135] In February 1818 workmen building a new parish church over the old choir found a vault with a decayed oak coffin with a skeleton covered in a gold shroud and lead; a crude crown of lead fashioned over the head.[136] The sternum had been sawed longitudinally from top to bottom, which would have been necessary for removal of the heart, a sign that they were indeed the remains of Robert the Bruce.[137] This was confirmed later by the discovery of a copper plate with the inscription *Robertus Scotorum Rex* (Robert, King of Scots) in the shape of a cross with a crown at the top.[138] A plaster cast was made of the skull by William Scoular before the remains were covered in pitch and reinterred. Several forensic sculptors have reconstructed the face from this plaster cast; the most recent recreation is the one by Christian Corbet.[139]

Modern research indicates that there may be a genetic basis for personality involving several genes, but more than likely cultural mores and values are also responsible for the consistency of strong character cascading down through the generations of Meinels, Pettits, Steeles, Bards, and the rest, leading to Aden and Marjorie Meinel. Throughout the centuries, those generations, as well as so many others, relied on their own resources for survival as they moved from the east to the west, or in the case of natural leaders such as Robert the Bruce, defended their country because they defined themselves by it. Their methods for coping with hardships that are unimaginable today endured in their children and their grandchildren, who encountered their own perils—oppression, war, disease, natural disasters—and had to devise ways to meet their own needs.

That pioneering spirit carried through to Aden and Marjorie, members of the Greatest Generation, called to the service of their country in war and peace without fanfare or self-aggrandizement. Like Edison Pettit, Aden built whatever he required in his home workshop, whether it was his own set of varifocal lenses for his eyeglasses, an f/1 camera, solar collectors, a model of the Multiple Mirror Telescope, or the Meinel tool—a way to polish aspheric mirrors without digging grooves. He always had the farsightedness to see the big picture, instead of simply its individual components, which made him the right person at the right time to establish a national observatory and a college of optical sciences. The narrative of how this developed, with Marjorie at his side, begins in Pasadena during the Great Depression.

Notes

1. Year: 1910; Census Place: Pittsburgh Ward 8, Allegheny, Pennsylvania; Roll T624_1301; Page 14A; Enumeration District: 0378; FHL microfilm: 1375314.
2. *Marriage Records. Pennsylvania Marriages.* Various County Register of Wills Offices, Pennsylvania.
3. Letter from Aden Meinel to Dr. Aden C. Irwin, January 16, 1973; Aden S. Baker death certificate.
4. Meinel, A. B., Meinel, M. P., and Jacobs, B. M. 2008. *The golden age of astronomy: In the beginning. . . .* Unpublished, 12.
5. The National Archives, Washington, DC; *Pension Payment Roll of Veterans of the Revolutionary War and the Regular Army and Navy, 3/1801–9/1815;* NAI Number: *2600769;* Record Group Title: *Records of the Department of Veterans Affairs, 1773–2007;* Record Group Number: *15;* Series Number: *M1786;* Roll Number: *1.*
6. *Find a Grave.* http://www.findagrave.com/cgi-bin/fg.cgi.
7. Select publications listed in endmatter.

8. Meinel, M. S. 2014. Pettit, Edison. In *Biographical encyclopedia of astronomers* (eds. T. Hockey et al.). Springer, New York.
9. Meinel, M. S. 2014. Pettit, Edison.
10. 1912 Alumni Directory for State Normal School of Nebraska.
11. 1912 Alumni Directory.
12. 1912 Alumni Directory; Meinel, M. S., Pettit, Edison.
13. Nicholson, S. B. 1962. Edison Pettit, 1889–1962. *Publications of the Astronomical Society of the Pacific* 74: 495–8.
14. Nicholson, Edison Pettit, 495–8.
15. Pettit, E., and Steele, H. 1918. The application of Schaeberle's method in the photography of the corona at Matheson, Colorado, June 8. *Popular Astronomy* 26: 466–80.
16. Meinel, M. S., Pettit, Edison.
17. .*American Physical Society News*, 26 (9).
18. Meinel, M. S., Pettit, Edison.
19. Meinel, M. S., Pettit, Edison; Meinel et al., *Golden age*, 6.
20. Meinel, M. S., Pettit, Edison; Meinel et al., *Golden age*, 6.
21. Meinel, M. S., Pettit, Edison; Nicholson, Edison Pettit, 496; Miller, J. A., Pitman, J. H., and Steele, H. B. 1920. The parallaxes of fifty stars determined at the Sproul Observatory (second list). *Sproul Observatory Publications* (5): 1–65.
22. Meinel et al., *Golden age*, 6.
23. Nicholson, Edison Pettit, 496–7.
24. Nicholson, Edison Pettit, 496–7; Meinel, M. S., Pettit, Edison.
25. Meinel, M. S., Pettit, Edison.
26. Meinel, M. S., Pettit, Edison; select publications listed in endmatter.
27. Pettit, E. 1932. Characteristic features of solar prominences. *Astrophysical Journal* 76: 9–43.
28. Pettit, personal correspondence.
29. Letter from A. O. Howard to Edison Pettit, July 21, 1939.
30. Memoir of Walter Jones.
31. Nicholson, Edison Pettit, 497; Meinel, M. S., Pettit, Edison.
32. Meinel, M. S., Pettit, Edison.
33. Nicholson, Edison Pettit, 497.
34. Mumford, G. S. 2014. McMath, Robert Raynolds. In *Biographical encyclopedia of astronomers* (eds. T. Hockey et al.). Springer, New York.
35. Mumford, McMath, Robert Raynolds; Meinel, M. S., Pettit, Edison.
36. Letter from Aden Meinel to W. Patrick McCray, November 9, 1998.
37. Letter from Edison Pettit to Robert R. McMath, May 31, 1939, Box 6, Edison Pettit Papers. The Huntington Library, San Marino, California.
38. Letter from Edison Pettit to Dr. C. D. Shane, July 3, 1945, Box 6, Edison Pettit Papers. The Huntington Library, San Marino, California; Meinel, M. S., Pettit, Edison.
39. Butler, C. 1962. The light of the atom bomb. *Science* 138: 485.
40. Aden and Marjorie Meinel, personal correspondence.
41. Meinel et al., *Golden age*, 7.
42. Nicholson, Edison Pettit, 498.

43. Meinel, M. S., Pettit, Edison.
44. Van Wyck, K. L. W. 1936. *Genealogy of Pettit families in America: Descendants of John Pettit 1630–1632 first of that name in America.* Reprinted 2013. Isha Books, New Delhi, India, 9.
45. Vivian, J. L. 1897. *The visitations of Cornwall comprising the heralds' visitations of 1530, 1573, & 1620.* William Pollards & Co., Exeter, England, 494.
46. Vivian, *The visitations of Cornwall*, 494, 495; Van Wyck, *Genealogy of Pettit families*, 9.
47. Vivian, *The visitations of Cornwall*, 494.
48. Berry, W. 1830. *County genealogies: Pedigrees of the families in the county of Kent.* Sherwood, Gilbert and Piper, London, 157.
49. Cooke, R. 1924. *The visitations of Kent taken in the years 1574 and 1592.* John Whitehead and Son, London, 20–1.
50. Van Wyck, *Genealogy of Pettit families*, 15, 17.
51. Van Wyck, 21; Huntington, E. B. 1868. *History of Stamford, Connecticut, from its settlement in 1641, to the present time, including Darien, which was one of its parishes until 1820.* Self-published, Stamford, Connecticut, 59.
52. Van Wyck, *Genealogy of Pettit families*, 94–112.
53. Van Wyck, *Genealogy of Pettit families*, 29.
54. Van Wyck, *Genealogy of Pettit families*, 29.
55. Van Wyck, *Genealogy of Pettit families*, 61.
56. Van Wyck, *Genealogy of Pettit families*, 62.
57. Stratton, E. A. 1986. *Plymouth Colony: Its history and people, 1620–1691.* Ancestry Publishing, Salt Lake City, Utah, 21, 280–1.
58. Genealogical profile of Philip Delano/De la Noye (collaboration of Plimoth Plantation and New England Historic Genealogical Society), www.plimoth.org/media/pdf/delano_phillip.pdf; accessed September 15, 2018.
59. Roberts, G. B. (comp.). 1989. *Ancestors of American presidents.* New England Historic Genealogical Society, Boston, 31.
60. Roberts, *Ancestors*, 70.
61. Roberts, *Ancestors*, 227; however, Van Wyck, *Genealogy of Pettit families*, 66, cites a child by Mary named Daniel born in 1656, the same year that Mary died, perhaps in childbirth.
62. Roberts, *Ancestors*, 227; Van Wyck, *Genealogy of Pettit families*, 66.
63. Stratton, *Plymouth Colony*, 265.
64. Stratton, *Plymouth Colony*, 312.
65. Van Wyck, *Genealogy of Pettit families*, 29.
66. War of 1812 Pension Applications. Washington, DC: National Archives. NARA Microfilm Publication M313, 102 rolls. Records of the Department of Veterans Affairs, Record Group Number 15.
67. Lewis Publishing Company. 1904. *A biographical and genealogical history of southeastern Nebraska.* Chicago, Illinois, vol. II, 981.
68. Van Wyck, *Genealogy of Pettit families*, 57.
69. Letter from Leonard Knox Pettit to Edison Pettit, December 17, 1959.
70. Morris, R. 1883. *William Morgan: or Political anti-Masonry, its rise, growth and decadence.* Robert Macoy, New York, 9.

71. Morris, *William Morgan*, 30.
72. Versluis, A. 2006. North American esotericism. In *Introduction to new and alternative religions in America* (eds. E. V. Gallagher and W. M. Ashcraft), vol. 3, 92–148. Greenwood Publishing Group, Westport, Connecticut, 94.
73. Lewis Publishing Co., *A biographical and genealogical history*, 981; Hallenbeck, P. P. Undated. The life of Samuel Morrow. Unpublished, deposited with Meinel papers, Special Collections, University of Arizona, Tucson, Arizona.
74. Lewis Publishing Co., *A biographical and genealogical history*, 982; an obituary for Martha Knox Pettit states that her father Leonard was a scout for "Boone, Duncan, McArthur, and Gen. Clark" [presumably William Clark who explored the West with Meriwether Lewis], but the last three names could not be confirmed elsewhere.
75. Chernow, R. 2017. *Grant*. Penguin Press, New York, 730.
76. Hallenbeck, The life of Samuel Morrow.
77. Hallenbeck, The life of Samuel Morrow.
78. Farley, M. L. 1930. Much early history is found in memories of pioneer Pettit family. *Peru Pointer*, deposited with Meinel papers, Special Collections, University of Arizona. Tucson, Arizona.
79. Farley, Pettit family.
80. Dundas, J. H. 1902. Nemaha County. Digitized by Google, 14; Farley, Pettit family, *Peru Pointer*.
81. Hallenbeck, The life of Samuel Morrow; Farley, Pettit family.
82. Farley, Pettit family.
83. Farley, Pettit family.
84. Farley, Pettit family.
85. Hallenbeck, The life of Samuel Morrow.
86. Farley, Pettit family; Historic floods on the Missouri River, http://www.dnr.ne.gov/flo odplain/mitigation/mofloods.html; accessed September 22, 2018.
87. Hallenbeck, The life of Samuel Morrow.
88. Hallenbeck, The life of Samuel Morrow.
89. Hallenbeck, The life of Samuel Morrow.
90. Hallenbeck, The life of Samuel Morrow.
91. Hallenbeck, The life of Samuel Morrow.
92. Hallenbeck, The life of Samuel Morrow.
93. Hallenbeck, The life of Samuel Morrow.
94. Stone, F. D., Jr. 1896. *The descendants of George Steele of Barthomley, Cheshire, England, and Chester County, Pennsylvania*. Self-published, Philadelphia, 5, 12.
95. Narain, Jaya. "Author puts medieval home in trust to protect it from footballers invasion of village," *Daily Mail*, January 3, 2008.
96. Stone, *The descendants of George Steele*, 12–13.
97. Fea, A. 1908. *Secret chambers and hiding places: Historic, romantic, & legendary stories & traditions about hiding-places, secret chambers, etc.* 3rd ed. Project Gutenberg, eBook # 13918; accessed September 23, 2018.
98. Stone, *The descendants of George Steele*, 7.
99. Steele, J. D. 1930. *Recollections of my life in England and America. With introduction by his great granddaughter Loraine S. McKinstry*. Self-published, 25.

100. Steele, *Recollections of my life*, 31, 33, 40.
101. Steele, *Recollections of my life*, 44–8.
102. Stone, *The descendants of George Steele*, 8.
103. Steele, *Recollections of my life*, 58–9.
104. Steele, *Recollections of my life*, 59.
105. Steele, *Recollections of my life*, 12.
106. Steele, *Recollections of my life*, 12.
107. Downingtown Area Historical Society Hist-O-Gram, vol. 5, no. 43. October 23, 2014.
108. www.southernchestercountyweeklies.com/three-men-remember-charles-a-lind-bergh/article_de84a1cc-73e1-5afa-91a6-c4454ba50f6b.html; accessed September 20, 2018.
109. Berg, A. S. 1998. *Lindbergh*. Berkley Books, New York, 170.
110. www.southernchestercountyweeklies.com/three-men-remember-charles-a-lind-bergh/article_de84a1cc-73e1-5afa-91a6-c4454ba50f6b.html; accessed September 20, 2018.
111. www.southernchestercountyweeklies.com.
112. https://pabook.libraries.psu.edu/literary-cultural-heritage-map-pa/bios/Rains_Cla ude; accessed September 20, 2018.
113. Seilhamer, G. O. 1908. *The Bard family: A history and genealogy of the Bards of "Carroll's Delight" together with a chronicle of the Bards and genealogies of the Bard kinship.* Kittochtinny Press, Chambersburg, Pennsylvania, 10.
114. Seilhamer, *The Bard family*, 145–6.
115. Seilhamer, *The Bard family*, 158.
116. Seilhamer, *The Bard family*, 159–85.
117. Chernow, R. 2010. *Washington: A life.* Penguin Press, New York, 91.
118. Seilhamer, *The Bard family*, 188.
119. Seilhamer, *The Bard family*, 189.
120. Seilhamer, *The Bard family*, 209.
121. Seilhamer, *The Bard family*, 210.
122. Seilhamer, *The Bard family*, 224.
123. Seilhamer, *The Bard family*, 224–5, 246.
124. Seilhamer, *The Bard family*, 225.
125. Press Reference Library, 163–4.
126. Seilhamer, *The Bard family*, 251.
127. Seilhamer, *The Bard family*, 251–4.
128. Seilhamer, *The Bard family*, 254–5.
129. Press Reference Library (Western edition). 1915. *Notables of the West. Being the portraits and biographies of the progressive men of the West.* International News Service, 165.
130. Press Reference Library, *Notables of the West*, 163.
131. Press Reference Library, *Notables of the West*, 255–6.

132. Harrison, T. 1997. Archibald Philip Bard. *Biographical memoirs of the National Academy of Sciences*. Vol. 72: 14–26. National Academy Press, Washington, DC.

133. Scott, R. M. 1982. *Robert the Bruce: King of Scots*. Carroll & Graf, New York, 10.

134. Scott, *Robert the Bruce*, 159.

135. Scott, *Robert the Bruce*, 226, 228.

136. Jardine, H. 1821. *Report relative to the tomb of King Robert the Bruce, and the cathedral church of Dunfermline*. Hay, Gall, and Co., Edinburgh, 28–30..

137. Jardine, *Report relative*, 37.

138. Jardine, *Report relative*, 36.

139. Jardine, *Report relative*, 44.

2
Rising Stars

It is understandable how the daughter of a world-class astronomer, who assisted him in his solar research at their home observatory for many years, could excel in her coursework, from elementary school through a master's degree in astronomy and become a leading exponent of solar energy. How the son of a house-painter of modest means could emerge from the Depression as a literal rocket scientist, fulfill dangerous undercover assignments in Germany at the end and aftermath of World War II as a 22-year-old Navy ensign, receive both a BA and PhD at Berkeley within the same three-year period, help plan and direct the first national observatory, pioneer solar energy systems with his wife, establish and secure funding for the renowned College of Optical Sciences, consult on classified aerial surveillance during the Cold War at a high security level, and assist in repairing the Hubble Space Telescope and designing visionary terrestrial telescopes around the world is less obvious, if not staggering.

Through a confluence of fateful encounters, limitless ingenuity, and incredible opportunities afforded to few, both Aden Baker Meinel and Marjorie Steele Pettit were nurtured in an environment rich in scientific experimentation and achievements by Nobel Prize winners, one that placed a high premium on academic excellence and figured prominently in the development of military and space technology. That environment was the growing city of Pasadena, California, which attracted not only tourists by the thousands and the nation's most honored scientists, but also snowballing corporate investment and endowments. The fruits of such largesse would eventually include Mt. Wilson Observatory and Caltech. To understand the atmosphere in which Aden and Marjorie were born and raised, it is necessary to consider the founding of Pasadena and how it became the academic hot spot that would influence their future together.

Origins and settlement

Spanish explorer Gaspar de Portolá and Father Juan Crespí, accompanied by a troop of soldiers, traveled by land and established the Presidio of San Diego in 1769, and then headed north toward Los Angeles. At a fertile watershed that came to be known as Arroyo Seco, they erected a cross, sprinkled holy water, and thereby founded Mission San Gabriel Arcángel, all under the watchful eyes of

With Stars in Their Eyes. James B. Breckinridge and Alec M. Pridgeon, Oxford University Press. © Oxford University Press 2022. DOI: 10.1093/oso/9780190915674.003.0002

the Tongva, the Native American people of southern California.[1] Their language was Uto-Aztecan, pointing to a historical if not genealogical relationship with the Hopis and Comanches of the American Southwest.[2] The peaceful Tongva, also called Gabrieleños and Gabrielinos, were forced by the Spaniards to help build the chapel, by some accounts only after brutality prior to their conversion to Christianity.[3] The San Gabriel Mission prospered until after the 1822 Mexican War of Independence, when California passed into Mexican jurisdiction. Eleven years later, Mexico passed the Secularization Act of 1833, which dismantled the mission system and divided and distributed the land as *ranchos*. Two of those ranchos, Rancho del Rincon de San Pasqual and Rancho Santa Anita, comprised much of present-day Pasadena and Altadena.[4] Both ranchos changed hands several times until the Mexican-American War of 1846–1848. In the Treaty of Guadalupe Hidalgo in 1848, Mexico ceded California to the United States, as well as all or portions of New Mexico, Arizona, Nevada, Utah, and Colorado. California was granted statehood two years later.

The title to Rancho San Pasqual, which spanned almost 14,000 acres, included Pasadena, Altadena, South Pasadena, Alhambra, San Marino, and San Gabriel. It passed first to trader/trapper Benjamin Davis Wilson (1811–1878) in 1859 and then jointly to Wilson and surgeon John Strother Griffin, MD (1816–1898) in 1860.[5] Wilson had survived attacks by Indians, Mexicans, and even a grizzly bear. Notably, he was the grandfather of General George S. Patton, Jr., in whose 3rd Army Aden would briefly serve in World War II (Chapter 4). Wilson was elected mayor of Los Angeles, Los Angeles County supervisor, and California state senator. His role in the development of his namesake, Mt. Wilson, will be discussed later. John Griffin was ranking surgeon under General Stephen Watts Kearny and Commodore Robert F. Stockton during the Mexican-American War.[6] Beginning in the 1870s, both Wilson and Griffin sold off large parcels of their properties, thousands of acres at a time, to buyers from the East, anxious to escape cold winters and respiratory diseases such as tuberculosis and to start new lives in an ideal climate on fertile land with abundant water.[7] The completion of the First Transcontinental Railroad in 1869 facilitated quicker shipment of goods over the rails at 22 miles per hour in Pullman palace-cars and gave visitors easier and safer transportation to scout out new properties in California. One of the curious was Daniel M. Berry from Indianapolis, Indiana.[8]

Berry had read a book by British-born novelist Charles Nordhoff titled *California for Health, Pleasure, and Residence: A Book for Travellers and Settlers*, which heralded the new state as the land of milk and honey:

> There are no dangers to travelers on the beaten track in California; there are no inconveniences which a child or a tenderly reared woman would not laugh at; they dine in San Francisco rather better, and with quite as much form and

a more elegant and perfect service, than in New York; the San Francisco hotels are the best and cheapest in the world; the noble art of cooking is better understood in California than anywhere else where I have eaten; the bread is far better, the variety of food is greater; the persons with whom a tourist comes in contact, and upon whom his comfort and pleasures so greatly depend, are more uniformly civil, obliging, honest, and intelligent than they are anywhere in this country, or, so far as I know, in Europe; the pleasure-roads in the neighborhood of San Francisco are unequaled anywhere; the common country roads are kept in far better order than anywhere in the Eastern States; and when you have spent half a dozen weeks in the State, you will perhaps return with a notion that New York is the true frontier land, and that you have nowhere in the United States seen so complete a civilization—in all material points, at least—as you found in California. Moreover, the cost of living is to-day less in California by a third than in any Eastern State; it is, at this time, the cheapest country [sic] in the United States to live in.[9]

In the course of his promotional works, Nordhoff described the pleasure of the train journey, with suggested stops in St. Louis and Salt Lake City, the orange orchards and olive and almond groves in southern California, sightseeing throughout the state, opportunities for hunting and fishing, the finest climate in the world, and the costs of purchasing land and how to farm it for every crop from alfalfa to wine grapes.

By chance, Daniel Berry happened to meet Myer J. Newmark from Los Angeles at a New York hotel and persuaded him to stop in Indiana on his way back west and speak to Berry's group of friends at the home of surgeon Thomas Elliott, MD. Newmark, who then owned Rancho Santa Anita, was anxious to sell his property and extolled its merits. Elliott and others then formed the California Colony of Indiana, consisting of 27 families, and authorized four men to go on a reconnaissance mission to California and report back. Only Berry ended up going and ruled out San Diego and Rancho Santa Anita for different reasons. He settled on Rancho San Pasqual, then owned by Wilson and Griffin, and acting as agent for the California Colony, found other investors and purchased 3,962.35 acres for $6.31/acre, deeded to the newly named San Gabriel Orange Grove Association at the end of December 1873. The settlers amicably apportioned the land among themselves, planted 1,000 orange and lemon trees for future income, and beautified the streets with ornamental trees.[10] An orange orchard of only 10 acres would provide at maturity a net annual profit of $10,000.[11] They named their agricultural community "Muscat" after the grape variety they hoped to grow, but to some it sounded too much like "muskrat," and they preferred a Native American name.[12] Over the years it had been called *llave del valle* ("key of the valley") and "crown of the valley"; in 1875 the settlers decided on the Chippewa phrase meaning "the valley" or "of the valley"—Pasadena.[13] The first school was built that year for 17 students; it was also used for an interdenominational church and as meeting room for a literary society. In the next few years the settlers built Presbyterian and Methodist churches and a post office and extended a water line from the reservoir to serve the residences.[14]

The next decade saw exponential growth of the new city of Pasadena. In 1883 a weekly newspaper, the *Pasadena Chronicle*, published its first issue in August, announcing a drug store, a second blacksmith, newsstand, shoemaker, two hardware stores, harness shop, and restaurant in the business district at Fair Oaks Avenue and Colorado Street.[15] To accommodate the rapid influx of tourists from the East, two hotels were erected, the Pasadena House and the more imposing, five-story Raymond Hotel.[16] To build houses and barns, lumber and tools had to be transported from Los Angeles over rough roads and swollen streams, so when the Los Angeles and San Gabriel Valley Railroad was finally completed in September 1885, Pasadena was no longer isolated in many respects, even less so when the rail line was bought out two years later by the Atchison, Topeka and Santa Fe Railroad with its connections to Chicago and eastern cities.[17] Two new schools were built in 1885 to join the Central School—the Monk Hill School and Wilson School. By 1887–1888, even six schools were not enough for the 1,800 students; by 1900 there were 3,000 students. Fred Sears, who would become assistant director at Mt. Wilson Observatory, was one of six students in Pasadena High School's first graduating class.[18]

The housing boom collapsed temporarily in 1888 for a variety of reasons, not the least of which were real estate speculation and a water shortage. Emphasis was then placed on tourism and restoring the garden image that Pasadena once had but lost in the building frenzy. The Board of Trade emphasized the city's educational and cultural facilities—schools, churches, a new Grand Opera House, a public library with thousands of accessions, three newspapers (*Pasadena and Valley Union*, *Pasadena Star-News*, and the short-lived *Pasadena Standard*).[19] Millionaires from the East now moved to Pasadena again for the lowered real estate prices, climate, and scenery, including in 1891 the new Tournament of Roses.[20] In that same year, Throop (pronounced "troop") University opened its doors.

Amos Gager Throop (1811–1894), born in upstate New York, had come to Pasadena from Los Angeles via Michigan and then Chicago, and in his final years decided to donate his fortune first to the Universalist Church (1886). Two years later he was elected mayor of Pasadena. While serving in that position he personally funded and opened a university in 1891 that his friends named Throop University in his honor. Trustees were chosen to oversee the coeducational and nonsectarian school, which was first located at the corner of Fair Oaks Avenue and Green Street. An additional building to be known as Polytechnic Hall was erected a few blocks north on Fair Oaks to Chestnut Street, and the name of the school was soon changed to Throop Polytechnic Institute. What set Throop apart from other colleges in the nation was its emphasis not only on the classical liberal arts, but also on trades that could guarantee employment after graduation, such as typewriting, stenography, woodworking, machine shop, cooking, sewing, and weaving.[21] In 1910 it was moved to the southeastern part of town and two years later was renamed the Throop College of Technology.[22] In an address at Throop the following year, former president Theodore Roosevelt recognized the one person who could take the school to the next level of excellence:

> I want to see institutions like Throop turn out perhaps ninety-nine of every hundred students as men who are to do given pieces of industrial work better than any one else can do them; I want to see those men do the kind of work that is now being done on the Panama Canal and on the great irrigation projects in the interior of this country—and the one-hundredth man I want to see with the kind of cultural scientific training that will make him and his fellows the matrix out of which you can occasionally develop a man like your great astronomer, George Ellery Hale.[23]

That Roosevelt should single out George Ellery Hale (1868–1938) was prescient, for it was Hale who, more than any other, shaped not only Throop into Caltech but also Mt. Wilson Observatory, the Huntington Library, and for that matter much of modern-day Pasadena. All these institutions would figure prominently in the careers of Edison Pettit, his daughter Marjorie, and Aden Meinel.

Mt. Wilson Observatory

In the late 1880s, former mayor of Los Angeles and trustee of the University of Southern California, Edward F. Spence, decided to endow a 40-inch refracting telescope on the summit of Mt. Wilson at 5,710 feet in the San Gabriel Mountains. University president Marion M. Bovard consulted Edward Pickering, director of Harvard College Observatory, who then sent his brother William and optician Alvan Clark to Mt. Wilson with a 4-inch telescope to test its observing conditions, its "seeing."[24] After making it up the eight-mile, primitive trail cleared to the summit in 1864 by Benjamin Wilson, Pickering and his expedition companions were excited by the visibility. That excitement spread throughout Pasadena and then back to Boston upon Pickering's report to his brother Edward.[25]

One of Edward Pickering's volunteer assistants at the observatory was MIT student George Ellery Hale, whose wealthy father and astronomer William had built an observatory at their home in Kenwood on the south side of Chicago.

Before graduating in 1890, Hale invented the spectroheliograph, which produces monochromatic photographs of the Sun at a single wavelength. He then married, traveled extensively to Europe and the Lick Observatory in California, and spent several years at his Kenwood Astrophysical Observatory. In 1892, at the age of 24, Hale was invited to become associate professor of astrophysics at the University of Chicago. He and University president William Harper persuaded financier Charles Yerkes to endow a new observatory bearing his name, and Yerkes Observatory was dedicated in 1897 with young Hale as its first director (see Chapter 6).[26]

Hale went to Pasadena in December 1903 with other staff members from Yerkes and a coelostat to explore simply whether Mt. Wilson was suitable for a temporary solar observatory. The expedition was funded by a grant of $10,000 from the Carnegie Institution of Washington.[27]

Hale himself lucidly explained how a coelostat works: "It consists . . . of a plane mirror 30 inches in diameter, rotated by clockwork at such a rate as to keep the beam of sunlight, reflected from its silvered (front) surface, in a fixed position on a second plane mirror standing above and south of it. From this mirror the beam is reflected nearly horizontally to a point 100 feet north, where it falls on a 24-inch concave mirror of 60 feet focal length, which forms a solar image about 6½ inches in diameter on the slit of the spectrograph or spectroheliograph."[28] Within 14 months, a semi-permanent solar observatory had been established there, staffed by three astronomers from Yerkes (Walter Adams, Ferdinand Ellerman, and George Ritchey) whose salaries were personally paid by Hale. This was a professional and personal gamble, as the Carnegie Institution had agreed to support only a study for a temporary solar observatory, whereas Hale, always the visionary, wanted a permanent solar facility with a large reflector to study stellar evolution as well. The Carnegie Board of Trustees met in December 1904 and awarded Hale $150,000 (about $4.9 million in 2021) for two years.[29]

The next task was to move the coelostat telescope donated to Yerkes by Helen Snow up the Wilson Trail to the summit of Mt. Wilson; Edwin Frost, who replaced Hale as director of Yerkes, made a permanent gift of the Snow telescope to Pasadena.[30] The logistics of transporting telescope parts weighing as much as 350 pounds required Hale to devise a steel truck with rubber tires to bear the weight. A man led a mule pulling the truck, with other men at each end to steer, followed by another mule as a reserve or to pull the truck back down the trail if necessary.[31] Eventually materials to build a 60-foot tower and a 150-foot tower, which were necessary to overcome warm air currents rising from Earth that blurred the solar image, were hauled up the same way. Using the Snow telescope, Hale and Adams photographed sunspot spectra, and with Henry Gale showed that sunspots have lower temperatures. Hale and Ellerman then photographed the sun in the light of H alpha and noted that bright hydrogen filaments (flocculi) aggregated around sunspots in a spiral pattern that resembled iron filings around magnetic poles; Hale called these solar vortices. Using the 60-foot solar tower, Hale discovered that, indeed, there are strong magnetic fields in sunspots, earning him a nomination for the Nobel Prize in Physics.[32]

In time Hale recruited and retained some of the brightest astronomers in the country (if not the world) for the Mt. Wilson Observatory, drawn in part by his addition of the 60-inch reflector telescope in December 1908 and, most of all, the 100-inch reflector completed in 1918, at that time the largest and most powerful telescope in the world, with a double yoke designed by Ritchey, three different secondary mirrors, two different flat mirrors, and three observing stations. Los Angeles hardware mogul John D. Hooker had donated $45,000 for a 100-inch mirror blank, which was cast in France, shipped to New Jersey and then New Orleans, and finally transported to Pasadena by train; the only condition of the donation was that Hooker wanted the telescope named after him.[33] To transport both telescopes up to the Observatory, the Wilson Trail had to be widened in two stages—one for each telescope—first by hand and mule-drawn scrapers and plows for the 60-inch, and then by power shovels and drills (and still by hand also) for the 100-inch.[34] Francis Pease, Ritchey's engineering assistant, joined the staff from Yerkes about a year later, after Ritchey, Ellerman, and Adams, and was responsible for designing the 60-inch telescope and its dome.[35] Other permanent appointments followed in short order up to 1920: Arthur King, superintendent of the physics lab; Charles St. John, solar physics; Harold Babcock, physics laboratory; Frederick Seares, stellar photometry; Edward Fath, stellar spectroscopy; Charles Abbot, research director; Adriaan van Maanen, solar physics; Harlow Shapley, stellar photometry; Alfred Joy, stellar spectroscopy; John Anderson, physics laboratory; Seth Nicholson, solar physics; Gustaf Stromberg, stellar spectroscopy; Paul Merrill, stellar spectroscopy; Edwin Hubble, nebular photography; and Edison Pettit, solar physics (father of Helen and Marjorie Pettit; see Chapter 1, section "Marjorie Steele Pettit"). Most were recruited directly from Yerkes (or had worked there at one time) or the University of California at Berkeley/Lick Observatory on Mt. Hamilton near San Jose. Two other staff members were added in the 1920s: Milton Humason, nebular photography, and Theodore ("Ted") Dunham, Jr., stellar spectroscopy. In the 1930s, six others were hired: Walter Baade, nebular photography; Rudolph Minkowski, nebular photography; Harold Babcock, stellar spectroscopy; Olin Wilson, stellar spectroscopy; Ira Bowen, stellar spectroscopy, and Don Hendrix, optics.[36] By 1939 Hale had assembled an impressive staff of astronomers who would help to define and describe the known universe (Figure 2.1). Although Aden interacted with John Anderson and Alfred Joy as a young student, Baade and Hendrix would play larger roles in his training in astronomy and optics, respectively, so some space is devoted here to an account of their contributions.

Figure 2.1. Mount Wilson Observatory astronomers assembled at their Pasadena office in 1939. From left to right, front row: Joseph O. Hickox, Adriaan van Maanen, Gustaf Stromberg, Harold D. Babcock, Frederick H. Seares, John A. Anderson, Walter S. Adams, Paul W. Merrill, Alfred H. Joy. Back row: Ralph E. Wilson, Milton L. Humason, Robert S. Richardson, Seth B. Nicholson, William H. Christie, Arthur S. King, Theodore H. Dunham, Jr., Edwin Hubble, Robert B. King, Rudolph Minkowski, Walter Baade, Edison Pettit, Olin C. Wilson, Edison R. Hoge, Roscoe F. Sanford.
Courtesy of the Huntington Library, San Marino, California.

Walter Baade (1893–1960) received his PhD from the University of Göttingen just after World War I ended in 1919, and then worked at the Hamburg Observatory in Bergedorf until 1931.[37] While there, he photographed globular clusters (most with cluster-type variable stars known as RR Lyrae variables today) and "nebulas" (mostly galaxies), discovered a new comet, and began work on the Andromeda Galaxy (M31) that would dominate his work at Mt. Wilson. He received a fellowship grant to visit several American observatories and finally arrived there in July 1926 when Walter Adams was director, having succeeded an ailing Hale.[38] There Baade continued his work on variable stars, specifically Cepheid variables, which vary in temperature and diameter with stable pulsation periodicity. He wrote a paper that described a way to determine the radius and absolute magnitude of a pulsating variable, now known as the Baade method.[39] He returned to Hamburg after his fellowship abroad and in the next few years earned an international reputation for his photometry, enough for Walter Adams to persuade his colleagues that Baade should be invited to join the permanent Mt. Wilson staff at an annual salary of $3,300 to work on "nebulae" and globular clusters.[40] Baade accepted the offer and arrived in September 1931 and remained there until 1958.[41] While there, he collaborated with Fritz Zwicky of Caltech to classify novas into two types; the more luminous type they called supernovas, which they claimed were the sources of cosmic rays, soon to be studied by Robert Millikan and William Pickering at Caltech with Aden's assistance.[42] Baade identified the Crab Nebula (M1) as a supernova remnant, one of the first nebulas to be recognized as such.[43] He would later identify many others with Rudolph Minkowski, using the Hale 200-inch telescope at the Palomar Observatory in San Diego County (Figure 2.2).

One of Baade's most profound discoveries was that there are two different populations of stars in galaxies. Population I stars, such as our Sun, are metal-rich and include extremely luminous O-type (blue-white) and B-type (blue)

Figure 2.2. 200-inch Hale telescope pointing to zenith, seen from east.

Photo by Mt. Wilson-Palomar Observatories. Courtesy of the Archives, California Institute of Technology.

stars, mostly in the disks of galaxies; in other words, they are hotter, bluer, and younger. Population II stars, in the galactic halo and bulge and in globular clusters, are metal-poor yellow giants rather than blue stars; they are cooler, redder, and older. He dealt with the origins of the two-population concept in his book *Evolution of Stars and Galaxies*.[44] The filters that Aden brought back from Jena after World War II (see Chapter 4, section "Jena") at Baade's request likely improved Baade's observations.

An even more profound discovery by Baade was that Cepheid variables are also of two types—classical and type II. Classical Cepheids are younger and larger Population I stars, whereas type II Cepheids are older and fainter Population II stars. The luminosity of classical Cepheids is about 1.5 times greater than that of type II Cepheids, and Baade calculated from the difference that the distance to the Andromeda Galaxy should be doubled.[45] Edwin Hubble had used classical Cepheids as a measure of distance in the universe, but Baade showed that calibrations of absolute magnitudes from motions and radial velocities of classical Cepheids in the Milky Way were wrong, as then also were the distance scales by a factor of two.[46] Baade had doubled the size of Hubble's universe. At the dedication of the 200-inch telescope funded by the Rockefeller Foundation, Baade stressed the need for measuring Cepheid variables at the faintest limits of the telescope and using those measurements to calibrate the absolute magnitudes of the brightest stars in measuring distances to other galaxies.[47]

Don Hendrix (1905–1961) did not have the academic pedigree of the other members of the Mt. Wilson staff but made equally momentous contributions at not only Mt. Wilson but also the Palomar, Kitt Peak, and Lick Observatories. He was born in Fort Worth, Texas, as Don Osgood Hendricks. His father later shortened the surname to Hendrix. After two years of high school he moved with his parents to the Los Angeles area in 1921 and worked installing radios for a music company until it went bankrupt in 1931 during the Depression. Through his father's connections with the support staff at Mt. Wilson, he learned of an immediate opening as an optics apprentice to Walter Adams. He was hired that same year and was trained by opticians W. L. Kinney and John Dalton, who had ample construction and maintenance experience with the 60-inch and 100-inch telescopes and other optical equipment. He was a quick study and would soon become the go-to optics specialist for all the observatories (Figure 2.3). In the course of his career he had hundreds of patents to his name, including one for the proximity fuze[48] for rockets and bombs, which explode not on impact or at a preset time but when near the target; he donated all his patents to the US government. Because of the top-secret nature of much of his work, he never openly received credit for his inventions.[49]

Walter Baade's friend Bernhard Schmidt had devised a fast, wide-field camera with a corrector plate that removed the coma and astigmatism of paraboloidal mirrors, which the 60-inch and 100-inch telescopes had. When Baade came to

Figure 2.3. Don Hendrix polishing the 200-inch mirror.
Courtesy of the Archives, California Institute of Technology.

Mt. Wilson Observatory in 1931, he shared this new design with John Anderson in the optics laboratory. Hendrix was probably the one who made the first Schmidt camera there in 1932 for spectroscopy.[50] Soon Hendrix was in demand by other observatories. Over the course of 21 weeks in 1939–1940, he and one assistant produced the 72-inch spherical primary mirror for the very fast (f/2.5) 48-inch Schmidt telescope at Palomar. In 1948 he corrected the mirror of the 200-inch Hale telescope by removing excess glass from the outer 20 inches and aluminized it. From 1951 to 1955 he was seconded to the Lick Observatory to grind, polish, and figure the 120-inch mirror of the Donald Shane reflector.[51] While at Lick, Hendrix trained Don Loomis, who would be the chief optician at Kitt Peak National Observatory and close colleague of Aden and Marjorie.[52] At the time of his death in 1961, Hendrix was helping plan the 80-inch reflector on Kitt Peak and the astrometric reflector at the US Naval Observatory Flagstaff station.[53]

Early years of Caltech

In the first part of the 20th century, George Ellery Hale seemed to be everywhere, not just on Mt. Wilson and not just in Pasadena. He persuaded railroad industrialist Henry Huntington to build a research library and art gallery; Huntington's architects designed a mansion for him (now the Art Gallery) and a separate building to house his rare books (Huntington Library).[54] Hale was elected a trustee of Throop Polytechnic Institute in 1907 and envisioned the school as an "MIT of the West." He lobbied his fellow trustees to appoint Lutheran minister James A. B. Scherer as its first president; although Scherer was not a scientist, he was a widely respected and accomplished fundraiser.[55] Next Hale persuaded lumber magnate Arthur Fleming to donate money to purchase 22 acres of orange groves at the corner of Wilson Avenue and California Boulevard and build what would ultimately become Throop Hall, dedicated in 1910.[56] Three years later, Throop Polytechnic Institute was renamed Throop College of Technology.

How better for Hale to begin to create an "MIT of the West" than by recruiting his contemporary there while a student, Arthur Amos Noyes (1866–1936). Noyes received his Master of Science degree in 1887 and was appointed assistant in analytical chemistry. He struck up a friendship with Hale (one of his students) and served as professor of theoretical chemistry at MIT from 1899 to 1919.[57] He was the foremost physical chemist of the era and established the Research Laboratory of Physical Chemistry on campus in 1903. He and MIT split the funding for salaries 50:50; the Carnegie Institution supported Noyes's own research until 1927. He then served as acting president of the school for two years, beginning in 1907. Six years later, Noyes was invited by Hale to be advisor for three months each year at the Throop College of Technology.[58] During World War I, Noyes chaired the National Research Council in Washington, DC, which had been set up as a committee (Hale, Noyes, and Robert A. Millikan) within the National Academy of Sciences to bring science to bear on solving military problems.[59] After the war ended, he resigned from MIT and joined Hale full-time as director of chemical research at Throop, which through the fundraising efforts of Scherer and/or Hale received sizable endowments from Arthur Fleming ($1 million), Norman Bridge, MD ($250,000 for a physics lab), and retired lumberman Robert R. Blacker ($50,000).[60] Adjusted for inflation, the total of those donations would be almost $19,758,000 in 2021. The research endowments, a tantalizing salary, and previous, fulfilling collaboration with Hale and Noyes on the National Research Council were sufficient incentives for the third man of this triumvirate (Figure 2.4) to join them from the University of Chicago—Robert Andrews Millikan.[61]

Millikan (1868–1953) was born in Morrison, Illinois, and was raised chiefly in the Iowa countryside. His chief interests at Oberlin College were mathematics and classical Greek; he showed no interest or talent in physics. To his surprise, he was invited to teach elementary physics in his junior year, so over the summer he worked all the problems in the physics textbook and required his students to do likewise. He received his bachelor's degree in 1891 and his master's degree two

Figure 2.4. The "Triumvirate" in front of the Caltech gates. Undated. From left: Arthur A. Noyes, George Ellery Hale, Robert A. Millikan.
Courtesy of the Archives, California Institute of Technology.

years later at Oberlin, followed by a summer at the University of Chicago doing research with Professor Albert A. Michelson, who invented the interferometer, measured the precise speed of light, and would be awarded the Nobel Prize in Physics in 1907 for his "optical instruments used for spectroscopic and metro-logical observations."[62] Upon receiving his doctorate in physics at Columbia University in 1895, Millikan studied in Europe with Max Planck, among others, and then accepted an offer of an assistant professorship at the University of Chicago from Michelson. After 12 years of working six hours a day on teaching and writing textbooks for high schools and colleges and another six hours on research, he decided to devote more time to research and specifically to developing his elegant technique of measuring the value of e, the electric charge of a single electron, using a charged oil drop falling between two plates between which an electric field can be applied. If the weight of the drop and the viscosity of the air are known, then the electric field can be applied to counteract gravity and hold the droplet at rest; once the upward force equals the weight of the drop, the charge can be determined. His other major contributions were to confirm Einstein's particle theory of light and measure the value of Planck's constant (h) in his photoelectric research. For all of these Millikan was awarded the Nobel Prize in Physics in 1923.[63]

Millikan was also a hard-nosed negotiator and agreed to join Hale and Noyes in Pasadena only if he had limited administrative duties and was granted most of the school's financial resources as director of the Norman Bridge Laboratory of Physics and chairman of the Executive Council (effectively president). After his terms were accepted, he arrived in late 1921 to succeed Scherer and reshape the school, which had been renamed the California Institute of Technology the year before, into one of the leading universities in the world.[64] He immediately set about recruiting a few of his colleagues from Chicago: instrument-maker Julius Pearson, research assistant Ira Bowen, and one of his former graduate students, Earnest Watson. Bowen would go on to receive his doctorate in physics under Millikan and join the staff of Mt. Wilson Observatory. Watson supervised the construction of the new physics laboratory and taught kinetic theory and thermodynamics.[65] Apart from expanding the faculty with national scientists, Millikan initiated a visiting scientist program comprising a breathtaking list of international luminaries in atomic structure and quantum theory: Niels Bohr, Max Born, Paul Dirac, James Franck, Erwin Schrödinger, and Werner Heisenberg.

Millikan's coup, though, was in bringing Albert Einstein for visiting professorships at Caltech for a few months in three successive years, beginning in 1931, with the hope that Einstein would accept a permanent position at Caltech; Einstein eventually accepted a more lucrative offer from Princeton University. While in Pasadena he discussed his general theory of relativity with physics professor Richard Tolman and Edwin Hubble and soon came to accept Hubble's idea of an expanding universe based on the proportional relationship between redshift in the spectra of stars in galaxies and distance from the observer, forcing him to abandon his "cosmological constant," which he had never liked.[66] Today, it is now believed that the expansion of the universe is accelerating, requiring a nonzero cosmological constant after all; the value of the Hubble constant H_0 is still widely debated, depending on the method used to determine it. Millikan would go on to augment Caltech's faculty and prestige in the years ahead by hiring famous geneticists Thomas Hunt Morgan and Theodosius Dobzhansky, aerodynamicists Theodore von Kármán and Clark Millikan (Robert's son), and physicist Charles C. "Tommy" Lauritsen. Kármán, Clark Millikan, and Arthur Klein designed a wind tunnel to test lift, drag, and stability on scale models of airplanes.[67] Lauritsen achieved fame by designing X-tubes for radiation therapy of cancer patients, but later developed proximity fuzes and rockets for the war effort at the Kellogg Radiation Laboratory, under strict secrecy, with his student William "Willy" Fowler. Fowler shared the Nobel Prize in Physics in 1983 with Subrahmanyan Chandrasekhar.[68] Aden would earn the respect of Fowler at Caltech and Chandrasekhar at Yerkes Observatory/University of Chicago. By 1939, the faculty and research fellows included the preceding individuals, along with what were (or would be) other impressive, critical figures in American science, such as Paul Epstein (theoretical physics), Robert Oppenheimer (theoretical physics), Linus Pauling (chemistry), William Sears (aeronautics), Alfred Sturtevant (genetics), Tsien Hsue-shen (aeronautics), Fritz Went (plant physiology), and Fritz Zwicky (theoretical physics).

Youth and education

Into this *sui generis* world, within easy pedaling or walking distance from their childhood homes, Marjorie and Aden were introduced as they were growing up. Standing like Isaac Newton on the shoulders of giants, it is hard to believe that they could have ended up any other way than they did, namely as polymathic visionaries who helped to harness solar energy and revolutionize optics again and again as opportunities arose. There were certainly obstacles and political adversaries along the way, but they overcame them as a devoted husband-wife team for over 63 years. As for many of what Tom Brokaw called the "Greatest Generation," namely those who lived and worked through the Great Depression and World War II and then rebuilt the US economy to become the envy of the world,[69] Marjorie and Aden learned to be inventive, resourceful, and unflappable in the face of adversity. Coping with hardships bracketed by poverty and war built character and forged an early maturity. Women in particular had to prove that they were capable of being more than human computers for their male counterparts, which was how Marjorie's mother Hannah was treated at Yerkes Observatory, even with her PhD in astronomy from the University of Chicago. They had to show that they could solve problems and make meaningful contributions to any task set before them, whether in wartime or peace, in the public or private sector. The rest of this book details how they did just that, drawing on their autobiography in three parts: *The Golden Age of Astronomy: In the Beginning* (Meinel et al.,

2008), *Echoes from a Simpler Time* (Meinel and Meinel, 2002), and *The Solar Odyssey: Adventures Along the Way* (Meinel et al., 2003), hereafter referenced in the endnotes by shortened titles. There is some repetition in *Golden Age* and *Echoes*, as also in long emailed documents that Aden sent to William Patrick McCray. McCray was a research associate at the University of Arizona who was writing the history of the Steward Observatory Mirror Laboratory with sponsorship by the National Science Foundation. He wanted to interview Aden in that regard, but Aden opted to answer the questions in writing.

Marjorie was born on May 13, 1922, in Pasadena to Edison and Hannah Steele Pettit, whose astronomical careers had taken them from Yerkes Observatory to Pasadena and Mt. Wilson Observatory at the invitation of George Ellery Hale. While Edison was employed as a solar astronomer at the salary of $2,000/year until 1945, Hannah worked as his assistant at their Alvan Clark 6-inch refractor at home until a chronic heart ailment left her confined to bed or otherwise disabled most of the day. She was able to translate papers from foreign languages, presumably German or French. At that point Marjorie would spend time at the home telescope with him or in his laboratory in the city offices of the Mt. Wilson Observatory at 813 Santa Barbara Street in Pasadena (Figure 2.5). Marjorie recalled that her father built much of his own instrumentation and made his own eyepiece crosshairs for telescopic observations from spider webs. Black widow spiders were common in Pasadena in the late 1920s and early 1930s, and Edison "cultivated" them for their webs in hidden, dark places at the house, with dire warnings about them to his two daughters. Every night before going to bed Marjorie would peek under the covers to make sure there were no black widows there to bite her toes.[70] She would travel with the family every summer for several years to the Yerkes Observatory and the McMath-Hulbert Observatory in Lake Angelus, Michigan, in northern Oakland County. While in Michigan they lived in a vacant faculty house on the campus of private Cranbrook College (now Cranbrook Schools) in Bloomfield Hills, which is where McMath lived until his death in 1962. While there, she listened to Detroit Tigers baseball on the radio and would become a lifelong fan of the game.

Beginning in 1910, George Ellery Hale suffered from a series of nervous breakdowns, limiting his scientific activity and travels, and eventually he spent more and more time at his home near Caltech.[71] Edison was the only person allowed to speak with Hale there, as Hale seemed to like Edison's movies of solar prominences and his soothing voice in discussions of solar work in general; meanwhile, Marjorie and her sister Helen sat in the car. According to Marjorie, Walter Adams, who succeeded Hale as director of Mt. Wilson Observatory, resented Edison's privileged access to Hale and never gave him priority in the instrument shop.[72]

Figure 2.5. Marjorie Pettit with her parents, Hannah and Edison, at their home observatory ca. 1937. Note their Alvan Clark refractor telescope extending above the roof line.

Courtesy of David Drach-Meinel.

As a result, Edison developed his own instrument shop at home and built a solar monochromator to observe prominences in the absence of an eclipse. The monochromator and Alvan Clark refractor sealed Marjorie's interest in astronomy. She spent hours after school, on weekends, and in the summer at the telescope, which was housed in a roll-off roof observatory (Figure 1.2 in Chapter 1) behind their house at 963 East Villa Street and only a few blocks from the Mt. Wilson offices on Santa Barbara Street. That was the same telescope that she would use with a visual photometer for her master's thesis at Claremont Colleges on the long-period variable star RT Cygni.

Marjorie's formal schooling likely began near her home, at Jefferson Elementary School at 1500 East Villa Street, which was erected in 1909 but then replaced by another building on the same site in 1926. Aden probably attended Hamilton Elementary School at 2089 Rose Villa Street.[73] Edison Pettit's salary at Mt. Wilson Observatory was sufficient to help his family weather the Depression, but Aden remembered that when President Franklin D. Roosevelt closed all the banks on March 6, 1933, his family had only 27 cents to their name (Figure 2.6). They did have pet rabbits, chickens, and a few ducks, plus some fruit trees, which helped them survive; as a 10-year-old he scavenged in trash containers behind grocery stores to feed the animals.[74]

Contrary to Roosevelt's oft-quoted sound bite in his first inaugural address two days before the bank closures, there would in fact be much more to fear than fear itself in the years ahead. In fact, as soon as March 10, the magnitude 6.4 Long Beach earthquake struck southern California, killing up to 120 people and causing millions of dollars in damage. Schools in particular were heavily damaged, which led to passage of the Field Act that established earthquake construction codes throughout the state.[75] By the end of the decade, tourism and building permits in Pasadena had all but vanished. Hotels and restaurants were shuttered or demolished. Thousands were unemployed and starving until New Deal programs such as the Works Progress Administration, Public Works Administration, and State Employment Relief Administration put them back to work on renovating the Rose Bowl and Brookside Park, laying utility

Figure 2.6. John and Gertrude Meinel with young Aden ca. 1930.
Courtesy of David Drach-Meinel.

lines underground, engineering flood control, constructing the Arroyo Seco Parkway, etc.[76]

Both Aden and Marjorie excelled in their classes early on, enough to skip a grade, and entered John Marshall Junior High School, which had been severely damaged in the earthquake and then rebuilt piecemeal by the New Deal's Public Works Administration over the next seven years. They graduated in 1937 and enrolled in Pasadena Junior College (PJC; now Pasadena City College). Students were required to adopt either a college track curriculum (languages, math, sciences) or trades track (architectural and engineering drafting, math, sciences). Marjorie chose the former, but Aden never expected to go to college (for lack of money) and opted to learn a trade so that he could land some sort of job after graduation in those lean years. The skills in drawing and modeling that he learned would serve him well in the years ahead when he would flesh out his design concepts for telescopes. He attended scientific talks at the Pasadena Public Library and planned to become an aeronautical engineer.[77]

His interest in aeronautics blossomed as a member of the Airplane Club at Marshall Junior High when he constructed an ornithopter, propelled by a rubber band, as a science project. He launched it and watched it flap bird-like over the heads of his classmates; he received an A for it. His older half-brother, Mark Perry Meinel, was another influence. Mark was then taking an aeronautical engineering course at PJC and asked Aden if he would like to observe wind tunnel tests there, evaluating airplanes of the students' own construction. Aden learned how to balance the angle of tilt with weight controls from Professor Albert Merrill (1875–1952), the first instructor of aeronautics at Caltech.[78]

Later, Aden was present at the PJC wind tunnel for the testing of a monoplane called the PJC-1, designed by the students of instructor Max Harlow. There he met Ernest E. Sechler (misspelled by Aden as "Schlecter" and "Schlicter"), who was visiting from the Guggenheim Aeronautical Laboratory at Caltech and invited an enthusiastic Aden to tour the wind tunnel there, attend weekly seminars, and use the library. The next year (1937) Aden watched the H-1 Racer of Howard Hughes fly over his house toward the Burbank airport after setting a transcontinental speed record of 7 hours, 28 minutes.[79]

Another of Aden's first mentors was his physics teacher at PJC, Bailey Howard, who organized a pass for Aden to the Rare Book Room at the Huntington Library. Aden pored over the works of Isaac Newton, especially an English translation of his *Philosophiae Naturalis Principia Mathematica* and *Opticks: or, A Treatise on the Reflexions, Refractions, Inflexions and Colours of Light*. He was deeply impressed by Newton's experiment that used a prism to separate light into its spectrum and a second prism in reverse to restore the white light. As Howard's laboratory assistant the following autumn, Aden repeated Newton's experiment for the classroom of students. He was likewise inspired to construct his own reflecting telescope following Newton's diagram and instructions for polishing the parabolic mirror.[80]

His chemistry class at PJC was taught by O. G. Dressler. He had already read the textbook and received all As on class exams. On the basis of Aden's grades, Dressler was impressed enough to nominate him as one of four to compete on the PJC team on May 13, 1939, in the 25th Annual High School Contest sponsored by the American Chemical Society, Southern California Section. One member of the PJC team placed second overall and was admitted as a freshman to Caltech the following year. Aden placed 27th overall but second on the PJC team.[81]

Although he did not do as well in the contest as he would have liked, 16-year-old Aden had continued to impress Bailey Howard, who recommended him to Earnest Watson, chair of the Department of Physics at Caltech, as a volunteer lab assistant there that summer. This would be a critical connection for Aden because Watson was the administrator of Caltech's rocket program prior to 1945, spending a million dollars a week.[82] Watson agreed to take Aden on, led him through the Norman Bridge Physics Laboratory, and introduced him to Luke Chia-Liu Yuan (1912–2003), grandson of Yuan Shikai, the first president of the Republic of China from 1912 to 1916. Luke was working on a doctorate in millimeter wave science under Millikan; with Aden's help, he set up a transmitter on Mt. Wilson and dipole receivers at several locations around Pasadena to compare apparent radio direction with the true visual angle.[83] Luke Yuan would go on to become a senior physicist at Brookhaven National Laboratory.

One day Yuan introduced Aden to Robert Millikan, along with Victor Neher and William Pickering, who were studying cosmic rays at the time. In 1902 reports of atmospheric radiation were announced at a meeting of the American Physical Society, but it was not until 1915 that it was firmly established that the radiation originated from outside Earth's atmosphere instead of from its own geological strata. Millikan coined the phrase "cosmic rays" to describe high-energy atomic particles entering Earth's atmosphere from outer space and colliding with particles there; he then set out to measure the radiation at different altitudes and in different locations. Before airplanes were in common use, an early method was to launch balloons with a Geiger counter, radio, and parachute. Millikan asked Aden if he would like to assist. Aden rarely said "No" to any request made to him throughout his life. Whether he was qualified was irrelevant. If he didn't know how to proceed, he would teach himself. If he didn't have the necessary equipment, he would build it. His task in this instance was to hold six, hydrogen-filled balloons on the roof of the laboratory and then release them on signal. When he did, two of the balloons burst only a few thousand feet above them, and the whole apparatus was found an hour later in a tree in Altadena. Another such balloon flight ended up in the ocean by Long Beach, and still another in a municipal swimming pool in Arcadia. If it accomplished nothing else, the experiment was

Aden's introduction as a teenager not only to the "great and powerful" Robert Millikan, but also to William Pickering, who would later be a colleague to both him and Marjorie in rocketry.[84]

A double bout with influenza and eyesight problems, as well as spending too much time at Caltech instead of on his PJC homework, forced to Aden to take a leave of absence in the spring semester of 1939. Bailey Howard told him about a Caltech graduate student named Jim Edson who with three others had organized the Planet Group, dedicated to planetary observations and photography. Jim had been a night assistant at the Lowell Observatory in Flagstaff, Arizona, where Clyde Tombaugh had discovered Pluto; he later married Tombaugh's sister, Lil. The Group borrowed a 6-inch reflector from Roger Hayward at Caltech and a 6-inch refractor from Mt. Wilson to use as a guide telescope, and with a sun shield successfully observed the inferior conjunction of Venus. After that, they transported the equipment to the Table Mountain station of the Smithsonian Astrophysical Observatory overlooking Oroville in north-central California. Aden and Edson established a lifelong friendship. Hayward also told Edson of a 20-inch replica of the 100-inch on Mt. Wilson that needed repairs. The Group refurbished it, built a fold-down shelter for it, and set it up on Table Mountain in the summer of 1940.[85]

In his early years Aden had dreamed of becoming a concert pianist and had taken lessons from the renowned piano teacher Ora Leola Caldwell. Those dreams were shattered when his girlfriend, Dorothea Behm, was awarded a music scholarship to Juilliard; after she left, his future abruptly shifted from a musical career to astronomy, all because of a matchmaker. Nonetheless, his musical talent remained until late in life, when he was able to play a movement of a Beethoven piano sonata skillfully from memory.[86]

Although Aden and Marjorie had attended the same schools to that point, they did not meet formally until both were students at PJC, and not from shared classes but from the efforts of a mutual friend named Barbara. Barbara lived near Aden and often walked to school with him. She thought that Marjorie would like Aden and invited her to the next meeting of the Math Club, of which Aden was president. In the same time frame, Barbara asked Aden if he had heard about Edison Pettit of the Mt. Wilson Observatory. Yes, he said, he was aware of Edison's measurement of the low temperatures on Mars. Barbara told him that his daughter was in her math class and would be attending the next Club meeting. When that day came and Marjorie opened the door, her eyes locked onto those of a tall fellow and followed his glasses as they slipped slowly, inexorably, down his nose. For Aden anyway, it may have been love at first sight. He asked about her class schedule, and more importantly, invited her to the PJC football game that Friday night in the Rose Bowl to see PJC's superstar in football, baseball, basketball, and track and field: Jackie Robinson, who was scoring an average of over 30 points a game on the gridiron. Aden resolved to learn to drive—the sooner the better—because his father had to chauffeur them back and forth.[87]

Aden changed his class schedule to be in Marjorie's astronomy class, but continued to work after school with Luke Yuan at Caltech. As a result his grades began to suffer, and Marjorie (who always got straight As) happened to notice the D on his physics experiment paper while they were in the school library and encouraged him to do better. That was all the motivation he needed, because the next week he received an A on a paper.[88] More importantly, his overall grades rose dramatically such that after he and Marjorie graduated from PJC in June 1941, he was accepted as a sophomore at Caltech for the fall, receiving credits for freshman mathematics, physics, English, and drawing. At the same time, Marjorie began a Bachelor of Arts curriculum at the University of California at Berkeley.

Aden's father was able to pay the $200 tuition per semester at Caltech, but Aden still needed to look for part-time employment. One day he wandered into the Astrophysics Building and along the hallway saw drawings of the 200-inch telescope on Palomar Mountain. An elderly lady named Miss Gianetti emerged from an office and asked if she could help him. He explained that he was a volunteer assistant in physics and a member of the Planet Group. He went on to say that he was interested in the 200-inch because he was in the crowd when the mirror blank for it arrived at Arcadia train station. Perhaps by no coincidence, Marjorie and her parents were also in the crowd. Miss Gianetti introduced him to John Anderson, who supervised the 200-inch Optical Shop. He escorted Aden there and onto the Shop floor, while both were wearing white protective suits and boots. When he came within inches of the mirror, he was transfixed. He knew that optics would be his future. Anderson said that he would make some inquiries on Aden's behalf and for him to come back next week.[89]

Upon Aden's return, Anderson was pleased to say that he had found a position for him, not at Caltech but at Mt. Wilson Observatory, which paid $100 per month. Walter Adams needed an apprentice optician for National Defense Research Committee (NDRC) projects under Theodore ("Ted") Dunham, specifically for roof prisms and the Schmidt camera. With Walter Adams, Dunham discovered that the amount of oxygen on Mars is less than one-tenth of 1% of that in the Earth's atmosphere over the same surface area.[90] Dunham told Aden that he would be working with Roger Hayward, Rudolph Langer, George Mitchell (inventor of the Mitchell cameras), and Don Hendrix. First, Aden was to assist Alfred Joy and Walter Baade. On his first night on the 100-inch, Joy asked Aden if he would like to guide the telescope, which entailed climbing out onto the narrow Cassegrain platform in total darkness. He could hear the humming of electric motors as he manipulated the guidance controls. On his second night, the "seeing" was excellent as he worked with Baade photographing galaxies and a recent nova, Nova Herculis 1934, with a ring developing around it. Baade complained about the nightglow and the need for larger Schott filters and also one to eliminate atomic emission lines. Aden was to procure those filters for him in 1945 (see Chapter 4, section "Jena").[91]

The next night, Aden watched Rudolph Minkowski photograph suspected planetary nebulae using the 60-inch telescope. Minkowski had been born to a prominent Jewish family in Strasburg, Germany, in 1895, and although he had converted to Christianity later in life, he knew that the Nazis emphasized genetic heritage rather than beliefs. Baade, who had left Germany in 1931, encouraged his friend and collaborator from Hamburg University to do likewise and lobbied Walter Adams to take on Minkowski at Mt. Wilson. Adams agreed to hire him on the basis of his expertise in theoretical physics and applied optics, and over the next 20-odd years Minkowski worked with Baade to identify two classes of supernovae. They published two joint papers on optical identifications of radio sources, such as elliptical galaxy M87, radio galaxy Cygnus A, and supernova remnants. Minkowski is equally well known for contributions to the Palomar–National Geographic Sky Survey, which assembled photographs from Palomar of the entire sky to a declination of –30° taken with a wide-field, 48-inch Schmidt camera.[92]

With his apprenticeship in the Mt. Wilson Optical Shop, Aden had to discontinue his participation with Jim Edson's Planet Group and devote more time to his studies at Caltech. Nonetheless, he went to Table Mountain on a few weekends to keep up to date with observations of Mars. Because of a persistent dust storm on the planet, it was difficult to see any surface features, such as the so-called canals previously described by Percival Lowell.[93]

But his association with Edson would take a new turn as war clouds gathered. Aden remembered listening to Sunday concerts by the NBC Symphony Orchestra while working on Table Mountain. On the morning of December 7, 1941, he was at home and tuned in the radio for that day's concert when he heard a news bulletin about the Japanese attack on Pearl Harbor. Like millions of others around the world, his life and Marjorie's would change forever. She returned home after that year at Berkeley and transferred to the prestigious Pomona College in Claremont, California, about 25 miles east of Caltech, to complete her Bachelor of Arts degree. She had a scholarship in astronomy and taught classes for a Professor Whitney, one of which was navigation for Army Air Force men. A student in the class turned in a navigation report that incorrectly placed his latitude and longitude in Baja California instead of the Brackett Observatory at Pomona College. With her help he repeated the experiment and ended up closer to Claremont. Later she learned that he was the navigator in one of the B-25 bombers in Jimmy Doolittle's raid on Tokyo that made it safely into China and then back home again.[94]

Vannevar Bush (1890–1974), president of the Carnegie Institution and professor at MIT, was appointed by President Roosevelt to chair the NDRC, which Roosevelt approved on June 15, 1940. His Dutch given name (pronounced Vuh-NEE-ver) was a constant source of irritation to him and others, who simply called him "Van." His "screwball first name" notwithstanding, Bush was a brilliant engineer and an effective leader and motivator before, during, and after the war. Educated at Tufts and MIT, he cofounded Raytheon with Laurence Marshall and Charles Smith as an electronics firm that provided magnetron tubes for radar and parts for proximity fuzes in the war effort. It evolved into a leading contractor for US defense and homeland security. Bush built the first analog computer—his differential analyzer—and developed the concept of the memex as a forerunner of hypertext. And, with support from Roosevelt, he originated the Manhattan Project.[95]

The purpose of the NDRC was to correlate and support scientific research on mechanisms and devices of warfare, finance that research in educational and scientific institutions and industry, and support the War and Navy departments. Researchers remained at their institutions, funded by government contracts and unencumbered by bureaucracy and the military. Bush organized the NDRC program at Mt. Wilson Observatory, headed by Ted Dunham. He had requested that Don Hendrix go to the Frankford Arsenal in Philadelphia to troubleshoot manufacture of roof prisms. As a result, Aden was assigned to make the aspheric plate for the Schmidt camera destined for the University of Bristol in the United Kingdom without Don's guidance. Aden did own a copy of John Strong's (1938) *Procedures in Experimental Physics*,[96] with a chapter on laboratory optical work partly written (and autographed) by Hendrix, and another on molding and casting by Roger Hayward, who had also illustrated the book to make ends meet in the Depression. In Don's absence, Hayward was to serve as Aden's supervisor.[97]

Roger Hayward (1899–1979) was a polymath in fields that spanned architecture, astronomy, illustration, optics, and even puppetry. His extraordinary illustrations appeared in two editions of Strong's *Procedures*, four books with Caltech Nobel Laureate Linus Pauling, including *General Chemistry* (1946) and *The Architecture of Molecules* (1964), *Scientific American*, and dozens of textbooks in the sciences. With staff from Mt. Wilson, he reverse-engineered the aspheric corrector plate for Schmidt optics that later contributed to the 18-inch and 48-inch Schmidt telescopes on Palomar Mountain. During the war he worked in the Mt. Wilson Optical Shop designing gun-sight optics, anamorphic lenses, roof prisms, wide-field binoculars, etc. To assist the British in spotting incoming German bombers, he patented a compact, two-mirror Schmidt telescope, as well as a simple device for grinding and polishing surfaces that could allow "anyone" to produce corrector plates (US Patent No. 2,399,924).[98]

That "anyone" became Aden as the guinea pig. His job was to finish polishing the 12-inch, f/2.5 corrector plate using Hayward's device with two co-geared, counter-rotating rings. Tile facets were waxed on the rings for grinding, replaced by pitch facets for polishing. Years later, Aden found the corrector plate that he had finished and took it with him to Yerkes Observatory; it eventually became the f/2.5 camera spectrograph delivered to Bill Petrie in Canada for auroral studies. The University of Bristol needed an f/1 Schmidt camera to spot Nazi bombers, so Hayward invented a simple but ingenious method for marking progress when grinding the Schmidt curve. It involved placing a transparent bar of glass, slightly tilted, over the workpiece. Then a hundred or so small ball bearings were fed into the upper gap between the bar and the workpiece. These balls then stopped where the depth difference between a flat and the curve equaled the ball diameter. A graph of the desired Schmidt shape was below the two pieces so that one could compare the curve defined by the ball bearings and the desired curve. For polishing, a flexible sheet of neoprene was applied to both sides of a fabric sheet. Tile facets were waxed to the fabric sheet over an elliptical area large enough that when it oscillated, the edge of the facets came just to the periphery of the workpiece. A circular piece of neoprene foam was placed on top of the flexible polisher. Then lead weights were placed on top of the disc, enabling differential grinding. Aden made one f/1 plate for the University of Bristol and two additional ones. He took the latter two with him to Yerkes Observatory, where one was incorporated in an auroral spectrograph and the other in a flat-field camera used by W. W. Morgan and Hugh Johnson to map the sky in H-alpha light. Aden used the same methods patented by Hayward to make a Schmidt plate for his dissertation and several others for the US Air Force in the 1950s. Upon Hendrix's return, Aden was taken off the Schmidt work and was assigned to polish roof prisms that Hendrix and George Alfred Mitchell had diamond-milled; although Hendrix and Mitchell contracted severe eye irritations from the fine glass dust mixed with water vapor during the milling, Aden was polishing at the other end of the lab and avoided the problem. Apart from his wartime work at Mt. Wilson Observatory, Mitchell founded the Mitchell Camera Corporation with Henry Boeger in 1919, which supplied most, if not all, of the Hollywood film production companies with various motion picture cameras for the remainder of the century and beyond. For example, the Mitchell BNC (Blimped Newsreel Camera) was used to film *Citizen Kane* in 1940–1941, and the Mitchell VistaVision camera was used for everything from *White Christmas* (1954) and *The Ten Commandments* (1956) to special effects for *Star Wars* (1977) and *Interstellar* (2014). Mitchell was awarded an Honorary Award for cinematography by the Academy of Motion Picture Arts and Sciences.[99]

Living at home instead of in the dorms while attending school saved the Meinels money, but for the same reason Aden was unable to discuss homework problems with his fellow students. Fortunately, Caltech mathematical physicist Rudolph Meyer Langer, who also participated in the NDRC project, helped Aden with his physics homework.[100] His physics grades improved from a C in the first term of his sophomore year (1941–1942) to a B in the second and third terms. He took physics courses from Earnest Watson, Victor Neher, and John Strong (and probably used Strong's textbook, mentioned earlier).[101] Aden received only one A in that first year (mathematics), and all the other grades were either B (history and geology), or C (chemistry, biology). He realized that he could no longer work in the Optical Shop and maintain the pace to achieve the grades he wanted, so he resigned from the Mt. Wilson research team, with an invitation from Walter Adams to resume work the following summer. He also took a leave of absence from Caltech at the start of his junior year (1942–1943), although he did study elementary French in the second semester.[102]

In the meantime, another phase of Aden's kaleidoscopic career was quickly spinning into focus. As the war raged on in Europe and the Pacific Theater, he learned that all Caltech undergraduates had to either (1) join the Navy's V-12 College Training Program and receive a commission as ensign upon graduation, or (2) face being drafted. Nationwide, 125,000 young men in 131 colleges enrolled in the program. Of those, 60,000 would become officers in the Navy or Marine Corps. Although the stated purpose of the national program was to serve the Army and Navy, it would pay for using the college's facilities and staff; the formulas for those payments, while convoluted, were acceptable to the vast majority of the 131 colleges participating.[103] Regardless of the perks, Aden was not enamored of either option, so he asked his Planet Group friend, Jim Edson, if there was a job for him in the rocket project working out of Caltech's secret High Voltage Research Laboratory (now the Alfred P. Sloan Laboratory of Mathematics and Physics) because it came with an automatic draft deferment.[104] Edson found a place for him. Little did he know at the time that his experience and training with rockets and other ordnance would probably spare him a horrific death by shark attack in 1945.

Notes

1. Page, H. M. 1964. *Pasadena: Its early years.* Lorrin L. Morrison, Los Angeles, California, 1–2.
2. Lund, A. S. 1999. *Historic Pasadena: An illustrated history.* Historical Publishing Network, San Antonio, Texas, 8.
3. Page, *Pasadena,* 3.
4. Lund, *Historic Pasadena,* 11–12.
5. Lund, *Historic Pasadena,* 14–15.

6. Page, *Pasadena*, 8, 9.
7. Page, *Pasadena*, 16.
8. Page, *Pasadena*, 18.
9. Nordhoff, C. 1873. *California: For health, pleasure, and residence. A book for travellers and settlers*. Harper & Brothers, New York. https://www.loc.gov.item/14022123/, 17.
10. Page, *Pasadena*, 18–27; Lund, *Historic Pasadena*, 17–21.
11. Lund, *Historic Pasadena*, 22.
12. Sandage, A. 2012. *Centennial history of the Carnegie Institution of Washington*. Volume 1. *The Mount Wilson Observatory*. Cambridge University Press, 8; Babcock, H. D. 1938a. Address of the Retiring President of the Society in announcing the award of the Bruce Gold Medal to Dr. Edwin Hubble. *Publications of the Astronomical Society of the Pacific* 50: 87–96; Page, *Pasadena*, 38.
13. Lund, *Historic Pasadena*, 7, 29.
14. Page, *Pasadena*, 39–47.
15. Page, *Pasadena*, 54.
16. Page, *Pasadena*, 54, 56; Lund, *Historic Pasadena*, 31.
17. Page, *Pasadena*, 63–64; Lund, *Historic Pasadena*, 32.
18. Page, 67; Lund, *Historic Pasadena*, 38.
19. Page, *Pasadena*, 107, 109, 135–6; Lund, *Historic Pasadena*, 40, 45.
20. Page, *Pasadena*, 147.
21. Page, *Pasadena*, 153–4; Lund, *Historic Pasadena*, 44; Goodstein, J. R. 1991. *Millikan's school: A history of the California Institute of Technology*. W. W. Norton & Company, New York, 27.
22. Page, *Pasadena*, 152–4; Lund, *Historic Pasadena*, 44.
23. California Institute of Technology, Caltech Catalog 2010–11, 13.
24. Page, *Pasadena*, 126; Simmons, 2020. https://www.mtwilson.edu/bringing-astronomy-to-an-isolated-mountaintop/.
25. Page, *Pasadena*, 131–2.
26. Adams, W. S. 1938. George Ellery Hale (1868–1938). *Astrophysical Journal* 87: 372–4; Babcock, H. D. 1938b. George Ellery Hale. *Publications of the Astronomical Society of the Pacific* 50(295): 156.
27. Sandage, *Centennial history*, 158–9.
28. Hale, G. E. 1915. *Ten years' work of a mountain observatory: A brief account of the Mount Wilson Solar Observatory of the Carnegie Institute of Washington*. Reprinted in 2013 by HardPress Publishing, Miami, Florida, 14–15.
29. Sandage, *Centennial history*, 10–12.
30. Sandage, *Centennial history*, 57.
31. Adams, W. S. 1947. Early days at Mount Wilson. *Publications of the Astronomical Society of the Pacific* 59: 228.
32. Hale, *Ten years' work*, 15–29; Sandage, *Centennial history*, 58, 63, 65–6.
33. Hale, *Ten years' work*, 86; Sandage, *Centennial history*, 170–1, 177; Babcock, George Ellery Hale, 160–1.
34. Adams, Early days at Mount Wilson, 292–3.
35. Adams, Early days at Mount Wilson, 219.

36. Sandage, *Centennial history*, 85–7, 198–200.
37. Osterbrock, D. E. 2001. *Walter Baade: A life in astrophysics*. Princeton University Press, 5, 7.
38. Osterbrock, *Walter Baade*, 9–11, 15, 21.
39. Osterbrock, *Walter Baade*, 22.
40. Osterbrock, *Walter Baade*, 45, 46.
41. Osterbrock, *Walter Baade*, 49.
42. Osterbrock, *Walter Baade*, 59.
43. Osterbrock, *Walter Baade*, 91; Sandage, *Centennial history*, 388.
44. Osterbrock, *Walter Baade*, 102; Sandage, *Centennial history*, 389.
45. Baade, W. 1956. The period-luminosity relation of the Cepheids. *Publications of the Astronomical Society of the Pacific* 68: 10.
46. Osterbrock, *Walter Baade*, 175–6; Babcock, H. D., Address of the Retiring President, 91–2.
47. Osterbrock, *Walter Baade*, 122.
48. The preferred spelling for mechanical or chemical components of an explosive device.
49. Osterbrock, D. E. 2003. Don Hendrix: Master Mount Wilson and Palomar Observatories optician. *Journal of Astronomical History and Heritage* 6: 1–2; Sandage, *Centennial history*, 474–5.
50. Osterbrock, Don Hendrix, 3.
51. Sandage, *Centennial history*, 475; Osterbrock, *Walter Baade*, 121, 123; Osterbrock, Don Hendrix, 6.
52. Osterbrock, Don Hendrix, 9.
53. Osterbrock, Don Hendrix, 10.
54. Lund, *Historic Pasadena*, 106; Babcock, George Ellery Hale, 10.
55. Lund, *Historic Pasadena*, 55; Goodstein, *Millikan's school*, 31, 50.
56. Lund, *Historic Pasadena*, 56.
57. Pauling, L. 1958. Arthur Amos Noyes: September 13, 1866–June 3, 1936. Biographical Memoirs of the National Academy of Sciences, Washington, DC, 323.
58. Pauling, Arthur Amos Noyes, 325–6.
59. Pauling, Arthur Amos Noyes, 329.
60. Lund, *Historic Pasadena*, 56.
61. DuBridge, L. A., and Epstein, P. A. 1959. Robert Andrews Millikan: March 22, 1868–December 19, 1953. Biographical Memoirs of the National Academy of Sciences, Washington, DC, 246.
62. Goodstein, *Millikan's school*, 88–90; Pauling, Arthur Amos Noyes, 242.
63. Pauling, Arthur Amos Noyes, 243–4, 252–3; Goodstein, *Millikan's school*, 91–2.
64. Goodstein, *Millikan's school*, 94; Pauling, Arthur Amos Noyes, 246.
65. Goodstein, *Millikan's school*, 94; Watson, E. C. 2012. Interview by Larry Shirley. Pasadena, California, January 20, 1969. Oral History Project, California Institute of Technology Archives. Retrieved July 4, 2019 from http://resolver.caltech.edu/Caltech OH_Watson_E.
66. Goodstein, *Millikan's school*, 101–2; Isaacson, W. 2007. *Einstein: His life and universe*. Simon & Schuster, New York, 355.

67. Goodstein, *Millikan's school*, 164–6.
68. Goodstein, *Millikan's school*, 230.
69. Brokaw, T. 1998. *The greatest generation*. Random House, New York.
70. Meinel, M. 1984. *Optical Center OSCillations*, no. 318, May 25, 1984.
71. Sandage, *Centennial history*, 171.
72. Meinel, A. B., Meinel, M. P., and Jacobs, B. M. 2008. *The golden age of astronomy*. Unpublished, 8.
73. Personal communication, Anuja Navare, Pasadena Museum of History; the Pasadena Unified School District was unable to supply any enrollment records for the two of them.
74. Meinel et al., *Golden age*, 12.
75. https://earthquake.usgs.gov/regional/states/events/1933_03_11.php.
76. Lund, *Historic Pasadena*, 73–6; Phillips, C. I., and Pasadena Museum of History. 2008. *Images of America: Early Pasadena*. Arcadia Publishing, Charleston, South Carolina, 109.
77. Meinel et al., *Golden age*, 13–14.
78. Meinel et al., *Golden age*, 14–15.
79. Meinel et al., *Golden age*, 16–17; Parker, D. T. 2013. *Building victory: Aircraft manufacturing in the Los Angeles Area in World War II*. Self-published. Cypress, California. Kindle ed.
80. Meinel et al., *Golden age*, 17–20.
81. Meinel et al., *Golden age*, 20.
82. Watson, Interview.
83. Meinel et al., *Golden age*, 21–3.
84. DuBridge and Epstein, Robert Andrews Millikan, 263–7; Meinel et al., *Golden age*, 23; Pickering, W. H. 1978. Interview by Mary Terrall. Pasadena, California. November 1978–December 19, 1978. Oral History Project, California Institute of Technology Archives. Retrieved July 4, 2019, from http://resolver.caltech.edu/Caltech OH: OH_Pickering_1.
85. Meinel et al., *Golden age*, 27–8; email from Aden Meinel to Wally Meinel et al., September 9, 2008.
86. Email from Aden Meinel to Lisa Fiorenza; personal communications, David Drach-Meinel and Ed Meinel.
87. Meinel et al., *Golden age*, 23–4.
88. Meinel et al., *Golden age*, 24–5, 27.
89. Meinel, A. B., and Meinel, M. P. 2002a. *Echoes from a simpler time*, unpublished, 22.
90. Adams, W. S., and Dunham, T., Jr. 1934. The B band of oxygen in the spectrum of Mars. *Astrophysical Journal* 79: 308–16.
91. Meinel et al., *Golden age*, 30; letter from George R. Harrison, NDRC, to Sylvia Burd, Mt. Wilson Observatory, September 5, 1941.
92. Meinel et al., *Golden age*, 30–1; Sandage, *Centennial history*, 533–5; Osterbrock *Walter Baade*, 153–6.
93. Meinel et al., *Golden age*, 32.
94. Meinel et al., *Golden age*, 32–3.

95. Zachary, G. P. 1997. *Endless frontier: Vannevar Bush, engineer of the American century*. The Free Press, New York, 20, 49, 112, 197, 261–3.

96. Strong, J. 1938. *Procedures in experimental physics*. Prentice-Hall, New York.

97. Meinel et al., *Golden age*, 22; Goodstein, *Millikan's school*, 241; Zachary, *Endless frontier*, 115–16, 121, 124.

98. Email from Aden Meinel to W. Patrick McCray, November 9, 1998; Bell, 34–37; http://scarc.library.oregonstate.edu/omeka/exhibits/show/hayward/item/3006. Accessed July 18, 2019.

99. Email from Aden Meinel to W. Patrick McCray, November 9, 1998; http://www.mit chellcamera.com/forum/viewtopic.php?t=77, accessed July 19, 2019.

100. Email from Aden Meinel to W. Patrick McCray, November 9, 1998.

101. In 1939 Strong invented coating telescope mirrors with aluminum instead of silver, which influenced Aden's designs in later years. Using an optical laboratory held aloft in the stratosphere by balloons, Strong was also the first to discover water vapor on Venus.

102. Meinel et al., *Golden age*, 34; official transcript, Aden Meinel, California Institute of Technology.

103. Herge, H. C. 1996. *Navy V-12*. Turner Publishing, Nashville, Tennessee, 6, 20–1, 30.

104. Meinel et al., *Golden age*, 34–5.

3

The Rocketeers

As early as 1933, the Navy foresaw another international crisis developing, but it was only after Germany's invasion of Poland in September 1939, and then of western Europe a year later, that the US Congress appropriated significant funds for naval ordnance. The first priorities were expanding production facilities and training additional personnel, particularly in leadership roles, and it was that gap that Caltech scientists took the initiative to fill. Robert Millikan, a conservative Republican, promoted military preparedness for national defense among not only his own faculty, but other universities as well. He and Max Mason, a professor of mathematics at the University of Wisconsin who had developed a successful submarine detection device at the Naval Research Laboratory in New London, Connecticut, were members of the Advisory Committee of the National Academy of Sciences to the Engineer in Chief of the Navy, Rear Admiral Harold Bowen. That committee recommended involving university faculty members in defense research but took no concrete steps to implement it.[1]

Earnest Watson (Figure 3.1), who had been a graduate student of Millikan and followed him to Caltech, now disliked, even resented, Millikan's benevolent dictatorship (e.g., any Caltech expenditure over $50 had to be approved by Millikan) and seized the reins. He invited Mason and Richard Tolman (professor of physical chemistry and mathematical physics and then dean of the Graduate School) to form an ad hoc group, the CIT Council on Defense Cooperation, led by Tolman. After speaking to their colleagues on campus in the spring of 1940, they prepared a report on what Caltech could offer to national defense, namely 95 professors and instructors and 126 others representing 27 fields of research, who were willing to work part-time or full-time toward that effort.[2] Tolman soon thereafter went to Washington and helped to form the NDRC, chaired by Vannevar Bush.[3]

Tolman was a close colleague of Charles C. Lauritsen (1892–1968; Figure 3.2) at Caltech and, with Millikan's blessing, recruited Lauritsen to visit Washington and help out. "Charlie," as he was known, was born in Denmark and had come to the United States in 1917 with a degree in architecture, but not knowing how he would make a living. He tried designing boats, professional fishing off the Florida coast, designing radio receivers, and ended up producing popular 10-tube radios for the Kennedy Corporation. By chance he had heard Millikan give a lecture

With Stars in Their Eyes. James B. Breckinridge and Alec M. Pridgeon, Oxford University Press. © Oxford University Press 2022. DOI: 10.1093/oso/9780190915674.003.0003

Figure 3.1. Earnest Watson with 5-inch high velocity aircraft rocket ("Holy Moses")
at a 1944 celebration of the one-millionth rocket manufactured at Eaton Canyon.
Courtesy of the Archives, California Institute of Technology.

in St. Louis in 1926, and, inspired, moved to Pasadena with his wife Sigrid and
young son Thomas without acceptance for a degree program at Caltech. During
the course of his entrance interview with Earnest Watson, Lauritsen admitted
that he had limited training in mathematics and none in calculus whatsoever.
Watson had misgivings about how successful Charlie would be at the age of 34
without the necessary prerequisites, but agreed to let him try and enrolled him
in advanced calculus, which Watson thought would end Charlie's academic
aspirations quickly.[4]

Figure 3.2. Charles C. Lauritsen. Undated.
Courtesy of the Archives, California Institute of Technology.

Instead, Charlie excelled in calculus and took courses from Tolman, Ira Bowen, Fritz Zwicky, and others. Millikan's suggestion for a research proposal for him led to his development of the first million-volt X-ray. He received his doctorate in physics in 1929 from Caltech, joined the faculty as assistant professor in 1930, and became director of the new Kellogg Radiation Laboratory on campus. He received tenure the next year and was awarded the school's Gold Medal; he became full professor in 1935. Until his retirement in 1962, Lauritsen made several significant contributions to nuclear physics, such as radiative capture; when carbon captures protons, a new isotope is formed that then releases gamma rays as it decays back to its normal low-energy state. He participated in the scientific development of the Manhattan Project and helped to establish the Naval Ordnance Test Station (NOTS) at China Lake, the Office of Naval Research (ONR), and the Aerospace Corporation. His work on various aspects of ballistic missile research was pervasive.[5]

When Lauritsen arrived in Washington in August 1940, Tolman appointed him vice chairman of the Armor and Ordnance Division of the NDRC. They met with Commander Gilbert Hoover of the Bureau of Ordnance, who emphasized the need for research on proximity fuzes. Gauging the correct range for anti-aircraft guns was difficult because the lag time between firing and reaching the target was not easily corrected for. The proximity fuze, with its radio-controlled detonator, exploded near its target, not upon impact or at a predetermined time. It screwed into the front of an artillery shell and emitted a steady radio wave, while the nose cone was a receiving antenna. When the fuze was within a few wavelengths of the target, it set off the detonator. One section of the NDRC, headed by Merle Tuve, developed fuzes for shells, and another section, led by Alexander Ellett and supported by Lauritsen, worked on fuzes for rockets. By the end of war, the Navy was producing up to 70,000 proximity fuzes per day, which were highly successful against Japanese aircraft such as kamikazes and especially in defense of Britain against V-1 rockets (see Chapter 4). After the Battle of the Bulge beginning on December 16, 1944, General George Patton commented on the devastation caused by "the new shell with the funny fuze" and was glad the Americans thought of it first. The official historian of the Office of Scientific Research and Development (OSRD), James Phinney Baxter, listed the proximity fuze among the three or four most extraordinary scientific achievements of World War II.[6]

Lauritsen was asked to go to England in the spring of 1941 and observed British antiaircraft rockets targeting German bombers almost every night during the Blitz, noting that "I don't think they ever shot down any bombers . . . but they made a beautiful fireworks display over Hyde Park. . . ." However much he regretted the lack of accuracy in hitting the German targets, he did prefer the British use of a propellant powder mix of nitrocellulose and nitroglycerin (cordite) over the American solvent-extrusion method. Upon his return to the States, he prepared a memorandum to Vannevar Bush that set the future course of rocket research and production at Caltech, much to the relief and delight of Millikan, who wanted to keep the physics faculty intact in Pasadena instead of farmed out to Washington and elsewhere.[7]

Lauritsen continued to buttonhole anyone he could—military and civilian—to promote the new rocket program, which by word of mouth reached the Office of Scientific Research and Development (OSRD). The OSRD had been created in 1941 and funded by Congress to supersede the NDRC, which had depended on the president's emergency funds to support projects. The OSRD, with Vannevar Bush as director, would receive funding directly from Congress, while the NDRC, headed by James Conant, would become the de facto operating unit of the OSRD. Within a year the NDRC had recruited about 2,000 scientists, including 75 percent of America's premier physicists and half of its most renowned chemists. By 1944, the OSRD was spending $3 million a week on 6,000 researchers at more than 300 industrial and university labs; at war's end, it had spent almost half a billion dollars and had signed 2,300 contracts with 321 companies and 142 academic institutions and nonprofits. Both the Army and Navy recommended several projects to be overseen by the NDRC. The memo was also circulated among the armed services to gauge interest in rockets, and the chief of the Coast Artillery stressed the need for target rockets that could match the speed and course of enemy aircraft. Lauritsen's son Thomas, along with a former doctoral student and now faculty colleague in the Kellogg Radiation Laboratory, William Fowler, flew to Washington to develop target rockets at the National Bureau of Standards.[8]

Several contracts with OSRD were eventually awarded for rocket research, the largest two to George Washington University in the East and Caltech in the West. The former started in October 1941 and focused on propellants extruded with solvents, leading to the bazooka, the 4.5-inch rocket, and the recoilless gun; there was little interaction between East and West. The contract at Caltech for $200,000 was not scheduled to begin until February 1942, but Caltech trustees advanced funds so that the project could start September 1, 1941. By the next February, Caltech's indebtedness had soared to half a million dollars, so as a matter of some urgency Earnest Watson flew to Washington to inquire about the promised government check. Although the full amount was approved, he was summarily told that a check for it would be sent to Pasadena by ground mail because unnecessary use of airmail was being discouraged. Stunned, Watson asked if he could pay for the (50-cent) airmail stamp himself. His offer was accepted only because there were no rules at the time prohibiting that. Caltech received its check soon thereafter.[9] It was that sort of no-nonsense approach to dealing with the federal bureaucracy that made Watson so integral and invaluable to the administration of Caltech's rocket project, leaving more time for Charles Lauritsen and his assistant Fowler to direct project research. The other scientists never had to justify what they wanted and simply wrote a memo to Watson, who would type up the requisitions and send them to Washington.[10]

Tom Lauritsen and Fowler returned to Pasadena to start the project being administered by Watson. Other faculty and students from various disciplines, including 18-year-old Aden Meinel, joined them. The learning curve was steep. Faculty members, including even Charles Lauritsen, knew little about rockets, and the longer it took to master the science of constructing and launching them, the more Allied lives and property would be lost.[11]

Fortunately, Tom Lauritsen was on board to lead the group that designed and developed the rockets, along with Fowler to test them.[12] Thomas "Tommy" Lauritsen (1915–1973) was the son of Charles and Sigrid Lauritsen. He graduated from Caltech in 1936, received his PhD from Caltech three years later, and joined the physics faculty. He was skilled in developing devices to answer physics questions, such as the pressurized electrostatic accelerator that he built with his father and Fowler. The three of them later designed the proximity fuze and solid-fuel rockets in the Kellogg Radiation Laboratory. After the war, they recast the Kellogg Lab as a center for research in nuclear physics and nuclear astrophysics, and Tom himself was particularly interested in nuclear spectroscopy and energy levels of light nuclei. His profound knowledge guided federal agencies and institutions in nuclear physics throughout the 1960s and 1970s. He was elected as a Fellow of the Royal Danish Academy of Sciences and Letters in 1965 and to the National Academy of Sciences in 1969.[13]

William "Willy" Fowler (1911–1995; Figure 3.3) graduated from Ohio State University in 1933 with a major in engineering physics and completed his doctorate in 1936 at Caltech under Charles Lauritsen. Together they pioneered particle-induced reactions at low energies and radiation capture reactions. He applied what he learned to measuring rates of hydrogen, helium, carbon, and oxygen fusion reactions inside stars and deducing their evolution. After the war he studied thermonuclear power in the Sun and stars with the Lauritsens and then nucleosynthesis in stars with British astrophysicist Fred Hoyle. He received

Figure 3.3. William Fowler with equipment in 1956.
Courtesy of the Archives, California Institute of Technology.

the Nobel Prize in Physics in 1983 for this work, sharing the Prize that year with Subrahmanyan Chandrasekhar, who published significant papers on physical processes associated with the structure and evolution of stars. (At Yerkes Observatory, Chandrasekhar was one of Aden Meinel's most devoted advocates; see Chapter 6.) Fowler was elected to the National Academy of Sciences in 1956 and received numerous awards in his long career, including the National Medal of Science (1974) and the *Légion d'honneur* of France (1989).[14]

One of the first orders of business was to decide on a propellant for the rockets and the method used to extrude it, that is, press it through a die of fixed cross-section profile into a shape (e.g., sticks, ribbons, grains) that could be inserted around the other components inside the rocket casings. Most of the grains for later rocket models were cruciform in shape.[15] Lauritsen was impressed by the British method of dry extrusion instead of wet extrusion (using a solvent), and he preferred to use ballistite instead of cordite because it was a hotter-burning powder and had a higher percentage of nitroglycerin and a higher percentage of nitrogen in the nitrocellulose. As a result, it had a faster reaction rate, but was more dangerous to process and use. Nonetheless, because of the exigencies of war, they opted to use the ballistite until a more effective weapon was developed.[16]

Tom Lauritsen assembled a small extrusion press and took it to Eaton Canyon in the foothills of the San Gabriel Mountains, where he successfully extruded grains of 15/16 of an inch in diameter (about 2.3 cm) totaling 180 pounds. Soon thereafter other presses were designed that could extrude grains with diameters ranging from 1.75 inches to 4.50 inches, and these were moved to Eaton Canyon, where about 160 acres were eventually leased. Up until the time that operations were transferred to China Lake, some 4,700,000 pounds of ballistite grains were produced there by 800 employees under the supervision of Bruce Sage (associate professor of chemical engineering) and William Lacey (professor of chemical engineering)—enough, as Watson said, "to have blown Pasadena off the map."[17]

Handling even small amounts of ballistite had inherent dangers, as proven by two serious explosion events, one of them in the Kellogg Radiation Laboratory on campus. On March 27, 1942, worker Raymond Robey was busy machining propellant grains on a lathe in room 106, where about 200 pounds of propellant were being stored. Most of it was in closed metal containers, but 40 sticks of it, weighing over 50 pounds, were in an open box. Another worker, named Carl B. Sanborn, walked in to deliver an empty powder can and did not notice any propellant in or around the lathe. After Sanborn left and closed the door behind him, however, Frank Crandell across the hall heard a hissing from room 106 and then saw flames arising from the lathe, but did not see Robey and assumed that he had moved out of the line of fire. Then he heard a series of small explosions followed by a larger ones, after which Robey ran down the hall screaming and out onto the pavement. Sanborn tried to smother the flames on Robey with his apron. Just about that time Aden was riding his bicycle behind the chemistry building and saw yellow smoke arising from the High Voltage Research Laboratory (now the Sloan Laboratory adjacent to the Kellogg Laboratory). He hurried over to see a man on the ground with most of the clothes burned off his body. At first he thought it was Jim Edson, who had been working on a classified project in the lab, but then Edson came around the corner and asked what had happened. Later that day Earnest Watson went to the hospital and asked Robey if he had been working on the lathe, and Robey said no. Robey died soon thereafter. After that incident, all propellant work was moved off campus to Eaton Canyon.[18]

But that did not remove the hazards. On June 16, 1942, Sanborn himself was at Station 1 in Eaton Canyon, mixing magnesium powder and potassium perchlorate in a mortar for a primer charger. As careful and experienced as Sanborn was, the resulting detonation threw pieces of a tabletop 20 feet away, and the heat from it ignited exposed ballistite propellant. Carlton Horine heard the explosion and ran to help; he found Sanborn in shock and suffering from shattered arms and significant internal injuries. Sanborn was not wearing the prescribed fireproof overalls, and he died the next day.[19]

As the need for larger and larger testing areas emerged, Eaton Canyon became important mainly as a 24/7 production facility for propellant while more appropriate proving or training grounds were selected in California: Goldstone Lake northeast of Barstow, the Marine Corps Base at Camp Pendleton, the Naval Training Center in San Diego, Morris Dam (a flood control dam, 180 feet deep, in the mountains about 20 miles from Pasadena), and China Lake near Inyokern. Caltech itself was transformed as the rocket project extended beyond the Kellogg Radiation Laboratory to the Bridge Laboratory of Physics, Astrophysical Laboratory, optical shop, and hundreds of businesses in Los Angeles, outsourced to produce parts.[20]

Apart from propellant research, the Caltech Navy projects fell into one of five general categories: target rockets for training gunners, barrage rockets to soften up beach landings to save Allied lives, aircraft rockets, packing shells up to 5 inches in diameter, and antisubmarine rockets.[21] Aden participated in most of these projects, as did Marjorie to some extent in her role in the editorial section of the projects, even while she was working on her Master of Arts degree at Claremont Colleges.[22] Rocket launchers and fuzes were designed from scratch by the Caltech team, tested by the Navy, and then mass-produced by Caltech. The first launchers and projectiles used in the North African campaign (1940–1943) and against Axis submarines were in fact manufactured by Caltech.[23] In August 1943, Caltech received orders for 10,000 forward-firing 3.25-inch rockets for use against submarines every month and to equip at least 200 Grumman TBF Avenger torpedo bombers with launchers for those rockets.[24]

Aden's involvement in the Caltech rocket work began when Jim Edson invited him to join the Navy rocket project and thereby take advantage of the associated draft deferment. Aden preferred work as a research assistant but was satisfied when the head of the Physics Machine Shop took him on as Journeyman Machinist and briefed him on the next day's project. Hours later, he was in the passenger seat of a van ominously bearing red warning signs front and back that read "DANGER—EXPLOSIVES." The route took them on the Angeles Crest Highway over the San Gabriel Mountains through Palmdale to Barstow and then toward Camp Irwin (now the Fort Irwin National Training Center), ending at the dry lake bed of Goldstone Lake. After unloading the munitions, Aden and the driver went to the "Ranch House" for dinner a few miles away. Seated at the tables were Charles Lauritsen, Willy Fowler, and Ira Bowen, eating their dinner from Army mess kits and talking about the next day's tests of antisubmarine rockets (ASR).[25]

Target rockets

Aden and Marjorie would collaborate in the ASR research later, but Aden's first assignment was target rockets (aka rocket targets), work which Fowler and the Lauritsens had begun in 1941 before the bombing of Pearl Harbor. Fin-stabilized rockets, traveling at up to 450 miles per hour, were used because "dummy" airplanes could not be manufactured fast enough to give the Marines enough practice.[26] They simulated fast, low-flying aircraft when fired at low angles and steeply diving aircraft at high angles.[27] In the fall of 1941, Aden's chore at Goldstone Lake was to teach 22,000 gunners how to "lead" aircraft so that their shots hit it, instead of whizzing ahead or behind it (Figure 3.4). Scorers such as Aden were positioned below the trajectory of the target rocket to test the effectiveness of the lead; when the gunner had the correct lead, he would hear a buzzer through headphones. On the basis of tests that Aden demonstrated for officers from the Bureau of Ordnance and US Marine Corps on September 6, 1944,[28] the Navy Bureau of Ordnance ordered 10,000 3T4 target rockets (with a velocity of up to 425 mph) from Caltech.[29] After a stint at Goldstone, Aden was transferred for five weeks to Camp Dunlap near the Salton Sea, northeast of San Diego, to train Navy and Marine gunners in firing aircraft guns of up to 90 mm at the desolate and uninhabited Chocolate Mountains (leaving him with some hearing loss) as well as mobile 40-mm guns. He also participated in several naval practice cruises, where he was in charge of target rocket crews for experimental firings at sea.[30]

Although 21,000 gunners were successfully trained using target rockets, the Commander in Chief of the US Fleet ordered Caltech to cease the research on

Figure 3.4. Aden Meinel (right) at gunnery range. Undated.
Courtesy of David Drach-Meinel.

them as of January 1, 1944, because they were less important than rockets to fill other needs.[31]

Barrage rockets

Aden was then assigned a desk in Room 200 of the High Voltage Engineering Building, the floor above where the explosion had killed Raymond Robey. Len Richards, who was leader of the Launcher Group and later served as technical advisor in the South Pacific, asked Aden to assist him designing a self-propelled launcher for barrage rockets called a "Sandy Andy." It was named after a popular children's sandbox toy patented in the early 20th century. Sand was placed into a hopper at the top, below which was a ramp on an incline. A wheeled cart was attached to the ramp by a pulley and counterweight. When sand was released from the hopper into the cart at the top of the ramp, the added weight of the cart would carry it downward to dump the sand into a waiting tin reservoir. Then, when the cart was empty, the counterweight would pull it back up the ramp automatically for the next fill. The child could learn some simple pulley physics while enjoying the play action.[32]

In much the same way, Richards had developed a method of stacking rockets vertically (like filling the hopper with sand) into two racks, each holding 12 rockets. As one 4.5-inch rocket dropped down onto the launching rails and fired, another gravity-fed rocket replaced it. An entire rack could be emptied in 12 seconds. Such a weapon was to be fired from landing craft to soften up beachheads and thereby save Allied lives. Two versions were made. One was a 4.5-inch fin-stabilized rocket with a maximum range of 1,200 yards. The other was a 5.0-inch spinner rocket stabilized by canted holes with a range up to 5,000 yards. The earliest rockets used by US armed forces were fin-stabilized following the British examples, but they were only accurate when launched at high velocity from aircraft, limiting their usefulness on the ground. Spin-stabilized rockets, however, were more accurate, less bulky, and easier to load into launchers. As a result, they were in high demand by the US Navy Bureau of Ordnance. In September 1944, orders were placed for 10,000 3.5-inch spinners at 100/day, starting immediately; 24,000 5.0-inch spinners with high explosive bodies at 5,600/month; and 9,500 5.0-inch spinners with semi-armor-piercing bodies at 1,800/month. Production shops at Caltech ran 24 hours a day. Scientific personnel worked as expediters. When their regular work was completed, they and the office staff put in an extra shift as inspectors or assemblers. No one reaped extraordinary profits because the project's hourly pay rates were well below those of airplane factories and shipyards in southern California; all were happy knowing that they had made a material contribution to the war effort.[33]

In the spring of 1943, Aden was assigned to teach Marines at Tent Camp No. 3 at Camp Pendleton, under the command of Major V. P. Hoffman, how to use the 4.5-inch rocket as assistant range engineer. The first barrage caused a grass fire in the mountains that burned for days. One Marine recruit was killed when he suddenly ran behind the launchers and was hit by the blast of burned propellant gases. Aden was called on several times to remove unexploded rocket projectiles and destroy them, admitting that he was scared each time. He was reprimanded for these hazardous actions by the Caltech Safety Committee (chaired by W. N. Arnquist) and was forbidden to handle unexploded ordnance thereafter. While he was at Camp Pendleton, he was assigned to the desert for two more weeks to assist in the final maneuvers of the Special Weapons Group of the 4th Marine Division prior to their embarkation for combat at Roi-Namur, Marshall Islands, on January 13, 1944. After that, they fought at the battles of Saipan, Tinian, and Iwo Jima. During the five-week Battle of Iwo Jima, 40.4 percent of the 4th Marine Division suffered casualties; many of those trained by Aden are likely buried there.[34] About 1,600,000 barrage rockets were produced and installed on landing craft, patrol boats, and large ships in both the European and Pacific theaters beginning at Casablanca on November 8, 1942, and Arawe, an island off New Guinea, 13 months later. Thereafter they were launched in every amphibious landing, often by the thousands in the first few minutes, and killed or dazed enemy forces on islands such as Okinawa and Iwo Jima. By the end of the war, every Marine division typically had 12 one-ton trucks, each with three 12-round Sandy Andy launchers. The 4.5-inch barrage rocket was so effective that the Bureau of Ordnance nicknamed it "Old Faithful."[35]

Antisubmarine rockets [bombs]

In 1940, U-boats sank 56 merchant ships in August and 59 in September. Out of 83 merchant ships that embarked from Nova Scotia for Liverpool in October 1940, only 20 reached their destination; not one of the eight U-boats that torpedoed the ships was lost. More than 2.8 million tons of shipping were destroyed in the first half of 1941, mainly in the Atlantic; over 1,000 men were lost. Even after the US Congress passed the Lend-Lease Act on March 11, 1941, to provide Britain with tens of billions of dollars' worth of food, oil, steel, lumber, and military equipment, the convoys that were the lifelines of Britain were still being attacked by "wolf packs" of U-boats. By April 1942, there were U-boats patrolling the entire eastern seaboard of North America and sinking supply ships at will, from the St. Lawrence River south to the Florida coastline. Activity peaked early in 1943, when U-boats sank 108 Allied ships in February and 107 in March. Britain was being starved into submission. Meat and even clothing were being heavily rationed. Winston Churchill told Parliament on February 11, 1943, "The waste of precious cargoes, the destruction of so many noble ships, the loss of heroic crews, all combine to constitute a repulsive and sombre panorama.... Defeat of the U-boat and the improvement of the margin of shipbuilding resources is the prelude to all effective aggressive operations."[36]

The tide began to turn with developments at Caltech, one of them involving Aden and Marjorie directly. Charles Lauritsen's group designed and tested antisubmarine rockets Mark 5 (ASRs) 35 inches long and 7.2 inches in diameter, using a Mark 5 recoilless launcher called the "Mousetrap." They had a range of 325 yards and a peak velocity of 165 feet/second.[37] The tests off Key West, Florida, on "tame" submarines in mid-April 1942 were so successful that the Chief of Naval Operations ordered 367 Mousetrap projectors and 45,424 ASRs. By the fall they complemented depth charges against U-boats along the Atlantic coast and in the Caribbean, and then six months later in the Pacific against the Japanese. These were the first Caltech rockets fired against the enemy and therefore the first in the US Navy's rocket program used in battle.[38]

The MIT Radiation Laboratory developed the 10-cm-wavelength radar, the SCR-584, from the British magnetron oscillator. The cavity magnetron had been improved by British physicists John Randall and Harry Boot at the University of Birmingham in 1940. It was small enough to be installed in antisubmarine aircraft but still had problems, and so the scientists gave the Americans all their notes to optimize radar devices. Bush and the NDRC established a new Radiation Lab at MIT, directed by Lee DuBridge, who at the time was chair of the physics department at the University of Rochester. Within two months the Rad Lab team, numbering 140, had produced an airborne microwave radar system that could detect a submerged submarine from three miles away.[39] It was later adapted for use by the US Navy so that the conning tower of a U-boat could be pinpointed and "locked on" from up to 12 miles away; a bomber pilot could then release bombs or depth charges with increased accuracy.[40] But there was still a problem—if the depth charges did not sink the submarine, the shock wave caused the detectors on the destroyer or bomber to "blank out," allowing enough time for any U-boat that survived to change course and escape.[41]

Meanwhile, in February 1942, the Caltech rocket group, headed by Charles Lauritsen and Willy Fowler, was conducting tests at the dry lakebed of Goldstone Lake. When he arrived at Goldstone Lake with the explosives, Aden overheard Charles Lauritsen and Fowler discussing PBYs (Patrol Boat-Y; "flying boats" or seaplanes), generators, and loop antennas. In the next morning's tests, a PBY flew over a set of wires that Aden was told simulated the magnetic field of a submarine.[42] The objective was how to slow down the depth charges so that they would sink directly above the submarine in combination with a magnetometer called the magnetic anomaly detector (MAD), which had been adapted from instruments to detect mineral ore deposits. Although the MAD could identify the location of the submarine, its range was limited to the immediate area above the U-boat, so that by the time the depth charge was dropped by a PBY, the forward motion of the PBY would put it ahead of the U-boat. Beginning in July 1942, Caltech's rocket project, which involved 250 personnel including Aden and Marjorie and Conway Snyder, solved this problem by developing an ASR that could be fired <u>rearward</u> from the PBY at a velocity matching the airplane's forward speed—a retrorocket. With a depth charge attached to it, the retrorocket would then fall vertically over the submarine.[43] The MAD/retrorocket system with a Mousetrap launcher was operational by the autumn of 1942 along the Atlantic Coast and in the Caribbean, but soon the Germans learned that if they installed an anti-gaussing coil system around the submarine they could eliminate its magnetic field and render it almost invisible from above. After that, Navy destroyers would drop a pattern of depth charges to encircle the U-boat. By the end of 1943, 243 U-boats had been sunk using one or both of these methods.[44] The number of merchant ships lost worldwide fell from 629 in July–December

1942 to 314 in January–June 1943, 149 in July–December 1943, and 67 in January–June 1944. The loss of U-boats also helped to make possible the safe movement of men and supplies across the English Channel on June 6, 1944, and ensure an Allied victory in Europe.[45]

Aircraft rockets and NOTS

In the spring of 1943, Vannevar Bush and OSRD were promoting the increased use of aircraft rockets, based on earlier successes with barrage rockets and ASRs. The Army had tested 4.5-inch aircraft rockets using a solvent ballistite with disastrous results, mainly blowups. On one occasion, the rocket motor burst inside the firing tube, and the tube and rocket flew a loop around the wing of the airplane that was used for the test. Meanwhile, commanders of U-boats under attack by depth charges and retrorockets had decided to go on the offensive and surface with antiaircraft guns firing. Although some British planes were shot down or damaged, nine submarines were sunk by the British using 3-inch, fin-stabilized rockets. Based on these that Charles Lauritsen had seen on his visit to England in 1940 and on the 50 rockets that were later sent to him, the team at Caltech soon developed the 3.5-inch forward-firing aircraft rocket (FFAR) with a solid steel, hemispherical head and a speed of 1,180 feet/second (plus the airplane's speed) that could continue in a straight line after hitting the water, penetrate a submarine's hull, and force it to surface. After successful tests at Goldstone Lake in July 1943, the Navy ordered 10,000 of these per month for 4–6 months. The contract cemented the growing partnership between the Navy and Caltech.[46]

Len Richards asked Aden to allow others to continue the training of gunners at Camp Pendleton so that he could join the Launcher Group beginning on July 15, 1943.[47] They and the Projectile Group, about 25 men in all, had desks in Room 200 of the Kellogg Laboratory. Aden's primary task was designing a way to store a thousand or so Sandy Andys in the hull of a large landing craft. But as a team, the Launcher Group was trying to improve the accuracy of unguided rockets. To stabilize the tail fins for the first few feet of travel, the launchers were as long as the rockets themselves. The Launcher Group came up with a zero-length wing launcher that guided the 3.5-inch (and later 5-inch) aircraft rockets for only the first inch of travel and therefore reduced aerodynamic drag.[48] Aden acquitted himself well in the Launcher Group as a research assistant and then research associate, earning praise from Richards in a memo to Earnest Watson dated November 30, 1943, to support the extension of Aden's draft deferment: "Mr. Meinel is one of the most useful men we have in the Land/Amphibious Launcher Group. He has particular ability in originating devices, drawing the initial sketches and plans, and getting test units constructed. He has, in addition, had considerable field experience in the testing of such devices that makes him particularly useful."[49] Among his assignments was to work on target rocket launchers, truck launchers, barrage rocket blast shields, spinner launchers, and barrage rocket gravity quadrants as integral parts of the launcher.[50]

To be used with the zero-length launcher was another critical prototype that issued from Room 200: the 5-inch-high velocity aircraft rocket (HVAR), designed principally by Tom Lauritsen and Conway Snyder. It was the most formidable rocket to date, with a 5-inch motor, 5-inch warhead, and a velocity of 1,375 feet/second. At 73 inches long and weighing 140 pounds, the rocket could penetrate 1.5 inches of armor and 4 feet of concrete. At its first ground testing at Goldstone Lake, Snyder named it "Holy Moses," as it was more awesome than anything seen to date. The epithet stuck, and by the end of the war they were the most effective rockets of all. Carl Anderson and Willy Fowler took them to Europe to equip P-47 Thunderbolt fighter-bombers. The success of these rockets in knocking out tanks, armored cars, and pillboxes was so overwhelming that Major General E. R. Quesada of the Ninth Air Force requested thousands of the rockets and dictated, "We want Caltech rockets, repeat, we want Caltech rockets, not Army Ordnance." Later on, the Holy Moses was equally effective against Japanese transports, antiaircraft gun emplacements, submarines, destroyers, ammunition and storage depots, and locomotives. By the end of the war, the Navy had 1,200 manufacturing plants for it, at a cost of $100 million per month, and more than 1 million were stockpiled. It remained in use for 11 years until it was superseded by the 5-inch Zuni Folding Fin aircraft rocket.[51]

The testing of the HVAR illustrated the pressing need for a larger range, away from populated areas, for air-to-air and air-to-ground rocket trials, larger than Goldstone Lake, Salton Sea, and Camp Pendleton. About the time that the Navy ordered 10,000 3.5-inch FFARs from Caltech in July 1943 (raised to 100,000 later that year),[52] Captain Sherman E. Burroughs visited Charles Lauritsen at Caltech. Burroughs was the new aviation assistant to the director of research and development (Captain William M. Moses) of the Bureau of Ordnance and was touring ordnance facilities, first Eglin Air Force Base in the Florida Panhandle and then Caltech. Burroughs was impressed by the Caltech teams and, like Lauritsen, stressed the urgency of a large area to conduct aviation weapons research. The following month, Lauritsen and Commander Jack C. Renard, who had expedited the production of 3.5-inch FFARs, toured the Mojave Desert in a one-engine Beechcraft airplane. To their delight they looked down on a large, hard-topped landing strip with two runways in an expanse near the village of Inyokern, California, with only a few ranches and mercury mines nearby. Later, Lauritsen gave Burroughs and other officers of the Bureau of Ordnance an aerial tour of the Inyokern airstrip and the dry lake bed of China Lake, so-called for the Chinese workers who had once collected borax from it. The infrastructure for a new weapons testing center was already nearby: the Los Angeles Aqueduct from the Mono Basin, electric and telephone lines, the Southern Pacific Railroad, and US Highway 6. As an added bonus, there were only a few residents who would have to be displaced. In October 1943, Fowler, Ira Bowen, and Wesley Hertenstein (Caltech's construction and maintenance chief) set out by jeep to survey the area and agreed that it would be an ideal proving ground. That same month, the Navy was able to lease or acquire 650 square miles in Kern, Inyo, and San Bernardino counties and install Quonset huts, a mess hall, and ammunition storage units. The Secretary of the Navy, Frank Knox, issued a memo establishing the Navy Ordnance Test Station (NOTS) on November 8, 1943. A few weeks later, Captain Burroughs assumed command of the quickly evolving station, the mission statement of which emphasized research, development, and testing of weapons, especially aircraft weapons and particularly rockets. Within six months from proposal to realization, Caltech and the Navy began to transform desert scrub and hardpan lake beds into what would become the largest weapons-testing naval facility in the country and the source of rocket armaments until the end of the war.[53] As assistant range engineer, Aden was asked to fly into Inyokern and travel by jeep to China Lake and then eastward along the length of the valley. He noted that the lake bed was perfect for impact metrology and that some rock outcroppings would be suitable for housing of Command and senior staff.[54] Further developments, supervised by Willy Fowler, would include spin-stabilized rockets ("spinners") and Tiny Tim, ironically named given its length of 10.25 feet, diameter of 11.75 inches, gross weight of 1,285 pounds, warhead gross weight of 590 pounds, and armor-piercing head.[55]

Willy Fowler, one of Aden's mentors, spent most of 1944 and 1945 helping to establish NOTS and serving as the director of research there.[56] To obtain direct feedback on the rockets from military personnel in the Pacific Theater, he flew 21,000 miles in 66 days and to visit (1) the Commander in Chief of the South Pacific, Admiral W. F. "Bull" Halsey, Jr.; (2) the Third Amphibious Command at Guadalcanal; (3) the Fourteenth Army Corps and Thirteenth Air Force at Bougainville in the Solomon Islands; (4) Headquarters of the US Army Forces in the Far East (USAFFE) in Australia; (5) the Second Engineer Special Brigade (amphibian unit of the US Army Corps of Engineers) in New Guinea; and several combat areas. He asked admirals and generals down to enlisted men how the rockets were being used and what problems had arisen that the scientists could solve. He learned that the 3.5-inch aircraft rockets and 5.0-inch aircraft rockets were ineffective against large vessels but were highly successful against antiaircraft emplacements. He was told that 4.5-inch barrage rockets were effective in amphibious operations. Admiral Chester W. Nimitz, Commander in Chief of the Pacific Fleet, wanted 100,000 rockets per month from Caltech, which Fowler knew was possible. The main deficiencies, Fowler learned, were in overall rocket policy, guidelines, installations, supplies of ammunition and launching equipment, and training materials, resulting in too many wasted man-hours, misfires, and accidents. Another problem was shortage of ammunition and rockets. Fowler reported that Admiral Halsey "literally cussed a blue streak because he can't get enough rockets out here."[57] Operating manuals and procedures were assembled by technical writers at Caltech and read by the G.I.s, even in foxholes, jungles, and under mosquito netting.[58] In October 1944, the rocketeers celebrated the production of their one-millionth rocket. It should be clear that Caltech's contribution to the war effort cannot be overstated. In fact, after the war ended, Caltech's Contract OEMst-418 received the Naval Ordnance Development Award from the US Navy Bureau of Ordnance.[59]

Meanwhile, in the European Theater, a call from London to the Caltech teams inquired about rockets that could reach an altitude of 30,000 feet to supplement British antiaircraft defenses. Aden was asked to design such a rocket. He knew Tsien Hsue-shen from the Guggenheim Aeronautical Laboratory across the alleyway from the Kellogg Radiation Laboratory. Tsien had prepared Memorandum No. 8 on drag coefficients of rocket projectiles. Aden used the information and devised a three-stage, solid propellant rocket with delay fuzes so that each successive stage would ignite after the previous one ran out of fuel. Although he constructed the rocket in Edison Pettit's machine shop, it is doubtful that one was ever tested, much less put into service.[60] Beyond Aden's adoption of some of Tsien's equations in designing the multistage rocket, there is little evidence of "cross-fertilization" between the Kellogg Radiation Lab working on Navy ordnance rockets and Theodore von Kármán's Guggenheim Aeronautical Lab (GALCIT) designing much larger rockets for the Army right next door. Indeed, Charles Lauritsen did not encourage his rocket group to have any dealings with them. Fowler noted that ". . . as everyone knows, the Navy doesn't tell the Army what it's doing, and the Army doesn't tell the Navy what it's doing."[61]

Nonetheless, it is incumbent here to recount briefly the transformation from the GALCIT Rocket Research Project founded by a handful of graduate students and two local enthusiasts into the Jet Propulsion Laboratory, where both Aden and Marjorie were hired in 1982 when it was NASA/JPL (National Aeronautics and Space Administration/Jet Propulsion Laboratory). The graduate students were Frank Malina, Tsien Hsue-shen, William Bollay, Apollo Smith, and Weld Arnold, joined by mechanic Edward Forman and chemist John Parsons. Theodore von Kármán agreed to be their faculty sponsor but could not offer any financial support. Arnold promised to raise $1,000 (= $15,585 in 2021) and somehow came up with $100 (= $1,558) in one-dollar and five-dollar bills as the first installment. The mathematical skills of Tsien, Malina, Smith, and Bollay, as well as Tsien's reports on rocket motors, liquid fuels, rocket planes, and rocket shells, gave the group its intellectual foundation. Forman and Parsons were the engineers who built the rockets with scrap metal from junkyards and second-hand stores. They were allowed to conduct their first experiments in the basement of the Guggenheim Lab, but after two major accidents indoors (without casualties) Kármán forced them to work outdoors on the Lab porch. It was not surprising that they became known as the "Suicide Squad."[62]

In 1938, Kármán funded Malina's trip to New York to present a paper at a meeting of the Institute of Aeronautical Sciences (IAS) titled "Flight Analysis of the Sounding Rocket," the first on rocketry ever presented at one of its meetings and the first to describe launching a rocket higher into the atmosphere than ever before. The paper garnered publicity in national and local media, including *Time* magazine. The Associated Press took special interest, even tailing Malina around campus. He and the others continued to experiment with rocket motors in the Arroyo Seco until one spring day in 1938 when General Henry "Hap" Arnold visited the campus and was intrigued by the possibility of employing rockets for national defense just as the Third Reich was annexing Czechoslovakian territory. He invited Kármán and Robert Millikan to attend meetings of the National Academy of Sciences Committee on Army Air Corps Research. The Committee asked Caltech to prepare a proposal for rocket-assisted takeoff for heavy bombers. Malina's report resulted in a $10,000 grant from the NAS to develop solid and liquid rocket fuels for allowing takeoffs on short runways. The grant, administered by the Army Air Corps beginning in 1940, was titled the Air Corps' Jet Propulsion Research Project. Work was carried out in the Arroyo Seco near the present site of the Jet Propulsion Laboratory (JPL), about six miles from Caltech. Critical to the success of the project was the productive partnership between Malina and Tsien. Outside of the academic environment, Malina and radically liberal graduate students such as Sidney Weinbaum invited Tsien to their social gatherings to discuss current events and topics such as socialism in Russia and the inevitable collapse of capitalism. Those connections, despite the lack of irrefutable evidence that Tsien was a member of the Communist Party, would ultimately lead to his harassment, humiliation, and deportation with his family to China in 1955 toward the close of the McCarthy era.[63]

Tsien Hsue-shen (1911–2009), also known as Qian Xuesen, originally intended to be a railway engineer. Beginning in 1929 he took classes toward that goal in mechanical and electrical engineering, mechanical design, and the basic sciences at Jiaotong University in Shanghai, the top engineering school in China. He graduated first in his class four years later. By then his interests had changed to aeronautical engineering, and for that he had to study in the Great Britain or the United States. He won a Boxer Rebellion Indemnity Scholarship in 1934 and boarded a steamship bound for MIT in 1935. He spent an unhappy year there because his talents lay in theoretical, not applied, aerodynamics. Rather than return to China so soon, he flew to Caltech in 1936 to ask Kármán to take him on as a PhD student. They would become an inseparable team, with Kármán the visionary and Tsien the gifted mathematician to make those visions realities. One of their chief contributions together was the Kármán-Tsien pressure-correction formula, which solved the problem of planes stalling at subsonic speeds.[64] Upon receiving his doctorate, Tsien was invited in 1943 to remain at Caltech as assistant professor of aeronautics, about the same time that Aden was working in the Kellogg Laboratory next door.

With Kármán, Malina, and Clark Millikan, Tsien received a $3 million-dollar grant from Army Ordnance to develop long-range rockets as soon as possible. The project had four divisions: ballistics, materials, propulsion, and structures. Tsien headed the propulsion division and shared supervision of the ballistics division with Homer Joseph Stewart. Not long after that, in September 1944, Kármán met secretly with General Henry "Hap" Arnold in a parked car at LaGuardia Airport in New York City. General Arnold wanted him to organize other scientists in Washington, DC, to prepare a study on the future of aerial warfare for the next 20–50 years. Kármán agreed to be an Army Air Force consultant and was granted a leave of absence from Caltech. He invited William Bollay and Tsien to join him there as members of the Scientific Advisory Group. In March 1945, General Arnold asked them to fly to Germany and assess progress there in research and development, both by interrogation and site inspection. This would be part of Operation LUSTY (LUftwaffe Secret TechnologY), tasked with the capture and study of German aeronautical technology. Tsien personally interrogated Wernher von Braun upon his surrender and arrest and asked him to write a report describing his past rocket work and future plans. In response, von Braun submitted a detailed paper titled *Survey of Development of Liquid Rockets in Germany and Their Future Prospects*. There in fact were Nazi plans to produce the A-9, an intermediate-range missile that could strike the Eastern Seaboard of the United States, and the A-10, a booster stage for the A-9 to convert it into an intercontinental ballistic missile.[65]

Tsien returned to Caltech in late June 1945 and was soon promoted to associate professor. He edited and contributed to the 1946 book titled *Jet Propulsion: A Reference Text Prepared by the Staffs of the Guggenheim Aeronautical Laboratory, GALCIT*. That and other publications secured him a place in the pantheon of the world's aerodynamicists, along with Kármán. He was lured back to MIT briefly by the promise of a full professorship. But he did not get along with the faculty and students, and so when Lee DuBridge, who succeeded Millikan as president of Caltech, offered him a position as Robert Goddard Professor of Jet Propulsion, including postdoctoral fellowships and research funds, Tsien leapt at the opportunity, returned to Caltech in the summer of 1949, and applied for US citizenship. That would be the start of his nightmare.[66]

About a year later, two FBI agents visited Tsien's office and confronted him with his participation in the social meetings at the home of Sidney Weinbaum as a graduate student. These events, the agents claimed, were in fact meetings of the Pasadena Communist Party, and Tsien's name was listed on the membership list under an alias. Tsien denied being a communist and said that he was opposed to communism and that Weinbaum was loyal to the US government. Regardless of how Tsien answered, his security clearance had already been revoked. That same day, Caltech administrators received a letter from the headquarters of the Sixth Army saying that Tsien was no longer allowed to work on classified military projects, which constituted 90% of all research at JPL at the time. He and the Caltech faculty members were incredulous. Tsien felt that he had lost face and offered to return to China, where his father was ill. Dubridge intervened, and in August, Tsien flew to Washington, DC, only to learn that his return to China posed a threat to national security and was therefore disallowed. His luggage, unclassified books, and papers, which he said were only drawings and logarithm tables (not code books or blueprints), were confiscated and examined. The episode made the national headlines, and many in the media were already labeling him a spy. He was arrested in September 1950 and imprisoned for 15 days in an Immigration and Naturalization Service (INS) detention center in San Pedro, California. Caltech officials were finally able to arrange bail for $15,000 (= $170,732 in 2021). But by April 1951, the INS decided to deport him under the Subversive Control Act of 1950.[67]

While he waited under partial house arrest, the FBI tailed him, broke into his house and office, and opened his mail. Friends who telephoned him were interrogated by the FBI. In June 1955, President Eisenhower released Tsien to return to China. On September 17, 1955, Tsien and his family boarded the *President Cleveland* in Los Angeles Harbor Region, bound for China. Upon his arrival three weeks later, he was lauded as the most important scientist in China and in a few months submitted a proposal to the party leadership for aeronautics and missile development. He later helped to create the first Chinese satellite, Dong Fang Hong-1, launched in 1970. In 1991, the Chinese government named Tsien a State Scientist of Outstanding Contribution, its highest honor.[68] Today, China ranks only behind the United States and Russia in atomic weaponry.

Was Tsien Hsue-shen a member of the Communist Party when he lived in America? Dan Kimball, who was executive vice president and general manager of Aerojet Corporation and then undersecretary of the Navy in 1950, interviewed Tsien in Washington in August of that year. He became convinced of Tsien's innocence and sent him to a Washington attorney for representation.[69] About Tsien's deportation, he said, "It was the stupidest thing this country ever did. He was no more a Communist than I was, and we forced him to go."[70] On January 3, 1999, a committee of the House of Representatives chaired by Christopher Cox released classified Report 105-851 in three volumes on US national security and military/commercial concerns with the People's Republic of China (PRC), principally dealing with how the Chinese acquired US missile and satellite technologies; it was declassified on May 25 of that year.[71] In December, four highly credentialed professors from Stanford University, Harvard University, and Lawrence Livermore National Laboratory evaluated the Cox Report in a study edited by Michael May.[72] They concluded that the Cox Report was riddled with inaccuracies, shallow scholarship, and speculation.[73] For example, the Cox Report stated that Tsien (Qian) worked on the Titan Intercontinental Ballistic Missile.[74] However, Tsien was denied security clearance in 1950, and work on the Titan ICBM did not even commence until 1955 when he was deported.[75] No new evidence was adduced to support the assertion that Tsien was a spy, and the authors of the Cox Report simply wrote, "The allegations that he was spying for the PRC are presumed to be true."[76] The battle between the INS (which wanted to deport Tsien) and the Department of State (which tried to keep him in the United States as a matter of national security) was lost by the latter and catapulted the PRC decades ahead in developing ICBMs and satellites.

Given his collegial relationship with Tsien while at Caltech and with physicist C. Y. Fan at the University of Arizona, it is hardly coincidental that after President Richard Nixon's diplomatic flight to China in February 1972, Aden and Marjorie would be invited to Taiwan and the People's Republic of China (Nanjing, Shanghai, Beijing), all expenses paid, seven years later. Over the course of a few weeks in the fall of 1979, they gave a series of lectures and toured observatories and optics factories.[77]

Following his construction of the three-stage, solid propellant rocket with delay fuzes, Aden submitted a Caltech memorandum to Ira Bowen on March 20, 1944, titled "High-altitude Spectroscopic Sounding Rocket." His goal was to design a multistage rocket or "step rocket" carrying an automatic recording spectrograph to obtain the far-ultraviolet solar spectrum. He computed the deceleration coefficient and the trajectories using formulas published by Tsien in CIT Memorandum No. 8. He calculated that with fins sharpened properly, the eddying could be reduced such that the drag could be explained mainly by turbulent skin friction. To reach altitudes of 84 km, a five-step rocket would be required and would be launched from 4.8 km (Mt. Whitney). He requested permission from Bowen to conduct experiments with the two-step rocket from low elevations first. Bowen doubted the calculations of an undergraduate and said that Aden was wasting his time on something not that important. Besides, Bowen continued, from his own research and that of others, he was already certain of what wavelengths would be found in the solar ultraviolet. As Aden's report was not war-related and not even important to astrophysics, he was reprimanded by his Launcher Group leader, Len Richards, and it likely jeopardized his deferment.[78] Nonetheless, he may have been one of the first to design multistage rockets. When he went to Germany in 1945 (see Chapter 4), he discovered that the Germans were also planning a multistage rocket, not for spectrographic research but for leveling cities in the West.

During Aden's work at Salton Sea, China Lake, and Camp Pendleton, Marjorie was living at home helping her father Edison take care of her mother Hannah and working toward her Master of Arts degree in astronomy at Claremont Colleges east of Pasadena. Her thesis, titled *A Study of the Long-period Variable Start RT Cygni,* was directed by her faculty advisor Walter Whitney, who was director of the Brackett Observatory at Pomona College. The visiting examiner was Alfred Harrison Joy (1882–1973), who had also helped Aden guide the 100-inch telescope in the darkness at Mt. Wilson (see Chapter 2, section "Youth and education"). Joy was hired by George Ellery Hale in 1915 from the stellar parallax program at Yerkes Observatory. He collaborated for 20 years with Walter Adams on stellar spectroscopy, particularly radial velocities of thousands of stars. His major contribution was determining a new class of variables, T Tauri stars, the name based on one of the brightest of the 11 stars in the group lying in or near the Milky Way dark clouds. These stars with unique features showed that star formation is continuous and that young stars originate in interstellar clouds. He was elected to the National Academy of Sciences in 1944. Two years later, he fell from the observing platform of the 100-inch telescope and sustained serious injuries; however, he recovered in less than a year and continued to publish until his death.[79]

Marjorie used the 6-inch Alvan Clark refractor at home with a wedge photometer to study RT Cygni, a red giant like all the other long-period variable stars. Its magnitude varies from 7 to 11 or 12 within a period of 190 days. Her research was designed to determine an accurate light curve of the star every night for at least one period, study the light curve for the previous 43 periods to spot any irregularities and construct a mean curve, and find out as many physical properties of the star as possible. She made 137 observations between June 21, 1943, and March 16, 1944, using the 6-inch refractor, and Edison Pettit supplied some data using the 20-inch telescope on Mt. Wilson. Magnitude maxima occurred on July 23, 1943 (magnitude 7.47) and January 1, 1944 (6.32), and minimum occurred on October 4, 1943 (11.04). She noted that within two days in September 1943 the magnitude dropped from 10.15 to 10.41. Alfred Joy and Paul Merrill of Mt. Wilson helped her determine that the spectrum of the star at maximum light is like that of another red giant, Alpha Orionis (Betelgeuse). Its absolute magnitude is about –4.0, making it a super-giant star, at its maximum about 16,600 times

Figure 3.5. Wedding portrait of Marjorie and Aden Meinel, 1944.
Courtesy of David Drach-Meinel.

brighter than the sun. Using the absolute and visual magnitudes, Marjorie was able to compute the parallax and therefore the distance of RT Cygni as 2,500 parsecs or 8,100 light years from Earth.[80] She also calculated its average diameter as up to 940 times larger than that of our Sun and the highest temperature as 2,999°K.[81] Claremont Colleges awarded her a Master of Arts degree on August 4, 1944.

While Aden was at Goldstone, he asked Marjorie to marry him, although the circumstances are unknown. Apparently she did not reply. One day he had just finished the day's test firings in the desert when he was called to the field telephone. It was Marjorie. The phone line stretched mostly on the Mojave Desert floor, so the reception was extremely poor. Neither he nor she could hear the other well at all as they shouted into the receiver. Finally the operator in Barstow interrupted and asked if she could help. She relayed Marjorie's answer to Aden: "She accepts and so do her parents, thanks to her sister Helen's assistance." Then the operator repeated Aden's reply for Marjorie: "Whoopee! See you to-morrow night."[82]

Only a few weeks after Marjorie received her master's degree, they were married on September 5, 1944 (Figure 3.5). Director Walter Adams granted them permission to honeymoon on Mt. Wilson in Kapteyn Cottage, built in 1910 and named after the Dutch astronomer Jacobus Kapteyn (1851–1922), who made regular visits to Pasadena as a research associate from 1909 to 1914. He had persuaded Hale to survey limited parts of the sky ("Selected Areas") to work out the structure of the Milky Way and even the universe by photometry, spectroscopy, proper motions, and radial velocities. Work would be parceled out to different observatories around the world. Hale convinced the Carnegie Trustees to adopt Kapteyn's plan, which today has helped to answer fundamental questions about stellar evolution and the origin of the Milky Way.[83]

In those years Kapteyn Cottage was limited in creature comforts, but it does command a clear view of the valley below and even the Pacific Ocean on a clear day. Just outside the door was the 50-foot interferometer that Albert Michelson and Francis Pease had constructed in 1928. Earlier they had mounted a 20-foot interferometer on top of the 100-inch telescope and had measured the angular diameter of Alpha Orionis and five other stars by 1925. The 50-foot instrument was constructed as a stand-alone unit, but it never produced any usable data. Aden knew the problems with it because he had built a four-armed interferometer as an honors project while a student at PJC.[84]

A week after they returned from their honeymoon, Aden received notice that his deferment for rocket research had not been renewed, and he was ordered to report for duty on November 10, 1944.[85]

Notes

1. Christman, A. B. 1971. *History of the Naval Weapons Center, China Lake, California.* Vol. 1. *Sailors, scientists, and rockets.* Naval History Division, Washington, DC, 19, 71–2, 80.
2. Watson, E. C. 2012. Interview by Larry Shirley. Pasadena, California. January 20, 1969. Oral History Project, California Institute of Technology Archives, https://core. ac.uk/download/pdf/32977014.pdf; accessed July 25, 2019; Christman, *History of the*

Naval Weapons Center, 80–1, 115; Goodstein, J. R. 1991. *Millikan's school: A history of the California Institute of Technology*. W. W. Norton & Company, New York, 241–2.

3. Goodstein, *Millikan's school*, 242–3.

4. Christman, *History of the Naval Weapons Center*, 86–8; Fowler, W. A. 1975. Charles Christian Lauritsen. 1892–1968: A biographical memoir. National Academy of Sciences, Washington, DC, http://www.nasonline.org/publications/biographical-memoirs/memoir-pdfs/lauritsen-charles.pdf; accessed July 22, 2019, 221–2.

5. Christman, *History of the Naval Weapons Center*, 88–9; Fowler, Charles Christian Lauritsen, 222–3, 227, 230–31.

6. Christman, *History of the Naval Weapons Center*, 93, 95; Zachary, G. P. 1997. *Endless frontier: Vannevar Bush, engineer of the American century*. The Free Press, New York, 174, 175, 182–3.

7. Christman, *History of the Naval Weapons Center*, 108.

8. Christman, *History of the Naval Weapons Center*, 107–10, 112; Goodstein, J. R., *Millikan's school*, 250; Zachary, *Endless frontier*, 129–30, 138, 183; Gibson, E. 2020. NSF and postwar US science. *Physics Today* 73: 41–2.

9. Christman, *History of the Naval Weapons Center*, 113–14.

10. Fowler, W. A. 1987. Supplemental interview by Carol Bugé. Pasadena, California. October 3, 1986. Oral History Project, California Institute of Technology Archives, http://resolver.caltech.edu/CaltechOH:OH_Fowler_W, 175; accessed July 4, 2019. By 1944, Caltech was spending $2 million/month on the rocket project. Horine, C. 1998. WWII rocket project at Caltech's Eaton Canyon. Historical files, Y1.15. Caltech Archives, 9.

11. Christman, *History of the Naval Weapons Center*, 112, 114; Price, E. W., Horine, C. L., and Snyder, C. W. 1998. Eaton Canyon: A history of rocket motor research and development in the Caltech-NDRC-Navy Rocket Program, 1941–1946. 34th AIAA/ASME/SAE/ASEE Joint Propulsion Conference and Exhibit, July 13–15, 1998, Cleveland, Ohio. American Institute of Aeronautics and Astronautics, 3.

12. Fowler, Supplemental interview, 176.

13. Fowler, W. A., and Ajzenberg-Selove, F. 1985. Thomas Lauritsen. 1915–1973: A biographical memoir. National Academy of Sciences, Washington, DC, http://www.nasonline.org/publications/biographical-memoirs/memoir-pdfs/lauritsen-thomas.pdf; accessed July 22, 2019, 385–9.

14. Clayton, D. D. 2014. Fowler, William Alfred. In T. Hockey et al. (eds.), *Biographical encyclopedia of astronomers*. Springer, New York, http://doi.org/10.1007/978-1-4419-9917-7, 7.

15. Horine, WWII rocket project, 8.

16. Christman, *History of the Naval Weapons Center*, 117–18; Goodstein, *Millikan's school*, 248; Price et al., Eaton Canyon, 8.

17. Snyder, C. W. 1991. Caltech's *other* rocket project: Personal recollections. *Engineering & Science* 54: 4; Christman, *History of the Naval Weapons Center*, 118; Goodstein, *Millikan's school*, 255; Price et al., Eaton Canyon, 12.

18. Christman, *History of the Naval Weapons Center*, 126–7; Snyder, Caltech's *other* rocket project, 4; Meinel, A. B., Meinel, M. P., and Jacobs, B. M. 2008. *The golden age*

of astronomy: In the beginning. . . . Self-published, 34; Price et al., Eaton Canyon, 13; Horine, WWII rocket project, 9–10.

19. Christman, *History of the Naval Weapons Center*, 127–8; memo to W. A. Fowler from V. Anthony, William Fowler Papers, Box 57, Folder 11, Caltech Archives; Horine, WWII rocket project, 10.

20. Goodstein, *Millikan's school*, 255; memo to Admiral R. S. Holmes from Earnest Watson, June 12, 1943, William Fowler Papers, Box 99, Folder 10, Caltech Archives.

21. Goodstein, *Millikan's school*, 255.

22. Meinel et al., *Golden age*, 37.

23. Memo to Earnest Watson from J. Foladare on rocket history for Navy release, May 30, 1944, William Fowler Papers, Box 95, Folder 1, Caltech Archives.

24. Letter to Irvin Stewart, OSRD, from Earnest Watson, August 19, 1943, William Fowler Papers, Box 99, Folder 10, Caltech Archives.

25. Email from Aden Meinel to W. Patrick McCray, November 9, 1998.

26. Rocket Demonstrations and Files, William Fowler Papers, Box 61, Folder 11, Caltech Archives.

27. OSRD, Contract OEMsr-418, Gunnery and Tactical Training with Rocket Targets, December 14, 1943, William Fowler Papers, Box 88, Folder 12, Caltech Archives.

28. Memo from A. B. Meinel to W. A. Fowler regarding Dam Neck rocket target tests, September 14, 1943, William Fowler Papers, Box 110, Folder 8, Caltech Archives.

29. Memo to the Chief of the Bureau of Ordnance from the Coordinator of Research and Development, November 10, 1943, William Fowler Papers, Box 110, Folder 8, Caltech Archives.

30. Christman, *History of the Naval Weapons Center*, 113, 131; Fowler, Supplemental interview, 174; Meinel, A. B., and Meinel, M. P. 2002a. *Echoes from a simpler time*, unpublished, 27; memo from A. B. Meinel to W. A. Fowler regarding target launcher tests at Niland, California, November 1, 1943, William Fowler Papers, Box 110, Folder 8, Caltech Archives; letter of recommendation for Aden Meinel from James Edson to Director of Naval Officer Procurement, January 5, 1944; letter from Aden Meinel to Lt. Commander Boroughton, US Naval Training Center, December 28, 1944.

31. Letter to F. L. Hovde, Chief of Division 5, NDRC, November 18, 1943, William Fowler Papers, Box 99, Folder 10, Caltech Archives.

32. Meinel et al., *Golden age,* 28.

33. OSRD, Contract OEMsr-418, 3.5″ and 5.0″ Spin-stabilized Rockets, William Fowler Papers, Box 77, Folder 7, Caltech Archives; letter to Captain G. L. Schuyler, Chief of the Bureau of Ordnance, from C. C. Lauritsen, December 27, 1943, William Fowler Papers, Box 96, Folder 2, Caltech Archives; memo to the Office of the Coordinator of Research and Development from the Chief of the Bureau of Ordnance, September 8, 1944, William Fowler Papers, Box 102, Folder 1, Caltech Archives; Huse, W. W. 1957. The California Institute and rocket production. Historical files, Y1.5. Caltech Archives, 8, 14.

34. Chapin, J. C. 1945. *The Fourth Marine Division in World War II.* History and Museums Division, Headquarters, US Marine Corps, Washington, DC, 82E.

35. Email from Aden Meinel to W. Patrick McCray, November 9, 1998; Veysey, Victor V. Interview by Shirley K. Cohen. Pasadena, California. July 14 & 21, 1993, and February 4, 1994. Oral History Project, California Institute of Technology Archives, http//resolver.caltech.edu/CaltechOH:OH_Veysey_V, 15; accessed May 31, 2018; Snyder, Caltech's *other* rocket project, 5–6; Christman, History of the Naval Weapons Center, 138–40; Meinel et al., *Golden age*, 39; Price et al., Eaton Canyon, 4; Gerrard-Gough, J. D., and Christman, A. B. 1978. *History of the Naval Weapons Center, China Lake, California*. Vol. 2. *The grand experiment at Inyokern*. Naval History Division, Washington, DC, 114; letter from Aden Meinel to Lt. Commander Boroughton, US Naval Training Center, December 28, 1944; memo to Commanding General, Camp Pendleton, from Earnest Watson, May 10, 1943, William Fowler Papers, Box 99, Folder 10, Caltech Archives.
36. Dimbleby, J. 2016. *The Battle of the Atlantic: How the Allies won the war*. Oxford University Press, 102, 111, 212; Snyder, Caltech's *other* rocket project, 6; Zachary, *Endless frontier*, 142; Manchester, W., and Reid, P. 2012. *The last lion: William Spencer Churchill, defender of the realm. 1940–1965*. Little, Brown, and Company, New York, 243–4; https://api.parliament.uk/historic-hansard/commons/1943/feb/11/war-situation.
37. William Fowler Papers, Box 61, Folder 1, Caltech Archives.
38. Christman, *History of the Naval Weapons Center*, 131–4; Snyder, Caltech's *other* rocket project, 5.
39. Zachary, *Endless frontier*, 132–5.
40. Dimbleby, *The Battle of the Atlantic*, 386.
41. Email from Aden Meinel to Edward Meinel, October 26, 2009. As a result, the success rate was only about 5 percent; Snyder, Caltech's *other* rocket project, 5.
42. Email from Aden Meinel to W. Patrick McCray, November 9, 1998.
43. Snyder, Caltech's *other* rocket project, 7; Christman, *History of the Naval Weapons Center*, 133; Veysey, Interview, 12; email from Aden Meinel to W. Patrick McCray, November 9, 1998; Price et al., Eaton Canyon, 4–5; email from Aden Meinel to Edward Meinel, October 26, 2009; memo to Earnest Watson from Thomas Lauritsen, May 11, 1944, William Fowler Papers, Box, 97, Folder 1, Caltech Archives.
44. Dimbleby, *The Battle of the Atlantic*, 429; Goodstein, *Millikan's school*, 257; Christman, *History of the Naval Weapons Center*, 133; Veysey, Interview, 12.
45. Dimbleby, *The Battle of the Atlantic*, 431.
46. Christman, *History of the Naval Weapons Center*, 145, 147, 149, 163; Price et al., Eaton Canyon, 5; Snyder, Caltech's *other* rocket project, 8; press release from Industrial Incentive Division, Navy Department, undated, William Fowler Papers, Box 99, Folder 9, Caltech Archives.
47. William Fowler Papers, Box 57, Folder 5, Caltech Archives.
48. Meinel et al., *Golden age*, 29.
49. William Fowler Papers, Box 57, Folder 5, Caltech Archives.
50. Memo to W. A. Fowler and C. C. Lauritsen from L. A. Richards, October 30, 1943, William Fowler Papers, Box 108, Folder 6, Caltech Archives; memo to W. A. Fowler and C. C. Lauritsen from L. A. Richards, November 16, 1943, William Fowler Papers,

Box 108, Folder 6, Caltech Archives; memo to W. A. Fowler and C. C. Lauritsen from L. A. Richards, December 11, 1943, William Fowler Papers, Box 108, Folder 7, Caltech Archives; memo to W. A. Fowler and C. C. Lauritsen from L. A. Richards, December 29, 1943, William Fowler Papers, Box 108, Folder 7, Caltech Archives; memo to W. A. Fowler from A. B. Meinel regarding 3R1 firing from M. C. mount, Camp Pendleton, January 3, 1944, William Fowler Papers, Box 110, Folder 5, Caltech Archives; memo to W. A. Fowler from A. B. Meinel regarding range data of 3R1 automatic spinner launchers, January 13, 1944, William Fowler Papers, Box 108, Folder 8, Caltech Archives; memo to W. A. Fowler from A. B. Meinel regarding 3R1 firing from machine gun tripod at Camp Pendleton, January 22, 1944, William Fowler Papers, Box 108, Folder 8, Caltech Archives; memo to W. A. Fowler and C. C. Lauritsen from L. A. Richards, January 26, 1944, William Fowler Papers, Box 108, Folder 8, Caltech Archives; memo to W. A. Fowler from A. B. Meinel regarding 3R1 dispersion from launcher CIT Type 32A, Goldstone, January 28, 1944, William Fowler Papers, Box 108, Folder 8, Caltech Archives; memo to W. A. Fowler from A. B. Meinel regarding 3R counter flow launcher test, Goldstone, February 7, 1944, William Fowler Papers, Box 108, Folder 8, Caltech Archives; memo to W. A. Fowler from A. B. Meinel regarding 3R launcher tests, Camp Pendleton, February 22, 1944, William Fowler Papers, Box 108, Folder 8, Caltech Archives; memo to W. A. Fowler from L. A. Richards, February 23, 1944, William Fowler Papers, Box 108, Folder 8, Caltech Archives.

51. Email from Aden Meinel to W. Patrick McCray, November 9, 1998; Snyder, Caltech's *other* rocket project, 8, 10; Latimer, J. 2019. Tank killer—5-inch High Velocity Aircraft Rocket (HVAR) "Holy Moses." *The China Laker* 22: 2–3; Price et al., Eaton Canyon, 5–6; Gerrard-Gough and Christman, *History of the Naval Weapons Center*, 84–5; Martin, W. T., and Navy Department (eds.). 2013. *Arming the fleet: Providing our warfighters the decisive advantage*. 3rd ed. Department of the Navy, Washington, DC, 12.

52. Letter to Irvin Stewart, OSRD, from Earnest Watson, November 18, 1943, William Fowler Papers, Box 99, Folder 10, Caltech Archives.

53. Christman, *History of the Naval Weapons Center*, 167–71, 182–3, 187–8, 194; Gerrard-Gough and Christman, *History of the Naval Weapons Center*, 190.

54. Meinel et al., *Golden age*, 29.

55. Bowman, N. J. 1957. *The handbook of rockets and guided missiles*. Perastadion Press, Whiting, Indiana, 204.

56. Fowler, Supplemental interview, 171.

57. Memo to C. C. Lauritsen from W. A. Fowler in Noumea, New Caledonia, March 8, 1944, William Fowler Papers, Box 111, Folder 9, Caltech Archives.

58. Gerrard-Gough and Christman, *History of the Naval Weapons Center*, 72, 74–7; memo to Joseph Foladare from W. A. Fowler, April 9, 1944, William Fowler Papers, Box 111, Folder 9, Caltech Archives; memo to Chief of Field Services from William Fowler, May 17, 1944, William Fowler Papers, Box 95, Folder 1, Caltech Archives.

59. Memo to employees and former employees of Contract OEMsr-418 from Earnest Watson, April 1946, William Fowler Papers, Box 100, Folder 2, Caltech Archives.

60. Meinel et al., *Golden age*, 30.

61. Fowler, Supplemental interview, 177, 179.
62. Chang, I. 1995. *Thread of the silkworm*. BasicBooks, New York, 71–4; Goodstein, *Millikan's school*, 262.
63. Chang, *Thread of the silkworm*, 75–6, 80–1, 83; Goodstein, *Millikan's school*, 263.
64. Kármán, T. von, and Tsien, H. S. 1938. Boundary layer in compressible fluids. *Journal of the Aeronautical Sciences* 5: 227–32.
65. Chang, *Thread of the silkworm*, 23, 33, 38, 64, 93, 104–5, 108–9, 111–13; Simons, G. M. 2016. *Operation Lusty: The race for Hitler's secret technology*. Pen and Sword Books, Barnsley, South Yorkshire, UK, 12, 105–7, 111; Bowman, *Handbook of rockets and guided missiles*, 76–7.
66. Chang, *Thread of the silkworm*, 117, 130, 141–3.
67. Chang, *Thread of the silkworm*, 149–51, 154–7, 160, 162, 165–6, 171.
68. Chang, *Thread of the silkworm*, 173, 189, 201, 212–13, 225, 260.
69. Chang, *Thread of the silkworm*, 154–5.
70. Nolan, https://www.latimes.com/nation/la-me-qian-xuesen1-2009nov01-story.html.
71. Cox, C. 1999. Report of the select committee on U.S. national security and military/commercial concerns with the People's Republic of China. 3 vols. U.S. Government Printing Office, Washington, DC, ii–xxxvii.
72. May, M. M. (ed.). 1999. The Cox Committee Report: An assessment. https://carnegieendowment.org/pdf/npp/coxfinal3.pdf; accessed August 18, 2019.
73. May, Cox Committee Report, 38.
74. Cox, Report of the select committee, 178.
75. May, Cox Committee Report, 36.
76. Cox, Report of the select committee, 178.
77. Meinel, A. B., and Meinel, M. P. 1980. Aden and Marjorie Meinel's China trip October–November 1979. *Applied Optics* 19: 2666–9.
78. Meinel, A. B. 1944. C.I.T. memorandum to I. S. Bowen: High-altitude spectroscopic sounding rocket. March 20, 1944. Walter S. Adams correspondence, Box 46, Huntington Library and Archives, 1–11.
79. Herbig, G. 1974. Obituary: Alfred Harrison Joy. *Quarterly Journal of the Royal Astronomical Society* 15: 527–30; Joy, A. H. 1945. T Tauri variable stars. *Astrophysical Journal* 102: 168.
80. We now know it is only 1,149.43 parsecs and 3,749 light years away; [https://www.universeguide.com/star/rtcygni#distance.
81. Pettit, M. S. 1944. *A study of the long-period variable star RT Cygni*. Master's thesis. Claremont Colleges, Claremont, California, 5, 7, 24–5, 36, 37, 41, 45–8.
82. Meinel et al., *Golden age*, 42–3.
83. Sandage, A. 2012. *Centennial history of the Carnegie Institution of Washington*. Vol. 1. *The Mount Wilson Observatory*. Cambridge University Press, 211–12, 214–15, 422.
84. Meinel et al., *Golden age*, 41; Sandage, *Centennial history*, 424–5.
85. Meinel et al., *Golden age*, 41.

4

Foreign Intelligence across the Rhine

Aden wanted no part of slogging through the mud and snow of Europe as an infantryman. Rear Admiral Ralston Smith Holmes, who had served as the director of Naval Intelligence from May 1937 to June 1939, was leading the Caltech rocket project at that time and wrote a successful supporting letter for Aden's request to be assigned to the Navy.[1] Although Aden did not mention the admiral by name, Victor Veysey identified him in an oral interview with Shirley Cohen for the Caltech Archives.[2] Such requests were rarely granted.

In 1944 there were four Navy boot camps: San Diego, California; Bainbridge, Maryland; Newport, Rhode Island; and Great Lakes, Illinois. As the chill of winter was bearing down up north, he was relieved to be inducted in the relatively mild climate of San Diego on November 10.[3] The US Naval Training Center at the north end of San Diego Bay had been commissioned 21 years earlier and during World War II housed as many as 25,000 recruits under Lt. Cmdr. Simon L. Shade. Apprentice Seaman Meinel was assigned to Company 44-576 and issued his uniform, sleeping gear, *The Bluejackets' Manual*[4] with everything a young sailor needed to know (ranging from personal hygiene and first aid to ships and aircraft, gunnery, and seamanship) and his canvas sea bag that would contain all his essentials rolled in prescribed ways to minimize wrinkles and save space. He was given the standard buzz cut and vaccinations for almost every disease imaginable (Figure 4.1). The Navy General Classification Test determined levels of aptitude for different military roles; it was similar to today's SATs and ACTs for college, evaluating verbal and math skills. More specialized tests also assessed abilities in administration, language skills, and technical pursuits such as electronics. Aden wrote that his scores on all the exams were almost perfect.[5] On his Enlisted Personnel Qualifications Card under Duties, Skills, Machines, he listed "rockets and rocket launchers; supervised preliminary design and construction and service tests with various armed forces; trained anti-aircraft gunners with rocket targets—Camps Callen, Pendleton, Dunlap, Pacific Beach & Dam Neck, VA; used lathe shaper, milling machine in experimental construction of parts."

For about six weeks, over 10 hours a day were spent on calisthenics and fitness tests, one drill after another, firearms training, preparing for inspections,

With Stars in Their Eyes. James B. Breckinridge and Alec M. Pridgeon, Oxford University Press. © Oxford University Press 2022. DOI: 10.1093/oso/9780190915674.003.0004

Figure 4.1. Aden Meinel, enlistment photo.
National Archives and Records Administration.

scrubbing uniforms, seamanship, and batteries of academic tests. Among the training exercises was learning how to jump from a sinking ship while gripping a life jacket under your chin so that your neck would not be broken when you hit the water. This was repeated over several days in a large swimming pool. A Marine instructed the new recruits how to repel sharks if the ship sank in warm tropical waters: Beat on the water to scare the sharks away." Aden disagreed: "Sir, that's the way to attract them. I know because beating the water with a long bamboo pole was how I attracted them when I was fishing for sharks." The Marine glared at him and shouted, "Don't you ever contradict me again. What I told you is Navy policy. If you do that again, I'll confine you to the brig." He then told the other recruits to ignore what Aden said.[6] Before experimentation with shark repellents, that was indeed Navy policy.[7]

Following boot camp, Aden's company was to be assigned to a spectacular "new" Portland-class heavy cruiser across the bay for their first mission (Figure 4.2).[8] It was a sight to behold with nine 8-inch guns along with 34 antiaircraft guns: eight 5-inch guns, two Hotchkiss 3-pound guns, and 24 Bofors guns. At

Figure 4.2. USS *Indianapolis* docked at the pier in San Diego harbor (1935).
Photo by Harry A. Erickson. Courtesy of the San Diego History Center.

610 feet long and 133 feet high from the waterline to radar antennae, the cruiser was a behemoth that, despite its size, could reach speeds of 32.75 knots. The explanation (and ultimate price) for such speed was that its belt armor amidships was only 3–4 inches, whereas a battleship had about 13 inches; this was to be its downfall in July 1945.[9] Aden and his fellow "boots" were excited by the prospect of having bunks and showers, a far cry from slogging through the rain, snow, and mud that GIs would have to endure.[10] But the cruiser was not as new as Aden assumed. It had been commissioned in 1932 and had served as Roosevelt's ship of state on a goodwill tour of South America before the attack on Pearl Harbor. Afterward, it had engaged Japanese ships and shore installations in New Guinea, the Aleutian Islands, and other islands in the Pacific: the Gilberts, Marshalls, Marianas, and Western Carolines. It would later participate in the attacks on the Japanese "Home Islands" in February–March 1945 and then bombarded Iwo Jima and Okinawa to prior to landings there. On the bow was hull number 35. It was the USS *Indianapolis* (CA-35).

Whether because of his scores on the exams and/or intervention by influential people such as Admiral Holmes and Willy Fowler, Aden was called to

Headquarters after only a few weeks of boot camp. He was ordered to report back to Caltech and would not be sailing on the *Indianapolis* after all. He would later learn of its fate in July 1945 while stationed in Paris. So he ended up behind the same desk in Room 200 of the High Voltage Laboratory. Marjorie, who as of January 1945 was working on the editorial staff of OSRD,[11] had an office across the alleyway in the Annex.[12] Marjorie told him that the secretaries there loved "the rocket engineer wearing a white Navy cap" and the uniform of a Seaman Second Class.[13]

On January 2, 1945, Aden was reassigned to China Lake and applied for commission as Ensign on the recommendation of Rear Admiral Holmes and Willy Fowler, Jim Edson, and W. R. Van Auken.[14] Captain Van Auken (BuOrd AD4) argued that Aden's two years of experience as a research associate at Caltech in connection with the design and testing of rocket-ranging equipment made his services extremely valuable to the BuOrd Rocket Program. If the application was approved, Van Auken continued, he recommended that Aden report on February 4, 1945, to the Commanding Officer at the Naval Ammunition Depot at Hingham, Massachusetts, for temporary duty. The Chief of Naval Personnel, Lieut. (JG) John F. Geis, handling the application wrote, "Recommend waiver on lack of college degree and appointment because of importance and urgent need of personnel with rocket experience. Am in doubt about this one but recommend we go along with bu ord [*sic*] due to importance of program. (signed J. F. G.)."[15]

After temporary additional duty in Pasadena to study the design of rocket launchers, he returned to Inyokern on January 28. A secret effort was then well underway at NOTS: Project CAMEL. Aden recalled that before he was drafted, he had been asked to measure the exact timing of fuzes, specifically how long it took a fuze to detonate after an electrical current was sent to it.[16] He soon began to understand why.

Project CAMEL

Project CAMEL, as explained to Aden by Fowler, referred to the observation that once a camel (in this case, Caltech) gets its nose through the tent flaps, it will soon occupy the whole tent. It represented Caltech's growing and leading role in research at NOTS for the atomic bombs of the Manhattan Project. Charles Lauritsen, cofounder of NOTS with Commander Sherman E. Burroughs of the Navy's BuOrd, and Fowler headed the rocket team there and worked closely with fellow Caltech professor and theoretical physicist J. Robert Oppenheimer at the Los Alamos Laboratory. Project CAMEL specifically encompassed the development of detonators and also drop tests of the Fat Man bomb using different fin configurations from the Boeing B-29 Superfortress. The most effective bomb shape was determined using instrumentation designed by Gerald E. Kron, also of Caltech and head of the Special Devices Group at NOTS.[17]

The problem with designing working detonators (known as "sockets") was that the action had to be on the order of milliseconds in activating the high-explosive block around the nuclear core. Lauritsen's son, Thomas, and Fowler were in charge of developing and testing of the sockets made in Pasadena at China Lake.[18]

As the Manhattan Project was top secret, Aden could not reveal his work on detonators to Marjorie, nor could she tell him that she had prepared and edited reports about the Project for the government. One of those reports was a message with the code name Fat Man. A drop test departed Alhambra Airport (now a shopping mall east of Los Angeles) for the China Lake Range with a dummy bomb loaded under the fuselage. When the bomber reached the drop site, a voice over the radio from ground control asked the pilot where Fat Man was. When the pilot checked the arming circuits, there was no signal. His secret payload had been dropped prematurely; everyone hoped that it was inaccessible high in the mountains between Alhambra and NOTS. There were no reports of an object parachuting to the ground.[19] It is not known if it was ever found. It may have simply penetrated the desert floor to a depth of a 10-story building when dropped from 25,000 feet, as occurred in the first dummy bomb drop.[20]

Detonators and other parts for the Little Boy and Fat Man bombs were covertly loaded aboard the USS *Indianapolis* for transport to Tinian in the Northern Mariana Islands. From there the *Enola Gay* took off for Hiroshima with Little Boy; three days later the B-29 bomber *Bockscar* was bound for Nagasaki with Fat Man, which ended World War II. Although Aden was not aboard the *Indianapolis*, the results of his detonator research were.

New orders

In February 1945, only three months after he had been inducted, Aden was appointed Ensign S(04) and received orders to transfer for temporary duty to the Hingham Naval Ammunition Handling Depot at Hingham, Massachusetts, which produced munitions for the war effort.[21] He delayed reporting until March 4. That year the Depot, spanning almost a thousand acres, consisted of about 90 buildings (barracks, houses, workshops, storage facilities, PX, etc.) and employed close to 2,400 civilians and military personnel. Aden's tasks were related to deployment of barrage rockets on LSTs (Landing Ship, Tank) to soften up Pacific landings, including at Iwo Jima and Okinawa. Because there was no recoil from them, they could be launched from any boat or ship. Some were fin-stabilized, whereas others were spin-stabilized. What they still lacked in accuracy was compensated by the sheer numbers that could be launched at once.[22] He was to teach officers how to use them and would likely be assigned as Chief Ordnance Officer on one of them.[23] Only two weeks later, on March 26, 1945, the Commandant called Aden into his office at 5 p.m. and conveyed orders for him to report by commercial air to Navy Headquarters BuOrd in Washington, DC. Aden said that he would be ready the next morning. The Commandant shouted back, "They mean *now!*"[24] He quickly gathered his gear, loaded it into a chauffeured station wagon, and headed for the Boston airport. The car with a broken gas gauge sputtered out of gas along the way, so Aden hitchhiked the rest of the way and barely made his flight to Washington, DC, at 10:35 that night.[25]

Arriving at Navy HQ at 7:30 the next morning, he was surprised to see people on roller skates gliding down the long halls and thought to himself, "So that's how they expedite things here."[26] He reported to Lieut. (JG) Flook, the BuOrd Commander who had earlier assigned him to the Pacific theater. Flook told Aden that he had been designated the new rocket and optics expert for the Naval Technical Mission in Europe (NavTechMisEu). His assignment was twofold: "(1) provide specialized technical experience and detailed knowledge of foreign ordnance and prepare reports under supervision of NavTechMisEu, and (2) inspect and evaluate this ordnance, the design and manufacture, and any features thereof that will be of value to U.S. Navy and prepare complete reports thereon."[27]

NavTechMisEu had been activated on January 20, 1945, led by Commodore Henry Adrian "Packy" Schade, who was directly responsible to the Commander in Chief of the US Fleet and Chief of Naval Operations.[28] Schade had been ordered to form teams to work behind the front lines to seek out German documents, equipment, and weaponry just behind the front lines in Europe and ship them back to the States. He had "maximum freedom of action consistent with operational Naval and Military requirements" and had complete oversight of a fleet of Navy aircraft to transport captured Axis personnel, technology, and documents to Great Britain and the United States before the Soviets could lay their hands on them.[29]

On the team from Caltech were aerodynamicists Clark Millikan and Tsien Hsue-shen (also known as Qian Xuesen). Millikan helped to engineer the construction of the wind tunnels at Caltech in which Aden had trained. He was the son of Robert Millikan, winner of the Nobel Prize for Physics in 1923, with whom Aden had launched balloons carrying cosmic ray counter systems as a laboratory assistant. As discussed in Chapter 3, Tsien Hsue-shen was a protégé of Hungarian aerodynamicist Theodore von Kármán and would later become Goddard Professor at Caltech and cofounder of the Jet Propulsion Laboratory. On May 5, 1945, Tsien and others from the team interrogated leaders of the German rocket group led by Wernher von Braun and Walter Dornberger, two days after their surrender in Bavaria.[30] There were also civilian technical specialists selected by Navy contractors, including Charles Lindbergh, who later followed in Aden's footsteps at Mittelbau-Dora.[31]

Aden was to accompany General Patton's Third Army, part of the 12th Army Group, staying as near the front as possible.[32] Aden would be a member of the T(echnical)-Force mission (Combined Advance Field Teams, or CAFT), gathering and assessing intellectual resources and seizing or destroying technology before the Soviets arrived on the scene.[33] His own principal targets would be the optics works in Wetzlar (Leica) and Jena (Zeiss and Schott). In addition, he was to locate the underground factory where V-2 rockets were being assembled and bring back anything of scientific value.[34] Curious about why he was chosen, Aden learned that the BuOrd Commander had recommended him after a DC-3 carrying the first group of personnel chosen from MIT had crashed outside Paris, killing everyone aboard. They were meant to be there to participate in an earlier crossing of the Rhine River. Now it was up to Aden, and he was to go immediately to Headquarters of the Office of Strategic Services (OSS) for personal defense training in case of ambushes or booby traps.[35] This was probably his first of many direct dealings with national intelligence agencies, although as mentioned earlier, he had certainly worked on classified projects at China Lake in years prior.

The Office of the Coordination of Information (OCI) was established by President Roosevelt in 1941 and chose Major General William J. "Wild Bill" Donovan as director. For his brave and fearless actions as commander of the 165th Infantry Regiment of the US Army in World War I, Donovan had been awarded not only two Purple Hearts but also the Congressional Medal of Honor, Distinguished Service Cross, Distinguished Service Medal, and the National Security Medal; he was the only American to have received all four of the nation's highest awards.[36] In 1942, the OCI became the OSS with Donovan still in charge. Eight major departments coordinated the gathering and analysis of intelligence, weapons and equipment development, counterespionage, code-breaking, sabotage (e.g., railroad tunnels), guerrilla warfare, subversive propaganda by the Morale Operations branch, and organizing resistance groups in Europe.[37]

At the height of its influence during World War II, the OSS employed almost 24,000 people from all walks of life, including such notables as chef Julia Child, historian Arthur Schlesinger, Jr., and actor Sterling Hayden.[38] At OSS Aden learned in a one-day crash course how to detect and avoid booby traps and also how to kill a person with a pencil.[39] President Truman dissolved the OSS after the war ended, but it would eventually be redefined, combined with other agencies, and renamed the Central Intelligence Agency (CIA), or simply "the Agency." Donovan would become known as the "Father of Central Intelligence," and Aden's name would later become "synonymous" with the CIA.[40] That night he called Marjorie to say that he was going across the Atlantic, not the Pacific, on a secret mission and that he would write as soon as possible.[41]

The next day Aden flew to RAF Prestwick in Scotland aboard an unpressurized Douglas DC-4 with benches along the wall and small, round holes in the center of each window to insert a machine gun should German Messerschmitts approach. After refueling in Gander, Newfoundland, it was a grueling 16-hour flight to Scotland, but Aden managed to get some sleep atop parachutes on the floor. After a day in Scotland, he overflew London directly to Paris because of pressing new developments at the Rhine.[42] He spent that night at Le Royal Monceau Hotel at 37 Avenue Hoche in the 8th arrondissement, only 500 meters from the Arc de Triomphe. The hotel had been stripped of interior decorations by the Germans before they retreated, but it had been de-loused, and the food was decent. The next morning at breakfast Aden noticed some civilians among otherwise military personnel and asked about them. He was told simply that they were members of the top secret Alsos mission, which had something to do with determining the extent of German development of atomic weapons.[43] One of those members, Gerard Kuiper, would become Aden's lifelong friend and colleague at Yerkes Observatory and the University of Arizona (UA), so it is relevant to provide more detail on Kuiper and Alsos at this point.

Alsos mission

Alsos, the Greek word for "grove," was the code name referring to Brigadier General Leslie Groves of the US Army Corps of Engineers, who directed the Manhattan Project. Hitler repeatedly claimed that Germany had devised secret weapons, and there were also intelligence reports that Germans were following the nuclear research of French physicist Frederic Joliot-Curie and producing heavy water (deuterium oxide) at the plant in Rjukan, Norway, and delivering an estimated 120 kilograms of it to the Nazis every month. This was enough for Project administrators to be concerned that Germany could be developing an atomic bomb or that uranium and other bomb materials could fall into Soviet hands. German physicist Werner Heisenberg, a pioneer in quantum mechanics and famous for his "uncertainty principle," and Otto Hahn, who had conceived of the principle of nuclear fission with his assistant Fritz Strassmann in 1938, were prime targets for capture. Both were leading researchers in the Nazi nuclear power project *Uranverein*.[44]

The mission, led by Colonel Boris Pash (1900–1995) with physicist and science advisor Samuel Goudsmit (1902–1978), was staffed by the Office of Naval Intelligence (ONI), OSRD, the Manhattan Project, and Army Intelligence (G-2). Pash and others recruited 13 military personnel and civilian scientists (by

V-E Day they numbered 114) who were focused on gathering intelligence about mainly nuclear physics but also bacteriological warfare, aeronautical research, proximity fuzes, and guided missiles.[45] Among the scientists were the Dutch-American astronomers Peter van de Kamp from Swarthmore College and Gerard Kuiper from Yerkes Observatory—whom Aden would meet upon his return to Paris and who would play a major role in Aden's career in astronomy and optics.[46] Goudsmit had formulated the concept of electron spin in 1925 with George E. Uhlenbeck as doctoral students, which had major implications for atomic theory and quantum mechanics.

In a memo dated April 1, 1945,[47] Goudsmit outlined for Kuiper his assigned duties in order of importance:

1. Assist the senior scientific member, if so requested, on special tasks of high priority.
2. Contact enemy astronomers and investigate their homes and institutes in order to uncover their contributions to war research.
3. Visit other research institutions with the working radius of the Aachen Headquarters, concentrating as much as possible on fundamental research to the exclusion of development and production. Of special interest to Alsos is intelligence about the organization of German war research.
4. Contact our colleagues in Holland, when liberated, in order to obtain information on German research.

In a handwritten memo to Kuiper dated nine days later, Goudsmit (who was of Dutch-Jewish descent), added a task of a more personal nature. Goudsmit knew that his parents Isaac and Marianne had been held in the Westerbork transit camp in the Drenthe province in the northeast Netherlands until they were removed from there in January 1943. Goudsmit asked Kuiper to search the records to find out where they were sent.[48] It is not known when Goudsmit learned that they had been murdered in an extermination camp; detainees in Westerbork, such as Anne Frank, were sent by train to Auschwitz.

Fourteen German scientists were located and interned; four of them were sent to the United States, and the other 10 were held in Versailles before being moved to an estate outside of London as part of Operation Dustbin interrogations.[49] Two of the scientists interrogated were Nobel Prize winners Heisenberg and Hahn. By the end of the questioning, it had become clear that Germany was nowhere close to producing atomic weapons and indeed had concentrated on developing an atomic engine only as a power source. The Germans had not been able to separate U-235 and had never achieved a sustained chain reaction. They "had not even reached first base," said Vannevar Bush.[50] As a result of questioning by Goudsmit and others, Alsos team members recovered 68 tons of uranium blocks in Belgium and 30 tons in Toulouse, France, which were sent through England to the United States.[51] On April 26 the heavy water was removed from the cellar of an old mill near Haigerloch and was shipped to Paris. From a field just outside the town, one and a half tons of uranium cubes were dug up and also sent there.[52] Just as important, teams working within and out of Heidelberg in the spring of 1945 confiscated nuclear equipment such as cyclotrons and centrifuges, discovered significant documents, and took prominent atomic scientists into custody. To the Germans it was impossible to separate pure U-235, and they had never thought of incorporating plutonium in the bomb.[53] In fact, when the German scientists heard the news about Hiroshima on August 6, they were incredulous that the Americans had succeeded in achieving so quickly what they could not; it shattered their belief in their own scientific superiority.[54] Goudsmit later wrote that the team "wondered if our government had not spent more money on our intelligence mission than the Germans had spent on their entire project."[55] On October 15, 1945, the Alsos team of 28 officers, 43 enlisted men, 19 scientists, 5 civilian employees, and 19 Counterintelligence Corps (CIC) agents was officially disbanded.[56]

Across the Rhine

Aden and two other Navy officers fluent in German were given a Jeep and a road map. One of the officers was fluent in French, which was useful in trying to navigate the confusing streets out of Paris toward Verdun, the scene of one of the longest and bloodiest battles of World War I with a total of about 700,000 casualties. Aden had taken a semester each of French and German for reading knowledge at Caltech, but he soon discovered that speaking languages is an entirely different matter, so he was grateful for having interpreters along.[57] Along the way, tall poplar trees in bud lined narrow, winding French roads. It was raining, and although the Jeep had a canvas roof, the mud splashed up from the front wheels and soon began to cake on everyone's uniforms. They slept that night in the old World War I barracks in Verdun and were briefed on the crossing of the Rhine at Ludwigshafen,[58] where an estimated 40 to 50 percent of Germany's output of chemicals and chemical weapons was produced by the IG Farben facilities.[59]

One of those chemical weapons was the nerve agent *tabun*, developed first as a cyanide-based insecticide by chemist Gerhard Schrader at the Farben laboratory north of Cologne. Finding that it was 100 percent effective on lice, Schrader tested it in a gas chamber on an ape, which died in minutes. Karl Krauch, the president of Farben's board of directors, collaborated with Hermann Göring to drop *tabun* from airplanes on the enemy.[60] Otto Ambros, who smiled while giving soldiers in the Third Army free bars of soap, was in fact Farben's chief of chemical weapons production. Farben also manufactured Zyklon B, the gas used to murder millions at Auschwitz and other concentration camps.[61]

Ludwigshafen was taken by the 94th Infantry Division by March 24, 1945, assisted by the XIX Tactical Air Command.[62] The rails of the bridge were twisted and impassable after Germans had planted explosives two days earlier. Aden remarked that fortunately there were no German airstrikes, as there were at the Ludendorff Railway Bridge at Remagen farther north.[63]

A few German planes appeared, but they were driven off by North American P-51 Mustang fighter-bombers.[64] General Patton ordered the Third Army to begin crossing the Rhine at 2200 on March 22, a day earlier than scheduled, thereby depriving British Field Marshall Bernard Montgomery and his 21st Army Group of bragging rights for being the first to cross on the 23rd.[65] However, the legitimate (and richly deserved) honors for the first crossing of the Rhine are reserved for the US First Army at Remagen on March 7.

Once across, Aden and his two fellow Navy officers headed north to Frankfurt, which was captured by the Third Army on March 29. Finding fuel for the jeep was a continuing problem, and they stayed on the safe "Red Ball Route," which had fuel depots.[66] Gasoline was delivered to the Rhine in railway tank cars and then pumped across the bridges into pipelines.[67] It was also airlifted to four airfields, so that by April 8, 1945, the Third Army had 1,344,670 gallons of gasoline at depots east of the Rhine.[68]

The sight of a US Navy jeep sloshing along muddy roads must have made more than one soldier scratch his head. US Army Colonel Frederick P. Field of the 12th Armored Division, which had been temporarily assigned to Patton's Third Army in crossing the Rhine, saw a jeep with US Navy printed on it and wondered, "What in the hell is the Navy doing so far from the sea?" and described "a tar all done up in a pea jacket and a white sailor cap" who explained that he was on a highly classified mission.[69]

On arrival in Frankfurt, Aden saw a city still in ruins from heavy bombing, which had been especially pummeled in March 1944. As he arrived, the townspeople were emerging from underground bomb shelters. He was stationed in a bunker of a bombed-out building along with some Army Intelligence officers. One evening when they were sitting around a fire eating C-rations (canned meats and vegetables, jam, crackers, powdered drinks, etc.), they told Aden that frontline troops had liberated death camps and showed him a photo of victims in an issue of the US Armed Forces newspaper, *Stars and Stripes*. He was thunderstruck by the realization that, because his father's mother was Jewish, he would still be considered a Jew by the Nazis and could have been one of those pictured in the photo.[70] This is perhaps why he was told never to discuss his background (see Chapter 1, section "Aden Baker Meinel"). He was soon to witness remnants of the horrors that would later bring tears to his eyes when talking about it.

Aden's next targets were rocket factories near Munich, which were difficult to locate behind hedges of evergreens. The search for one of them took him to a town just northwest of Munich: Dachau. He saw a few Army trucks beside the road and ignored a soldier waving him down. He continued on in the Jeep through a double wall of barbed wire and pulled up to what he figured was the main building. An MP stopped him and told him that he could not enter, even over Aden's protests that he was with Naval Intelligence and therefore had authority to search for rocket-related targets. "This was not a factory," the MP said. "It was an extermination camp." Aden could see the gas chambers beyond and smelled death.[71] Shaken to his soul, Aden returned to Paris for his new orders: First, get to Dresden and Jena before the Soviets and collect significant documents and materials. Then drive to Nordhausen, locate the underground factory for V-1 and V-2 rockets in the vicinity, and bring back anything of scientific value.[72]

The reason for the urgency was that Franklin D. Roosevelt, Winston Churchill, and Joseph Stalin had agreed to a postwar division of Germany at the Yalta Conference in February 1945, reaffirmed in more detail at the Potsdam Conference five months later by Stalin, Churchill, Clement Attlee, and Harry Truman, who assumed the presidency on the death of Roosevelt in April of that year. The states of Thuringia, Saxony, Saxony-Anhalt, Brandenburg, and Mecklenburg-Vorpommern were to be incorporated into the Soviet zone of occupation.

Dresden

Destruction from Allied bombings throughout Germany had been pervasive, from Bonn to Stettin (now Szczecin, Poland) and from Hamburg to Munich. In 1945, Jena was hit on several days in February, March, and April. By April 13, when the Americans occupied Jena, hundreds had been injured, killed, or reported missing there; among the targets had been the university district, the city center, the Carl Zeiss and Schottwerk optics workshops, museums, banks, and churches.[73] Between February 13 and 15, Allied forces dropped more than 3,900 tons of high-explosive bombs and firebombs on Dresden, killing an estimated 25,000–35,000 people, leveling most of the city, and leaving the vast majority of surviving civilians homeless and hopeless; debate continues today about whether it was justified.[74]

By April 1945, General Eisenhower had given strict orders that Allied forces remain west of the Elbe River, not for political reasons as commonly assumed, but for military reasons such as minimizing casualties.[75] However, when the Third Army arrived at the river, Aden's Army colleagues saw no German defenses and asked Patton to order them to cross. Aden supported their request because he had targets in the east.[76] Finally that order came on April 11, 1945. The XX Corps, comprising the 80th and 756 Infantry Divisions and 6th and 13th Armored Divisions, was to advance to the Elbe and capture Dresden.[77] Aden's target there was a factory where guidance equipment was made for the V-2 rocket and reportedly other weapons.[78]

On arrival there, Aden and his interpreter discovered that the factory had been nearly destroyed. He crawled under some concrete slabs and burrowed his way to the likeliest former site of the executive offices. Among the scattered papers marked *Geheim* [Secret], mostly details of operations and reports to Berlin, was one labeled *Geheime Kommandosache* [Secret Commando Operation] and listing the five most urgent weapon development programs. It was signed "A. Hitler."[79] One of them was *Wasserfall* [waterfall], a supersonic antiaircraft missile with infrared homing and propelled by the combustion of vinyl isobutyl ether and SV-Stoff (mostly nitric acid). It was named *Wasserfall* most likely because it was designed to explode in the middle of a bomber formation, bringing down several planes at once in small pieces. In the postwar interrogation of General Dornberger, who was in charge of all ground-to-ground, air-to-air, and ground-to-air missiles, he revealed *Wasserfall* would use one radar to track the target and by a second similar radar bring the missile into the beam of the first radar. The missile could then "ride" the beam to within 3,000 meters or less, when the homing system would take over. This deadly missile, conceived by General Dornberger and Wernher von Braun, was tested but never implemented, primarily because the Germans lacked effective centimeter radar equipment.[80]

Another of the antiaircraft weapon programs on the list that Aden saw was a subsonic missile with sweptback wings named *Schmetterling* (butterfly), which was radio-controlled and burned nitric acid and R-Stoff, a hydrocarbon, self-igniting propellant (Figure 4.3). Only a few were produced before the end of the war. Half the test firings failed, including some from aircraft as well as from the ground.[81] Göring had authorized development of both of these on September 25, 1942, along with three other high-altitude, surface-to-air missiles: *Taifun* (typhoon; ground-to-air, unguided rockets) as well as *Enzian* (gentian flower) and *Rheintochter* (Rhine Maiden), which were remote radio-controlled rockets.[82] Aden recalled two other projects on the list. One was "Jaguar," a 3 km/sec liquid-fueled rocket, and the other was a rocket that could be launched from a submarine.[83] The last of these was mentioned in later interrogations as being tested in 1942, launched from a U-boat at a depth of 40 feet.[84]

Aden found a safe on the floor below that he thought might hold other secret documents. An Army team guarding the site set an explosive charge on the lock. Unfortunately it set the contents of the safe on fire, making them irretrievable.

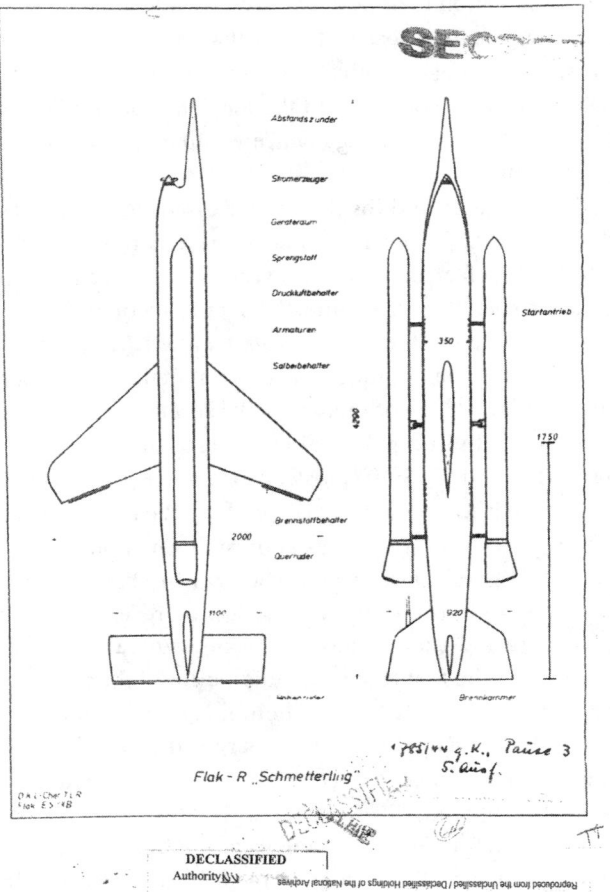

Figure 4.3. Design of the subsonic, radio-controlled missile *Schmetterling*.
National Archives and Records Administration, RG 38, LR 41-45 (German Controlled Missiles),
Identifier 4345900, Container 11.

Still, the code names that he had discovered would prove useful in interrogating
German scientists and engineers about those projects.[85]

Nordhausen/Mittelbau-Dora

Backdrop

In order to document the significance of Aden's exploits at Nordhausen, we must
first summarize the evolution of German rocketry in the 1930s and 1940s and

why Hitler was forced to rely on it in his attempt to crush Great Britain and later Holland, both physically and psychologically.

At the end of 1935, 23-year-old rocket enthusiast Wernher von Braun had conceived of an experimental rocket establishment that was funded by the Luftwaffe. In the search for a suitable site, his mother, Emmy von Quistorp, suggested Peenemünde at the northern tip of the Baltic island of Usedom, where her father once went duck-hunting near the family's properties there. With the political backing and guidance of Major (later General) Walter Dornberger, von Braun moved his rocket-development operation from the test range at Kummersdorf, about 15 miles south of Berlin, to Peenemünde in 1937 and would soon have a staff of 5,000 there.[86]

By the time that Aden had been inducted on November 10, 1944, the war in Europe had been raging for just over five years, following Germany's invasion of Poland on September 1, 1939; Great Britain entered the war two days later when Germany had not responded to the British ultimatum to withdraw from Poland by the deadline of 11 a.m. on September 3; France followed suit.[87] Almost 1.5 million people—children, pregnant women, the elderly and infirmed—were evacuated to the countryside.[88] Following the surrender of Holland on May 15, 1940, Belgium on May 27, and France on June 22, Adolf Hitler offered British prime minister Winston Churchill a treaty in his Reichstag speech on July 19, but Churchill refused to negotiate—over the objections of some in Parliament (even Lord Halifax in the Conservative party) and a handful of celebrity pacifists from the arts.[89]

Without a fleet of landing craft other than river barges at the time, Nazi Germany had to rely on air and sea attacks to bring Great Britain to its knees and the negotiating table; about 24 out of about 90 British destroyers were sunk in the first months of the war by U-boats, mines, and the Luftwaffe.[90] The Battle of Britain began on July 10, 1940, when the Luftwaffe attacked a convoy off Dover, only to lose twice as many aircraft as the Royal Air Force.[91] Hitler then decided on a new strategy, namely to bomb major cities, especially London. On September 15, 200 German bombers heading toward London encountered 300 British Spitfires and Hurricanes; the Luftwaffe sustained losses of 25 percent.[92] By the end of the Battle of Britain, generally recognized as October 31, 1940, the Luftwaffe had lost 1,887 aircraft and the RAF 1,547; the Luftwaffe lost 2,698 airmen and the Fighter Command of the RAF 544.[93]

Overlapping with the Battle of Britain, the Blitz began on September 7, 1940, with large-scale nighttime (and even daytime) raids on major ports such as Plymouth, Portsmouth, Southampton, Liverpool, Bristol, and Glasgow, as well as industrial centers, including Belfast, Manchester, and Birmingham. London itself endured 57 consecutive nights of bombings.[94] Coventry, raided on November 14–15, was a factory city that had been converted to munitions production, making it a significant target. Overnight 500 tons of bombs and 30,000 incendiaries were dropped on the center of the city. By the end of the siege, 568 people had been killed and 1,200 injured; most homes were destroyed or damaged.[95] Throughout Britain, about 2.25 million people were made homeless during the Blitz, two-thirds of those in London.[96] By the end of the Blitz, over 43,000 civilians throughout Great Britain had been killed and over 159,000 injured,[97] although these numbers do vary from source to source.

Revenge weapons

Seeing the devastation caused by incendiaries, the Bomber Command of the RAF used them to bomb the medieval town of Lübeck in Schleswig-Holstein, on March 28, 1942, destroying over 200 acres, including three churches and the town center. In retaliation, Hitler immediately wanted revenge, and Reich Minister of Propaganda Josef Goebbels originated the name *Vergeltungswaffe*, revenge weapons, which were to be aimed at British civilian targets.[98] Rockets were already in development at Peenemünde but not yet ready to be weaponized and launched toward Britain. In the meantime, Fritz Gosslau of Argus Motorenwerke showed designs for a pilotless aircraft propelled by a pulse-jet engine to an impressed Field Marshal Erhard Milch, deputy of Reichsmarschall Hermann Göring. By November 1942 it was being tested at Peenemünde.[99]

The Fieseler 103 became known as *Vergeltungswaffe Eins*, the V-1 (Figure 4.4), much cheaper and easier to produce than rockets and without the loss of pilots. These steel flying bombs were 25 feet long with a wingspan of 17 feet, a rear rudder, and a warhead containing TNT. It ultimately had a range of 250 miles and flew at an altitude of 2,000–3,000 feet and an airspeed of up to 40 mph. Guidance was governed by a magnetic compass, two gyroscopes with an accelerometer, a barometer, and a vane anemometer on the nose to measure distance. The windmill was preset to rotate a specific number of times based on the distance and air speed between the launch site and target. When that distance was reached, the V-1 went into a nosedive. They were usually launched from mobile ramps but also from atop Heinkel 111s.[100]

Because the pulse-jet engine alternately sucked air and fuel, it sounded like a grating thrump-thrump-thrump to those below.[101] A late British colleague of one of us (AP), Tom Reynolds, who lived through the bombings, heard that sound many times and said that "as long as you heard it passing overhead you were safe, but when it stopped, you prayed." The first salvo of these "buzz bombs" or "doodlebugs," as they were called, were launched from France and struck Kent, Sussex, and London on June 13, 1944, a week to the day after D-Day, twisting up a railway, destroying houses, claiming the lives of six and severely injuring 30.[102] Within three days of V-1s striking London, 647 of them had killed

Figure 4.4. Cutaway drawing of a V-1 showing fuel cells, warhead, and other equipment.
Public domain.

499 people, severely injured more than 2,000, and damaged or destroyed 137,000 buildings.[103] As time went on, 100–150 V-1s were aimed at London every day, taking their deadly toll and making thousands homeless just as autumn was setting in. By mid-September 1944, 2,622 bombs had fallen on Kent, 886 on Sussex, 512 on Essex, and 295 on Surrey; from June 15 to August 1 there were 6,046 launched at London.[104]

The first line of defense comprised RAF fighters and a 5,000-yard-deep gun belt from Beachy Head to Dover that could reach 10,000 yards out to sea; at its peak, the belt included 1,600 guns of all calibers and 200 rocket projectors. Between the fighters and gunners, up to 56 percent of the V-1s were shot down before they could make landfall.[105]

The final line of defense was a line of barrage balloons, which held up steel cables to prevent the V-1s from flying over them. By July 8 there were 1,750 balloons over London.[106] The barrage altogether covered an area of 260 square miles from Surrey to Kent, flying from 1,000 to 5,000 feet.[107] By the time the last of the effective 10,000 doodlebugs was launched at England, on September 1, 7,488 made it across the English Channel. Of those, about 3,900 were intercepted by defenses. Despite that 52 percent success rate, 2,419 V-1s had fallen on London, killing over 6,000 and seriously injuring more than 18,000.[108] There were no such defenses against the next revenge weapon, the V-2, which began to hit London a week after the last V-1.

British and American forces had advanced farther into Germany and Italy in April 1945, taking prisoners and liberating concentration camps. From August 1, 1944, until May 13, 1945, the Third Army alone captured 1,280,688 prisoners of war, but at a cost of 27,104 killed, 105,242 wounded or injured, and 28,237 missing in action.[109] According to Fuller, 47,500 German soldiers were killed and 115,700 wounded.[110] Benito Mussolini was executed April 28, and Adolf Hitler committed suicide two days later in his Berlin bunker. Grand Admiral Karl Dönitz, who had commanded the U-boat fleet, assumed command as head of state of the Reich and would later order the unconditional surrender of all German forces on May 8. As resistance was crumbling and fighting waned, roads were filled with wartime refugees returning to what was left of their homes or heading west in advance of the Soviets. German POWs were marched with their hands folded on top of their heads or transported in trucks to be interned in what had once been concentration camps.[111]

For von Braun, the V-2 rocket was the end product of years of testing earlier iterations in the so-called *Aggregat* series, beginning in Kummersdorf in 1933 with the A-1. It was a relatively simple, liquid-fueled rocket, 4.5 feet long, 1 foot in diameter, with a takeoff weight of 330 pounds; all tests failed on the launch pad.[112] A year later, two A-2 rockets named after the Katzenjammer Kids were launched from the island of Borkum in the North Sea. They were not much larger than the A-1, but improved on the A-1 by separating the fuel and liquid-oxygen tanks, elongating the combustion chamber, and adding a gyroscope for better initial control. The rockets were successfully launched and reached an altitude of 1.4 miles.[113] In 1937 the A-3 was launched from the Baltic island of Griefswalder Oie. It was 21 feet 8 inches long, 2 feet 4 inches in diameter, and had a takeoff weight of 1,650 pounds. Stability was provided by a new guidance system, four fins, and molybdenum vanes. All tests crashed when the guidance platform could not correct pitch or yaw greater than 30 degrees, which released the parachute prematurely to burn up in the exhaust.[114] The A-4 was already designed as a military rocket, so an improvement on the successful A-3, much smaller than the A-4, was named the A-5. But the design of the A-3/A-5 had its limitations and could not be used without significant modifications to achieve supersonic flight.[115] To test the A-4, a long-distance firing range was needed, and Peenemünde fit the bill; by 1937 most of von Braun's/Dornberger's staff had moved there from Kummersdorf. Laborers were still needed to work on the production plant, power grids, rail systems, and liquid oxygen plant; they were drawn from surviving Soviet POWs, civilians from occupied countries, and concentration camp prisoners.[116]

The A-4, later to be renamed the V-2, was 46 feet long and 5.5 feet in diameter at its widest point (Figure 4.5). Its warhead, containing amatol, weighed about a ton. Fueled by ethyl alcohol and liquid oxygen in the combustion chamber, it

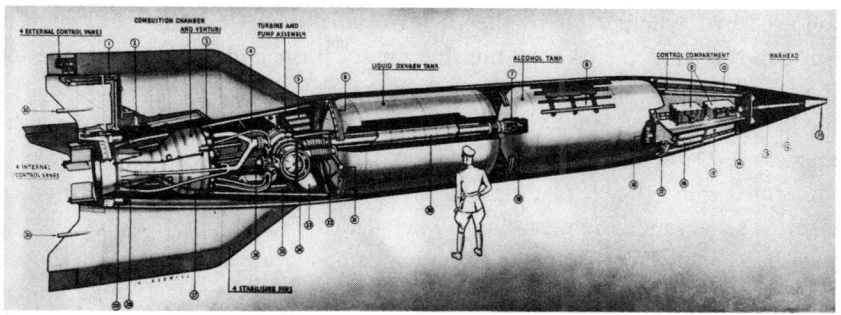

1 CHAIN DRIVE TO EXTERNAL CONTROL VALVE
2 ELECTRIC MOTOR
3 BURNER CUPS
4 ALCOHOL SUPPLY FROM PUMP
5 AIR BOTTLES
6 REAR JOINT RING AND STRONG POINT FOR TRANSPORT
7 SERVO-OPERATED ALCOHOL OUTLET VALVE
8 ROCKET SHELL
9 RADIO EQUIPMENT
10 PIPE LEADING FROM ALCOHOL TANK TO WARHEAD

11 NOSE PROBABLY FITTED WITH NOSE SWITCH, OR
   OTHER DEVICE FOR OPERATING WARHEAD FUZE
12 CONDUIT CARRYING WIRES TO NOSE OF WARHEAD
13 CENTRAL EXPLODER TUBE
14 ELECTRIC FUZE FOR WARHEAD
15 PLYWOOD FRAME
16 NITROGEN BOTTLES
17 FRONT JOINT RING AND STRONG POINT FOR
   TRANSPORT
18 PITCH AND AZIMUTH GYROS
19 ALCOHOL FILLING POINT
20 DOUBLE WALLED ALCOHOL DELIVERY PIPE TO
   PUMP

21 OXYGEN FILLING POINT
22 CONCERTINA CONNECTIONS
23 HYDROGEN PEROXIDE TANK
24 TUBULAR FRAME HOLDING TURBINE AND PUMP
   ASSEMBLY
25 PERMANGANATE TANK (GAS GENERATOR UNIT BEHIND
   THIS TANK)
26 OXYGEN DISTRIBUTOR FROM PUMP
27 ALCOHOL PIPES FOR SUBSIDIARY COOLING
28 ALCOHOL INLET TO DOUBLE WALL
29 ELECTRO-HYDRAULIC SERVO MOTORS
30 AERIAL LEADS

Figure 4.5. U.S. Army cutaway of V-2 rocket, showing engine, fuel cells, guidance unit, and warhead.

Courtesy of Niday Picture Library, Alamy Stock Images.

could reach an altitude of 50–60 miles and travel at 3,600 mph over a range of up to 225 miles. Guidance was provided by (1) gyroscopes connected to two sets of vanes that corrected for any deviation from the preset course, (2) an accelerometer that cut off the fuel at a given speed, or (3) a radio beam.[117] Unlike the V-1, it was supersonic and therefore gave no warning and was next to impossible to shoot down. The fatal flaw of the A-4 was its instability on launching, and an estimated 100 of them blew up on the launchpad.[118] In fact, it was not until the fourth attempt, on October 3, 1942, that it was successfully launched, quickly followed by a string of failures. But with losses in the Battle of Britain and defeats in Africa and Russia, Hitler ordered mass production of the A-4 on December 22, 1942, up to 5,000 rockets per year. Armaments Minister Albert Speer and General Dornberger recommended they be aimed at England, and Hitler gave top priority to building concrete bunkers as launch sites.[119]

How the Allies discovered Peenemünde most likely had multiple sources— German POWs, Polish laborers there, the Polish resistance, and others. The information was passed along to Churchill, who then assigned Member of Parliament (and his son-in-law) Duncan Sandys to oversee the investigation. Later photo-interpretation of the site by Professor R. V. Jones revealed long objects that were clearly rockets.[120]

In the bombing of Peenemünde that occurred under a full moon on August 17–18, 1943, there were 596 RAF bombers and about 4,000 crew members, followed by daylight raids by the US Army Eighth Air Force.[121] Although it was Sandys's idea to strike the housing for scientists and technicians, most of the casualties were forced laborers, including those passing along intelligence to the British, although key scientists were also killed. Before the night was over, there were 732 dead and over 800 injured. Dornberger and von Braun escaped unharmed with valuable plans and records. Reichsführer SS Heinrich Himmler visited Peenemünde the next day and persuaded Hitler to move the A-4 work elsewhere, using labor from concentration camps to prevent further sharing of information with the Allies. Himmler appointed as his deputy Hans Kammler, who had helped design the extermination camps and gas chambers.[122]

A new factory site

In less than a week, Kammler decided to transfer V-1 and V-2 production to an underground site in the Harz Mountains in Thuringia, less than two miles from the city of Nordhausen. In 1936, one hill in particular, Kohnstein, began being mined for calcium anhydrite, which when hydrated becomes gypsum and can be used in making plaster and fertilizer. The *Wirtschaftliche Forschungsgesellschaft* (Company for Economic Research; WIFO for short) was to manage the digging of two parallel tunnels running north to south, Tunnel A on the east and Tunnel B on the west. Each would be a little over a mile long with 46 transverse galleries or halls, each 490 feet long, dug between them so that the end product resembled a ladder. By the time WIFO abandoned the project, the tunnels were being used to store gasoline.[123] Hitler approved the converted use of Kohnstein on September 10, 1943, and decided to call it the *Mittelwerke*, or Central Works. Kammler ordered the first transport of prisoners from Buchenwald to Mittelbau-Dora—soon to be known as the hell of all the camps—on August 23, 1943. The deportees came from France, Belgium, Holland, Italy, Czechoslovakia, Hungary, Yugoslavia, Russia, Poland, and Germany. When they arrived, the prisoners discovered that Tunnel B was the only one completed and also too low for the production of V-2s.[124] By the end of September there were 3,000 at work drilling and digging with their bare hands, hauling heavy rocks and machinery, for 18 hours a day. Ammonia dust burned their lungs.[125] At its peak, 9,000 people worked in the Mittelwerk.[126]

Jean Michel, a survivor of Dora, described what life was like for those first prisoners:

> The deportees saw daylight once a week at the Sunday roll-call. The cubicles were permanently occupied, the day team following the night team and then vice versa. Very faint electric bulbs lit this nightmarish scene. There was no

drinkable water. You had to make do with any water you could find, for example where condensation formed, you lapped up liquid and mud as soon as the SS had their backs turned, for it was forbidden to drink "undrinkable" water.

The cold and damp in the tunnel were intense. The water that oozed from the rock caused a disgusting, permanent clamminess. Chilled to the bone, we felt as if our emaciated bodies would go mouldy. Some prisoners went mad, others had their nerves shattered as the installation progressed: the constant din was one of the causes—the noise of machines, of pick-axes, the bell of the locomotive, continual explosions, and all of it echoing mercilessly in the closed world of the tunnel. No heat, no ventilation, not the smallest pail to wash in: death touched us with the cold, the sensation of choking, the filth that impregnated us. As for the latrines, they were barrels cut in half with planks laid across. They stood at each exit from the rows of sleeping cubicles. The SS took great sport and laughter in walking up to a man seated on the plank and pushing him into the excrement (made even worse because all the men had dysentery). Without water, the only way to cleanse himself afterwards was to roll in the dust.[127]

The daily diet consisted of a quart of rutabaga soup, a piece of bread, a few tablespoons of butter, and either jam or ersatz cottage cheese or ersatz sausage.[128] Some men saved the bread for later, only for it to be stolen during the night by gangs of young Russians.[129]

Parts for the missiles arrived by rail from the south and entered camouflaged Tunnel A on the east side. It was in three halls adjacent to the southern entrance that V-1s were assembled. In early 1945 *Taifun* antiaircraft missiles were also being assembled in the factory.[130] The northern half of Tunnel B was devoted to the manufacture of airplane engines by the Nordwerke Company. Under the direction of chief engineer Arthur Rudolph, the V-2s were assembled in Tunnel B and transported out through the same entrance. Hitler wanted 900 V-2s produced each month so that 5,000 could be launched at London at the same time, but between September 1944 and March 1945, an average of only 600 per month were completed.[131] General Dornberger was later to say that 5,400 V-2s were built, of which 3,600 were fired against the Allies.[132] Pressure was on the SS and Kapos (cruel overseers, usually German criminals) to meet Hitler's target number, and no hint of laziness, weakness, poor workmanship, or sabotage was permitted. Wernher von Braun was a witness to mass hangings,[133] but he later denied that he had seen any brutality to the prisoners while he was in the tunnels.[134]

One of those who did not survive was the Dutchman Cornelius Gehrels, older brother of Tom Gehrels (1925–2011) who was an astronomer, professor of planetary sciences at UA, and close friend of Aden. According to Professor Gehrels, "Cor" distributed illegal newspapers, collected intelligence for the British, took

Jews into his home, and worked on an escape route for downed Allied pilots. He was arrested and interrogated but revealed nothing. Then he was crammed into cattle cars with other men and transported from one camp to another. By the time he arrived at Dora in April 1945, shortly before Aden arrived, Cor was emaciated from dysentery and died there.[135]

The first two V-2s landed in Greater London from The Hague on September 8, 1944, with 35 more that month and 131 in October. After the last one exploded in Orpington, Kent, on March 27, 1945, 2,754 people had been killed by V-2s and 6,523 injured in Great Britain and one million homes destroyed.[136] Worse damage was inflicted on Belgium, especially in Antwerp, which had become an important port when it was liberated by the Allies on September 4, 1944. The single worst death toll caused by V-2s was a direct hit on the Cinema Rex in Antwerp on December 16, 1944, when military personnel and civilians were watching *The Plainsman* starring Gary Cooper. Altogether 561 people were killed and 291 seriously injured in that attack alone;[137] over 5,000 were killed throughout Belgium by V-1s and V-2s.[138]

The Allies first learned of the existence of the Mittelwerk from someone who had worked at the Technische Hochschule in Berlin and escaped to London on October 3, 1944. On the basis of his account, air reconnaissance mapped out the region and relayed its location to ground troops.[139]

On April 1, 1945, von Braun heard that the Americans would soon be arriving at Nordhausen, so over the next two days he packed up three trucks with all the important documents—14 tons worth—and had them buried in an abandoned mine about 50 miles NNW at Dörnten for two reasons: (1) to prevent SS General Kammler from destroying them, and (2) to use them as a bargaining chip upon his surrender to the Allies.[140] Five hundred scientists and engineers then traveled by train to Oberammergau in the Bavarian Alps while von Braun went there by car because he had suffered a broken arm and shoulder in a previous car accident. About 4,500 more remained around Nordhausen and Bleicherode to the southwest.[141] Despite the Allied bombing of Nordhausen on the April 4, able-bodied prisoners were evacuated by cattle car to other concentration camps such as Bergen-Belsen or on foot as "death marches" under heavy SS guard; the rest remained in the hospital.[142]

The most reliable figures for the death toll at Dora and during the evacuations were provided by Sellier, although Michel as an eyewitness to the horrors humanized the numbers with his own personal accounts. Out of a total population of 40,000 prisoners in the camp over a period of 20 months, 15,500 died in the camp and transports and another 11,000 during the evacuations.[143] Perhaps the most heinous atrocity was the massacre at Gardelegen, between Berlin and Hanover, on April 13, 1945. Up to 4,000 prisoners who had survived the evacuations by freight cars and on foot were moved into a former cavalry school there. Soldiers and local civilians moved 1,016 of the sickest and weakest from Gardelegen into a barn at Isenschnibbe. Gasoline-soaked straw on the floor was set alight and the doors bolted shut. The fire burned for seven hours. Some prisoners who tried to escape by digging under the wall were quickly shot. When the 102nd Infantry Division reached Gardelegen on April 15, they found that the locals had already buried half of the bodies to hide the evidence.[144]

Arrival of the allies at Nordhausen

The last of the able-bodied prisoners left the camp on April 5. Six days later, the Third Army arrived at Nordhausen and Dora, and they were stunned by the sight and smell of piles of corpses, the "indescribable human filth," and the extremely emaciated condition of the remaining prisoners in the hospital. Survivors were given medical treatment and food; by May 7 all of them had departed, most on stretchers.[145] American troops were also astonished to see freight cars in one tunnel with V-2s aboard, so they notified Chief of Army Ordnance Technical Intelligence Col. Holger Toftoy, who arranged to ship 100 V-2s and parts weighing 360 metric tons to Antwerp by rail and then to New Orleans and White Sands, New Mexico.[146] The first trainload left on May 22.[147] The troops gathered up thousands of blueprints and models that were later to be used in interrogating German scientists.[148] Among the interrogators were many faculty members from Caltech.

Aden did not mention seeing any prisoners or scientists at Dora in his auto-biography, nor did he give any dates of his work there, but it was clearly after the last of the prisoners had left and before the V-2s were shipped off.[149] When he and his driver arrived at Nordhausen by Navy truck, all was silent. He saw the wooden barracks that the prisoners had built for themselves and also a makeshift graveyard. There on the rail line were several flat cars loaded with V-2s, waiting to be shipped out to Antwerp. He entered one of tunnels and was deep in the complex when suddenly the lights went out. The flashlight he had could illumi-nate nearby machinery and equipment but not show the route out, so he fired his pistol in the hope that his driver, who had stood guard at entrance, could hear the shot. The driver fired back, but the echo made it impossible to locate the source. The lights came on again briefly, giving Aden his bearings, then dimmed again. He saw some light in a side tunnel and headed in that direction. Along the way he saw two large crates labeled *Schmetterling*. His driver backed the truck into the tunnel entrance, and they loaded the two crates for the trip to Jena, their tem-porary headquarters. Aden later learned that British intelligence had found the crates earlier that day and had left them for retrieval later; when they returned, the crates were inexplicably missing. In the end the US Navy dispatched one of them to England but kept the other.[150] It was important to retrieve them for two reasons: (1) If the war continued, the missiles could be used against the Japanese; (2) if the Germans had already passed the plans for *Schmetterling* on to the Japanese, then the Allies needed to develop countermeasures.[151]

On June 10, 1945, perhaps six weeks after Aden had left Dora, Navy Lieutenant E. H. Uellendahl gave a tour of the tunnels there to Charles Lindbergh, who 17 years earlier had been forced to land his monoplane at Stock Grange, the property of Marjorie Meinel's grandfather, John Dutton Steele. Lindbergh saw V-2s in different stages of assembly, the crematorium, a cadaver covered by canvas, and a pit filled with ashes, bone chips, and a knee joint. He wrote in his journal the next day, "Of course, I knew these things were going on, but it is one thing to have the intellectual knowledge, even to look at photographs someone else has taken, and quite another to stand on the scene yourself, seeing, hearing, feeling with your own senses."[152]

Jena

After leaving Nordhausen, Aden and his interpreter pushed on by jeep through the state of Thuringia to Jena, the world-famous optical center that was headquarters of Carl Zeiss and the Glassworks of Schott & Genossen. Both firms, united as the Carl Zeiss Foundation, were respected pioneers in optics over a century or more, and because the glass, lenses, and cameras they produced had scientific and military applications, their factories and laboratories were targets for both NavTechMisEu and Alsos. For example, among the many products of Schott were X-ray tubes used as capacitors for radios in fighter planes, binocular lenses, glass-to-metal feedthroughs for wireless communications, and illumination glasses for fighter bases.[153] In 1938 the Nazis designated the company an armaments plant and demanded that Erich Schott, the manager and son of founder Otto Schott, join the Nazi party. The firm's employment statute that people be "hired without consideration of their ethnic origin, religion and party membership" had to be scrapped.[154] Hundreds of employees were conscripted and sent off to war to die. They were replaced by forced workers from eastern Europe (chiefly Russia) and "guest workers" from Italy and Croatia.[155]

Jena had been bombed by the Allies on March 17, 1945, with loss of life and heavy damage to the glasswork factories (Figure 4.6). Less than a month later, beginning on April 11, American troops bombarded the city with artillery and then swiftly marched into the city, bringing relief to Erich Schott, who said to himself, "Now things can only get better."[156] Two US Air Force officers took command of the Carl Zeiss Foundation, and soldiers occupied first the Schott glassworks and Zeiss optical workshops and then the rest of the city.[157]

Figure 4.6. Carl Zeiss GmbH at Jena, destroyed buildings after the bomb attack on March 19, 1945.

Courtesy of Alamy Stock Images.

Specialists from NavTechMisEu, including Aden, combed through the factories, gathering documentation on glass composition, molten glass, production processes, patents, and samples of equipment to be shipped west by freight train.[158] While there, based at the Old Mill, he met Lt. Commander George Dimitroff (1901–1968), an astronomer from Harvard University and later Dartmouth College, who also had been assigned to question scientists and remove documents and military hardware for the US Navy.[159] Dimitroff solicited Aden's help in doing that and especially wanted two Soviet submarine periscopes that had been captured.[160]

The documents and/or hardware shipped back included: (1) a survey of optical manufacturing methods at Zeiss, Jena Glass Works, and E. Leitz (in Wetzlar, which Aden later visited), most already in use in the States;[161] (2) aerial lenses and aerial cameras made by Zeiss for military and mapping purposes;[162] (3) a survey of optical design at Zeiss and E. Leitz, which noted in particular the use of projected reticules to keep glass surfaces out of the focal planes and thereby minimize the dirt problem;[163] and (4) filters and high-refractive index glass.[164] Among these optical glass filters were UG1 (dark purple) that transmits ultraviolet and extreme red and also RG2 (pure red, with a sharp spectral cutoff), which Aden somehow "procured" for Walter Baade (1893–1960) at the

Mt. Wilson Observatory.[165] With them Baade was able to resolve individual stars in the center of the Andromeda galaxy (M-31) for the first time. Aden eventually led a convoy of six trucks filled with optical technology (including the two Soviet periscopes) all the way to Dover.[166]

All in all, 44 tons of valuable equipment, such as range finders and fire control telescopes, were removed and shipped to the United States from the Zeiss plant.[167]

Operation Paperclip

A secondary assignment given to Aden was to persuade the top scientists and industrialists at Zeiss and Schott to move to the West within two weeks rather than remain and thereby fall under Soviet control; most decided to relocate (knowing the conditions in the USSR and the way the Soviets treated their scientists).[168] The underlying reason was to restrict the postwar scientific expertise flowing to the Russians. This was part of Operation Paperclip, adopted by the Joint Chiefs of Staff in July 1945, first as the highly classified Operation Overcast to evacuate about 350 German rocket scientists and place them in weapons or intelligence projects in the military; in November 1945 it was renamed Paperclip in reference to the paper clips attached to the folders of those with superior knowledge and skills.[169] The most famous of these recruits was the architect of the V-2 rocket, later director of NASA's Marshall Space Flight Center and architect of the Saturn V rocket that took NASA's Apollo astronauts to the Moon—Wernher von Braun. It was later expanded to German scientists and technicians in other fields and approved by the State-War-Navy Coordinating Committee as Project Paperclip on March 4, 1946, overseen by the Joint Intelligence Objectives Agency (JIOA), itself created by the Joint Chiefs of Staff.[170] President Truman secretly approved Operation Paperclip on September 3, 1946, allowing for the entry of up to 1,000 German and Austrian scientists and their families, the selection of which was directed by the Departments of War, Navy, and Commerce.[171]

The rationale for Paperclip was spelled out succinctly by Kuiper in a memorandum titled "The Future of German Science" to Major Fisher of Alsos: "Germany has many hundreds of leaders in fields of war research. The most prominent ones and most dangerous to our security *should probably be moved to Allied territory in a sort of German enclave for at least 5 to 10 years* [his emphasis]. They could do research of interest to the Allies under decent living conditions for them and their families. This would liquidate the most important part of German 'Scientific General Staff.' The alternative would be to let such powerful men go to countries where they would be welcomed but potentially dangerous to us."[172]

Applicants for visas would need to prove that they sincerely wished to migrate and state why they wanted to become Americans and what connections that they wanted to keep with German persons, companies, institutions, and organizations. Each would need to be thoroughly interrogated and be willing to submit to additional investigations by the State Department.[173]

Teams of engineers and scientists from all parts of the States were dispatched to Germany to help Air Technical Intelligence teams interrogate their German counterparts, study technical documents, and visit factories and laboratories—in other words, to gather "exploitation intelligence." Among these groups was a strong contingent from Caltech: Theodore von Kármán, Tsien Hsue-shen, Clark Millikan, and Fritz Zwicky. Representatives from the defense industry joined them: Boeing Aircraft, Bell Aircraft, Lockheed, Douglas, Bendix, Curtiss-Wright, General Electric, General Motors, Honeywell, and others.[174]

There were initially three principles under which the scientists could be brought to the States: (1) certain German specialists were to be used to augment war-making capacity against Japan and aid postwar military research; (2) no known or alleged war criminals would be brought to the United States; and (3) they would be returned to Europe at the conclusion of the work.[175] The problem was that many scientists would be brought to the States without prior denazification, in direct violation of the Law for Liberation from National Socialism and Militarism, which Germany had passed on March 5, 1946, meant to exclude from influence in public life those who supported National Socialism in any way, not only committing crimes against humanity.[176] Nevertheless, the JIOA not only overlooked certain negative security reports, but even changed the documents to allow a "pass" in other agencies and departments.[177] Theirs was a prolonged cost-benefit analysis: How much do we have to gain in exploiting their scientific expertise for national defense, and how much can we stand to lose in terms of public opinion when their contributions to the Nazi war effort are brought to light?

Between May 1945 and December 1952, the State Department admitted 642 German scientists under Operation Paperclip, according to a report by the Joint Chiefs of Staff.[178] Perhaps because of the difference in dates, Crim (2018) listed the number of "Paperclippers" from 1948 to 1952 as between 492 and 555 based on a German source. Contrary to original intent when they entered, most of them became permanent residents and citizens, and most of them had been members of the Nazi party or an affiliate.[179]

Several of those who were associated with Mittelbau-Dora traveled by jeep from Witzenhausen to Paris and then by air to Wright Field in Dayton, Ohio, to Fort Bliss outside El Paso, Texas (along with the White Sands Proving Ground in New Mexico), and finally in 1950 to Redstone Arsenal at Huntsville, Alabama. Among them were (1) Wernher von Braun, who had been a member of the SS with the final rank of *Sturmbannführer* (Major) and who directed that prisoners from Buchenwald be sent to work on the V-2 at Dora;[180] (2) his brother Magnus, and (3) his close friend and advisor, Ernst Stuhlinger. Stuhlinger had been trained as a nuclear physicist but spent two years as a soldier on the Russian Front before being assigned to Peenemünde.[181]

Tom Gehrels, professor of astronomy at UA and whose brother died at Dora, sent his book *On the Glassy Sea: An Astronomer's Journey* to Stuhlinger. In that book he asked why Wernher and Magnus von Braun, who did nothing to help the prisoners and reduce the rocket assaults on England and Antwerp, should be lionized, instead of those such as Count Klaus von Stauffenberg who resisted Hitler and the SS.[182] Stuhlinger wrote back from Huntsville:

> I lived through all 12 years of Hitler's tragic regime, it was an experience that will certainly remain with me throughout my life. If we got to talk about it, you might perhaps like to tell [me] what you would have done if you had been a 20-year-old German in 1933, and I could tell [you] what the result of each action would have been (for example, you might have found yourself marching through the snow fields of the Ukraine as a Pfc., although you had a Ph.D. in physics). Also, I would be glad to tell you what I would have done as a 17-year old Dutchman in occupied Holland: Escape to England, get trained as a saboteur, parachute behind the enemy lines, and blow up their supplies!—There was an underground in Germany; in fact, there were numerous different activities, from simple sabotage to the very sophisticated and clever attempts by persons such as Col. Stauffenberg, General Rommel, Prof. Max Planck, Prof. Werner Heisenberg. The fact that none of them worked will remain one of the most tragic circumstances of the Hitler period.[183]

Werner Heisenberg, best known for his uncertainty principle and as winner of the Nobel Prize for Physics in 1932, had openly opposed the Nazis by writing papers on the proton-neutron model of the nucleus and defending Einstein's theories. Theoretical physicists like Heisenberg and Stuhlinger were considered "white Jews" by the Nazis and therefore were marginalized in some way or were sent to the Russian Front until their value in developing weaponry such as atomic bombs and rockets was realized in the closing years of the war.[184] On the other hand, Kuiper reported to Major Fisher that he learned from interviewing German scientists that they "appear *devoid of moral responsibility for the consequences of their work* [his emphasis] . . . [and] appear little concerned about the use to which their results are put." In 1943 Heisenberg himself told Professor Hendrik Casimir in Eindhoven, Holland: "History legitimates Germany to rule Europe (and later the world). . . . Only a nation which rules ruthlessly can maintain itself."[185]

The optical scientists left by different routes. According to one account, 213 technicians and their families were evacuated from Jena to Heidenheim in Württemberg.[186] SHAEF (Supreme Headquarters Allied Expeditionary Force) approved 41 Schott employees with their closest family members and belongings for evacuation by army truck, giving rise to the saga known as "The Odyssey of the 41 Glassmakers."[187] In Heidenheim they were temporarily housed in former barracks of Russian prisoners-of-war, fenced in and guarded by American soldiers (as it was located in the American zone).[188] The Americans had also transported scientists and their families as well as specialists from other parts of central Germany to Heidenheim, bringing the camp population to 1,200 under appalling conditions of bedbugs and a limited water supply.[189] Once released, most of the glassmakers eventually found work in glass factories in Landshut, Zweisel, and Mitterteich, all in Bavaria.[190] American troops left Jena on June 30, 1945, and the Soviet army arrived one day later.[191]

Aden assisted several of those optical scientists in their relocation to the States in 1945. One of them was Lou Bruckner, most likely from Zeiss. In the 1960s he and other Germans worked in the Optics Section, Photographic Branch, Reconnaissance Division, Air Force Avionics Lab of Wright-Patterson Air Force Base in Dayton, Ohio. He served on the Air Force Optical Patents Evaluation Board and helped Aden to draft a curriculum for the Optical Sciences Center at UA. Lou had a 35-mm Zeiss Contarex camera, as did NASA Mercury Seven astronaut Gus Grissom, who always took it on his space missions. When it developed a problem, he sent it to Lou and Jim Mayo for repair.[192] Grissom was later killed with Edward White and Roger Chaffee in a pre-launch test of Apollo 1 on January 27, 1967.

Some 20 years after the war ended, Aden and Marjorie were attending a dinner with engineers from Schott at "Wiesbaden Castle," now known as Wiesbaden City Palace, which had been headquarters for Allied occupation forces during the war and also Aden's home base in Germany in 1945. As it happened, two of the men remembered that he had asked them to pack up the two Soviet submarine periscopes for shipment to the United States. Their memories were fresh because they had been among those that he helped to evacuate to Heidenheim.[193]

Wetzlar and Toplitzee

A few weeks after leaving Nordhausen and its horrors behind him, Aden accompanied George Dimitroff to Wetzlar in the state of Hesse, specifically to the Ernst Leitz Company.[194] The optical unit of the Office of Technical Services, organized by the Department of Commerce, converged on Leitz and came away with photographs of 180,000 pages that describe the manufacture of Leitz equipment, advanced color film, and the magnetophon (a reel-to-reel tape recorder that had secretly been used in German radio broadcasts). This treasure trove of technology is believed to have saved American industries billions of research dollars.[195] For example, one report included microfilms of lens formulae diagrams, curves, and charts of the principal lenses produced by Leitz. In addition to the BuOrd, copies were disseminated directly to Bausch & Lomb and Eastman Kodak, both in Rochester, New York, which at that time was the epicenter of optics research in the United States.[196] A Technical Report from NavTechMisEu investigated processing equipment and materials used by four German firms, two in Wetzlar and two in Munich, in the manufacture of lenses and prisms for military optical instruments. The two in Wetzlar were Leitz and M. Hensoldt Söhne, and the two in Munich were G. Rodenstock and Steinheil Söhne. None had been severely damaged by bombs. Most documentation was provided for the manufacturing procedures at Leitz, such as glass-cutting; prism-grinding; polishing of prism blocks; lens grinding, polishing, and cementing with Canada balsam; coating with sodium silicate solution and cryolite; and production of reticules and scales, mainly for Leica cameras and 7×50 binoculars with Porro prisms.[197]

While at the Leitz factory, Aden met an elderly woman scientist who supervised the testing of camera optics. She offered to take his photograph, and the next day gave him the print, marred only by a slight light leak at the top right. After apologizing, she likened it to a ray of sunshine, which the Germans could now anticipate with glee. Aden had not been given a camera before he was sent overseas, so with a month's pay he bought a Leica from Dimitroff, who had scavenged it and wanted to be paid for his efforts.[198] However, earlier Aden had written from Paris that Dimitroff gave him the camera.[199]

By now it was August 6, 1945. Aden was having breakfast in Wiesbaden and listening to President Truman speaking on Armed Forces Radio:

> A short time ago an American airplane dropped one bomb on Hiroshima and destroyed its usefulness to the enemy. That bomb had more power than 20,000 tons of TNT. The Japanese began the war from the air at Pearl Harbor. They have been repaid many fold, and the end is not yet. With this bomb we have now added a new and revolutionary increase in destruction to supplement the growing power of our Armed Forces. In their present form these bombs are now in production, and even more powerful bombs are in development. It is an atomic bomb. It is a harnessing of the basic power of the universe. The force from which the sun draws its power has been loosed against those who brought war to the Far East. . . . We have spent more than two billion dollars on the greatest scientific gamble in history, and we have won. But the greatest marvel is not the size of the enterprise, its secrecy or its cost but the achievement of scientific brains in making it work. . . .

Suddenly it all became clear to Aden when he recalled Project CAMEL: "That's why I was asked to measure detonation times for fuzes!"[200]

At about the same time he heard Truman's announcement of dropping "Little Boy" on Hiroshima, Aden heard from the Navy radio operator of the sinking of the cruiser USS *Indianapolis*.[201] The components of that atomic bomb had been delivered to the island of Tinian in the Northern Mariana Islands northeast of Guam by the USS *Indianapolis*, the heavy cruiser on which Aden had once been slated to serve. Unlike a battleship with an average of 13 inches of steel amidships, a cruiser had only 3–4 inches, which is where one of two Japanese torpedoes struck her starboard and exploded on the night of July 30, 1945, as it was headed for the Philippines.[202] The ship sank in 12 minutes. As it was going down, a distress signal with the ship's coordinates was transmitted on frequencies monitored at sea and onshore, but it was ignored three times on the island of Leyte, 650 miles away.[203] Of the 1,195 crewmen aboard, about 300 died instantly from the explosions and 900 went into the oily sea. Of those 900, half had the foresight to grab a life vest or life belt, but the oil dissolved the seams, resulting in leaks.[204] Four days later, Lieutenant Chuck Gwinn was flying his Lockheed PV-1 Ventura bomber over the site and noticed an oil slick. At first he thought it was a Japanese submarine and was preparing to bomb it when he saw hands waving and floating bodies.[205] When the rescue finally came, 321 of the 900 who went in the water were still alive, and four died later in hospital. The rest succumbed to exposure, dehydration, suicide, salt water poisoning, and relentless attacks by sharks, mainly oceanic whitetip and tiger sharks. The wreckage of the ship was eventually located in August 2017 at a depth of 18,000 feet in the Philippine Sea by an expedition led by Paul Allen, cofounder of Microsoft.[206]

Aden later wrote that the few who were saved told their rescuers that they had held onto pieces of debris, which were few and far between. "And be assured," he observed, "that if someone gentle, like me, got onto a piece of debris, some bully would have knocked that person off and taken it himself. So I would have died just like so many of my former friends died."[207] It is noteworthy if not ironic that even though Aden was not aboard the *Indianapolis*, the results of his research on Project CAMEL were delivered at Tinian aboard her.

On August 7, 1945, Aden received written orders from the Chief of NavTechMisEu, Henry "Packy" Schade, to report to NavTechMisEu headquarters in Munich and then to Third Army headquarters in Freising to obtain clearance for executing verbal instructions concerning Toplitzsee (Lake Toplitz) in the Austrian alps, east-southeast of Salzburg.[208] He had been told to follow up leads about the underwater rocket project that the German Navy had begun at Toplitzsee not long before the German surrender.[209] The chief scientist on that project, Dr. Dettermann, was meanwhile interned at the former concentration camp at Dachau. Aden was to pick him up and interrogate him on the way to Bad

Aussee. The last time that he stopped by there looking for other targets, the camp had just been liberated, and an MP had prevented him from entering. Now there were Germans inside the barbed wire. Aden and Dettermann drove through Bad Aussee to their base at Grundlsee, where Dettermann's family lived. From there they drove east to Toplitzsee, where the naval testing station had been under construction, but found nothing of interest because they had no equipment to explore the depths of that lake down to 300 feet. There were reports of Nazis who had arrived in black limousines carrying boxes to the edge of the glacial lake but returned empty-handed, giving rise to rumors that they had dumped Nazi gold or documents that revealed the numbers of Swiss bank accounts comprising funds confiscated from Jewish victims. Fourteen years later, in 1959, the lake gave up some of its secrets. A team financed by the German magazine *Stern* discovered a printing press and metal boxes with £72 million in forged British currency, which had been part of Hitler's Operation Bernhard to undermine the British economy.[210]

Back in Paris in July before heading home, 22-year-old Aden was writing up reports while staying in Le Royal Monceau Hotel. One of those reports concerned a high-altitude, German naval 21-cm rocket-propelled flare. Its development had been under the control of the A Wa A Lab of the Oberkommando der Kriegsmarine in cooperation with Rheinmetall-Borsig A.G., Dusseldorf. The rockets were tested at the ranges at Unterlüss. Aden described in detail the ballistics, rocket motor data, and flare unit data. Its maximum altitude was over 5,000 m with a range of almost 8,000 m and maximum velocity of 560 m/sec.[211]

While in Paris he explored the city for books on astronomy and optics and was fortunate to meet the famous authority in optics and spectroscopy, Charles Fabry (1867–1945), who died just six months later. Fabry and Alfred Pérot invented their namesake interferometer in 1899, and Fabry and Henri Buisson discovered the ozone layer in 1913. In the hotel itself Aden wrote that there were "enough astronomers to hold a convention!" He mentioned Gerard Kuiper (Yerkes Observatory), George Dimitroff (Harvard), Fritz Zwicky (Caltech), Peter van de Kamp (Swarthmore), and physicists working under Wallace Brode of OSRD (Figure 4.7a–d).[212]

When Aden visited several French observatories with Kuiper, Kuiper asked him where he had received his PhD. Aden replied that he had dropped out of Caltech in his junior year to work on the Navy rocket project. Kuiper was surprised that someone so young and without a doctorate had been assigned to such a responsible position in naval intelligence. After Aden said that he hoped to attend Berkeley to finish his undergraduate work and undertake a PhD, Kuiper suggested that he consider the University of Chicago and Yerkes after that, especially because both of Marjorie's parents had worked there.[213]

(a)

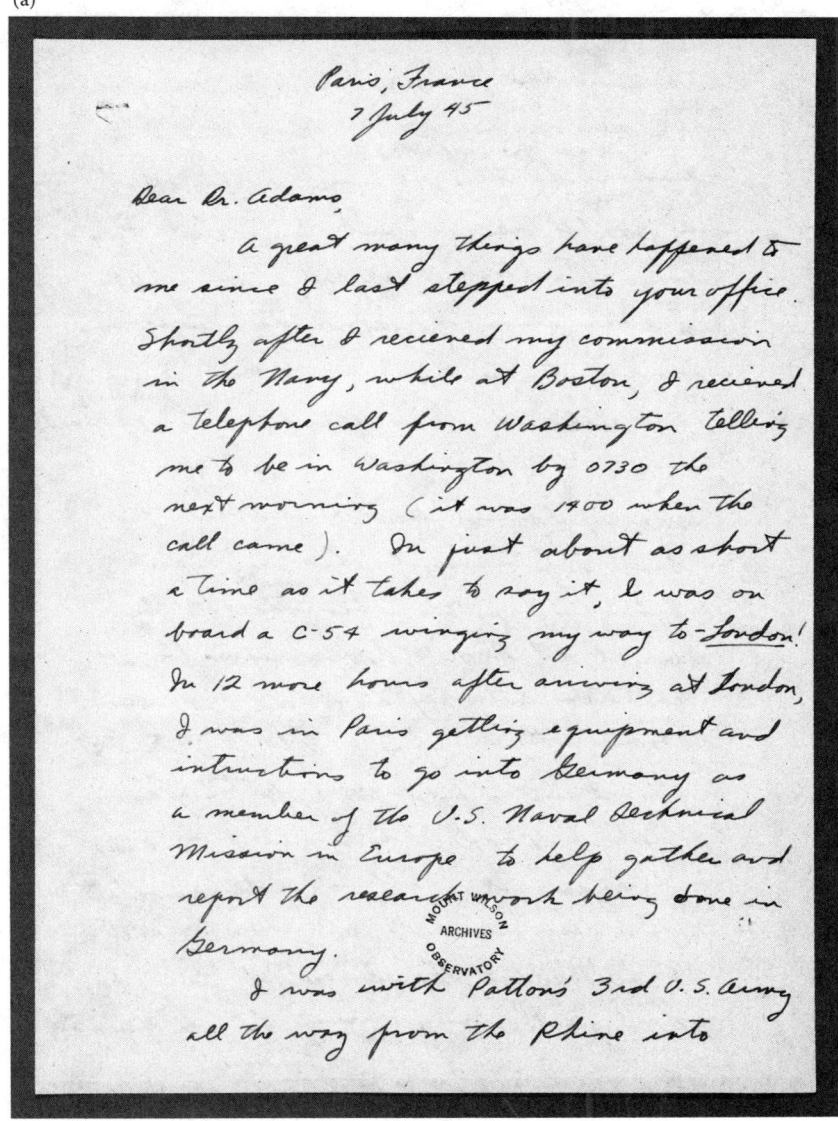

Paris, France
7 July 45

Dear Dr. Adams,

A great many things have happened to me since I last stepped into your office. Shortly after I received my commission in the Navy, while at Boston, I received a telephone call from Washington telling me to be in Washington by 0730 the next morning (it was 1400 when the call came). In just about as short a time as it takes to say it, I was on board a C-54 winging my way to London! In 12 more hours after arriving at London, I was in Paris getting equipment and instructions to go into Germany as a member of the U.S. Naval Technical Mission in Europe to help gather and report the research work being done in Germany.

I was with Patton's 3rd U.S. Army all the way from the Rhine into

Figure 4.7a–d. Letter dated July 7, 1945, from Ensign Aden Meinel, USN, in Paris to Walter Adams, director of Mount Wilson Observatory, Pasadena, California.

On July 10, 1945, Aden drafted a memo to the Bureau of Naval Personnel asking that he be assigned to the NOTS in Inyokern, California, upon completion of his temporary additional duty with NavTechMisEu via the Chief of NavTechMisEu and the Chief of the BuOrd.[214] Commodore H. A. "Packy" Shade

(b)

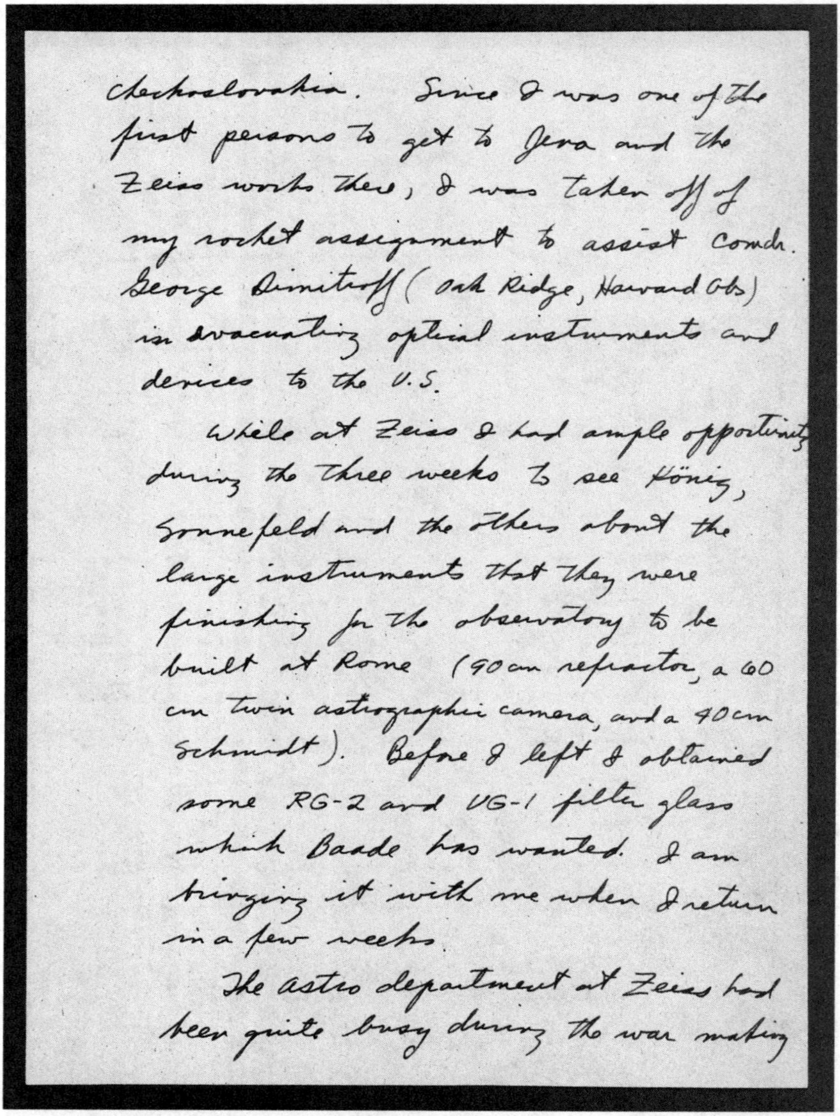

Checkoslovakia. Since I was one of the first persons to get to Jena and the Zeiss works there, I was taken off of my rocket assignment to assist Comdr. George Dimitroff (Oak Ridge, Harvard Obs) in evacuating optical instruments and devices to the U.S.

While at Zeiss I had ample opportunity during the three weeks to see Hönig, Sonnefeld and the others about the large instruments that they were finishing for the observatory to be built at Rome (90 cm refractor, a 60 cm twin astrographic camera, and a 40 cm Schmidt). Before I left I obtained some RG-2 and UG-1 filter glass which Baade has wanted. I am bringing it with me when I return in a few weeks.

The astro department at Zeiss had been quite busy during the war making

Figure 4.7a–d. Continued.

wrote a memo on Aden's behalf, which read: "Ensign Meinel is a highly intelligent young officer who had contributed to the success of this Mission to a high degree during the period from 2 April to 19 September 1945. While attached to the Naval Technical Mission he was engaged in exploitation of German technical

(c)

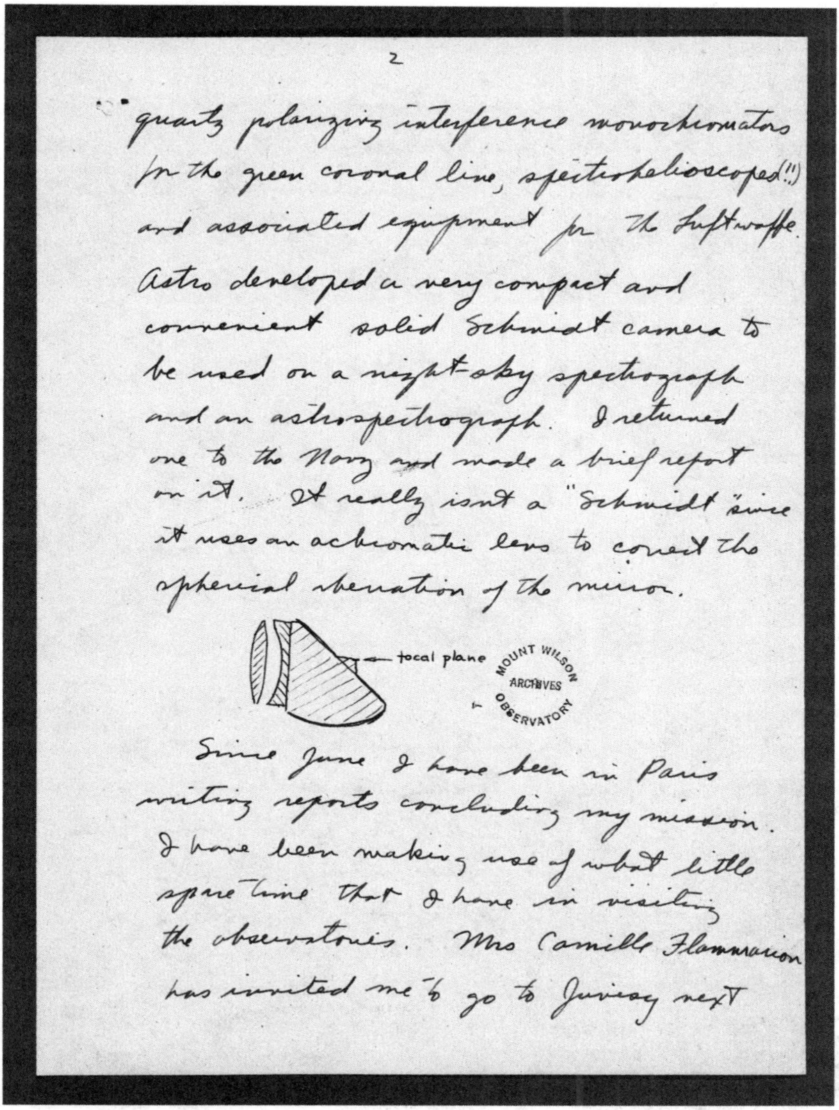

2

quartz polarizing interference monochromators for the green coronal line, spectrohelioscopes(!!) and associated equipment for the Luftwaffe. Astro developed a very compact and convenient solid Schmidt camera to be used on a night-sky spectrograph and an astrospectrograph. I returned one to the Navy and made a brief report on it. It really isn't a "Schmidt" since it uses an achromatic lens to correct the spherical aberration of the mirror.

Since June I have been in Paris writing reports concluding my mission. I have been making use of what little spare time that I have in visiting the observatories. Mrs Camille Flammarion has invited me to go to Juvisy next

Figure 4.7a–d. Continued.

intelligence in the field of rockets. He has been responsible for obtaining technical information and ordnance specimens through the means of searching for material throughout Germany and interrogating German scientists. All of his duties have been performed cheerfully and in a very efficient manner, frequently

(d)

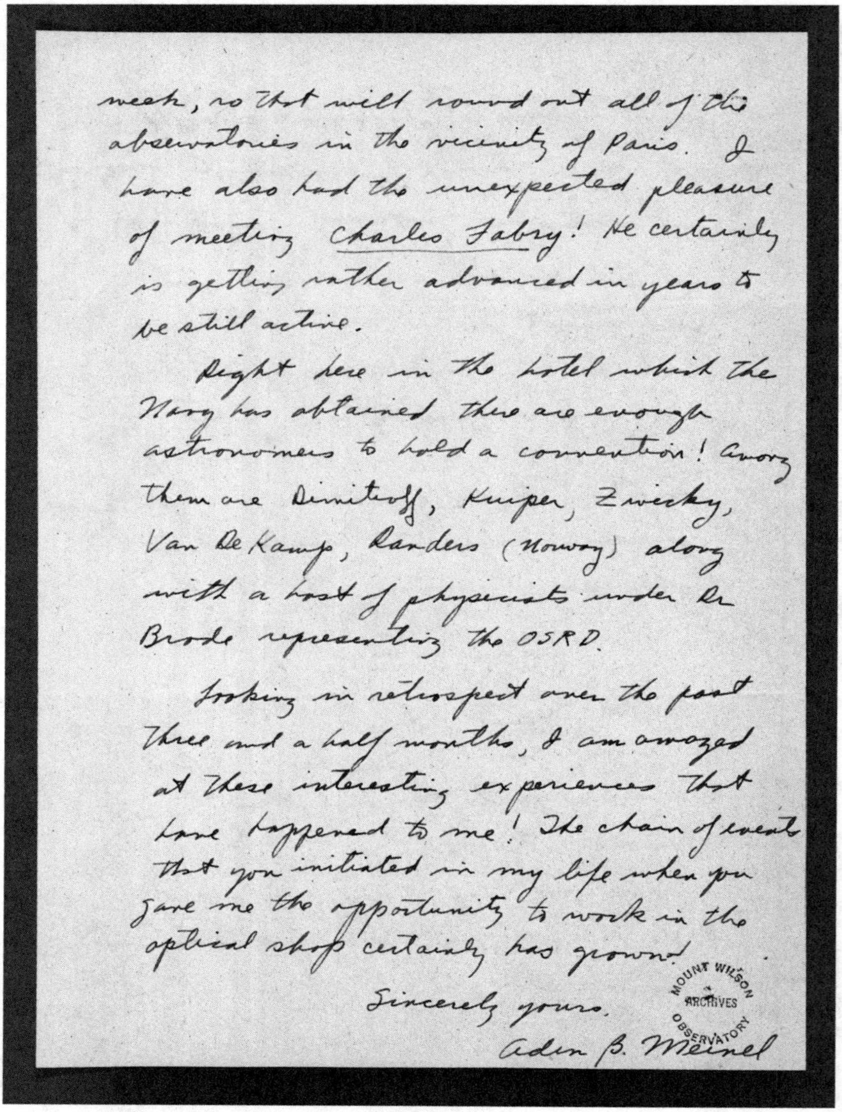

week, so that will round out all of the observatories in the vicinity of Paris. I have also had the unexpected pleasure of meeting Charles Fabry! He certainly is getting rather advanced in years to be still active.

Right here in the hotel which the Navy has obtained there are enough astronomers to hold a convention! Among them are Dimitroff, Kuiper, Zwicky, Van De Kamp, Randers (Norway) along with a host of physicists under Dr Brode representing the OSRD.

Looking in retrospect over the past three and a half months, I am amazed at these interesting experiences that have happened to me! The chain of events that you initiated in my life when you gave me the opportunity to work in the optical shop certainly has grown.

Sincerely yours.

Aden B. Meinel

Figure 4.7a–d. Continued.

under adverse conditions in the field."[215] As of August 30, 1945, Aden was authorized to wear the European-African-Middle Eastern campaign ribbon, having completed 30 consecutive days in the European Theater prior to July 1, 1945.[216] However, he received no medals because, as he later learned, he was classified only as a temporary visitor and had never been formally assigned to the Third Army or any naval facility in Europe. However, he was surprised to receive a substantial check as temporary additional duty (TAD) for expenses he would have submitted if he had slept in hotels rather than bombed buildings and had eaten restaurant meals instead of C-rations.[217]

In mid-September he returned to Washington, DC, where he received news that he was to return to NOTS at China Lake to be commissioned Assistant Rocket Project Officer. He telephoned Marjorie to give her the good news and heard that she had some of her own—she had resigned from her Caltech position when the war ended. His officer rank and foreign duty qualified them for a house on the base, a duplex, their first home together. He shared a taxi to the train station and hopped on the Santa Fe *Super Chief* to Pasadena where Marjorie, her parents, and his parents were waiting for their emotional reunion. He surprised her with the news of the TAD check, which they used to buy their first car and take a two-week vacation to Arizona before he was to report for duty.[218]

On New Year's Day 1946, Aden returned to the Special Launcher Group supervised by Al S. Gould as project engineer with the Projects Engineering Division, Explosives Department, for the design and testing of experimental ordnance equipment. At that time the Naval Ordnance Test Station at Inyokern was ground zero for revolutionary changes from guns to rockets as primary Navy weapons because Caltech's major rocket programs had been transferred to the Navy when the war ended.[219] In an Annual Qualifications Questionnaire for the Bureau of Naval Personnel, Aden wrote that he invented rocket launchers MK17 and MK18 along with others and had generated numerous publications for the OSRD and Explosives Department at NOTS.[220] Aden's group designed a launcher controlled by automatic tracking servomechanisms ("servos") with the range determined by radar. A computer would aim the launcher so that the spinner rockets and aircraft would intersect wherever the target aircraft was. In testing this new launcher one day on rockets with inert warheads and using a twin-engined drone operated remotely as a target, Aden's job was to watch the radar range and communicate the drone's position to the gunner who was managing the aiming mechanism close to the launcher. As Aden called out the ranges and heard the rockets being launched, suddenly the launch sounds stopped. The operators were behind him with the news that the drone was headed straight for them, so they managed several hits on it. Even though the warheads were inert, they pummeled the drone with enough rockets to blow off the wings and cause it to crash only a few yards in front of them. They had proved, nearly with casualties, that a barrage of spinner rockets with active warheads was more than adequate to defend Navy ships under attack.[221]

Aden applied to the University of California at Berkeley on the Servicemen's Readjustment Act of 1944 (the GI Bill of Rights), which would pay for up to four years of college tuition tax-free plus a small living allowance. In the spring of 1946 he received a reply accepting him as a student with the proviso that credits for some of his courses at Caltech were disallowed and he would be admitted as a sophomore.[222] Marjorie was happy because she was pregnant with their first child and planned to start a family while Aden received further training as an astronomer. Seven children in all would be born to them from 1946 through 1957.[223]

Notes

1. Meinel, A. B., Meinel, M. P., and Jacobs, B. M. 2008. *The golden age of astronomy*, unpublished, 41–2.
2. Veysey, V. V. Interview by S. K. Cohen. Pasadena, California, July 14 and 21, 1993, and February 4, 1994. Oral History Project, California Institute of Technology Archives. http://resolver.caltech.edu/CaltechOH:OH_Veysey_V; accessed June 20, 2019, 10.
3. Meinel et al., *Golden age*, 42.
4. United States Navy. 1944. *The bluejackets manual*. United States Naval Institute, Annapolis, Maryland.
5. Meinel et al., *Golden age*, 42.
6. Email from Aden Meinel to Barbara Meinel, dated April 6, 2009.
7. Stanton, D. 2001. *In harm's way: The sinking of the USS* Indianapolis *and the extraordinary story of its survivors*. Henry Holt and Company, New York, 166.
8. Meinel, A. B. 2009a. Reminiscences: Aden goes to war. *Optics & Photonics News* 20: 17.
9. Stanton, *In harm's way*, 28.
10. Meinel, Reminiscences: Aden goes to war, 17.
11. William Fowler Papers, Box 57, Folder 10, Caltech Archives.
12. Meinel et al., *Golden age*, 42; the High Voltage Laboratory was redesigned in 1960 and renamed the Alfred P. Sloan Laboratory of Mathematics and Physics. The Annex is now named the Sloan Annex.
13. Meinel, Reminiscences: Aden goes to war, 17.
14. Meinel, Reminiscences: Aden goes to war, 16; letter from James Edson to the Director of Naval Procurement, dated January 5, 1944; letter from W. R. Van Auken to the Chief of Naval Personnel, dated January 23, 1945, NARA.
15. Officer Application, Enlisted File No. 881-84-15, dated January 15, 1945, NARA.
16. Meinel et al., *Golden age*, 42.
17. Gerrard-Gough, J. D., and Christman, A. 1978. *History of the Naval Weapons Center, China Lake, California*. Vol. 2. *The grand experiment at Inyokern*. Naval History Division, Washington, DC, 218.
18. Gerrard-Gough and Christman, *History of the Naval Weapons Center*, 209.
19. Meinel et al., *Golden age*, 43.
20. Gerrard-Gough and Christman, *History of the Naval Weapons Center*, 219.
21. Meinel et al., *Golden age*, 45.

22. Veysey, Interview, 15.

23. Veysey, Interview, 45.

24. Meinel, Reminiscences: Aden goes to war, 17; Meinel et al., *Golden age*, 45; letter from A. B. Meinel to Jennifer Meinel, dated April 24, 1996.

25. Meinel et al., *Golden age*, 46; letter from A. B. Meinel to Jennifer Meinel, dated April 24, 1996.

26. Letter from A. B. Meinel to Jennifer Meinel, dated April 24, 1996.

27. Request for Assignment of Service Personnel, dated March 2, 1945.

28. Kohnen, D. A. 2015. Seizing German naval intelligence from the archives of 1870–1945. *Global War Studies* 12: 156; Simons, G. M. 2016. *Operation Lusty: The race for Hitler's secret technology*. Pen and Sword Books, Barnsley, South Yorkshire, UK, 109.

29. Kohnen, Seizing German naval intelligence, 156, 157.

30. Hall, R. C. 2015. Earth satellites, a first look by the U.S. Navy. https://ntrs.nasa.gov/search.jsp?R=19770026119 2019-02-05T18:30:50+00:00Z, Accessed February 5, 2019, 253; Neufeld, M. J. 2007. *Von Braun: Dreamer of space, engineer of war*. Random House, New York, 201.

31. Simons, *Operation Lusty*, 109; Berg, A. S. 1998. *Lindbergh*. Berkley Books, New York, 467.

32. By an odd twist of fate, George S. Patton, Jr., was raised in Pasadena, California, and attended Stephen Cutter Clark's School for Boys at 39 South Euclid Avenue. One of his friends there was Thomas G. Bard, son of US senator Thomas R. Bard. Later, while Patton was a student at Virginia Military Institute in Lexington, Virginia, he applied with 15 others to Senator Bard for an appointment to West Point and got it; Hirshson, S. P. 2002. *General Patton: A soldier's life*. Harper Perennial, New York, 25, 3. Senator Bard was the nephew of Marjorie's maternal great-grandfather, Richard Bard (1806–1867). See Chapter 1, section "Steele/Bard ancestry."

33. Memo from H. A. Schade to Ensign Aden Meinel, dated April 3, 1945, NARA.

34. Reminiscences: Aden goes to war, 17.

35. Letter from A. B. Meinel to Jennifer Meinel, dated April 24, 1996.

36. "A look back . . . Gen. William J. Donovan heads office of strategic services." News and Information, CIA.gov; accessed January 29, 2017.

37. O'Donnell, P. K. *Operatives, spies and saboteurs: The unknown story of the men and women of World War II's OSS*. Free Press, New York, xi.

38. "Chef Julia Child, others part of WWII spy network." Archived August 22, 2008, at the Wayback Machine, CNN, August 14, 2008.

39. Letter from A. B. Meinel to Jennifer Meinel, dated April 24, 1996.

40. Personal communication to the authors from Joseph Houston (retired engineer for both Perkin-Elmer and Itek), August 3, 2016.

41. Meinel et al., *Golden age*, 46.

42. Letter from A. B. Meinel to Jennifer Meinel, dated April 24, 1996; Meinel et al., *Golden age*, 47.

43. Meinel et al., *Golden age*, 47.

44. Jones, V. C. 1985. *United States Army in World War II. Special studies: Manhattan: The Army and the atomic bomb*. Center of Military History,

United States Army, Washington, DC, 280; Groves, L. M. 1962. *Now it can be told: The story of the Manhattan Project.* Da Capo Press, Boston, 192, 249.

45. Jones, *United States Army in World War II*, 280; Groves, *Now it can be told*, 192, 249.

46. Meinel et al., *Golden age*, 56; Goudsmit, S. A. 1947. *Alsos.* Tomash Publishers, Los Angeles, 17.

47. MS480, Box 57, Folder 153-1.17[2], University of Arizona Special Collections.

48. MS480, Box 57, Folder 16.3-1-17.1.

49. Goudsmit, *Alsos*, 123, 134.

50. Jones, *United States Army in World War II*, 289–91; Zachary, G. P. 1997. *Endless frontier: Vannevar Bush, engineer of the American century.* The Free Press, New York, 213.]

51. Jones, *United States Army in World War II*, 287; Groves, *Now it can be told*, 220.

52. Groves, *Now it can be told*, 242.

53. Goudsmit, *Alsos*, 176.

54. Goudsmit, *Alsos*, 134.

55. Goudsmit, *Alsos*, 108.

56. Groves, *Now it can be told*, 249.

57. Meinel et al., *Golden age*, 47.

58. Meinel did not specify the date, but the Third Army captured Ludwigshafen and Speyer, Germany, on March 24, 1945; World War II database, Crossing the Rhine; www.we2db.com, accessed February 1, 2017.

59. MacDonald, C. B. 2012. *The last offensive: United States Army in World War II, European Theater of Operations.* Whitman Publishing, Atlanta, Georgia, 237; Jacobsen, A. 2014. *Operation Paperclip: The secret intelligence program that brought Nazi scientists to America.* Little, Brown, and Company, New York, 28–9, 74.

60. Jacobsen, *Operation Paperclip*, 146–8.

61. Jacobson, *Operation Paperclip*, 150, 152–4.

62. Province, C. M. 1992. *Patton's Third Army: A chronology of the Third Army advance, August, 1944 to May, 1945.* Hippocrene Books, New York, 222; World War II Database, https://ww2db.com/battle_spec.php?battle_id=134; accessed February 3, 2019.

63. Meinel et al., *Golden age*, 47; letter from A. B. Meinel to Jennifer Meinel, date April 24, 1996. Meinel mistakenly wrote in *The golden age of astronomy* that the crossing was at Friedrichshafen, which is nowhere near the Rhine but on the north shore of Lake Constance in the state of Baden-Württemberg near the borders with Switzerland and Austria.

64. Email from A. B. Meinel to Barbara Meinel, dated April 6, 2009.

65. Province, *Patton's Third Army*, 221, 223; Forty, G. 1976. *Patton's Third Army at war.* Arms and Armour, London, 160.

66. Meinel et al., *Golden age*, 48.

67. MacDonald, *The last offensive*, 325.

68. Province, *Patton's Third Army*, 235, 247.

69. Kohnen, Seizing German naval intelligence, 158.

70. Letter from A. B. Meinel to Jennifer Meinel, dated April 24, 1996.

71. Letter from A. B. Meinel to Jennifer Meinel, dated April 24, 1996.

72. Letter from A. B. Meinel to Jennifer Meinel, dated April 24, 1996; Meinel et al., *Golden age*, 48.

73. Jena, *Lichtstadt mit moderner Verwaltung*, http://www.jena.de/de/stadt_verwaltung/ stadtportraet/chronik/1945/236315; accessed February 7, 2017.

74. Bombing of Dresden in World War II, http://www.newworldencyclopedia.org/entry/ Bombing_of_Dresden_in_World_War_II; accessed February 7, 2017.

75. Pogue, F. C. 1952. Why Eisenhower's forces stopped at the Elbe. *World Politics* 4: 356.

76. Letter from A. B. Meinel to Jennifer Meinel, dated April 24, 1996.

77. After Action Report, Third U.S. Army, 1 August 1944–9 May 1945. Volume II. Parts 1–9. Staff section reports. U.S. Army Military History Institute, Washington, DC, 13.

78. Meinel et al., *Golden age*, 49.

79. Meinel et al., *Golden age*, 49.

80. Samuel, W. W. E. 2004. *American raiders: The race to capture the Luftwaffe's secrets*. University of Mississippi Press, Jackson, Mississippi, 139; Naval Technical Mission in Europe, Letter Report 41-45, RG 38, Identifier 4345900, National Archives at College Park, Maryland; Naval Technical Mission in Europe, Letter Report 81-45, RG 38, Identifiers 4345939, 4345940, 4345942, National Archives at College Park, Maryland; Christopher, J. 2013. *The race for Hitler's X-planes: Britain's 1945 mission to capture secret Luftwaffe technology*. The History Press, Stroud, Gloucestershire, UK, 128–9.

81. Naval Technical Mission in Europe, Letter Report 41-45, RG 38, Identifier 4345900, National Archives at College Park, Maryland; Christopher, *The race for Hitler's X-planes*, 126–7.

82. Christopher, J. 2013. *The race for Hitler's X-planes*, 131–2, 142, 144–5; Naval Technical Mission in Europe, Letter Report 41-45, RG 38, Identifier 4345900, National Archives at College Park, Maryland.

83. Meinel et al., *Golden age*, 49.

84. Lasby, C. 1971. *Project Paperclip: German scientists and the Cold War*. New Saucerian Press, New York, 18.

85. Meinel et al., *Golden age*, 49–50.

86. Neufeld, *Von Braun*, 80, 88; DeVorkin, D. H. 1992. *Science with a vengeance: How the military created the US space sciences after World War II*. Springer-Verlag, New York, 23.

87. Gardiner, J. 2004. *Wartime: Britain 1939–1945*. Headline Book Publishing, London, 2; King, B., and Kutta, T. J. 1998. *Impact: The history of Germany's V-weapons in World War II*. SARPEDON, Rockville Centre, New York, 55.

88. Gardiner, *Wartime*, 17.

89. Manchester, W., and Read, P. 2012. *The last lion: Winston Spencer Churchill, defender of the realm 1940–1965*. Little, Brown, and Company, New York, 129; Bungay, S. 2015. *The most dangerous enemy: A history of the Battle of Britain*. Aurum Press, London, 9–10.

90. Manchester and Read, *The last lion*, 130–1.

91. Manchester and Reid, *The last lion*, 140.

92. Gardiner, *Wartime*, 328–9.

93. Bungay, *The most dangerous enemy*, 368, 373.
94. Gardiner, *Wartime*, 332.
95. Gardiner, *Wartime*, 352.
96. Gardiner, *Wartime*, 393.
97. Gardiner, *Wartime*, 434.
98. Haining, P. 2002. *The flying bomb war: Contemporary eyewitness accounts of the German V-1 and V-2 raids on Britain*. Robson Books, London, 13; King and Kutta, *Impact*, 2, 87.
99. Haining, *The flying bomb war*, 13; King and Kutta, *Impact*, 78–86.
100. Haining, *The flying bomb war*, 14–15; Manchester and Reid, *The last lion*, 854.
101. Manchester and Reid, *The last lion*, 854.
102. Haining, *The flying bomb war*, 37; Gardiner, *Wartime*, 638–9.
103. Gardiner, *Wartime*, 640.
104. Gardiner, *Wartime*, 643, 649; King and Kutta, *Impact*, 216.
105. King and Kutta, *Impact*, 205–6.
106. King and Kutta, *Impact*, 207.
107. Haining, *The flying bomb war*, 91, 92.
108. Manchester and Reid, *The last lion*, 870.
109. Province, *Patton's Third Army*, 294–5.
110. Fuller, R. P. 2004. *Last shots for Patton's Third Army*. NETR Press, Portland, Maine, 254.
111. Meinel et al., *Golden age*, 48.
112. King and Kutta, *Impact*, 32.
113. King and Kutta, *Impact*, 32–3.
114. King and Kutta, *Impact*, 34–6.
115. King and Kutta, *Impact*, 38–40.
116. Neufeld, *Von Braun*, 143.
117. Haining, *The flying bomb war*, 17, 18; Gardiner, *Wartime*, 652–3; King and Kutta, *Impact*, 52; Simons, *Operation Lusty*, 61.
118. Haining, *The flying bomb war*, 18.
119. King and Kutta, *Impact*, 62–5.
120. Sellier, A. 2003. *A history of the Dora camp*. Ivan R. Dee, Chicago, 28; Simons, *Operation Lusty*, 62; King and Kutta, *Impact*, 105–14.
121. King and Kutta, *Impact*, 70.
122. Sellier, *A history of the Dora camp*, 29; King and Kutta, *Impact*, 70–1; Michel, J. 1975. *Dora*. Holt, Rinehart, and Winston, New York, 60.
123. Sellier, *A history of the Dora camp*, 30–3; Naval Technical Mission in Europe, ETO Ordnance Intelligence Report No 267; Naval Technical Mission in Europe, ETO Ordnance Intelligence Report No 270.
124. Michel, *Dora*, 61; Sellier, *A history of the Dora camp*, 46.
125. King and Kutta, *Impact*, 72; Michel, *Dora*, 62.
126. Naval Technical Mission in Europe, ETO Ordnance Intelligence Report No 270.
127. Michel, *Dora*, 62–3.
128. Sellier, *A history of the Dora camp*, 62.
129. Michel, *Dora*, 75–6.

130. Christopher, *The race for Hitler's X-planes*, 117.
131. Michel, *Dora*, 15; Sellier, *A history of the Dora camp*, 403; Naval Technical Mission in Europe, ETO Ordnance Intelligence Report No. 267; Naval Technical Mission in Europe, ETO Ordnance Intelligence Report No. 270.
132. Naval Technical Mission in Europe, Letter Report 81-45, RG 38, Identifiers 4345939, 4345940, 4345942, National Archives at College Park, Maryland.
133. Michel, *Dora*, 97.
134. Berg, *Lindbergh*, 161, 409.
135. Gehrels, T. 1988. *On the glassy sea: An astronomer's journey.* American Institute of Physics, New York, 29–31.
136. Gardiner, *Wartime*, 653; King and Kutta, *Impact*, 309.
137. King and Kutta, *Impact*, 281.
138. Haining, *The flying bomb war*, 219.
139. Sellier, *A history of the Dora camp*, 314.
140. Sellier, *A history of the Dora camp*, 289; Simons, *Operation Lusty*, 128.
141. Sellier, *A history of the Dora camp*, 290.
142. Sellier, *A history of the Dora camp*, 313–15.
143. Sellier, *A history of the Dora camp*, 398.
144. Michel, *Dora*, 287–9.
145. Sellier, *A history of the Dora camp*, 313–16; MacDonald, *The last offensive*, 391.
146. DeVorkin, *Science with a vengeance*, 45, 48.
147. Simons, *Operation Lusty*, 127–8.
148. Brugioni, D. A. 2010. *Eyes in the sky: Eisenhower, the CIA and Cold War aerial espionage.* Naval Institute Press, Annapolis, Maryland, 63.
149. Meinel et al., *Golden age*, 50.
150. Meinel et al., *Golden age*, 51.
151. Lasby, *Project Paperclip*, 37.
152. Berg, *Lindbergh*, 466–8.
153. Kappler, D., and Steiner, J. 2009. *Schott 1884–2009: From a glass laboratory to a technology group.* Schott AG, Mainz, 92–3.
154. Kappler and Steiner, *Schott*, 93, 94.
155. Kappler and Steiner, *Schott*, 94, 95.
156. Kappler and Steiner, *Schott*, 98.
157. Kappler and Steiner, *Schott*, 98; Kiaulehn, W. 1959. *The odyssey of 41 glassmakers.* Jenaer Glaswerk Schott & Genossen, Mainz, Germany, 31.
158. Kappler and Steiner, *Schott*, 99.
159. Letter from A. B. Meinel to Jennifer Meinel, dated April 24, 1996.
160. Meinel et al., *Golden age*, 51.
161. Naval Technical Mission in Europe, Letter Report 152–45, RG 38, Identifier 4346015, National Archives at College Park, Maryland.
162. Naval Technical Mission in Europe, Letter Report 86-45, RG 38, Identifier 4345946, National Archives at College Park, Maryland.
163. Naval Technical Mission in Europe, Letter Report, 155-45, RG 38, Identifier 4346018, National Archives at College Park, Maryland.

164. Naval Technical Mission in Europe, Letter Report 93-45 (A), RG 38, Identifier 4345943, National Archives at College Park, Maryland.

165. Meinel et al., *Golden age*, 48.

166. Meinel, A. B. 2009b. Reminiscences: Aden returns from the war. *Optics & Photonics News* 20: 20.

167. Naval Technical Mission in Europe. Reports and Office Files. Historical data on U. S. Naval Technical Mission in Europe—First narrative. Record Group 38. Identifier 4345763. National Archives at College Park, Maryland.

168. Letter from A. B. Meinel to Jennifer Meinel, dated April 24, 1996.

169. Lasby, *Project Paperclip*, 85.

170. Gimbel, J. 1990. German scientists, United States denazification policy, and the "Paperclip Conspiracy." *International History Review* 12: 448.

171. Jacobsen, *Operation Paperclip*, 229; Gimbel, German scientists, 454.

172. Memorandum from Gerard Kuiper to Major Russell A. Fisher dated June 30, 1945, Alsos, Box 57, File 15, Special Collections at the University of Arizona Libraries.

173. Lasby, *Project Paperclip*, 86.

174. Samuel, *American raiders*, 125–7; Simons, *Operation Lusty*, 112–20.

175. Jacobsen, *Operation Paperclip*, 176; Gimbel, German scientists, 454.

176. Gimbel, German scientists, 458.

177. Crim, B. E. 2018. *Our Germans: Project Paperclip and the national security state.* Johns Hopkins University Press, Baltimore, Maryland, 43.

178. Lasby, *Project Paperclip*, 2.

179. Crim, *Our Germans* 3, 41, 42.

180. Neufeld, *Von Braun*, 179; Jacobsen, *Operation Paperclip*, 281.

181. Neufeld, *Von Braun*, 162.

182. Gehrels, *On the glassy sea*, 226.

183. Letter from Ernst Stuhlinger to Tom Gehrels, June 5, 1987, Tom Gehrels Papers, Special Collections at the University of Arizona Libraries.

184. Goudsmit, *Alsos*, 113–14.

185. Memorandum from Gerard Kuiper to Major Russell A. Fisher, dated June 30, 1945, Alsos, Box 57, File 15, Special Collections at the University of Arizona Libraries.

186. Lasby, *Project Paperclip*, 24.

187. Kappler and Steiner, *Schott*, 100.

188. Kappler and Steiner, *Schott*, 103; Kiaulehn, *The odyssey*, 34.

189. Kiaulehn, *The odyssey*, 34; Kappler and Steiner, *Schott*, 103.

190. Kiaulehn, *The odyssey*, 41; Kappler and Steiner, *Schott*, 108.

191. Kappler and Steiner, *Schott*, 110.

192. Personal communications to the authors from Jim Mayo, April 7, 2019; June 10, 2019.

193. Letter from A. B. Meinel to Irene Howard, dated May 6, 2009.

194. Meinel et al., *Golden age*, 51.

195. Lasby, *Project Paperclip*, 89–90.

196. Naval Technical Mission in Europe, Letter Report 89-45 (A), RG 38, Identifier 4345949, National Archives at College Park, Maryland.

197. Naval Technical Mission in Europe, Technical Report 455-45, RG 38, Identifier 4346502, National Archives at College Park, Maryland.
198. Meinel et al., *Golden age*, 51–2.
199. Letter from Aden Meinel to John Anderson, June 13, 1945, Huntington Library, Mt. Wilson Collection.
200. Meinel et al., *Golden age*, 54.
201. Meinel, Reminiscences: Aden returns from the war, 21.
202. Stanton, *In harm's way*, 28.
203. Stanton, *In harm's way*, 130–4.
204. Stanton, *In harm's way*, 142.
205. Stanton, *In harm's way*, 211–216.
206. Buckley, C. 2017. Wreckage of U.S.S. *Indianapolis*, lost for 72 years, is found in the Pacific. https://nyti.ms/2vP87GF; accessed April 19, 2019.
207. Letter from A. B. Meinel to Barbara Meinel, dated May 6, 2009.
208. NavTechMisEu, Memo from H. A. Schade to Ensign A. B. Meinel, August 7, 1945, National Archives at College Park, Maryland.
209. Meinel et al., *Golden age*, 52.
210. Meinel et al., *Golden age*, 54–5; www.theguardian.com/world/2005/apr/06/austria.secondworldwar, accessed February 2, 2019.
211. Naval Technical Mission in Europe, Technical Report 282-45, RG 38, Accession #72A-5983, National Archives at College Park, Maryland.
212. Letter from A. B. Meinel to Walter Adams, July 7, 1945, Box 46, Folder 789, Huntington Library, Mt. Wilson Collection.
213. Meinel et al., *Golden age*, 56.
214. Memo from Ensign A. B. Meinel to the Chief of the Bureau of Naval Personnel, NavTechMisEu, File P16-3(95A/cw).
215. Letter Report on fitness of Ensign A. B. Meinel S(04) from H. A. Schade to the Chief of the Bureau of Ordnance, NavTechMisEu, File P20-2/00(10HDH/Ry).
216. Memo to Ensign A. B. Meinel, NavTechMisEu, File P15(95A/cw).
217. Meinel, A. B. 2009b. Reminiscences: Aden returns from the war, 20.
218. Meinel, Reminiscences: Aden returns from the war, 20–1; Meinel et al., *Golden age*, 57, 58, 60.
219. Gerrard-Gough and Christman, *History of the Naval Weapons Center*, 293, 297.
220. While he was a doctoral student at UC Berkeley he received one patent for a rocket launcher (number 2,478,774, filed January 20, 1947) and another (2,800,836, filed October 24, 1945) when he was working at Yerkes Observatory, but whether these are for the MK17 and MK18 is not known.
221. Meinel et al., *Golden age*, 60–1.
222. Meinel et al., *Golden age*, 61.
223. Meinel, Reminiscences: Aden returns from the war, 21.

5

Lights in the Night Sky

In the spring of 1946 Aden and Marjorie were living in Inyokern while Aden was wrapping up his work at NOTS at China Lake. Christian Elvey, chief scientist at NOTS and later director of the Geophysical Institute at the University of Alaska, asked Aden to remain at China Lake to work on studies of the atmosphere. Elvey wanted many of the young scientists and engineers to remain there to constitute the core of the scientific staff for the new Michelson Laboratory. Aden was contributing to the NOTS program at the graduate engineer level but had no degree. This fact would ultimately severely limit his chances for promotion, more responsibility, and higher salary, so he declined to stay.[1]

Aden and Marjorie considered graduate astronomy programs for Aden at three universities: Caltech, University of Chicago (at the invitation of Gerard Kuiper), and University of California (UC) at Berkeley. Caltech did not begin offering graduate degrees in astronomy until 1948 when the 200-inch telescope was complete and Jesse Greenstein (from Yerkes) was hired to establish the department. Aden imposed a condition on his 1945 application to Caltech that he graduate with the PhD under funding from the GI Bill, which provided three years of funding. That was insufficient at Caltech because he was only a sophomore and needed strong faculty support to execute a highly accelerated schedule. Edison Pettit encouraged Aden to attend UC Berkeley and used his influence to help Aden gain admission.[2] His application was accepted in mid-spring 1946 for the first summer session on the condition that he would complete his sophomore year before entering the astronomy department. Although credits for his English and humanities courses at Caltech were not accepted at UC Berkeley, he found faculty there who were willing to work with him in his special situation as a war veteran.[3]

Marjorie and her sister Helen spent the 1941–1942 academic year in the graduate astronomy program at Berkeley under the same professors who were to teach Aden later. Marjorie completed four astronomy classes, two semesters of mathematics, and four semesters of physics there.[4] In the spring of 1942, the nation was gearing up for a response to Pearl Harbor, and Marjorie returned to Claremont, California, to be closer to home, family, and boyfriend Aden. Helen dropped out of graduate school at UC Berkeley that spring and took a secondary school teaching certificate from Pomona College. Professor C. D. Shane commented that Helen was shy, quiet, and seldom asked questions in class or

With Stars in Their Eyes. James B. Breckinridge and Alec M. Pridgeon, Oxford University Press. © Oxford University Press 2022. DOI: 10.1093/oso/9780190915674.003.0005

after.[5] Marjorie transferred units from UC Berkeley and received her MA degree in astronomy at Claremont Colleges on August 4, 1944. She and Aden were married one month later, on September 4, 1944. Marjorie was well prepared to support Aden in his quest for a PhD in astronomy.

Graduate student

As May 1946 approached, Aden and Marjorie's thoughts turned to leaving the desert at China Lake and heading for Berkeley. It was easy for them to decide what to do next: school for Aden and homemaking for Marjorie. Their first child was on the way, and they wanted to get settled as soon as possible, so in late May 1946 they packed up the car they had bought earlier with Aden's Navy Temporary Additional Duty (TAD) pay and headed north. They passed over the 5,250-foot-high Walker Pass about 10 miles from Ridgecrest for one last look back at China Lake, then on down the winding Kern River to Bakersfield and onward onto Berkeley.[6]

Marjorie chose to end her academic career with her MA degree, although she could easily have pursued a doctorate. She and Aden had bought a beautiful oak-covered lot in Altadena beside Eaton Canyon wash even before they were engaged. They sold it to help with expenses, but clearly it was not enough. The GI Bill provided unemployed veterans with three years of support at a maximum of about $20/week. The typical student took six years or more to advance from entering sophomore to PhD. It seemed like an insurmountable goal both financially and intellectually.[7]

Upon arriving in Berkeley, they found crowded housing conditions. Many young men and women recently released from military service were also using their GI Bill subsidy to go to college, and the university was preparing up for an onslaught of veterans. The housing office assigned them an apartment in a former shipyard workers' project, Codornices (Quail) Village. Codornices is a creek that drains the Berkeley hills into San Francisco Bay. Their apartment was near the Bay, which smelled horribly at the time because of industrial pollution, but they called it home while they faced together the task of completing Aden's schooling, taking the required English and humanities courses in order to be enrolled as a junior in the upcoming fall 1946 semester and to join the Lick Astronomical Department.[8] Aden received his AB degree in astronomy at the end of summer school in 1947, 14 months after admission.

They didn't stay long at the Codornices housing development. One afternoon Aden came home from classes, and the house was empty. After the initial shock he remembered that they had helped Helen to get an offer from Frank Roach to join him in studies of the light of the night sky at China Lake. Helen had moved south to China Lake from a small cottage she had rented in the Oakland hills. The owner agreed that Aden and Marjorie could rent it, even though it was small for their growing family.[9]

At the university, Aden was assigned an office of sorts in an excavated room under one of the World War I redwood "temporary" buildings at Leushner Observatory, at that time close to the north edge of the campus. The office walls of this hole in the ground consisted simply of soil. Aden had three classmates to share the office—Harold Johnson, Victor Blanco, and George Herbig. Each went on to distinguish himself in astronomical research or observatory administration.[10]

Harold Johnson was a native of Colorado who completed his BS in mathematics at the University of Denver and spent the war years at the MIT Radiation Lab designing and building electronics. At UC Berkeley, he had the same professors as Aden and defended his dissertation in June 1948. Later, Johnson would join the Lunar and Planetary Laboratory at the University of Arizona under Gerard Kuiper. Johnson invented the first precision photoelectric system to measure the color-brightness of celestial objects through three filters (UBV: ultraviolet, blue, and visible), which became a standard tool for astronomical research for over 55 years. Colors are obtained by subtracting intensity measurements made through each filter and are directly related to stellar spectral type. William W. Morgan of Yerkes Observatory provided the spectral classification of many of the stars. Spectral classification is subjective, often depending on the astronomer's eye for almost undecipherable detail in an image of a spectrum. Photometry was unbiased, quantitative, and precise. Soon both measurement techniques taken together would enable astronomers to map stellar evolution and understand how our galaxy and its neighborhood evolved.

Victor Blanco was a native of Puerto Rico who obtained his BA in astronomy from the University of Chicago. He served in the US Army by building radar detectors during 1940–1946. After the war he attended UC Berkeley in astronomy with Aden and earned his MS in 1947. He defended his PhD dissertation eight days before Aden defended his.[11] After a brief career at the US Naval Observatory and a longer period in research and teaching astrophysics at Case Institute of Technology, Blanco became the second director of the Cerro-Tololo Inter-American Observatory (CTIO), a joint venture between the US (National Science Foundation) and Chilean governments. This Southern Hemisphere observatory was an offshoot of the Northern Hemisphere Kitt Peak National Observatory, which was established by AURA and Aden in 1957. He found financing for construction of a 4-meter telescope at CTIO, which is now named the Victor M. Blanco Telescope. He maintained excellent relationships between this Southern Hemisphere observatory and the Chilean astronomical and other scientific communities. Blanco is credited with opening up modern astrophysics of the Southern Hemisphere skies to the citizens of Chile.

George Herbig was a 1943 graduate of UCLA. His dissertation was on the spectrophotometry of variable stars in nebulosity. He defended his dissertation in June 1948, accepted a Lick Fellowship, and spent his career at Lick Observatory. He is the co-discoverer of Herbig-Haro objects, which are bright patches of nebulosity associated with newborn stars. Herbig had discovered an important phase of the life cycle of stars.

These four young astronomers who trained together at Berkeley did much to pioneer modern observational astrophysical sciences during the last half of the 20th century.

Marjorie read the fine print in the UC catalog and found something of vital importance to Aden. The university allowed a student to take courses by examination and receive full credit for a course if he or she earned a grade of A or B on the exam. If the student maintained a grade point average of a B or better, then units would be subtracted from the total number required for graduation. Taking advantage of these opportunities, Aden studied hard and tested out of two semesters of introductory astronomy during the second semester summer session of 1946. For the academic year 1946–1947, he enrolled in two astronomy classes (104a/b and 117ab), German, four semesters of physics classes (105a/108b and 115/121), and a math and a history class. He maintained an A– average before graduating in June 1947 with his BA in astronomy. By the fall of 1946 he already knew the general topic of his dissertation and started to think about the apparatus he would need to make the measurements.[12]

In March 1947, Joseph H. Moore, who was then director of Lick Observatory, greeted Aden as he came to astrophysics class and offered him a James Lick Fellowship to work as an assistant in the observatory.[13]

That summer he was accepted into graduate school. For the academic year 1947–1948 he took Physics 208a and 212:295, along with Astrophysics (Henyey), Astronomy 108:115, and seven semester units of research. His graduate school advisor was Nicholas Mayall, an observational astronomer. Mayall would later succeed Aden as director of Kitt Peak National Observatory in 1961 when Aden moved on to his illustrious career with the University of Arizona, first as director of Steward Observatory and then founding director of what is now the College of Optical Sciences in 1967.

In 1931, German optician Bernhard Schmidt developed innovative technology to create a wide-field, high-resolution telescope that added a new dimension to atmospheric science and observational astronomy. It was this innovative technology that Roger Hayward, with the help of 17-year-old Aden Meinel, brought to Mt. Wilson Observatory (see Chapter 2, section "Youth and education") and the US Army in 1939 that contributed to the war effort. By the fall of

1947, Aden had designed and built an early version of the Schmidt camera he would use to obtain data for his dissertation. By April 1948, Aden had a clear picture of the steps he would follow for his research and requested approval from the Observatory director for a dormitory room on Mt. Hamilton.

Classes were a pleasure, but getting to the next class after Robert Trumpler's superb astronomy lectures was sometimes difficult, especially when the next class was on the opposite side of campus. Marjorie had Trumpler as an instructor when she attended Berkeley in 1941–1942, and he was a good friend of the Pettits, so Aden knew many stories even before attending his classes. Trumpler would already be lecturing as he walked through the back door of the lecture hall, picking up the thread where he last left it. Another professor, Leland Cunningham, was quite different but also delightful. He was so involved in his computation of comet orbits that he scarcely knew anything else existed. He would fill the blackboard with equations and calculations that the students would dutifully copy. When the class convened the next time, he would wonder aloud, "Just where was I?" Students would remind him exactly where he ended, and he would continue without missing a beat.[14]

Courses in physics were daunting, and Aden received As, Bs, and Cs in them. Physics majors dominated classes because they invariably audited each course before they took it for credit. Aden did not have that luxury because he needed to finish and be on his way. Many Berkeley physics majors took six to eight years to complete graduate studies, supported by research assistantships at the Berkeley Radiation Laboratory up the hill. One of his contemporaries in physics was still a graduate student when Aden returned for a visit years later, even after he had become an associate professor at the University of Chicago.[15]

Dissertation challenge

The research topic Aden selected for his dissertation made it necessary for him to build a scientific apparatus from scratch, use it to make observations, record data, reduce, interpret the science in that data, and then write up his work and defend his conclusions to a committee of six senior professors. He conducted the research and wrote the dissertation while taking a full load of academic classes and graduating after only two years in graduate school.

A PhD in the sciences acknowledges that the recipient has completed a compelling research project that adds significantly to our knowledge about the universe. Most students select a topic on the recommendation of their academic advisor, and the topic often follows the interests of the advisor or an institute committee. However, Aden's infectious curiosity about nature and his independent spirit drove him to recognize a scientific topic of broad interest and pursue it to completion with little direction from others.

Aden had decided on his dissertation topic even before he went to Europe as a Navy ensign in 1945. In fact, he wrote that it dated back to a night in the early 1940s on the Mt. Wilson 100-inch telescope when Walter Baade complained about how fast his infrared (IR) photographic plates reached the night sky background. Baade had concluded that the IR night sky must be much brighter than in the visible. But the source of that IR light in the night sky was unexplored.[16] The physical processes that caused the IR night sky to be bright were unknown. But the invisible IR night sky was bright enough to fog sensitive astronomical photographic plates and mask important astronomical information. The light of the night sky is spread out over a wide area of the sky and difficult to analyze in detail. Astronomical telescopes of the period were designed to have a small field of view (FOV) with high magnification to examine details on planetary surfaces, galaxies, and star fields. But unfortunately, no suitable high-throughput, wide-field imaging telescope/spectrometer system had ever been designed or built until Aden's success.[17]

The top authority on the visible-UV light of the night sky, Polydore (Pol) Swings (spectroscopist at the University of Liege, Belgium) happened to be visiting Mt. Wilson Observatory when Aden and Marjorie were visiting the Pettits. Aden described his idea for a wide-field spectrometer to Swings and Pettit during one of the many family trips that Aden and Marjorie made to Pasadena from NOTS at China Lake before they left for Berkeley in June 1946.[18] The only wide-field camera that delivered high-quality images across the wide spectral band needed for the measurement was the Schmidt camera.[19] But for Aden's application an innovative modification to the design was needed. This idea was based on Aden's early work as a teenage volunteer at Mt. Wilson Observatory.

Schmidt cameras/telescopes

Classical Schmidt cameras have a doubly curved "hemispherical-shaped" focal surface, and the flexible optical film was typically bent to match that curvature. Because the light of the night sky was so faint, Aden needed to concentrate a large area of the sky onto a small image on the film. He calculated that he would require a camera with an f/ratio of less than 1 to obtain high enough energy density to expose the emulsion and thus record a clear image of the structure of the spectrum. The radius of curvature of the focal plane becomes smaller as the f/ratio is decreased, until in the case of f/<1 it is too small for any flexible film.

Aden needed to use photographic emulsions sensitive to the IR. Kodak had developed IR-sensitive photographic materials for the military during World

War I. These were commercially available only on flat glass surfaces. Therefore, he had to create a new design and build a completely novel flat-field Schmidt telescope/camera system. To collect enough light to record the faintest light in the night sky, he calculated that he needed at least a 6-inch diameter collecting aperture for the camera, which in turn would require a 6-inch diameter Schmidt corrector and an even larger primary mirror.

His innovation was to flatten the image field using a stop shift and a field-flattening lens of weak power in front of the focal plane.[20] Aden decided to ray-trace design his new flat-field Schmidt camera at China Lake for this spectrometer using the state-of-the-art tools at the time: logarithm tables. These printed tables enable the accurate multiplication of two several-digit numbers using the process of addition, which is less complicated than the more error-prone and tedious long-hand multiplication.

Aden had watched Frank Ross at Mt. Wilson doing the same, using a giant folio of Crelly's Log Tables.[21] A tiny man, Ross had to lean out over the book with a large magnifier in order to read the numbers that were at the top of the pages. Aden asked one day if that was how Ross designed his famous Ross lens, first at Yerkes and then his even more famous Ross correctors for the Mt. Wilson telescopes. Ross assented but also revealed that he used some tricks of the trade discovered by Alexander E. Conrady and advised Aden to read Conrady's book.[22] One can avoid the labor of tracing skew rays by means of simple relationships, said Ross. Today, first-order optical system design is more refined, and innovative solutions to optical aberration control are regularly used.[23] In addition, modern digital computers enable thousands of individual rays to be traced surface-to-surface through an optical system. Aden applied Ross's advice and the contents of Conrady's book to explore the design for a fast, flat-field Schmidt camera.[24] His design closed with excellent results, but he would not be sure if those simple relationships held for fast cameras until he actually built one.

Armed with a design that looked excellent, Aden went to present his ideas to Ira Bowen at Caltech. Bowen listened and related that based on his experience it was impossible to achieve a flat field with excellent images over the field of view Aden needed. Bowen said he designed and had added field-flattening lenses to the Schmidt cameras in the spectrographs at Mt. Palomar Observatory, and the residual, uncorrected coma was too large. Aden replied by showing him his design that had both the coma and color well corrected and under control by adding a new free variable: the spacing of the field-flattening lens from the image plane. Bowen was still skeptical even after the exchange of several letters and extended discussions.[25]

Aden now had to build his spectrograph in the face of an adverse opinion from the distinguished Caltech astronomer responsible for the success of the optics of the new 200-inch Mt. Palomar telescope. Aden's dissertation advisor, Nicholas

Mayall, however, was convinced that his student's calculations were correct and that a useful camera could result. Aden had no funds to order the necessary components from catalogs. In fact, there were no such catalogs for anything other than pedestrian optical components. He needed a fast spherical primary mirror of large aperture (a minimum of 6 inches), Schmidt corrector plate, and a special lens near the focus to flatten the field. All three optical elements would need to be made, but he knew he could do it because he had made several similar ones while working as an apprentice optician at the Mt. Wilson Observatory Optical Shop.[26]

Aden's dissertation was to identify the physical source causing auroras and explain why the night sky in the infrared was fogging the photographic plates that Baade exposed in order to image galaxies with the 100-inch telescope. To identify the atoms and/or molecules responsible for the lights in the night sky, Aden would need to integrate an optical element to disperse the light into his Schmidt camera for spectral analysis. The dispersive element was mounted over the top of the camera. Prisms were not practical because of the size and the necessary spectral resolution in the infrared. Aden's camera required one of the largest transmission diffraction gratings manufactured at that time. Where would he find the grating? Would he need to manufacture that also?

Edison Pettit suggested that he write to both R. W. Wood at Johns Hopkins University, who had made some excellent gratings, and John Strong. Pettit had worked with Wood during World War I, only because the Army did not know what to do with an astronomer and assigned him to the Signal Corps at Johns Hopkins to develop signaling techniques. While there, Pettit met Robert Goddard, who later built the first liquid-fueled rocket and made the Space Age possible. Aden decided to write to Strong. When Aden was at Caltech, Strong's laboratory was in the second basement of the Astrophysics Building, diagonally across the hall from where James Edson and the Planet Group had their laboratory. Aden had visited him often, once getting his autograph on his successful book, *Procedures in Experimental Physics* (see Chapter 2, section "Youth and education"). He was now a professor at Johns Hopkins. Aden asked Strong if he would ask Wood to locate a grating. On March 3, 1947, Aden received Strong's reply, asking whether a 14,000- or 15,000-line-per-inch grating bright in the first order red would be better for Aden's purposes. A box with grating 4 × 8 inches arrived soon thereafter.[27]

Aden still needed an optical polishing machine to complete his camera. There was none he could use in Berkeley, but then he remembered he had seen a machine in storage at Caltech, left over from the National Defense Research Council (NDRC) days. Bowen gave Aden permission to use it but stipulated that the work would have to be done elsewhere. Pettit then told Aden he could set up the polishing machine in a corner of his shop at his home on Villa Street in Pasadena over the summer of 1947. The glass for the Schmidt corrector plate was cut from a piece of 1/4 inch-thick window glass. It took about three weeks of polishing, figuring, and testing to finish a reasonably good Schmidt corrector plate, but Aden was disappointed with the image quality that the camera yielded. Still, it seemed adequate enough, and time had run out.[28]

The camera body (made of metal tubes turned by Aden) and brackets were made in Pettit's shop. It was an impressive, flat-field, coma-corrected Schmidt camera when Aden finally assembled it and examined the images. They were not the sharp points he had hoped for, but the field was flat, and the spectrum of an iron arc source looked reasonable. Aden could head for Berkeley to set up and test the spectrograph at Lick Observatory before classes started in the fall of 1947.[29]

Light of the night sky enters from the left to strike a 14,000 line-per-inch diffraction grating loaned to him by Wood (Figure 5.1). Light diffracted from the grating enters the Schmidt corrector of diameter 6.5 inches and passes to a spherical mirror of diameter 6.5 inches and radius of curvature 11 inches, with a highly reflective aluminum coating. The reflected light passes through a field-flattener

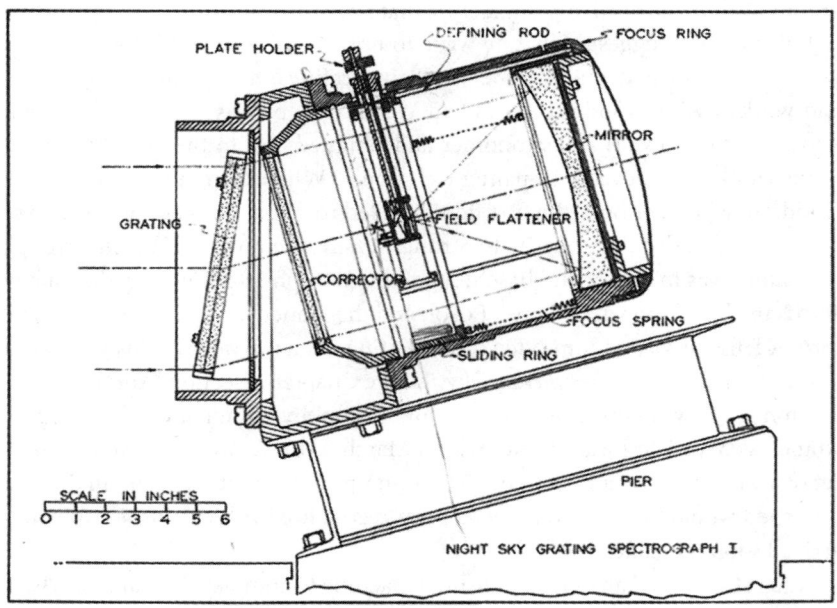

Figure 5.1. Flat-field Schmidt camera invented and built in 1948 by Aden Meinel for his PhD dissertation in 1949. He used this instrument to discover several physical processes in the upper atmosphere that limit astronomical telescope performance. Light enters the system from the left.
Drawing by Aden Meinel.

lens to converge to the focal plane at f/0.8. Aden designed, figured, and tested all the camera optical surfaces, including the zero-power, fourth-order figure on the Schmidt corrector. He designed and manufactured the metal mirror cells, support tube, structure, and the focusing mechanism using the lathe and small milling machine in Pettit's machine shop. The spectrograph dispersion was 257 Å/mm, and the size of the photograph plate was 0.75 × 1.25 inches. No one had ever made a Schmidt camera that was this fast (f/0.8), much less with a flat field.

While constructing the camera, Aden took a full load of classes during the 1947–1948 academic year, including three semesters of physics and six semesters of astronomy.

The qualifying exam

The qualifying exam is an examination of the student's retention of his classroom work, abilities for independent scientific investigation, and communication skills. It is typically given at the end of classroom work and before dissertation research begins. In Aden's case his dissertation research was well underway before this exam. His qualifying exam was scheduled for December 12, 1948, even before he had finished his coursework, but he was confident he could pass it.

The first two hours went well as Aden answered probing astronomy and physics questions, but at the opening question by the mathematics professor Aden suddenly felt his composure shatter and almost fainted at the blackboard. Although the committee members saw his problem and asked him to sit down for a minute, he still could not relay what he had learned about advanced mathematics. They excused him to go into the hallway and get a drink of water and await his fate. After a few minutes Harold Weaver, his astronomy advisor, called him back in to tell him that he had passed—with the recommendation that he retake the math course, and he resolved to do so. But when he saw the schedule for the next semester he discovered that the class conflicted with the atomic spectroscopy physics course that was vital to his dissertation. Aden requested and received permission to take the physics course; he never took that math course.[30] Time was running out. Funds from the GI Bill and the Lick Fellowship were not sufficient to support a growing family, which now included three young children. Aden needed to move on quickly to finish his research.

Eureka

His night sky spectrograph was finished, and Mayall told Aden to set it up at Lick Observatory as soon as possible. Observatory director Shane suggested he place it on top of the water tank on the highest rise just beyond the last house on Mt. Hamilton. Aden used the portable building that astronomers had used earlier to house a telescope on an early Lick expedition to Chile. At the end of summer all was ready, and a few test exposures with the comparison spectrum source were made. Multiple night exposures were needed because of the faintness of the light of the night sky, today called the airglow. Significant energy in the infrared night sky was revealed on several plates. However, a problem was that there appeared to be many emission lines blurred together into several groups, too blurred to identify any specific spectrometer features and assign them to atoms or molecules in the atmosphere. The camera/spectrograph lacked sufficient spectral resolution to resolve individual lines. Were these molecular emissions or atomic emissions? That question remained to be resolved later.[31]

Each exposure recorded the spectrum of the night sky along with a comparison spectrum with wavelengths that are well known: iron (Fe). Aden continued to work refining his camera and taking exposures of the sky from Lick Observatory in anticipation of finding anything of significance. But there were only blurred emission lines with a position that could not be measured. Not discovering something predicted makes for a subpar thesis.

Then, on the morning of August 8, 1948, when he developed the previous night's exposure, Aden saw something quite different—two sharp emission features. He walked over to the main building at the observatory and told Mayall, who had heard from the night assistant that there had been a faint aurora on the northern horizon (where Aden's spectrograph happened to be pointed). On the chance that Aden had captured that aurora, Mayall told him to measure the wavelengths and loaned him Charlotte Moore's wavelength tables for atoms of different elements. If it was an atomic emission, it might be in the tables; if not, it had a molecular origin.[32]

Aden measured the position of the lines on the photographic plate and found the wavelengths to be 7774Å and 8349Å. The closest candidates for those wavelengths were oxygen and nitrogen. He did not get as far as checking for N_2 bands, which had already been reported in auroras by Lars Vegard in Norway in 1930,[33] but two years later Aden did identify a new system of six bands of N_2 from an auroral storm on August 18–19, 1950 (see Chapter 6, section "Northern lights at Yerkes").[34] At the top of Aden's list for oxygen emissions were 7776Å and 8384Å, proving that auroras are caused in part by atomic oxygen emission (Figure 5.2; consult the color insert for this figure). Mayall told him to write it up and send the manuscript to the *Publications of the Astronomical Society of the Pacific* (*PASP*). Mayall sent an accompanying letter recommending immediate publication.[35] Aden prepared a full Grotrian diagram analysis of his spectra to show the allowed electronic transitions between atomic energy levels and thereby identify the atomic source of all the transitions indicated in his spectra. By now

Aden was at the end of his nervous energy. He doubted that he could have taken another course without having a nervous breakdown. And the family's bank account was approaching zero.[36]

Aden passed the defense of his dissertation exam at Lick Observatory and received his diploma from UC Berkeley on December 10, 1949. He needed a job—and soon. Mayall suggested that Aden apply for a grant from the newly established Office of Naval Research (ONR). The National Science Foundation was still years in the future.[37]

Office of Naval Research

Established after the war on August 1, 1946, and authorized under congressional Public Law 588, the ONR was the first federal agency to provide public support of merit-based science and technology research that benefits both the naval services and the nation. It issued a notification to academic departments saying they would be receptive to proposals for research work into the understanding of the physics of the atmosphere. The US Navy needed to understand the physical nature of Earth's atmosphere because changes in it caused communications blackouts. World War II had demonstrated the need for the US Navy to maintain secure and reliable communications system for all naval ships 24/7, no matter where on the globe, so predicting blackouts and forecasting their length was a high priority.

Aden had made many contacts with Navy scientists during his brief career at NOTS and as a member of the US Navy Caltech rocket program between 1943 and 1946. After his acceptance to the UC Berkeley astronomy graduate school and the award of his Lick Fellowship in the fall of 1947, Aden wrote to Shane that F. E. Roach from NOTS had approached him about his work on the night sky and had informed Aden that there was an "official directive and significant funds" to support an extensive program.[38] Aden's discovery and identification of emission lines in the atmosphere and the conclusion that they were caused by natural elements from the periodic table convinced the Navy to establish an aggressive program to understand their source. It would not be the first time that a discovery by Aden would lead to a major program in either science or national defense.

Marjorie and Aden were enthusiastic at Mayall's suggestion that they submit a proposal to continue work on lights of the night sky after Aden graduated. Together they wrote a proposal to ONR, the first of many proposals they would write together in their lifetime. It was approved by Mayall and then the university authorities, and their wait began. In the meantime, Mayall and Aden re-examined the spectra he had taken in detail, and both agreed that better images were needed to make further progress toward detailed identification. His Schmidt camera spectrometer needed to be upgraded or rebuilt.[39]

General Electric had just sent Mayall a 6-inch slab of fused quartz, a material being offered for the next large telescope mirror. Fused quartz has a much lower coefficient of expansion and better transmission in the ultraviolet than does Pyrex, the material used for the 200-inch mirror that was manufactured by Corning. Mayall had tested the substrate in transmission and found it had too much birefringence for the application he had in mind. But he mentioned to Aden that it might be an ideal material to make a new Schmidt corrector plate for his Schmidt camera/spectrometer. Aden decided to build his own optical figuring machine so that he could do the work at home and be closer to his family. The family was living in a small rental house on the estate (Lindcrest) of Earle Gorton Linsley in Oakland.[40] Linsley also had a small cottage down the hill from the Meinel house. The cottage was perfect for an optical shop. Progress on the optics was rapid, and a camera with superb definition resulted. They almost lost it, however, when a grass fire broke out farther down the hillside, and Aden was able to stop the flames just as they reached the cottage.[41]

To test the new camera, Mayall suggested putting it on the 36-inch Crossley reflector at Lick and recording a star field. Aden made a wooden frame to attach it to the top of the telescope, and they took a 10-minute exposure of M31, the Andromeda Galaxy. He developed the photographic plate and held it up to the darkroom light. After a cursory look, Mayall exclaimed that there were no stars, that the image was gray all over, and asked Aden if he had focused it. Aden assured him that he did, and it had to have star images on it. Then Mayall picked up a 10× hand loupe and saw thousands of stars, so small and so many that the overall image appeared gray and foggy.[42]

Aden's new spectrograph was ready for its next journey. Now with the replacement optics completed, it came time to contemplate leaving their home and moving on to their next adventure. Aden and Marjorie came to the same conclusion: Why not go to a university in the north where auroras are common occurrences?

At that time the University of Chicago operated Yerkes Observatory, which is 80 miles north of Chicago into Wisconsin, where auroras are frequently observed. Marjorie's parents earned their PhD astronomy degrees at Yerkes in the early 1920s. In 1947 Otto Struve had turned the directorship of Yerkes Observatory over to Gerard Kuiper, a 42-year-old Yerkes staff astronomer who was engaged in the first infrared spectroscopy of the planets.[43] But Struve remained at the observatory to continue his research. Aden wrote quickly to Kuiper, asking if Kuiper remembered him from Alsos and the Royal Monceau Hotel in the spring of 1945 (see Chapter 4, section "New orders") and if he and Marjorie could come to Yerkes to study auroras with their own spectrograph and

possible funding on an ONR grant. He enclosed a copy of his paper on the discovery of atomic emission lines in the infrared aurora.[44] Kuiper did remember.
The answer was an enthusiastic "Yes."[45]

Aden and Marjorie now finished their proposal to ONR, planning to go first
to Yerkes and then on to Saskatoon where Gerhard Herzberg had joined the
University of Saskatchewan.[46] While waiting for word about their ONR grant
application, exciting things began to happen at Lick Observatory; plans for a new
large telescope, rivaling that at Mt. Palomar, were moving forward. In 1946 the
State of California appropriated funds for a large astronomical telescope for the
University of California.[47] Aden believed he would find a role in the develop-
ment of the 120-inch telescope back at Mt. Hamilton after his one-year work at
Yerkes was over, but fate intervened.[48] He went on to serve as the first director of
Kitt Peak National Observatory and the Optical Sciences Center in Tucson.

It took several months after submission of the proposal before they would
learn whether or not ONR would fund the research work. Weather is good on
Mt. Hamilton in the fall, and a few nights were occupied with observing with the
new camera. But the excitement over the prospects of a 120-inch telescope would
soon occupy them.

Bowen suggested they use the 120-inch Pyrex disk left over from making the
200-inch mirror.[49] Shane asked Aden if on his next trip to Pasadena he would inspect that mirror blank stored at Caltech. It was intended to be a test flat for the
200-inch, but Ross designed a "null lens" that enabled testing of the required parabolic shape from the center of curvature, a great step forward in making large
telescope mirrors. Aden stood there, awed by the size of that lightweight, ribbed
mirror blank. Yes, it could be used, but the curve would need to be shallow because of the 4-inch thickness of the front plate and thus the f/ratio of the telescope, and the dome would need to be larger than usual.[50] Ten years later, Aden
would successfully develop technology that enabled re-heating a lightweight,
80-inch-diameter, fused silica mirror blank to slump the entire mirror assembly
to the concave "near net shape" needed for a short radius-of-curvature primary mirror, the KPNO 84-inch (see Chapter 7, subsection "The mirror") as an
example.

James G. Baker, a Harvard astronomer and leading expert in optical design,
was hired as a consultant to the 120-inch project by Shane. The pending telescope was exciting enough to compel Baker to move to Orinda, California, in
1948 to help Shane with both the design of the 120-inch and the red lens for
the 20-inch photographic astrograph, twin to the existing blue one. Shane decided to have a conference at Lick Observatory and invite visitors to review his
plans. Baker wanted to ride up with some of the VIPs and asked Aden if he would
drive Baker's car to the mountain. Aden was delighted because he had not been
invited as a participant but could come as an observer. He almost didn't make

the conference because it snowed during the night. Baker's car had worn tires, too smooth to get a grip on the road, so Aden parked it a mile below the observatory and trudged through the snow with Baker to attend the conference. When Aden indicated his hope to become more closely involved with the 120-inch project, Shane was noncommittal, but offered encouragement. Everything hinged on that proposal to ONR.[51]

One afternoon when Aden returned home, Marjorie was standing at the edge of the driveway holding the announcement that they were awarded the grant, with a salary that repaid all those years of carefully watching where every penny was spent. The car that they had bought with Aden's Navy TAD check was not roadworthy for the trip to Wisconsin, so they borrowed money to buy a new one. They loaded up the spectrograph, books, and few belongings into a trailer loaned by Aden's father, packed their three children into the car, and headed over the Donner Summit of the Sierras and eastward across the Great Basin Desert toward new horizons in Wisconsin.[52]

Notes

1. Meinel, A. B., and Meinel, M. P. 2002a. *Echoes from a simpler time*, unpublished, 50.
2. Personal communication from A. Meinel and M. Meinel to J. Breckinridge.
3. Meinel and Meinel, *Echoes*, 50.
4. Records of the Department of Astronomy, 1882–1960, CU-25, Box 10, University of California.
5. C. D. Shane (1942), Helen Bard Pettit graduate student file, Records of the Department of Astronomy, 1882–1960, CU-25, Box 10, University of California.
6. Meinel and Meinel, *Echoes*, 50.
7. Meinel and Meinel, *Echoes*, 51–2.
8. Meinel and Meinel, *Echoes*, 50. Aden began summer school on June 24, 1946 (UC Berkeley Summer Catalog, 1946, Bancroft Library).
9. Meinel and Meinel, *Echoes*, 50.
10. Meinel and Meinel, *Echoes*, 51.
11. Program of the Final Examination for the Degree of Doctor of Philosophy, Records of the Department of Astronomy, 1882–1960, CU-25, Box 5, University of California. Examiners: Trumpler, Henyey, Weaver, Birge, and Neyman.
12. Meinel and Meinel, *Echoes*, 52.
13. Letter from J. H. Moore, Lick Observatory, Students Observatory UCB, to John D. Hicks, Dean of the Graduate Division, justifying Lick Fellow for Aden Meinel; Records of the Department of Astronomy, 1882–1960, CU-25, Box 10, University of California; Meinel and Meinel, *Echoes*, 52.
14. Meinel and Meinel, *Echoes*, 51.
15. Meinel and Meinel, *Echoes*, 51.

16. Meinel and Meinel, *Echoes*, 52.
17. Throughput is a quantitative scientific term used by optical scientists to describe the light-gathering power of an optical system to record faint extended scenes. It is calculated as the area of the collecting aperture multiplied by the solid angle of the FOV. The units are cm² Steradians.
18. Meinel and Meinel, *Echoes*, 52.
19. Ingalls, A. G. 1996. *Amateur telescope making, vol. 2, Part E:* Schmidt cameras, pp. 421–517. Willmann-Bell, Inc., Richmond, Virginia.
20. Breckinridge, J. B. 2012. *Basic optics for the astronomical sciences.* SPIE Press, Bellingham, Washington, 95–7.
21. Morgan, W. W. 1967. *Frank Elmore Ross, National Academy of Sciences, Biographical Memoirs* 39: 391; Osborn, W. 2012. Frank Elmore Ross and his variable star discoveries. *Journal of the American Association of Variable Star Observers* 40: 133–40.
22. Conrady, A. E. 1929. *Applied optics and optical design.* Oxford University Press. Reprinted 1957, 1988, 1992 by Dover Publications, New York; Conrady, A. E., and Kingslake, R. 1960. *Applied optics and optical design*, Part Two. Dover Publications, New York..
23. Shack, R. V. 1974. The use of normalization in the application of simple optical systems. *Proceedings SPIE* 54: 155–62. https://doi.org/10.1117/12.954238; Gardner, I. C. 1927. Application of the algebraic aberration equations to optical design. *Scientific Papers for the Bureau of Standards* 22: 3–202. US Government Printing Office.
24. Meinel and Meinel, *Echoes*, 52–3.
25. Meinel and Meinel, *Echoes*, 53.
26. Meinel and Meinel, *Echoes*, 53.
27. Meinel and Meinel, *Echoes*, 54.
28. Meinel and Meinel, *Echoes*, 54–5.
29. Meinel and Meinel, *Echoes*, 55.
30. Meinel and Meinel, *Echoes*, 55.
31. Meinel and Meinel, *Echoes*, 55.
32. Meinel and Meinel, *Echoes*, 56.
33. Vegard, L. 1930. Die Spektren verfstigten Gase und ihre atomtheoretische Deutung. *Annalen der Physik* 6: 487–544.
34. Meinel, A. B. 1950a. A new band system of N_2^+ in the infrared auroral spectrum. *Astrophysical Journal* 112: 562–3.
35. Meinel, A. B. 1948. The near infrared spectrum of the night sky and aurora. *Publications of the Astronomical Society of the Pacific* 60: 373–8.
36. Meinel and Meinel, *Echoes*, 56.
37. Meinel and Meinel, *Echoes*, 57.
38. Letter from A. Meinel to C. D. Shane, December 24, 1947, Box 442, Lick Observatory Archives, University of California at Santa Cruz.
39. Meinel and Meinel, *Echoes*, 58.
40. Earle Gorton Linsley, a UC Berkeley PhD graduate (1938), was a world-renowned expert on the beetle family Cerambycidae. He was an emeritus faculty member at UC

Berkeley and emeritus director of the Chabot Science Center in Oakland during the time Aden Meinel was a graduate student.

41. Meinel and Meinel, *Echoes*, 58.

42. Meinel and Meinel, *Echoes*, 58.

43. Osterbrock, D. E. 1997. *Yerkes Observatory, 1892–1950: The birth, near death, and resurrection of a scientific research institution.* University of Chicago Press, 288.

44. Meinel, The near infrared spectrum, 373–8.

45. Meinel and Meinel, *Echoes*, 59.

46. Gerhard Herzberg was at that time one of the leading atomic spectroscopists in the world and had just completed writing *Atomic spectra and atomic structure.*

47. Shane, C. D. 1964. Lick Observatory: The first 75 years. *Publications of the Astronomical Society of the Pacific* 76: 77–87.

48. Meinel and Meinel, *Echoes*, 59.

49. Meinel and Meinel, *Echoes*, 59.

50. Meinel and Meinel, *Echoes*, 60.

51. Meinel and Meinel, *Echoes*, 60.

52. Meinel and Meinel, *Echoes*, 60.

6

Genius Blooms

Aden's preparation for the future was exceptional. At the age of 27 in 1949, Aden had apprenticed in both optical and machine shops, trained as a draftsman, researched and developed rocket technology, experienced military leadership during war firsthand on the front lines, demonstrated resourcefulness under extreme pressure, and completed a PhD in astronomy in record time. For his dissertation, Aden designed and built his own new innovative telescope/spectrometer system and applied it to discover the origin and characteristics of Earth's northern lights (Aurora borealis). He was well prepared to make profound contributions to telescopes and instruments for national defense reconnaissance and ground and space astronomy, as well as lay the foundation for modern solar energy engineering and establish the premier institution in the United States for education and research in optical sciences and engineering. Much later, at NASA/JPL, he would inspire a whole new generation of young scientists and engineers in space-based investigations for astrophysical and Earth sciences. Aden had extensive life experience, not only leading military troops on clandestine missions, but also working with skilled machinists, contractors, temperamental personalities, as well as Nobel Prize–winning physicists. He communicated with them all, and they each eventually had respect for his genius.

In the week of November 21, 1949, Aden and Marjorie and three children were driving their new car along the Wisconsin road from Delavan toward Williams Bay, Wisconsin, towing the small trailer that her father Edison Pettit had loaned them. In addition to family possessions, the trailer contained Aden's spectrograph, photographic plates, and small library. For Marjorie it was a homecoming. She had spent several summers in the mid-1930s as a young teenager with her parents in Michigan while they recorded motion pictures of solar prominences at the McMath-Hulbert Observatory, and her father built instruments for the observatory.

Robert R. McMath (1891–1962) was a businessman with years of experience working in industry leading commercial technical product developments. He held a BS degree from the University of Michigan in civil engineering but was an avid amateur astronomer whose work approached professional quality. He had developed an excellent business sense and how to motivate people (peers and employees) at all levels within a business environment. McMath achieved success in the astronomical community because of his work funding innovative

With Stars in Their Eyes. James B. Breckinridge and Alec M. Pridgeon, Oxford University Press. © Oxford University Press 2022. DOI: 10.1093/oso/9780190915674.003.0006

astronomical instrumentation for solar astronomy, recording and interpreting data on the dynamics of solar prominences, and collaborating with Edison Pettit, Marjorie Meinel's father, at Mt. Wilson Observatory. He received an honorary doctorate from Pennsylvania Military College in 1941 and served as president of the American Astronomical Society from 1952 to 1954 and was elected to the prestigious National Academy of Sciences (NAS) in 1958. While a member of the NAS he served on several influential committees, advising the government on scientific issues.[1]

As the Meinels traveled across Wisconsin by car through the November corn fields, suddenly the orange dome of the Yerkes Observatory's giant 40-inch refractor rose above the corn stalks, far different from the silver-domed observatories that Aden and Marjorie knew at Mt. Wilson and Lick Observatories. Staff members were expecting them and had reserved an apartment near the town center for them. The observatory is about 75 miles northwest of Chicago near the shore of Geneva Lake and close to the small town of Lake Geneva, Wisconsin, which in 1950 had a population of 4,300.[2]

After Aden was awarded his ONR grant to study auroras, the position at Yerkes Observatory where Marjorie's parents once worked was his first real professional job. His experiences at Yerkes framed his future career as founding director of first the nation's astronomical observatory and then the Optical Science Center at the University of Arizona. It is therefore important to understand the origins and culture of the University of Chicago's Yerkes Observatory and how these affected his career.

Yerkes Observatory, widely acclaimed as the birthplace of modern astrophysics and once home to Nobel laureates, was founded on October 7, 1892, when George Ellery Hale (see Chapter 2) and William Rainey Harper (then president of the University of Chicago) approached financier Charles T. Yerkes in Chicago to seek funding for the largest telescope in the world. Yerkes, who had made his fortune in streetcar and railroad franchises in Chicago and London, was only too happy to have his name attached to the grandiose project and gave them carte blanche that same day, probably to create a legacy of philanthropy and thereby counter his widespread reputation as a scoundrel. By its completion in 1897, the construction and telescope costs amounted to more than $500,000, equivalent to approximately $15 million today. A real estate speculator named John Johnston, Jr., donated 50 acres on the north shore of Lake Geneva, far from the oppressive smog of Chicago but within commuting distance.[3]

Hale learned of the availability of a 40-inch glass disk that Alvan Clark and Company had already begun grinding for another refracting telescope. He quickly seized the opportunity to surpass the 36-inch refractor at Lick Observatory on Mt. Hamilton near San Jose, California. Clark, his optician Carl Lundin, and one of the University of Chicago trustees personally escorted the well-packed lens in its own railroad car on a train designated for that sole purpose. It was as if they were guarding a special shipment of gold bullion. Hale and staff member Edward Barnard met them at the Williams Bay station and saw to it that the lens and its cell were unloaded into a horse-drawn wagon and carted over to the observatory. The next day Clark and Lundin installed the lens and its cell into the Warner & Swasey equatorial mount with assistance from optician George Ritchey, whom Hale had hired to grind, figure, and polish lenses.[4]

Before he left for Mt. Wilson, Hale hired some of the leading astronomers and astrophysicists of the day to work there: Sherburne Burnham, Edward Barnard, Edwin Frost, Walter Adams, and Frank Schlesinger. Edwin Frost succeeded Hale as Yerkes director, and during his tenure six eminent astronomers completed their doctorates at the University of Chicago/Yerkes: Edwin Hubble, Otto Struve, Christian Elvey, Nicholas Bobrovnikoff, William W. Morgan, and Philip Keenan. Frost suffered a detached retina in his right eye, followed by a cataract and hemorrhage in his left eye and was essentially blind by 1921. Nonetheless, he carried on his duties at a reduced level until retirement and his death in 1935. Struve succeeded Frost in 1932 and served until 1947; his tenure is considered by many the Golden Age of Yerkes (Figure 6.1). It was Struve who hired future Nobel laureate Subrahmanyan Chandrasekhar as well as Gerard Kuiper and Bengt Strömgren.[5]

Kuiper, who served as Yerkes director between the administrations of Struve and Strömgren (1947–1950) and then again after Strömgren (1957–1960), welcomed the Meinels with an invitation to have Thanksgiving dinner with his family at his home nearby. Marjorie was relieved that she did not need to cook the holiday meal for their family of five in an unfamiliar kitchen. Kuiper told Aden to come to the observatory on Monday to begin to settle into his office.

Gerard Kuiper (1905–1973), whom Aden first met in Paris toward the end of the war (see Chapter 4, section "New orders"), was born in northern Holland and attended Leiden University with Bart Bok. He then spent two years on a postdoctoral fellowship at Lick Observatory specializing in binary and white dwarf stars. After a brief stint first at Harvard University and then Bosscha Observatory in Java, he was hired in 1936 by Struve as assistant professor at the University of Chicago, where he taught courses on the solar system and statistics and dynamics among others. In Dutch the "ui" in his surname is pronounced "oui," a sound halfway between a 'u' and an 'i'. There is no true equivalent in English, and so his colleagues simply pronounced it with a long 'i' as KIE-per. His discovery that a majority of stars are actually two or more stars orbiting one another eventually led to groundbreaking work on stellar evolution and galactic structure. He became full professor in 1943. Up to the time when Aden joined Yerkes/McDonald, Kuiper had already determined from infrared spectroscopy that carbon dioxide is a principal part of the atmosphere of Mars, the rings of Saturn

Figure 6.1. Yerkes observatory staff, students, and visitors taken in the fall of 1950. Front row from left: Lyman Spitzer, Jr., Paul W. Merrill, Subrahmanyan Chandrasekhar, Otto Struve, Dirk Brouwer, Gerard P. Kuiper, Otto Heckmann, and Nicholas U. Mayall. Second row: William P. Bidelman (just behind Merrill), Guido Münch between Chandrasekhar and Struve, T. D. Lee between Struve and Brouwer, Karl-Otto Kiepenheuer between Kuiper and Heckmann, and W. Albert Hiltner at far right. Su-shu Huang is behind and to the right of Lee. Second row from rear from left: Stewart L. Sharpless, Irene L. Hansen (now Osterbrock), and Donald E. Osterbrock. Thornton L. Page is the tall man near the center of rear row, William W. Morgan is second to right of him, and Aden B. Meinel at far right of that row. Others in photo but not identified: J. Ramsey, R. Glenn Hall, A. Brown, L. Risino, R. Frerichs, L. Anderson, H. L. Helfer, F. Reno, R. Hardie, Frank N. Edmonds, J. G. Phillips, B. Stephenson, W. Fitch, Henry G. Horak, D. Duke, Charles Ridell, C. Robinson, Fred Pearson.

University of Chicago Photographic Archive [apf604331], Special Collections Research Center, University of Chicago Library. Some identifications by Helmut Abt.

are composed of ice, and Saturn's moon Titan has a methane-rich atmosphere. He had discovered the fifth moon of Uranus (Miranda) in 1948 and the second moon of Neptune (Nereid) in 1949 and predicted a vast region of icy objects extending from Neptune's orbit out to 50 AU from the Sun (one astronomical unit [AU] equals about 93 million miles/150 kilometers, the mean distance from

Earth to the Sun). It is now known as the Kuiper belt or the Edgeworth-Kuiper belt. While working at McDonald Observatory with the 82-inch telescope, he measured the diameter of Neptune and also began his benchmark studies of the lunar surface, culminating in the Photographic Lunar Atlas in 1960.[6]

Aden's office was on the first floor in the circular area under the 24-inch telescope, on the east side. He was one of three young astronomers who had desks in this area. Edison Pettit, Edwin Hubble, and some other students had used it for their bedrooms. According to Pettit, Hubble even slept with his head in the clock-well recess—until one night when he wasn't sleeping there the cable holding the weight that drove the 24-inch telescope came crashing down onto the bed. Hubble's blossoming career could have ended right there.[7]

The scientific strengths of Aden's dissertation were his construction of a new instrument that opened a new area of inquiry and his application of the principles of atomic spectroscopy to identify the chemical composition of upper Earth's atmosphere and its physical structure. Was there a possible a relationship between particles from the Sun and Earth's atmosphere?

When Aden was in Europe during World War II, Marjorie had sent him a copy of the new book by Chandrasekhar on stellar atmospheres. Now Aden, a new employee of the observatory, finally met Chandrasekhar himself.

Subrahmanyan Chandrasekhar (1910–1995), who was awarded the Nobel Prize in Physics in 1983 for his work on the structure and evolution of stars, was born in Lahore, Punjab, British India (now Pakistan), to a Brahmin family; he was the nephew of C. V. Raman, who won the Nobel Prize in Physics in 1930 for his work on light-scattering. Like his uncle, Chandrasekhar attended the University of Madras, after which he received a fellowship to study astronomy and physics at the University of Cambridge in the United Kingdom. Harlow Shapley, director of Harvard College Observatory, invited him to come to Boston as a visiting lecturer. Kuiper met him there and enthusiastically recommended to Struve that Chandrasekhar be invited to join the faculty as research associate.[8]

Struve then wrote to the president of the University, Robert Hutchins, saying that "Chandra is brilliant but his political views are pretty radical." Struve also mentioned what Shapley said about him: "You probably know that Chandrasekhar is extremely dark, but he has the bearing of an aristocrat."[9] Dean of Physical Sciences, Henry Gale, had been raised in segregated Chicago and thought of anyone with dark skin as Black. Consequently, he was opposed to the hiring of Chandrasekhar but did not object, knowing that Hutchins would ignore any protests.[10] Hutchins extended the offer in March 1936, which Chandrasekhar gratefully accepted in a Western Union telegram.[11]

Apart from winning the Nobel Prize in Physics with William Fowler of Caltech in 1983, Chandrasekhar overcame such covert racism and succeeded in becoming one of the foremost astronomers in the world for his brilliant work

on the evolution of stars, especially white dwarfs, but also for stellar structure and dynamics, radiative transfer, general relativity, and colliding gravitational waves.[12] Chandrasekhar was recognized as one of the greatest theoretical
physicists of all time, but his colleagues knew him to be rather inept as an instru-
mentalist and experimentalist. It is rare that a person is excellent at both. Aden
excelled in instrument development. Chandrasekhar appreciated Aden's energy,
experimental background, and his understanding of theory. He would also become one of Aden's most ardent advocates at Yerkes.

Chandrasekhar told Aden that Gerhard Herzberg, the world's leading expert
in atomic spectra, was no longer at the observatory.[13] But fortunately Herzberg
had sent the page proofs of his latest book, the second edition of di-atomic
spectra, to Chandrasekhar, who shared them with Aden. It was indeed a treasure,
so much better than the books Aden had read on the topic at Berkeley.[14]

The contents of the Herzberg book provided Aden with the analytical tools
he needed to interpret the wavelengths measured on the photographic plates
recorded by his spectrometer and convert them into an understanding of the
atomic and molecular structure of the source. This enabled him to identify the
unique atom or molecule in the atmosphere that was responsible for the signature in his recorded spectrum. Aden was highly adept at reading papers and
books and deriving skills useful to him in his research. He learned to do this
when reading textbooks and testing out of long classes at UC Berkeley.

Aden's spectrograph was temporarily in the basement hallway of the observatory so that it could be reassembled and checked out. He decided to put it up
on the roof to be pointed anywhere in the sky; only a tarp covered it from the
weather.[15] The new spectrograph enabled high-resolution spectra of the night
sky obtained with relatively short exposures. Because of Aden's ingenuity, the 32-
hour exposure at Lick became a 4-hour exposure at Yerkes. Aden attributed the
difference to the hypersensitivity of Eastman 1-N emulsion.[16]

Living in the small village of Lake Geneva, Wisconsin, was a pleasant change
from both Pasadena and Oakland in many ways. One could walk the length and
breadth of the village in a few minutes' time. Their apartment, which consisted
of four rooms in a large two-story house, was a two-minute walk to the grocery
store and post office, three minutes to church, and five minutes to the observatory.
Everyone was friendly and just a bit surprised to see that they were ordinary folks.[17]

They settled in and awaited the first snow, but it seemed slow in coming. Aden
and Marjorie were southern California natives, and in the first 27 years of their
lives they had experienced only "cold-wet" weather when living in Oakland. To
enjoy snow in California they would travel to the mountains, often for adventures
lasting only one day. They now anticipated living with snow and ice day after day.

The Wisconsin weather in the fall of 1949 was unusually warm, and Aden
and Marjorie began to wonder how snow could even stick to that warm ground.

Day after day they watched the weather program on Chicago television to see if a storm was finally coming. There was a sled outside their porch belonging to the landlord, so there was visual proof that snow must come sometime. It finally did, but not until after 1950 rolled around, when it kept falling and falling. The street next to their house had a gentle slope, just steep enough for sledding. The three children loved to be pulled on the sled and take turns going down the hill. After the children were sound asleep, Aden and Marjorie would often take the sled and go sledding themselves by the light of one streetlight. They even tried ice skating on Williams Bay, but the surface froze too rough that year and made skating difficult. The villagers thought these Californians were just a bit strange about snow.[18]

Living in Williams Bay was a change. Mountains are nowhere to be seen. From the top of the observatory the vista is farmland to the horizon and a wooded slope down to Lake Geneva. Townspeople kept referring to "the observatory up on the hill." Aden and Marjorie didn't see anything that could be called a hill—unless you viewed the observatory from down at the shoreline of Lake Geneva. Meanwhile Aden was establishing a routine at the observatory, and Marjorie was making a home for three children with a fourth on the way.[19]

The summer of 1950 they drove back to California for a brief period. Aden worked in Edison Pettit's shop to make minor repairs to his spectrometer. Upon returning to Yerkes that fall, they got a pleasant surprise. One of the three small "prefab" houses would soon be vacated, and they were offered the one where astronomers Glenn Hall and Guido Münch had lived, just a short walk to the observatory. It had only two bedrooms for their family of five, but at least it had a large basement. They decided to ship their goods, sell their car in California, and then take the train to Chicago and on to Williams Bay. They arrived six hours late in Chicago and missed their train connection but found a bus home, only to discover that their furniture had not yet arrived. Friends loaned them mattresses and bed clothes so that the five of them could sleep on the floor, but they were glad to be in their own home at last.[20]

Appointment to lecturer

Aden's salary for 12 months came from his ONR research grant, which ended December 31, 1949. As chairman of the Department of Astronomy, Struve called Aden into his office in the fall of 1949 and offered him a permanent tenure-track position as lecturer at the University of Chicago. Aden asked for and received the same salary as he had on his ONR grant.

Otto Struve (1897–1963) was born in Kharkov, Ukraine, and attended the university there before joining the army and fighting the Turks in the Caucasus, who were allied with the Germans, Bulgarians, and Austro-Hungarians as the Central Powers in World War I. After the Treaty of Brest-Litovsk was signed in 1918, Struve returned to university only to find himself fighting again in the Russian Civil War after Czar Nicholas II abdicated and was later murdered, along with the rest of his family, by the Bolsheviks. Struve enlisted as an officer in the White Army to fight the Red Army and was wounded before being evacuated to the Crimea and then Turkey. The successor to his uncle, who had been director of the Berlin Babelsberg Observatory, wrote to Edwin Frost and with Frost's support secured for Struve an assistantship in stellar spectroscopy at Yerkes, a field that would dominate his research career. By 1927 he had become an assistant professor at the University of Chicago and a naturalized citizen in the United States. Five years later he became director of Yerkes. From 1932 to 1947 he also served as editor of the *Astrophysical Journal*.[21] Among his many accomplishments and discoveries were ionized interstellar hydrogen, stellar rotation, and the correlation between rotational speed and temperature as part of stellar evolution theory.

Aden was happy to be a salaried astronomer after being a student for so many years. Prospects of any permanent appointment at an observatory in those days were remote because tenure positions only opened up when someone moved, retired, or died.[22] No one could have known how many new positions would sprout for young scientists after the launch of Sputniks 1 and 2.

It was so cold on the roof of the Observatory building that Aden's hands became too numb to do any useful work. He asked to put the spectrograph in the large transit room at the east end of the main hallway. In this room the north vertical shutter could open to expose the northern sky where auroras appear, and Aden could stand indoors out of the cold and wind.[23] He had built an instrument with unique capabilities. No one else had equipment that could record spectra of the night sky at the quality that his could (Figure 6.2). Throughout the early 1950s the instrument was unmatched, and it didn't take long to get exciting results or for others to build competing instruments.

The spring, summer, and fall of 1950 brought forth a torrent of scientific discoveries. Aden published nine peer-reviewed papers during 1950 in three journals: the *Astrophysical Journal, Physical Review*, and *Science*. One new paper was submitted to a journal about every six weeks.

The Meinel infrared bands

His first all-night exposure at Yerkes yielded a high-resolution spectrogram of the light of the night sky. The exposure showed that those groups of ill-defined emission lines he had obtained for his dissertation now were clearly resolved into an obvious molecular spectrum. He took the dry plate to the microdensitometer

Figure 6.2. Aden Meinel ca. 1952 shown in his office at Yerkes Observatory using a low-power binocular microscope to measure a spectrum of the night sky recorded by his Schmidt camera from the roof of Yerkes Observatory.

University of Chicago Photographic Archive [apf104375], Special Collections Research Center, University of Chicago Library.

(which, as the name implies, measures microscopic optical densities) and ran off a 10-foot tracing of the many molecular rotational lines. Showing it immediately to Marjorie, they could see that all of the many spectral lines clearly belonged to the same system.[24]

Referring to Herzberg's page proofs loaned to him by Chandrasekhar and seeing the wide spacing between the rotational lines showed that the source had to be a low-mass molecule: a hybrid. Measuring the wavelengths in the spectrum and knowing the molecular constants, it clearly had to be due to the hydroxyl radical, OH. Aden quickly drafted text that Marjorie edited, and they both hand-carried it to the editor (W. W. Morgan) of the *Astrophysical Journal* in an office just a few steps away on the main hallway at Yerkes. Morgan said he would insert it in the next issue, announcing the discovery and identification.[25] They sent copies to their friends in the airglow field and settled down to completing the identification: to what molecular transition system did these new spectral lines belong? Was it a new band system? It became apparent that there was no room for a new band system of the hydroxide molecule. All the theoretical ones had been discovered, yet the array of these spectral lines did not agree with the tabulated constants for any known bands. The structure said it had to be a doublet p to double p transition. But that seemed odd to them. They sent a letter off to Herzberg, then the director of physics at the Canadian National Research Council, asking his opinion. A letter back from Herzberg three days later cleared up the matter. It was indeed a p transition but within the same electronic transition, not a transition between electronic states, a rotation-vibration transition within the ground state of OH.[26]

A puzzle remained. The bands all seemed to originate from the ninth vibrational level, cascading downward. Aden sent this new interpretation to his friends. Then a letter came back from David Bates in England.[27] He had identified theoretically the chemical reaction in the high atmosphere that perfectly explained why the ninth level was the strongest. The reaction was between an ozone molecule and a hydrogen atom. The 3.2 electron volts of energy, which corresponded to the wavelength of the band-head, was exactly tuned to the ninth level of OH. The fact that it was only a two-body reaction explained the unusually strong intensity of these OH bands.

Northern lights at Yerkes

On a warm mid-August evening, a graduate student of Chandrasekhar named Don Osterbrock called to report that an aurora was beginning and brightening rapidly. It was the night of August 18–19, 1950, famous in the annals of upper atmosphere research. During the intense auroral activity on the nights of August 18–19 and 19–20, 1950, Aden recorded several spectra with his spectrograph. Aden went to the darkroom and cut several postage-stamp-size glass photographic plates from a larger fresh piece of Kodak 1-N emulsion for his Schmidt camera. He hyper-sensitized them and headed to the transit room. In the previous winter Aden's spectra indicated that protons aligned with the magnetic field of Earth were the source of energy driving the aurora glow. But to prove it, Aden needed not only spectra looking up into Earth's magnetic zenith along the magnetic lines of force, but also spectra recorded normal to these magnetic lines of force.[28]

The roof slit of the Transit Room was sealed. No transit instrument had ever been installed during the 56-year history of the observatory, and the doors over the opening were covered with roofing paper. Aden decided to force open that slit, hoping that over the years the tar paper had become weakened. Aden pulled hard several times to no avail until he heard a ripping noise. The slit doors slowly opened to reveal a bright auroral display just exactly where the spectrograph could record it, straight overhead.[29]

The spectrograph was pointed toward the magnetic zenith for three spectra and toward the north magnetic horizon for five spectra. Some spectra from both orientations showed H-alpha (Hα) in considerable intensity.[30] However, on one photographic plate which did not show Hα with appreciable intensity, they found a bright nearby spectral line. It did not take long for them to realize that the source of this feature was Hα but doppler-shifted to the blue. These spectra, recorded of the same auroral event, were obtained with respect to the geomagnetic coordinates. They enabled an unambiguous evaluation of the nature of auroral hydrogen. The profile of Hα observed from the magnetic zenith was seen to be asymmetrical, with the entire line shifted to the violet. The profile of Hα in the magnetic horizon is symmetrical but broadened. Aden reported that the velocity profile of Hα in the zenith cannot indicate a real velocity spread of the incident protons because of their nearly simultaneous arrival in the auroral zone. The profile, therefore, must be a consequence of the loss of energy of the protons upon entering the upper atmosphere.[31] The violet wing indicates that the velocity of the protons before entering the atmosphere is greater than 3,300 km/sec, corresponding to an energy of 57 keV. Although Aden did not directly state it, the obvious source of these charged particle protons at these energy levels was the Sun. This was the first direct measurement that proved that Earth is affected by particle fields from the Sun, a conclusion so controversial that Chandrasekhar had to write to the editor of *Science* twice to get the paper published.[32] He also wrote to the editorial secretary of the American Association for the Advancement of Science (AAAS) on October 2, 1950, saying, "I have no doubt that Meinel's discovery will be considered one of the most important in Geophysics and I have no doubts either that NATURE or the Physical Review would have given it the preference which is deserves and which you have not."[33] Chandrasekhar told the media of the discovery, and it was not long before *Time* magazine quickly published the news that hydrogen particles were racing toward Earth from the Sun at about 1,800 miles per second.[34]

When Aden had time to look at the microdensitometer tracing from that long exposure, he found that the usual molecular nitrogen bands were overexposed, but there were several other spectral bands that were only faintly seen before. Measuring their position on that long tracing with a meter stick, Aden and Marjorie saw the answer: a new set of molecular bands! The rotational lines were close together, so it was not a hydride. Aden and Marjorie knew that there was a missing molecular level in ionized molecular nitrogen, and it turned out that this was that missing band system, the Meinel bands of N_2^+.

At its annual meeting of the Optical Society of America held in October 1950, Aden received the distinguished Adolph Lomb Medal, which is presented to a person who has made a noteworthy contribution to optical sciences at an early career stage. Contributions from any area of optics, fundamental or applied, are considered. Candidates must be within 10 years of the completion of their highest degree earned in the year the medal is presented. In Aden's case, the medal was presented a short 13 months after the award of his doctorate, a record never broken to this day.

Management changes

Otto Struve resigned on December 31, 1949, and accepted an appointment to head the Department of Astronomy at UC Berkeley beginning June 30, 1950.[35] In 1947 Struve had promoted himself to chairman of the University's Department of Astronomy and recommended Kuiper as the new director.[36] Struve's departure from Yerkes was a shock to Aden, who had yet to come up for tenure and needed the votes of all the faculty to support his promotion.

It is well known that the professional astronomers at Yerkes did not get along well with each other. However, it appears that all had great respect for Aden, which accelerated his tenure. There had been some particular acrimony between Struve and both Kuiper and Chandrasekhar, but that was hardly news. Chandrasekhar wrote to Hutchins: "Mr. Struve, by his lack of confidence in the integrity of his associates and by his manner during the last two years (and particularly during the last months) succeeded in losing the regard of all of his colleagues and it was positively a relief to see him go."[37] Kuiper would have many more such stormy relationships during his lifetime, eventually even with Aden. But now it meant a major change at Yerkes in the person of Bengt Strömgren.

Bengt Strömgren (1908–1987) was a distinguished Danish astrophysicist who determined the chemical composition of stars, developed a photometric system using narrow-band filters, and discovered huge interstellar spheres of ionized hydrogen around stars, now called Strömgren spheres. He was born in Sweden, the son of astronomer Elis Strömgren, who served as professor of astronomy and director of the observatory at the University of Copenhagen. Raised behind a telescope, he graduated from that university at the age of 19 and received his doctorate at 21. He learned theoretical astrophysics and the new fields of relativity and quantum mechanics at the Institute of Theoretical Physics founded by Niels Bohr, now known simply as the Niels Bohr Institute. In 1932 he had already deduced the hydrogen content of the interior of stars using his theoretical coefficient of opacity. Struve offered him an appointment at the University of Chicago in 1936. A few years later, he followed up his 1932 paper with one discussing the helium and heavy element content of stars; additional helium is produced by nuclear transmutation of hydrogen, and the proportion of heavy elements can be used to determine stellar age. He returned to Copenhagen in 1938 and took over from his father as director of the observatory there until after the war.[38]

The staff didn't waste time when they learned of Struve's resignation. Chandrasekhar, Kuiper, and Hiltner heard that Strömgren was coming through Chicago after a stay at Mt. Wilson Observatory. The staff met quickly and decided on action. They sent a telegram to Strömgren on the train asking if he would accept the directorship of Yerkes. He had worked there from 1936 to 1938 and knew Yerkes, and they knew him. Struve sent a letter formally inviting him to join the Department of Astronomy and Astrophysics as assistant professor.[39] He accepted, with the incentive of knowing that Kuiper and Chandrasekhar were there. In the meantime Chandrasekhar served as acting chairman of the department.

Strömgren's arrival as director introduced quite a change in style. Struve had run the observatory as an autocrat. His secretary reinforced that attitude, so most faculty were glad when she announced that she also was going to UC Berkeley.[40] Strömgren was entirely different. He was a quiet, gentle man with high respect for others in all regards—personal, social, and professional. His wife Sigrid was a perfect match but not quite the socialite that some expected of a director's wife.[41]

They arrived from Copenhagen in June 1950 and moved into the refurbished director's house. They quickly became friends with Aden and Marjorie. Aden was asked to take charge of the optical shop and was welcomed by Fred Pearson and his young assistant, Chuck Robinson, who appreciated Aden's apprentice experience. The shop had an old project: revamping the large strip mirrors for the coudé spectrograph at the 82-inch McDonald Observatory telescope. The mirror surfaces had warped when the sides were cut off to accommodate a basically poor optical design. Aden had the opticians re-cement the material that had been cut off, repolish the surfaces until they were perfect, and replace the mirrors in the spectrograph, which was immediately moved to McDonald.[42]

Aden was placed in charge of crafting the optics for five 36-inch telescopes, one of which would go to the McDonald Observatory. He was also given the task of designing both that telescope and its building. In the meantime, he improved the efficiency of his auroral spectrographs by placing anti-reflection coatings on all of the transmission optics.[43]

Thornton Page, professor of astrophysics, decided to leave Yerkes, so his large home, the Parkhurst House, became available. The moment Marjorie heard this she was ready to move. It was as big as the Ross House where she had stayed when her parents traveled to Yerkes each summer. Now they had five bedrooms and a large dining room, but the basement was smaller than that of the prefab. Fortunately, the coal furnace was being replaced by an oil furnace, so Aden gained some room for his home office and optical shop.[44]

The US Air Force heard about Aden's spectrograph and asked if he would undertake to make several for their Cambridge Research Laboratories (AF-CRL).[45] Aden almost never declined any opportunity, even if it seemed impossible. If he did not know how to do or make something, he taught himself.[46] He added this assignment to his research and teaching responsibilities and set forth to deliver optical instruments to the Air Force. But the observatory declined to include this project in their shops. Aden decided to do it anyway, but with outside help. The head instrument-maker at Yerkes, Charles Ridell, retired and offered to manufacture the camera and spectrograph bodies, and he set up a shop at his 1900s-era home on the hill sloping down to the bay. Tinsley Optical in Berkeley took a contract to make the Schmidt corrector plates. When they arrived, they were useless. The surface contour met the point coordinates that Aden specified, but the surface shape between these specified points was not smooth. Today we use equation-driven, digital, numerically controlled polishing machines to prevent this from happening. But in 1950 these machines were still 30 years in the future. The surface contour on the delivered Schmidt plates (windows) met the specifications Aden provided, but there was almost a step function between points. How was he going to meet the delivery dates? He had to start over. He put an optical polisher in the basement along with a lathe and drill press and spent his evenings redoing the optics for the Air Force cameras. It was reminiscent of when he had made the optics for his auroral spectrograph back at Berkeley.[47]

Then he had a bright idea. Remembering some interesting geometrical relationships that he and Roger Hayward had explored when both were at Mt. Wilson almost 10 years earlier, Aden was able to diamond-generate a curve on those defective Schmidt plates that was close to the desired curve—and it was smooth. He also had discovered that, although one can generate accurate surfaces, the process of removing the diamond tool marks of an aspheric surface like a Schmidt plate resulted in the starting accuracy being lost. But if the mechanical accuracy is already close, why not cut the desired aspheric curve on an iron tool substrate and then use that shaped tool for the fine grinding process? One then only needed to hold the glass steady while the iron tool and abrasive turned underneath. It worked like a charm. Perkin-Elmer got wind of this, and Aden's long association as a consultant to Perkin-Elmer resulted. They used this same process in subsequent years to make a number of cameras having extremely steep aspheric surfaces. Aden used a membrane polisher for the final touch-up, the same one he had used earlier on the Schmidt camera for his thesis. The cameras were delivered on time, which established goodwill and led to contracts between the Air Force and Yerkes to continue upper atmospheric research for over a decade, long after Aden left Williams Bay.[48]

The spectrographs proved to be superb instruments and saw service for many years. The largest spectrograph, a 9-inch f/2.5, was sent to Canada, where Bill

Petrie of the Dominion Astrophysical Observatory in Victoria, British Columbia, used it for years (see Chapter 2, section "Youth and education"). The rest went to Cambridge, but the 9-inch f/0.8 spectrograph later returned to Yerkes to make some discoveries for Aden's associate and friend, Joe Chamberlain.[49]

National recognition

The Air Force asked Aden to start investigations to reproduce auroral phenomena in the controlled environment of a laboratory. And it was time for him to share some of the teaching load of the department. He was soon promoted to assistant professor and received a welcome raise along with it. Increased funding from the Air Force meant adding additional staff persons. The first was Dr. Charlie Fan, as soon as he finished his degree in physics from the University of Chicago. He and his wife June arrived at Yerkes on the day of the worst snowstorm in recent memory. He eventually followed Aden to the University of Arizona, where they collaborated on several projects. Joseph Chamberlain would leave AF-CRL and also join Aden in Tucson.[50]

One spring day in 1952, Strömgren called Aden into his office with some exciting news to report: Aden had been awarded the Warner Prize of the American Astronomical Society (AAS) in recognition of his developments of spectrographs and cameras and his discoveries with them. The Warner Prize was funded by Helen Warner, the daughter of Worchester R. Warner of Warner & Swasey, the company that built many of the telescopes in use at that time. Aden had been nominated for the Prize by J. J. Nassau of Case Institute of Technology, a long-time friend of the Pettit family. It would be the first time the AAS awarded the prize to a young scientist within five years of receiving a doctorate. The award cited his auroral and airglow discoveries and carried an award of $50, equivalent to almost $500 in 2020.[51]

Not long after that, Kuiper asked Aden to revise the chapter in Kuiper's book, *The Atmospheres of the Earth and Planets*, because the current chapter on airglow and auroras was now obsolete. The challenge was that the page count had to remain the same. Then Kuiper said he wanted a chapter for *The Earth as a Planet*, followed by a request to team with Sidney Chapman in a chapter on auroras for Springer's *Handbook of Physics*. It took some years for Chapman to write his part, by which time he had lost Aden's contribution. Aden rewrote it, and then Chapman said he was adding Syun-Ichi Akasofu as a coauthor. The article was finally published long after Aden left the field.[52] In addition to teaching and building new instruments, Aden was writing to publication deadlines along with maintaining a heavy observing schedule, recording spectra of auroras, and reducing data.

Air Force instruments

Diffraction gratings for the Air Force spectrographs that Aden was building were needed. The AF-CRL had a contract with Dave Richardson at Bausch & Lomb (B&L) in Rochester, New York, to make gratings along with a four-grating mosaic spectrograph that would need to be phased to produce its highest spectral resolution. Aden was asked to consult with B&L on behalf of their interests. He accepted the job and made a number of trips deep into the back areas of B&L where the grating laboratory was located. Aden also needed gratings for several new spectrographs he had designed for Yerkes and McDonald Observatories. It was an exciting time for rapid advancements in the art of making gratings.[53] The basic machine was old, but Richardson and his exceptionally skilled assistant had rebuilt the machine so that high-quality rulings could be obtained. The day of laser control was yet in the future, as was the laser itself. They tried interferometric control using atomic emission lines. Although the intensity of ghosts was reduced, they were still problematic. Richardson next faced a challenge from the Air Force program: Could he rule echelle gratings? These are a challenge to make because they have relatively few lines per millimeter in order to properly "blaze" the grating for a desired spectral region. And the grooves are relatively deep. The thick evaporated coating of silver or aluminum into which the grooves are drawn posed significant manufacturing challenges to avoid matte surfaces which scatter light and destroy diffraction efficiency.[54]

The diamond stylus used to shape the grooves in the soft metal coatings wore out quickly. But several good echelles were made, one which Richardson provided for Aden's use at Yerkes. He built a new auroral spectrograph using one of these echelles with a glass prism to provide cross-dispersion so that the many spectral orders were imaged side-by-side rather than overlapping. The purpose was to obtain enough dispersion in order to resolve the rotational lines of the blue and ultraviolet bands system in the airglow, thought to be due to a forbidden molecular oxygen transition: the Herzberg bands. Test spectra showed excellent resolution. Ghosts, caused by periodic errors in the screw thread that advanced the grating substrate from grove cut to groove cut, were faint, and there was no trouble removing the ghosts from the recorded data.[55]

To test the Air Force instruments properly, they would need direct exposures on faint airglow. Helen Pettit, Marjorie's sister, had joined Frank Roach's atmospheric group at Naval Ordinance Test Station (NOTS), China Lake, California, five years earlier. Frank paid Aden and Marjorie to travel to NOTS for some collaborative work. Joe Chamberlain would stay at Yerkes and record auroral spectra with Aden's 9-inch f/0.8 spectrograph that Aden had built for the Air Force. The Air Force permitted them to keep the 9-inch system at Yerkes. It would be a race to see who could first get the answer to these airglow bands.[56]

Aden and Marjorie arrived at NOTS and had a laboratory assigned where they tuned up the spectrograph. Aden placed his echelle spectrograph at the ranch home of one of the senior scientific staff members at NOTS, just west of Ridgecrest. They decided to use all of the dark of the moon that month during the spring of 1952 to make a two-week exposure. There were concerns about reciprocity failure of photographic emulsions, but they hoped that a pre-sensitizing exposure to diffuse light would keep the silver grains properly sensitized. It didn't, and no spectra were obtained. Midway in the long integration, Aden received a letter from Chamberlain, along with a microdensitometer tracing. A new snowfall had raised the ground reflection at Yerkes, so that Chamberlain obtained an excellent spectrogram resolving some rotational lines and asked Aden to help with the identifications. Aden analyzed the spectra and found that they did indeed appear to be from oxygen, but the precise values of the quantum levels could not be determined since the rotational number could only be ascertained to plus or minus 2. Aden and Marjorie suggested that they were the Herzberg bands. Chamberlain finished the analysis, confirmed they were the Herzberg bands, and published the results. That stay at NOTS was productive. Several reports to the Air Force were written describing a photometric method to determine of the altitude above Earth of the airglow emissions.[57]

The political situation was tense at NOTS between Frederick W. Brown, the chief scientist, and the Navy over research policies. Roach was in the middle of the fight, so when Brown announced that he was leaving to head the new National Bureau of Standards Laboratory at Boulder, Colorado, Roach, Helen, and a number of scientists also resigned to follow Brown to Boulder. Aden, Marjorie, and their now four children (their fifth on the way) then headed back to Wisconsin. The first signs of spring greeted them as they left the desert to return to Yerkes.[58]

Teaching responsibilities

Aden taught fall and spring semesters for the academic years 1951–1952, 1952–1953, and 1954–1955. The classes were attended by graduate students whose offices were at Yerkes, although he would occasionally lecture on the University of Chicago campus. During the three years he taught courses titled "Atomic Spectra" (applied wave mechanics; multiplet structure; intensities for permitted and forbidden lines; Pauli principle applications; Zeeman and Stark effects), "Molecular Spectra" (spectra of diatomic molecules; electronic states and transitions; rotational fine structure; electron configurations; dissociation, recombination, and pre-dissociation; polyatomic molecules; molecules in planetary, cometary, and stellar atmospheres; interstellar molecules), and "Basic

Astrophysics" (three courses). This was in addition to team-teaching classes in "Optics and Astronomical Instruments" with Strömgren (physical and geometrical optics of optical instruments; optical aberration; astronomical instruments and methods of observation). Such a busy load of teaching on top of full-time research was expected of all assistant professors.[59]

By the summer of 1955, other responsibilities forced him to spend full time on research, engineering, and administration for the new National Astronomical Observatory. It seems he relaxed outside the office and laboratory by designing and building astronomical telescopes and instruments for others.

Instrument development

One day Dave Richardson called Aden from Rochester, New York, to say that Alcoa wanted an all-reflective spectrograph without Schmidt plates but as fast as a typical Schmidt camera. The length of the focal plane needed to be about 10 cm long, not the 1 cm of his auroral spectrograph. The all-reflective requirement originated with the need to record high-acuity spectra over a broad wavelength bandpass.[60]

Aden recognized that the classical Schmidt camera design employs a transparent window (plate) which has a specially figured optical surface. This window is positioned at the center of curvature of the system's single spherical mirror and corrects for spherical aberration and fourth-order wavefront error. It also has zero optical power, so the system optical performance is first-order insensitive to tilt of this window. Aden's innovation was to replace the transparent window with a tilted reflecting surface with the fourth-order correction on the surface (Figure 6.3). This optical design principle was documented later by others and applied to an Earth-remote sensing application.[61]

Aden developed a new technique using graphical methods to solving the aberrations imposed by the tilted, reflecting Schmidt plate. He was asked to fabricate the optics for Richardson. On the last day, while edging the reflective aspheric window to final diameter, his fingernail touched some loose abrasive on the surface for just an instant. It left a short but surprisingly deep scratch on the asphere. Years later, Aden found that several such spectrographs had been made by B &L for the Fusion Physics Project at Princeton University. Aden was called by Lyman Spitzer to come and help realign the optics of what Princeton called the "Meinel Spectrograph." There was a surprise waiting. It was the Alcoa spectrograph. Sure enough, there was his "signature" scratch. Only it was on all three spectrographs. Bausch & Lomb had replicated the critical element, the reflection aspheric. Aden was paid once and wished that he had asked for royalties.[62]

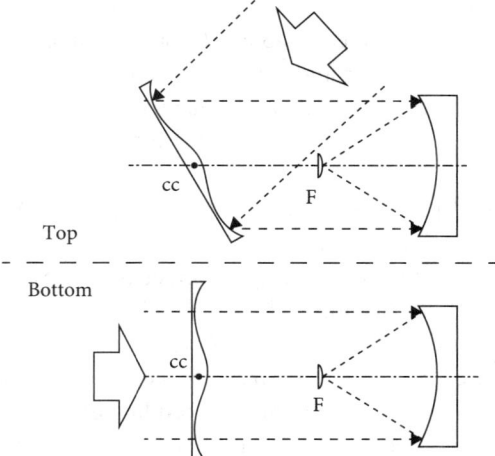

Figure 6.3. The difference between an all-reflecting Schmidt camera and the classic Schmidt designed and built by Bernard Schmidt. The center of curvature of the sphere at the right is given by "cc." The curved focal plane is shown as F. Top light is shown entering the system from the upper right to reflect from the fourth-order correcting mirror, pass to the sphere, and focus at F. Bottom light is shown passing straight through a transparent correcting window positioned at the center of curvature (cc) and into the spherical mirror, reflecting from the sphere to the focus.
Drawing by James Breckinridge.

Another aspheric optical task came from Bob Hopkins of the Institute of Optics at the University of Rochester. They had designed a wide-field anamorphic projection lens for Mike Todd's Todd-AO projection system but could not make the aspheric surface for it. Hopkins had heard about Aden's successful technique and challenged him to make the lens. So Aden manufactured the lens and suggested a test configuration. The results were well within their specifications. They paid Aden for his time, a total of $60. They also made replicas and sold the cameras for many, many times what Aden charged for the critical element. Aden commented that he should have had a business sense like that of Henri Chrétien, who received royalties from his projection camera anamorphic lenses for years.[63]

Another task that involved Aden's techniques for making Schmidt plates arose from a letter he received from Ed Carpenter, director of the Steward Observatory of the University of Arizona. Carpenter had an idea for a simple, inexpensive grating spectrograph for the 36-inch telescope at the observatory. It used a "grism," a transmission grating cemented onto the face of a glass prism. The net result was that the spectrum emerged along the same axis as the incoming light. It sounded like a good idea to Aden, and he sent back a design for a spectrograph using the grism and a flat-field Schmidt camera. Carpenter's grism had an effective beam diameter of 2 inches, about the same diameter as the Todd-AO aspheric. Aden made him the camera outside of Yerkes shops. The result was an excellent spectrograph used by students for many years.[64]

Astrophysics using the northern lights cameras

Aden investigated the utility of using his small (<20-cm) aperture Schmidt cameras for spectral classification with exceedingly low dispersion by replacing the diffraction grating with a small-angle objective prism. W. W. Morgan and Harold Johnson used one of Aden's small (10-cm aperture) Schmidt cameras with a 1-degree wedge objective prism to record low-dispersion, 30,000 Å/mm spectra to evaluate the system's utility for stellar spectral classification.[65] A low-dispersion spectrograph enabled recording spectra fainter with a limiting magnitude than the high-dispersion spectrograph. The key to building these optical systems was the low f/ratio uniquely achieved by Aden. The system's wide angle enabled the recording of a large number of faint stars with a short exposure, but the spectral dispersion was too low to be used reliably, and the approach was abandoned to the larger 24/36-inch Schmidt cameras with their 1-, 2-, and 4-degree wedge angle objective prisms at Case Institute of Technology and the University of Michigan.

Another important discovery was made by Helmut Abt using Aden's cameras. The Gum nebula was discovered in 1952 by Colin Gum, an Australian astronomer.[66] No one knew for certain how large the nebula was because its full extent had never been imaged. Abt received support from one of the early National Science Foundation (NSF) astronomy grants to mount one of the Meinel wide-angle f/1 Schmidt cameras onto the side of one of the smaller McDonald Observatory telescopes (to provide precision guiding).[67] Helmut discovered that the nebula was at least 15° × 22° and extends across 36° within the southern constellations Vela and Puppis, only 350 parsecs from Earth.[68] Hard to distinguish, it is widely believed to be the greatly expanded (and still expanding) remains of a supernova that took place about a million years ago.

Astronomical telescopes evolved over the centuries, driven by the desire to resolve details on the surface of planets and to penetrate deeper across the galaxy to fathom the depths of the universe. Larger apertures collect more light. But the technology for the optical fabrication and test of these large apertures limited them to longer focal lengths. These longer focal lengths give high angular resolution, which in turn leads to smaller fields of view. Telescope optical science and engineering evolved to satisfy the need for larger apertures and instruments for spectroscopy to support astrophysical studies. High angular resolution telescopes have evolved over the past 100 years, beginning with the 100-inch on Mt. Wilson (1917), to the 200-inch (1948) on Mt. Palomar, and to the 10-meter (~394-inch) Keck telescope in 1985. Today, 30-meter class telescopes are under construction. But the universe is not entirely narrow angle. For example, the Milky Way, when observed from Earth, extends across the entire sky, requiring an imaging system with a wide field of view to record it. The human eye does this well, but classic astronomical telescopes image it poorly without stitching together thousands of images at great expense.

University of Chicago 500-inch telescope

Big telescopes were often a conversation topic between Aden and Strömgren. Aden decided to see if he could develop a concept for a cost-effective, 10-meter class telescope. A whole new chapter in Aden and Marjorie's lives was about to open up. One March evening in 1953, while he walked around the wide circular lawn in front of the observatory, he thought about all those stellar photons falling uselessly into that grass. Was there a way to collect them and form an image? Can one create a primary mirror by excavating and paving a spherical bowl 10 meters in diameter with fixed spherical mirror segments and move only the observer's cage to reduce the cost? Could the wavefront arriving at the focus be corrected to obtain high acuity images? Aden went to his basement drafting table and put his ideas for a 10-meter telescope on paper. It looked like it would work if he used a big segmented Schmidt plate at the center of curvature of the bowl, the same point for the pivot for the observer's cage. There was room to put spectrographs and other large instruments in the long tube between the focus (halfway down to the big mirror) and the pivot point.[69]

The concept or design approach is described in a series of correspondence between Ira Bowen (director of the Mt. Wilson and Palomar Observatories) and Aden.[70] The ground is excavated into a bowl shape and the interior covered in concrete and finished to a near-net spherical concave shape, which is then overlaid with mirror segments with optically figured spherical surfaces. The segments are hexagonal in shape to fit or nest into a larger sphere to become the primary mirror of the telescope. The telescope is pointed not by tilting or moving the primary mirror across the sky; rather, telescope pointing and tracking are performed by tilting the telescope's much smaller secondary assembly and using different portions of the primary mirror sphere to reflect light to the secondary assembly. The secondary assembly consists of a structure with a segmented Schmidt plate to correct the spherical aberration at one end and the focal plane at the other. The Schmidt corrector is located at the center of curvature and the focal plane at the focus.

He was ready to present the concept (Figure 6.4) to Strömgren and Morgan. They went down to the basement of Parkhurst House with Aden and agreed that it was a great idea. But how big should it be? Aden said that the minimum would be 400 inches, but he preferred 500 inches (12.7 m). For perspective, the Keck 10-m telescope would not be operational until 58 years later, and Aden's visionary 12.7-m telescope will not be exceeded in size until 2025, 72 years later, when the planned ~39-m Extremely Large Telescope (ELT) of the European Southern Observatory comes online.[71]

Strömgren said that he would be happy with 200 inches. But Aden felt that would be too small to accommodate all the instruments at prime focus and noted that it would still require a segmented Schmidt plate, so aiming for the maximum affordable made more sense. Strömgren asked Aden for a construction plan and a budget to present to the University administration for their reaction. The response was favorable, but Strömgren was told to ask Aden to find funding for it himself, to the tune of a few million dollars.[72]

Aden then decided to poll other astronomers about the concept. He and Marjorie volunteered to go to both Lick and Mt. Wilson Observatories. Yerkes

Figure 6.4. Aden Meinel's concept for the X-inch telescope for McDonald Observatory. Light comes from above and strikes the large spherical, bowl-shaped mirror, reflects up from the stationary bowl, and converges to a focus where a smaller telescope pointed downward gathers the light, corrects for the spherical aberration with a large, segmented Schmidt corrector plate, and sends the light onto a focal plane. Scale is shown by the human figures shown standing on the rim in the foreground.
Photo courtesy of Ed Meinel.

would pay their train expenses, and the five children would go along at personal expense. It was an exciting time for all, especially Aden. Here was the student returning, but this time as a project manager and assistant professor at the University of Chicago.[73]

At Lick Observatory, Director Donald Shane listened with interest as Aden described the concept and went into the details of optical performance and construction of the most critical part: a segmented Schmidt plate. Shane was skeptical and wanted to see a more complete technical evaluation. Aden responded that he first wanted a chance to integrate Shane's thoughts and questions into that study. Then he was off to Pasadena and Mt. Wilson to show Ira Bowen the concept and get his reaction. When he arrived at Bowen's office, a portly white-haired gentlemen was sitting in a chair. Bowen introduced Robert R. McMath, who was experienced in manufacturing and telescopes. McMath's questions dealt with ground stability and whether the piers for the hundreds of mirrors in the spherical primary would remain stable even for a few days. Aden responded that that was one of the first experiments he would conduct. Bowen then described the monthly task that Aden faced in keeping four diffraction gratings properly aligned in the 200-inch coudé spectrograph. The gratings would be in an insulated room, but the spherical mirror segments would be in the open, exposed to the weather.[74]

Bowen raised a problem that Aden had not anticipated. Would not the chilled night air gather in the bowl-shaped depression and spoil the seeing? Aden suggested that perhaps the bowl could be placed where there is natural air drainage. Bowen then asked several questions related to the big segmented Schmidt plate. How would he make and test it? How would he keep it aligned? Where would he find sheets of glass large enough? And what about chromatic aberration? Aden had addressed each of these questions in the notebook he bought along, and conveyed his thoughts on each. When they parted, both Bowen and McMath made some encouraging comments but requested a report that addressed the many questions that they and Shane had raised. Aden built a scale model (Figure 6.4) for future use in demonstrating a concept that was so new. Then McMath asked something that presaged the future: Did Aden know that the Department of Defense was planning a radio telescope, likely in a natural bowl at Arecibo, Puerto Rico, following the same concept? It would have a focus and receivers placed at the end of a boom that pivots from the center of curvature of the spherical bowl—only that mirror will be one of wire mesh, not optical reflecting mirrors.[75]

Kuiper supported Aden's concept and sent a memorandum on March 26, 1953, to Chandrasekhar with a graph that plotted aperture against time, showing that a 400- or 500-inch telescope could be developed by 1985. He proposed siting the telescope at McDonald Observatory or in Prescott, Arizona, and hiring Aden,

Yerkes astronomers, Bowen, Donald Shane, or Lyman Spitzer to manage it. With a visual resolving power as low as 0.0094, it would allow ordinary spectroscopy and radial velocity measurements of faint stars (white dwarfs, old novae, stars in globular clusters); infrared spectroscopy of stars, nebulae, and planets; radiation and photoelectric measurements; visual investigations of close binaries, stellar disks, planets, and satellites; and direct photography with and without filters. He argued, as Aden often did, that waiting until large instruments reach their problem-solving ability before building even larger equipment "lacks boldness and misses the great opportunities for advance that come from pushing the technical facilities to the limit. . . . This generation of astronomers need not necessarily forego the advantages of a really powerful telescope, provided that a very good case can be made for its construction." Kuiper and others never followed up on the concept, perhaps taking into account an estimated annual operating budget of $50,000 at McDonald and $100,000 at Prescott (the equivalent of about $500,000 and $1 million today, respectively).[76]

Back at Yerkes, Aden gave more thought to the engineering questions about a quadrant-segmented Schmidt plate and decided to look at whether he could correct the spherical aberration with two small mirrors close to the focus. A few calculations, made laboriously in those days with mechanical calculators, showed that the field of view would be limited to only a few arc-minutes, not the 2 degrees desired.[77] Modern computer-aided, ray-trace design technology and detailed understanding of aberration balancing and adaptive/controlled optical surfaces have evolved until today such that Aden's concepts for a fixed spherical primary, innovative Schmidt corrector designs and downward-looking tracking optics have become a reality. In a nutshell, Aden was ahead of his time.

The concept of building an optical telescope having a primary mirror that did not track across the sky was finally implemented 43 years later in 1996 by the 10-meter Hobby-Eberly telescope of the University of Texas. This telescope was used to determine the mass of an extremely large black hole in the lenticular galaxy NGC 1277 and to characterize several exoplanets. It continues to be a highly productive instrument for its user community: University of Texas, Pennsylvania State University, Stanford University, Ludwig Maximillian University of Munich, Germany, and Georg-August University of Göttingen, Germany. This successful design concept was used again in 2005 for the 10-meter Southern African Large Telescope (SALT) near Cape Town, which is the largest aperture optical telescope in the Southern Hemisphere. Both telescopes are highly productive instruments, built at a small fraction of the cost of a fully tracking large primary mirror telescope (like the 10-meter Keck telescope)— exactly what Aden had envisioned.

McDonald Observatory

William Johnson McDonald, a prosperous banker in Paris, Texas, died on February 6, 1926, leaving over a million dollars to the University of Texas (UT) to build and equip an astronomical observatory but only $15,000 to each of eight relatives. Seven of the eight plus others left out of the bequest immediately contested it, claiming that he was of unsound mind. When the case was finally resolved after five court actions, $840,000 remained for UT to build an observatory to be named after its benefactor. The UT Dean of Science, Harry Benedict, sent letters to the directors of several observatories asking for advice about private endowments—and perhaps implicitly soliciting astronomers because there were none at UT.[78]

At about the same time, staff members at Yerkes were complaining about the growing limitations of their 40-inch, Alvan Clark refractor, as well as the often cloudy skies and bitter winters. Edwin Frost, who was then director of Yerkes, received one of Benedict's letters and struck an amiable friendship with him by correspondence, part of that to offer counsel during the legal proceedings related to the will. Otto Struve and George van Biesbroeck had been searching for a locale in Texas with better seeing, particularly Amarillo, but abandoned that town for the Davis Mountains in west Texas as talks began with Benedict and the presidents of UT and the University of Chicago. Struve drafted an agreement specifying a 30-year collaboration between the two universities, with shared responsibilities for staffing and operations. He made sure that the director of Yerkes Observatory would also be ex officio director of McDonald and that the new observatory would be built and equipped under his supervision. The budget for site selection and construction of the dome and an 80-inch telescope was not to exceed $375,000, with a completion date set for July 1, 1938. The agreement was ratified by the Texas Board of Regents in November 1932. Warner and Swasey of Cleveland, Ohio, submitted the successful construction bid, and the contract between it and the UT Board of Regents was signed by both parties in October 1933.[79]

Site selection was left largely to astronomers Christian Elvey from Yerkes and T. G. Mehlen from Amherst, Massachusetts, who crisscrossed the state, covering 8,000 miles from El Paso to Galveston and from Amarillo to Del Rio and San Antonio. They measured the "seeing," the atmospheric stability, by comparing the average oscillation of the image of a star (in this case Polaris) with the diameter of its diffraction pattern. In addition, they rated the sites for image quality and atmospheric transparency. They returned to Yerkes to discuss their recommendations; at the top of the list was Spring Mountain at 7,550 feet in the Davis Mountains. Struve went down to Texas himself and found that Mt. Locke at 6,809 feet, located 10 miles from Fort Davis, offered more usable rocky terrain for telescopes and outbuildings. Ground was broken in 1933, and the dedication was held on May 5, 1939. One of the most significant changes from the original budget was that the proposed 80-inch mirror ended up as 82 inches because the blank had to be remelted and recast at Corning Glass Works. When it cooled, the pressure of the glass had increased the diameter by 2 inches. It was still going to be the second-largest telescope in the world at that time, after the Hooker 100-inch at Mt. Wilson.[80]

Aden took his Schmidt camera/spectrometer to McDonald Observatory, bolted it onto one of the smaller telescopes there, and used the smaller telescope with its precise celestial tracking capability to guide long exposures and obtain low spectral resolution images of faint stars. He used the telescopes and instruments at McDonald frequently.[81]

Although the University of Chicago had agreed to operate and staff McDonald Observatory, later Yerkes director Strömgren believed that a first-class university such as UT should stand on its own with both an excellent department and telescopes to match. On one trip in early February of 1953 from McDonald to Austin, a drunken driver strayed over the center line at night and hit Strömgren's car head on.[82] It's surprising that Strömgren was not killed. After a long recuperation he did not pursue this dream for UT, because a palace revolt at Yerkes directed his attention to affairs at home.[83]

This revolt against the quiet ways of Strömgren resulted in a telephone call to Aden from McDonald Observatory, asking if he would consent to become associate director. Aden agreed, but with reservations. First, he was only an assistant professor, and if he expected to advance to a tenured position he would need the votes of the senior faculty at Yerkes, the same ones with whom Strömgren was having troubles. So Aden felt it would be necessary for him to be promoted to associate professor. Strömgren persuaded Dean Walter Bartky, and Aden became both associate professor and associate director of Yerkes and McDonald Observatories on June 5, 1953.[84]

Soon a new crisis arose from some Yerkes staff members who did not like Strömgren's style. Aden was put on the spot, being drawn aside several times, once for a stroll on the adjacent golf course that included the observatory lawn. The faculty wanted Aden to replace Strömgren as director. Aden liked the faculty, but he refused to undermine Strömgren and felt his talents were better served as an assistant to the director. Marjorie and Aden were private persons with personalities unlike those required of a good director and his wife if they are to keep the staff and others happy.[85]

Although he did not know it yet, Aden was soon to move into role where he would use his natural leadership skills, buoyant personality, optimistic can-do attitude, and infectious curiosity to lead national and international scientific investigations in geophysics, optical astronomy, and optical sciences on a global scale.

Radio astronomy

Aden received a call from John Kraus at Ohio State University inviting him to join a panel that would review plans for a radio telescope, the main dish of which was excavated within a valley at Arecibo, Puerto Rico. Aden wanted to see how they planned to correct for the huge amount of spherical aberration that a spherical bowl produces, so he agreed. There he saw the first concept for a distributed dipole array strung out along the boom to intercept pieces of the wavefront arriving at the optical axis of the boom. At radio frequencies, phase from astronomical sources can be measured directly, whereas at optical frequencies it can only be inferred, at least until the invention of the laser/maser as an optical local oscillator about 20 years later.[86]

Lloyd Berkner, chairman of the Board of Associated Universities, Inc. (AUI), invented a way to measure the height and electron density of the ionosphere, and so naturally had been following Aden's upper atmospheric work closely.[87] When he heard that Aden was heading the Yerkes Observatory development program for a 500-inch optical counterpart, he offered Aden a position on the AUI Board of Directors and asked him to join Bart Bok and Bernard Burke in advising him on both the site and development program for a radio astronomy observatory. Green Bank, West Virginia, was the prime site, as it was close to Washington, DC, and had a dark "radio" sky. Aden met with the advisory group at Green Bank to review plans and was especially impressed with Bok's ability to distill many facts into a prompt report, a habit that Aden followed as best he could in his career ahead. Green Bank was selected as the best site, and a 140-foot, high-precision reflector was chosen as the main instrument. Before that reflector was ready, a crude 300-foot dish was used so that the scientific staff could at least begin observations.

In 1992 that crude dish suddenly collapsed, but it had more than repaid the investment. The decision was to replace it with another, more versatile 300-foot dish at a cost of $24 million, instead of the original $300,000. To Aden it was just another sign that the Golden Age of Science and Big Science had indeed grown over three decades.[88]

Aden's association with Green Bank remained for a number of years, winding down about the time that Struve became its first director. This choice by AUI and NSF was a disappointment to Bok, who expected to have been asked to be its first director based on his deep involvement in its development. Aden never knew why Bok was passed over, but he was only beginning to understand that Big

Science has its political aspects. Aden didn't even realize that he was being drawn into the first of the Big Science projects, but his insatiable curiosity always drove him to keep accepting challenges and never decline an exciting opportunity.[89]

International Geophysical Year

Berkner had been a member of Byrd's Antarctic expeditions and had worked during the International Polar Year, 1932–1933, the 25th anniversary of which was approaching in 1957. Rapid advances in rocketry, geophysics, atmospheric science, and computing inspired Berkner and Sydney Chapman to propose to the International Council of Scientific Unions that an International Geophysical Year (IGY) be planned for 1957–1958, which happened to coincide with an approaching period of maximum solar activity.[90]

In 1952 the IGY was announced. Little did the planners know that the Russians would launch Sputnik, the first Earth-orbiting satellite, on October 4, 1957, during the IGY (see Chapter 11). Scientific findings from the IGY suddenly gained another dimension of importance. Broader international involvement was possible this time. Berkner asked Aden to plan the auroral and airglow studies. Aden went to Washington, DC, the following week to the new NSF headquarters at the Dolley Madison House. Aden created an ambitious plan for many observing stations around the world, including the South Pole. The Navy had agreed to provide support for the operation, but NSF would provide all the funds for scientific activities. Aden's plan totaled almost a million dollars, part of a $5-million-dollar package. Little did they dream that Antarctic operations would continue, not for just two years, but would still be going strong today after 50 years. Would Aden actually get to Antarctica? Fate had other things in store for Aden and Marjorie.[91]

Their thoughts turned to the idea of a home off the observatory grounds. They found a property just across the highway in front of the observatory. That would do nicely, and Aden and Marjorie could design the house to fit their needs. They had just about decided that this was the site for their future home when the unexpected happened. Aden was nearing the end of his days at Yerkes but didn't yet know it.[92] His plan was to exhaust discoveries with his Schmidt camera then move on to larger telescopes. He saw firsthand how powerful observational astronomy was. Edwin Hubble had discovered the expanding universe. It was observational astronomy that had distinguished galaxies from nebulae and had demonstrated that the Milky Way is our home galaxy, just one of many billions in the universe. What marvelous discoveries lay ahead for Aden and Marjorie?

Aden's plan included an Auroral Data Center at Yerkes, supported by the NSF. But it was derailed by Christian Elvey, who said it should properly be at the

University of Alaska Fairbanks, where he was then director of the Geophysical Institute. Joe Chamberlain wanted Aden to fight for the Center to be located at Yerkes, but Aden felt that because a member (Robert R. McMath) of the National Academy of Sciences was against him and because he was not a member of the NAS, he had little chance of making it happen. Aden almost became the director of the Geophysical Institute the year before when that post was vacant, but Chandrasekhar persuaded him that his future was bright at Yerkes, so he chose to withdraw his name from consideration. Elvey had told Aden back when he first proposed to write a dissertation on the airglow that he should leave the interpretation of his observations to experts such as Pol(ydore) Swings and Gerhard Herzberg. Aden disliked that statement at the time, and now he didn't like Elvey's power play to appropriate the planned Auroral Data Center, and Aden told him as much at that working session in the Dolley Madison House. After all, the most spectacular advances in both airglow and auroras had happened at Yerkes, and Aden was the person who made them. He still had the 500-inch project to give him more than enough to do, but he felt a special place in his heart for the airglow and auroras, even though Joe Chamberlain was assuming the lead in the Air Force program at Yerkes.[93]

Executive secretary

After World War II, several agencies were engaged in scientific pursuits. But these were mission-oriented and focused on achieving specific goals, such as energy, defense, industry, medicine, and standards. Vannevar Bush pointed out that it is basic or fundamental research that enables new discoveries in the applied research that is performed at the mission agencies. He observed that the nation had no agency that supported fundamental research, which is best performed in a university environment with students actively participating (see Chapter 11, section "Aden and the shifting role of science in the Cold War").

The NSF was founded in 1950 to provide funds for compelling, cutting-edge scientific research through modest individual research grants. By 1953, however, the Foundation was receiving research proposals that required access to facilities in nuclear physics, astronomy, and computing. Foundation leadership saw an opportunity to support the construction and operation of entire facilities for civilian-led, basic scientific research.

The astronomy community had already held several meetings on the subject of building new telescopes at sites in the western United States that had clear, stable atmospheres with constant high transparency for photoelectric photometry. In the fall of 1954, NSF management suggested to the president of the AAS (McMath) that astronomers study, provide a cost estimate, and prepare a proposal to build an astronomical observatory at the "best" site in the United States. Access to observing time at this observatory was to be merit-based and decided by peer review, just as awards of funds to individual researchers were managed by the NSF.

The National Science Board of NSF was going to meet in May 1955, and the subject of national facilities was on the agenda. An Advisory Panel for a National Astronomical Observatory (NAO) was formed under the auspices of the AAS, with McMath leading the effort and hosting the first meeting on November 4–5, 1954, at the University of Michigan in Ann Arbor. Those in attendance were Strömgren, Albert Whitford, Bowen, Frank Edmondson, and McMath. Minutes from that meeting were reported by Edmondson.[94]

The charge to the Panel was to develop plans for a national observatory and to recommend the appointment of an executive secretary. First choice for that position was Yerkes Observatory astronomer Aden Meinel. The panel recommended that $200,000 be included in the NSF's budget for FY 1956 to fund site and seeing investigations. Aden would continue to be an employee of the University of Chicago, but his salary would be paid by NSF under a grant. For administrative purposes, the University of Michigan would be the proposing organization.

Aden was still unhappy at the events in Washington, DC, where Elvey managed to take the NSF program on auroras to Alaska, when his train arrived at Delavan, Wisconsin, on November 5, 1954, from an IGY planning session at the NSF. Marjorie met him at the train station as she always did. She excitedly told him that Strömgren had called and wanted to know if he would accept the position of executive secretary for a planned ad hoc Committee for a National Observatory.[95] Aden would need to surrender his role with the IGY to have enough time to execute his responsibilities to site survey, telescope planning, facilities engineering, seeing measurements, staff development, etc.

Aden's acceptance ended the 500-inch project and their plans to buy a new home at Yerkes. The vision for a 500-inch telescope remained in what Aden listed as the "X-inch telescope" in the long-range plans he had prepared for the planned national observatory. Also evaporated was Aden's opportunity to see auroras from Antarctica and lead the Earth atmosphere investigations for the IGY. When Strömgren returned to Yerkes from Ann Arbor, he told Aden that he should develop a plan for a site survey, observatory, and its budget for the first few years. There would soon be a meeting at the University of Michigan, and Aden should make a presentation there. In the meantime, Aden sent a draft to Chairman McMath.[96]

Aden attended that meeting, and during lunch he sat next to McMath. As soon as Aden said, "My wife is Marjorie Pettit . . . ," he caught something chilling in McMath's glance, a chill that lasted as long as Aden dealt with him. Only after Aden left Kitt Peak did he learn why he had such a difficult relationship with

McMath, dating back to the bitter authorship issues between Edison Pettit and McMath (see Chapter 1, section "Marjorie Steele Pettit," subsection "Edison Pettit").[97] Later Chandrasekhar mentioned to Aden that Aden's membership in the NAS was being held up because of his association with Marjorie, the daughter of Edison Pettit, and Aden believed that McMath's grudge was the source of his unsuccessful nomination to membership.[98] He was to have many more confrontations with McMath as director of Kitt Peak National Observatory.

No one before Aden had ever faced the challenges of (1) siting and building an entire astronomical observatory facility with multiple stellar and solar telescopes, (2) engineering state-of-the-art telescopes and instruments, and (3) integrating the facility into the US national community of astronomers. It required someone charismatic, entrepreneurial, and high-energy as well as brilliant, modest, and patient. That person was Aden Meinel.

Notes

1. Mohler, O., and Dodson-Prince, H. 1978. Robert Raynolds McMath (1891–1962). *Biographical memoirs of the National Academy of Sciences* 49: 185–202. One of those issues was the so-called missile gap in the 1950s. Letter from R. McMath et al. to Allen White House Office, October 23, 1957, Office of the Special Assistant for Science and Technology, James R. Killian and George B. Kistiakowsky: Records, 1957–1961, Box 1, Folder: CIA (Oct. 1957–Oct. 1958, Dwight D. Eisenhower Presidential Library).

2. Meinel, A. B., and Meinel, M. P. 2002a. *Echoes from a simpler time*, unpublished, 62.

3. Osterbrock, D. E. 1997. *Yerkes Observatory, 1892–1950: The birth, near death, and resurrection of a scientific research institution.* University of Chicago Press, 1–2, 15.

4. Osterbrock, *Yerkes Observatory*, 9, 11, 20.

5. Osterbrock, *Yerkes Observatory*, 29–31, 33, 35, 58, 73.

6. Osterbrock, *Yerkes Observatory*, 170–2; University of Chicago Catalog of Courses 1950–51; Evans, D. S., and Mulholland, J. D. 1986. Big and bright: A history of the McDonald Observatory. University of Texas Press, 117–19; https://solarsystem.nasa.gov/people/720/gerard-kuiper-1905-1973/; Sears, D. W. G. 2019. *Gerard P. Kuiper and the rise of modern planetary science.* University of Arizona Press, 116.

7. Meinel and Meinel, *Echoes*, 63.

8. Osterbrock, *Yerkes Observatory*, 177–9, 180; Sears, *Gerard P. Kuiper*, 31–2.

9. Letter from O. Struve to R. Hutchins, January 13, 1936, Office of the President—Hutchins Administration Records 1892–1951, Box 280, Folder 1, University of Chicago Library.

10. Osterbrock, *Yerkes Observatory*, 180.

11. Osterbrock, *Yerkes Observatory*, 184; telegram from S. Chandrasekhar to R. Hutchins, March 24, 1936, Office of the President—Hutchins Administration Records, Box 280, Folder 1, University of Chicago Library.

12. Sears, *Gerard P. Kuiper*, 32.

13. Gerhard Herzberg was awarded the Nobel Prize in Chemistry in 1971 for his work on the structures of a large number of diatomic and polyatomic molecules, including the structures of many free radicals difficult to determine in any other way.
14. Meinel and Meinel, *Echoes*, 62–3.
15. Meinel and Meinel, *Echoes*, 63.
16. Meinel and Meinel, *Echoes*, 64–5; Meinel, A. B. 1950b. Hydride emission bands in the spectrum of the night sky. *Astrophysical Journal* 111: 207–8; Meinel, A. B. 1950d. OH emission bands in the spectrum of the night sky. *Astrophysical Journal* 111: 555–64.
17. Meinel and Meinel, *Echoes*, 63.
18. Meinel and Meinel, *Echoes*, 63.
19. Meinel and Meinel, *Echoes*, 63–4.
20. Meinel and Meinel, *Echoes*, 67.
21. Osterbrock, *Yerkes Observatory*, 77–9; Evans and Mulholland, *Big and bright*, 115–16.
22. Meinel and Meinel, *Echoes*, 69.
23. Meinel and Meinel, *Echoes*, 69.
24. Meinel and Meinel, *Echoes*, 64–5.
25. Meinel, A. B., Hydride emission bands, 207–8.
26. Meinel and Meinel, *Echoes*, 65.
27. Meinel and Meinel, *Echoes*, 65; Sir David Robert Bates (1916–1994) is often described as one of the greatest Irish physicists of the 20th century, an expert in atomic and molecular physics of the Earth's upper atmosphere. He was the 1977 recipient of the Gold Medal from the Royal Astronomical Society.
28. Meinel and Meinel, *Echoes*, 72.
29. Meinel and Meinel, *Echoes*, 72.
30. H_α is the visible hydrogen spectral line in the Balmer series created by a hydrogen atom when an electron falls from the third lowest to second lowest energy level, corresponding to a wavelength of 656.28 nm, or red light.
31. Meinel, A. B. 1950c. Identification of the 6560Å emission spectrum of the night sky. *Astrophysical Journal* 111: 433; Meinel, A. B. 1950e. On the entry into the Earth's atmosphere of 57-keV protons during auroral activity. *Physical Review* 80: 1096–7; Meinel, A. B. 1950f. On the entry into the Earth's atmosphere of high speed protons during auroral activity. *Science* 112 (2916): 590.
32. S. Chandrasekhar letters of August 21, 1950, and September 19, 1950, to the editor of *Science,* Chandrasekhar papers, Box 21, Folder 18, University of Chicago Library Archives.
33. Chandrasekhar papers, Box 21, Folder 18, University of Chicago Library Archives.
34. Time, Inc., September 18, 1950, 75.
35. Osterbrock, *Yerkes Observatory*, 300.
36. Osterbrock, *Yerkes Observatory*, 288.
37. Letter from S. Chandrasekhar to R. Hutchins, October 6, 1950, Office of the President—Robert Hutchins, Box 21, University of Chicago Library.
38. Osterbrock, *Yerkes Observatory*, 175–6; Evans and Mulholland, *Big and bright*, 127–8; Rudkjøbing, M. 1988. Bengt Georg Daniel Strömgren. *Quarterly Journal of the Royal*

Astronomical Society 29: 282–4; Strömgren, B. 1932. The opacity of stellar matter and the hydrogen content of the stars. *Zeitschrift für Astrophysik* 4: 118–52; Strömgren, B. 1938. On the helium and hydrogen content of the interior of the stars. *Astrophysical Journal* 87: 520–34.

39. Letter from O. Struve to B. Strömgren, January 6, 1936, Yerkes Observatory Office of the Director, Records, Box 162, Folder 5, University of Chicago Library.
40. Osterbrock, *Yerkes Observatory*, 291.
41. Meinel and Meinel, *Echoes*, 66.
42. Meinel and Meinel, *Echoes*, 69.
43. Meinel and Meinel, *Echoes*, 69.
44. Meinel and Meinel, *Echoes*, 69.
45. Meinel and Meinel, *Echoes*, 70.
46. Interview with David Drach-Meinel and Ed Meinel by A. Pridgeon, April 17, 2016.
47. Meinel and Meinel, *Echoes*, 70.
48. Meinel and Meinel, *Echoes*, 70.
49. Meinel and Meinel, *Echoes*, 71.
50. Meinel and Meinel, *Echoes*, 74.
51. Meinel and Meinel, *Echoes*, 74.
52. Meinel and Meinel, *Echoes*, 75.
53. When a diffraction grating is placed over a telescope objective lens, the spectrum of a star is shown much like a prism but at a higher resolution. Unlike a prism, the spectrum is not produced by refraction but by diffraction of light transmitted or reflected by the narrow lines in the grating, as many as 100,000 lines per inch.
54. Meinel and Meinel, *Echoes*, 75.
55. Meinel and Meinel, *Echoes*, 75.
56. Meinel and Meinel, *Echoes*, 76.
57. Meinel and Meinel, *Echoes*, 78.
58. Meinel and Meinel, *Echoes*, 78.
59. University of Chicago Catalog of Courses 51/53, 52/53, 54/55, 55/56, 56/57.
60. Meinel and Meinel, *Echoes*, 76.
61. Epstein, L. 1967. All reflecting Schmidt camera. *Publications of the Astronomical Society of the Pacific* 79: 132–5; Breckinridge, J. B., Page, N. A., Shannon, R. R., and Rodgers, J. M. 1983. Reflecting Schmidt imaging spectrometers. *Applied Optics* 22: 1175–80.
62. Meinel and Meinel, *Echoes*, 77.
63. Meinel and Meinel, *Echoes*, 77.
64. Meinel and Meinel, *Echoes*, 77–8.
65. Morgan, W. W., Meinel, A. B., and Johnson, H. M. 1954. Spectral classification with exceedingly low dispersion. *Astrophysical Journal* 120: 506–11.
66. Gum, C. S. 1952. A large H II region at galactic longitude 226°. *The Observatory* 72: 151–4; Gum, C. S. 1955. A survey of southern H_{II} regions. *Memoirs of the Royal Astronomical Society* 67: 155–77.
67. Morgan, W. W., Strömgren, B., and Johnson, H. 1955. A description of certain galactic nebulosities. *Astrophysical Journal* 121: 611–14.

68. Abt, H. A., Morgan, W. W., and Strömgren, B. 1957. A description of certain galactic nebulosities. II. *Astrophysical Journal* 126: 322.
69. Meinel and Meinel, *Echoes*, 79.
70. Letter from A. B. Meinel to I. S. Bowen, February 12, 1953, Papers of Ira Sprague Bowen, Box 25.413, Huntington Library.
71. Meinel and Meinel, *Echoes*, 79.
72. Meinel and Meinel, *Echoes*, 80.
73. Meinel and Meinel, *Echoes*, 80.
74. Meinel and Meinel, *Echoes*, 80.
75. Meinel and Meinel, *Echoes*, 80–1.
76. Confidential memo from G. P. Kuiper, March 31, 1953, Box 12, Folder 3[1], University of Arizona Special Collections; Sears, *Gerard P. Kuiper*, 165–7.
77. Meinel and Meinel, *Echoes*, 81.
78. Evans and Mulholland, *Big and bright*, 3–4, 11, 12–13, 20, 22.
79. Evans and Mulholland, *Big and bright*, 3–4, 11, 12–13, 20, 22, 25–7, 29.
80. Evans and Mulholland, *Big and bright*, 34–6, 44, 46, 69–70, 82.
81. Interview with Helmut Abt by J. Breckinridge, March 1, 2017.
82. Letter from A. B. Meinel to Ira Bowen, February 12, 1953, Ira Sprague Bowen Papers, Box 25.413 DP 53, Huntington Library.
83. Meinel and Meinel, *Echoes*, 81.
84. Meinel and Meinel, *Echoes*, 83–4; Evans and Mulholland, *Big and bright*, 128.
85. Meinel and Meinel, *Echoes*, 84.
86. Meinel and Meinel, *Echoes*, 81–2.
87. Hales, A. L. 1992. Lloyd Viel Berkner. *Biographical Memoirs of the National Academy of Sciences* 61: 2–24. https://doi.org/10.17226/2037.
88. Meinel and Meinel, *Echoes*, 82.
89. Meinel and Meinel, *Echoes*, 83.
90. Meinel and Meinel, *Echoes*, 83.
91. Meinel and Meinel, *Echoes*, 83–4.
92. Meinel and Meinel, *Echoes*, 83.
93. Meinel and Meinel, *Echoes*, 85–6.
94. Edmondson, F. K. 1997. *AURA and its US National Observatories*. Cambridge University Press.
95. Meinel and Meinel, *Echoes*, 86.
96. Meinel and Meinel, *Echoes*, 86.
97. Meinel and Meinel, *Echoes*, 86.
98. Personal communication between Aden Meinel and Jim Breckinridge in 1988.

7

The People's Observatory

Astronomy is a scientific field that depends in large part on observation and measurement of objects in a sky that is visible to every professional and amateur around the globe. It is a science of little or no commercial value, easily recognized by all, and captures the excitement and thrill of popular science fiction readers everywhere. Astronomy attracts a broad spectrum of talents, skills, and interests from around the globe. Indeed, there are millions of people who look at the stars each night, whether it be briefly to enjoy the beauty of the night sky or professional astrophysicists recording data to understand the physical composition of the universe. Clearly it is one of the oldest of the sciences and perhaps the most recognizable. Most important, it is a science familiar to taxpayers and their legislators. If NSF were to demonstrate success in its early initiatives, astronomy was a politically neutral place to start.

Weather is critical to the success of an observational astronomy program of discovery and science. The universities in the United States with highly productive astronomy departments and strong graduate programs were on the East Coast and in the Midwest, where clear skies were by no means guaranteed for much of the year. To compensate, some established university astronomy departments had established remote observing stations. For example, both Yale and Harvard had stations in the Southern Hemisphere, Yale in Australia, and Harvard in South Africa. As early as 1940, Otto Struve described the success of the Yerkes Observatory collaboration with the University of Texas. He reported on the advantages of clear skies and a stable atmosphere to observational astronomy and our understanding of the universe.[1]

Astronomers wanted access to observatories with large telescopes at sites with excellent, dependable weather. They banded together to propose a National Observatory to NSF. Of central importance to NSF was access to large, research-grade telescope facilities by any qualified scientist based on an assessment of the compelling merit of the research by a panel of peers. Federal funds were not to be used to support scientists at specific institutions which owned and operated exclusive large telescopes.

In 1950, the American Astronomical Society (AAS) appointed a special committee, chaired by C. D. Shane, director of Lick Observatory, Mt. Hamilton, California, to formulate a relationship between the astronomical community and the recently formed NSF. A draft report was written by Jesse Greenstein

With Stars in Their Eyes. James B. Breckinridge and Alec M. Pridgeon, Oxford University Press. © Oxford University Press 2022. DOI: 10.1093/oso/9780190915674.003.0007

and revised several times before it was published in late 1951. This document provided a plan for a national effort to develop observatories for observational astronomy.[2]

Optical astronomy through the atmosphere

A telescope aperture performs one basic function—to collect light. The larger the aperture the more light is collected, the brighter the image, and the higher the angular resolution. However, angular resolution is limited by what astronomers call "seeing," which is determined by atmospheric turbulence; the lower the turbulence, the better the seeing. On average, turbulence decreases with altitude, which is why optical astronomers select mountaintops for their telescopes.[3]

Ground-based telescopes for optical astronomy look up through a turbulent atmosphere, which distorts images and limits long-exposure angular resolution to about 1 arcsecond depending on location. Turbulence is caused by local temperature changes in the atmosphere, which in turn cause local changes in the index of refraction. The optical effect is as though one puts an array of small (10-cm diameter) weak optical lenses over the telescope aperture. The statistical size of these lens is determined by the temperature fluctuations. The wind blows this array of small lenses across the telescope aperture to cause the image to "boil" with time and destroy image quality. Good astronomical seeing occurs at a site which has small fluctuations in temperature and low wind. The only way to be sure that a site has a seeing worthy of the investment to locate a telescope is to measure seeing quality over a year, the longer the better.

The mountaintop site of the US National Observatory would be the first astronomical observatory to be sited based on years of survey work on several nearly inaccessible mountaintops. Aden invented several small, portable, optical instruments to measure seeing and predict the image quality that larger, more expensive telescopes would experience.[4] These instruments were carried on horseback to survey and evaluate several sites.

Precision photoelectric photometry: A new (1950) investigative tool

Visual photometry, the art of estimating the brightness of stars with the human eye as the instrument, has been used since humans first started looking to the sky. Astronomical photometry uses light-sensitive instruments to measure the brightness of stars. The earliest instruments were, of course, the human eye and brain. They grouped stars on the celestial sphere into constellations, recognized that stars have different levels of brightness, and established the well-known stellar magnitude scale used since prehistoric times. The eye is sensitive to a wide range of stellar brightness levels, but the eye/brain system response is nonlinear and provides a poor quantitative tool to support modern detailed scientific investigations.

Fundamental to the understanding of what makes stars shine and thus how our Sun provides light is the knowledge of how much power they radiate. This requires a quantitative measurement of the brightness of stars and estimation of their distance. Distances were well known using parallax studies and Cepheid variable stars. However, precision measurements of stellar brightness needed to wait for the photomultiplier tube developed during World War II to sense and accurately quantify faint light below the threshold of vision. By 1951, astronomers had developed techniques for making precision stellar brightness measurements. However, because of the highly variable transmission of the atmosphere at Midwest and East Coast observatories caused by weather conditions, the measurement process was slow, tedious, and error-prone. Observing sites with better weather were essential to further progress.

The vacuum photodiode is a device that converts light to electrons, then accelerates the stream of electrons to an anode to create an electric current. This current is measured with a current meter or voltmeter. Although a linear device (the number of electrons out is proportional to the intensity of incident light), it is unable to measure faint stars. Joel Stebbins, an astronomer at the University of Wisconsin, published the first quantitative data on an astronomical source by measuring the light from the phases of the Moon and later using the technique to make the first measurements of eclipsing variable stars.[5] Developments during World War II led to the development of electron amplifiers inside a vacuum that could achieve gains of 10^6 or more. Quantitative measurements of starlight far fainter than could be seen by the eye were recorded regularly by astronomers at large telescopes by increasing the integration time.

The brightness or intensity of a single isolated star could be measured accurately in relation to the brightness of nearby stars to follow the changes in variable stars such as Cepheids and eclipsing binaries. Colors for stars were accurately determined for the first time by measuring intensity through different-colored filters. There was interest in measuring stellar brightness for stars across the celestial sphere at different points in the sky. But atmospheric transmission added absorption and, not only that, the absorption of light was wavelength-dependent during periods when a weather front moved through and therefore changed during the night. It was obvious that the measured intensities and color of stars needed to be referenced to their brightness *outside* the atmosphere. That required calibration of the wavelength-dependent transmission of the atmosphere, as well as establishment of standard brightness for stars. If astronomers were to use measures of stellar intensity and color for astrophysical interpretation, it was clear that observing sites with little or no clouds, high atmospheric transmittance, and long-term, stable, clear weather were needed.

But this was not the situation at Midwest or Eastern universities, where most astronomers lived and worked. John Irwin of Indiana University proposed a national center for photoelectric photometry in October 1951 at a neighborhood meeting of astronomers from Ohio State University, Ohio University, and Indiana University.[6] It is not clear if Yerkes Observatory was part of this group or if Aden participated.

The need for a national astronomical observatory for photoelectric photometry was recognized again during a conference supported by NSF held at Lowell Observatory in Flagstaff, Arizona, on August 31 and September 1, 1953, on the subject of astronomical photoelectric photometry.[7] One outcome of this meeting was a recommendation that NSF provide financial support in the form of facilities and equipment at a suitable observing site to enable the effective application of this new investigative tool for all astronomy.

If federal funds were to be used to construct a National Observatory, it would need to be accessible to all. The fairest way to decide who would be allowed to use the telescope for research was to prioritize access based on the competitive merit of proposed research as decided by a group of peers. Such an approach was easily understood by funding agencies that received their funds through votes by members of Congress and state legislatures. Another purpose for a National Observatory was to provide access to large astronomical facilities to academically qualified students and faculty from small institutions and departments. The facilities at Mt. Wilson and Palomar Observatories, owned and operated by the private Carnegie Institution in Washington, DC, were inaccessible to all except students and faculty at Caltech, Caltech faculty collaborators, and Mt. Wilson Observatory scientists. The same was true for the facilities at Yerkes/McDonald (University of Chicago) and Lick Observatories (University of California), where telescope time was regularly over-subscribed.

National Astronomical Observatory Panel

Peter van de Kamp, a well-known astronomer from Swarthmore College, was serving as NSF program manager for astronomy in 1954. During a telephone call to van de Kamp from Leo Goldberg of the University of Michigan on September 21, 1954, the subject of a National Astronomical Observatory (NAO) was discussed. Meetings with the Foundation director, Alan Waterman, had led van de Kamp to believe that a proposal from the astronomy community to the National Science Board for an NAO facility would receive positive support.

The formation of an NAO ad hoc panel was recommended. During a telephone call from van de Kamp to Goldberg in September 1954, van de Kamp asked Goldberg to prepare and submit a proposal to cover the period of 18 months for travel by committee members to Ann Arbor, Michigan, for meetings. In a letter dated September 24, 1954, from Goldberg to van de Kamp, the ad hoc committee was charged in writing with the task of exploring the desirability and feasibility of a National Observatory under the leadership of NSF.[8] In that letter, Leo Goldberg confirmed that the members of the committee were Robert R. McMath (Committee Chair), Ira Bowen, Struve, Bengt Strömgren, and A. E. Whitford. The NSF also appointed five consultants to the committee: D. H. Menzel, Shane, James Baker, W. W. Baustian, and Bruce Rule. Representing NSF at Panel meetings were van de Kamp and Goldberg.

The purposes of the committee were (1) to investigate the national need for an optical astronomy observatory that would provide all astronomers with access to large-aperture telescopes based on merit, and (2) to plan for the construction and management of such a facility. It was this panel that established the US National Optical Astronomy Observatory (NOAO) and led its incorporation as a not-for-profit academic institution.

Waterman planned to have the NAO committee report presented to the National Science Board, along with plans for a National Radio Observatory facility (from a different committee) during the Board meeting in the spring of 1955 for a policy decision on national scientific facilities. Up until this time, NSF supported only individual scientists in their endeavors.

In academic circles, decisions are made by consensus, usually after much debate, considering the merits of multiple alternatives and pondering the impact of each. This approach is often inconsistent with creating the bricks and mortar necessary to create a physical institute. The NAO Panel differed in purpose significantly from the usual NSF peer-review, proposal-selection recommendation panel. Individual research proposals include fixed cost, schedule, and performance. Peer-review panels are convened to read, study, discuss, and review 15–20 proposals submitted. In the case of astronomy, access to telescope time is not awarded along with the funds to support data reduction and interpretation. Rather, the astronomer accepts his award from the government agency and must pass through an additional peer-review panel to obtain telescope time. The research panels are usually dissolved after a few weeks to be reconstituted with different members the following year.

The NAO Panel, however, evolved into a "board of directors" with responsibility for creating a facility. It would be three years before the nonprofit Association of Universities for Research in Astronomy (AURA) took over corporate responsibility for managing the continuing construction and operation of observatories. All of the Panel members, with the possible exception of Robert

McMath, had fundamental conflicts of interest. The academic members could expect NSF to manage astronomy as a "zero-sum game" where the Foundation assigned a fixed amount of dollars to astronomy and then divided these funds between support for facilities and support for merit-based research grants to individual scientists.

The first meeting of the NAO Panel took place at the University of Michigan in Ann Arbor on November 4–5, 1954. Eight conclusions emerged from the meeting, three of which eventually affected Aden directly: (1) The need for a National Observatory was answered in the affirmative by all members of the Panel, who felt it would stimulate interest in astronomy and be helpful in the training of young astronomers. (2) There was consensus that the site investigation should be the broadest possible and that seeing conditions for both day and night should be studied. There was no recommendation at this time regarding the Southern Hemisphere. The most important action was to recommend "the appointment of an executive secretary to act under the instructions and orders of the Panel for the purpose of making a broad study of possible sites and seeing." Aden Meinel, then of Yerkes Observatory, was the first choice of the Panel from a list of five names. The Panel recommended that NSF include $200,000 in its fiscal 1956 budget to fund the site and seeing investigations. (3) The Panel recommended that a 30-inch telescope be installed as soon as possible and that an 80-inch telescope be the second instrument. A decision on a larger telescope would come later. It was also agreed that Theodore Dunham's proposed 70-inch coudé telescope (possibly in the Southern Hemisphere) was not of interest to the National Observatory.[9]

One can speculate that the reason Aden was the committee's first choice to hold the position of executive secretary was because of his extensive experience with atmospheric optics. He was a well-recognized and enthusiastic professional astronomer and a widely published scientist who retained an engineering background to build telescopes/instruments. He had a natural infectious curiosity and was active in promoting a large telescope (500-inch) for Yerkes and McDonald Observatories. Aden was well-known to Struve and Bowen and had the strong support of the Yerkes director Strömgren. He would have been less well-known to McMath, whose interests were in solar astronomy instrumentation.

The appointment of Aden as executive secretary was a recommendation that the NAO Panel made to NSF, and it was up to NSF to respond with the official appointment. Van de Kamp was the program manager at NSF who was responsible for carrying forward the recommendation to NSF leadership and communicating action on the part of the Foundation back to the committee and Aden. Van den Kamp returned to Washington after the November 4–5 Panel meeting, but his report to NSF director Waterman was delayed by a month or so.

In the meantime, committee member Strömgren telephoned Aden and Marjorie and told them that the committee had recommended Aden to be the executive secretary of the NAO Panel.[10] But what did that mean? The NSF clearly defined the unpaid, volunteer role of a Panel member as providing the agency with peer scientific and technical review. But there was no role statement that defined the breadth and depth of the title of executive secretary. There seemed to be as many definitions as there were NAO Panel members. It was clearly more than "executive clerk," but was it a chief executive officer or a chief operating officer or director? What vision or strategic plan did the Panel have? Aden would spend the following 36 months working to define that role, sometimes with and sometimes without the support of the NAO Panel, but always with the focused goal of establishing a state-of-the art national astronomical observatory that would be available to all qualified scientists.

During the winter of 1954, Aden was extremely busy leading his team of auroral research workers and developing multiple ground stations for NSF's International Geophysical Year (IGY), 1957–1958. He was also building spectrometers for Air Force–Cambridge Research Laboratories, teaching graduate astronomy classes, sustaining his personal astronomical observing program in low-resolution stellar spectroscopy, and designing optics to increase the field-of-view of large telescopes.[11] In addition, for the past 17 months Aden had administrative responsibilities as associate director of Yerkes and McDonald Observatories. He was making commitments of his time and energy to these other projects and wanted to know as soon as possible of a formal offer. However, he quickly realized he could not handle both the role of NAO executive secretary and also lead the IGY aurora campaign. As soon as possible he would need to resign his IGY responsibilities, recommend new leadership, and pass off several tasks in progress to someone else.

On November 29, 1954, a little more than three weeks after the Ann Arbor meeting, Aden wrote a letter to committee chair McMath asking when he would receive the offer with its conditions.[12] In his reply on December 2, McMath summarized the recommendations of the Panel and observed that van de Kamp needed to report NSF's decision.[13] No Panel member could confirm this new leadership position because NSF had yet to approve his appointment. The details on how the funds were to reach Aden remained obscure.

Aden and Bowen had an unexpected meeting at Grand Central Station in New York City in mid-December 1954. Aden followed up this meeting with a letter to Bowen and a copy to McMath, dated December 15, saying that he was considering the Panel's nomination favorably but that he might not be the ideal choice if the Panel wanted someone merely to carry out its decisions rather than to bring a "driving leadership" and suggest modifications. He was also concerned about the "ominous aspects of government ownership of pure research facilities" and felt that the observatory "must not be relegated to the status of an observing station." McMath wrote to Bowen that he concurred with Aden's position and then back to Aden, saying that his concept of leadership was exactly why the Panel selected him. McMath continued: "And, of course, from NSF's standpoint your procedures and reports would naturally have to be approved by the government panel. Here, I am sure you would have no difficulty as the panel has no intention of giving you anything which might remotely resembled 'close supervision and direction.'"[14] The last two sentences contradict each other to a certain extent. Bowen realized this and responded to McMath's letter to Aden:

> The exact relationship of the executive director [*sic*] to the Panel as a whole has not been defined, and will doubtless change as the program from initial fundamental planning, to construction and then to final operation. Certainly, in the latter stages I would envision the relationship as somewhat similar to that of a college president and his board of trustees, namely most of the detailed planning and execution would be in the hands of the executive officer with major policy matters subject to review by the board. In all of the above it should of course be understood that I am giving my own thoughts and am speaking not for the panel. In any case, I very much hope you can see your way to help the committee on this problem. I am sure your experience with astronomical and optical problems would be of great assistance to getting the program underway.[15]

In December 1955, the NAO committee received a telephone request from NSF HQ for a proposal. On December 30, Goldberg responded by submitting a proposal to NSF through the University of Michigan on behalf of the NAO committee. The sum of $11,724.98 was intended to fund salaries of the executive secretary and an office secretary, plus funds for telephone, travel, postage, and drafting assistance for a five-month period beginning February 1, 1955. The executive secretary was authorized to act under the instruction and orders of the Panel for the purpose of making a broad study of possible sites and seeing.[16]

No specific instructions on how Aden was to interface with the Panel or NSF were given. Nor did the Panel provide any indication of the importance of strategic planning for development of the observatory. Word of Aden's new responsibility flowed quickly though the astronomy community. He visited several Panel members individually to discuss their perception of his role as executive secretary. In mid-January, McMath asked Aden to come to see him. On January 27, 1955, Whitford wrote to McMath that he wanted Aden to visit Ann Arbor as soon as he started work for a session with McMath and Goldberg. But Aden was traveling at the end of January to Arizona and California, talking with his colleagues and members of NAO about the impact and scope of this new appointment. No formal offer had been made, and Aden wanted to respond knowledgeably when it came. During his travel west, he met with Shane, Bowen, and others in California on engineering issues, telescope design, towers for testing the seeing, and the possibility of a new Southwest Astrophysics Institute. He then went to Flagstaff, Arizona, to speak with Alfred Wilson (director of the Lowell Observatory) and Lowell trustee, Roger Putnam, about a suitable peak in the Flagstaff area before going south to the University of Arizona at Tucson. While there he met with John Babbitt, member of the Board of Regents, to gauge interest in state support of the observatory. Finally, he spoke with University president Richard Harvill about the campus and faculty and Harvill's vision for the university.[17]

Aden was not yet "on contract." The title of executive secretary, an invention of McMath, had not been approved by NSF. The NAO Panel as a whole had not met to define his specific assignment nor produce a roles-and-responsibilities statement. His expenses for the trip were paid by Yerkes. Aden had a vision for the NAO and was formulating how he should proceed.

Aden's youthful and impetuous enthusiasm, genius, and resourcefulness baffled the NAO committee to a certain extent. His military training showed him how to get things done, and he was moving forward, ahead of any policy development, to make unilateral decisions that the committee would ultimately approve. In fact, it almost seemed that he was avoiding the committee members because he did not want to be held back. But in fact his official appointment was slow in coming, largely because of bureaucratic creep, and the Panel members had not agreed on how they were to interface with their new executive secretary.

For planning like this he did get static from McMath. His budget showed a bare-bones cost of $250,000 per year to operate this new observatory with a proposed 36-inch telescope and 80-inch telescope. McMath said Aden had told Waterman that it would only be $50,000. McMath demanded that the budget be cut back. Aden replied that the annual budget for operating McDonald Observatory with only one 82-inch telescope was over $200,000. That didn't matter. McMath demanded that Aden submit the budget as agreed. Recall that the NAO committee under the leadership of McMath advised NSF to plan an annual budget of $200,000 at their November 1954 meeting. Clearly McMath had an agenda in squeezing Aden into a $50,000 budget. Aden suggested that they leave those issues for the next committee meeting. McMath did not tolerate resistance from Aden, just as he had not from Aden's father-in-law, Edison Pettit (see Chapter 1, section "Marjorie Steele Pettit," subsection "Edison Pettit").[18]

Site survey

The Panel decided that Arizona should be given priority because its winter weather is better than that of California. Given the vagaries of weather fronts, the large telescope needed to be sited where the weather was out-of-phase as much as possible with coastal California and the East. No one had selected an observatory site based on detailed quantitative metrics for atmospheric turbulence, cloud cover, number of clear nights, and weather. Therefore, measurements would be needed at other sites, particularly locations of current long-term "reference" observatories such as Lick, Mt. Wilson, Mt. Palomar, and Lowell in order to locate the National Observatory in southern Arizona or New Mexico. Aden needed to examine options for designing and constructing the instruments and also make field measurements and fit theory to data in order to justify the recommendation of a mountaintop site for the observatory. Test towers rising 60 feet high were necessary to counter ground thermal effect that could affect the seeing measurements of the true quality of the site.[19]

Location requirements

By November 1955, one year after the first meeting of the NAO Panel, Whitford provided a clear summary of the rationale and a strategic plan for the construction of the National Astronomical Observatory in an address to the AAS in Troy, New York.[20]

Of paramount importance was the necessity that the site enable high-productivity, observational optical/IR astronomical research, and education into the foreseeable future. Aden had spent many years of his youth observing at Mt. Wilson and working in the optical and machine shops and in astronomical data reduction facilities and so fully understood the requirement for an optical/IR observatory. Several requirements, limited by technology in the mid-1950s, narrowed the search significantly. These were:

1. Atmospheric conditions permitting maximum number of cloudless nights per year, minimum atmospheric turbulence to blur images and reduce angular resolution, and maximum number of clear nights as a function of season (monsoons);
2. Minimization of background light scattered from nearby cities up to the atmosphere and then back down into telescopes and cameras;
3. Altitude of less than 8,000 feet for ease of sleeping and operations by astronomers, engineers, and support technicians; a national airport within an hour of the telescopes to allow East Coast astronomers to use them and maintain teaching and research responsibilities at their home institutions;
4. A flat area on top large enough to accommodate several telescopes for stellar astronomy (solar telescope sites had requirements for daytime

seeing; budgetary constraints eventually moved solar and stellar to the same mountaintop);

5. Adequate space for living quarters with kitchen and dining hall and also instrument shops for electrical and mechanical emergency repairs;
6. A low latitude in the continental United States in order to observe Southern Hemisphere objects;
7. Legal access to the mountaintop and highway right-of-way at minimum cost and schedule;
8. Sky background not affected by nearby cities and towns for over 50 years;
9. Easy access for visiting astronomers with amenities attractive to staff and their families.

These criteria limited the choice to a narrow strip of southern Arizona and New Mexico.

At the time, several additional factors were considered. The copper smelters were in full operation, blowing smoke into the sky. It was felt that visiting astronomers coming from sea level would find it hard to work at an elevation above 8,000 feet. For example, astronomers who use the observatory on Mauna Kea, Hawaii, today are required to spend a night at a lodge at 9,000 feet before traveling to the top at 13,800 feet to observe.

Before 1950, an East Coast astronomer would obtain his research data by scheduling three travel days on the train, a week to observe, and then three days back to his/her home institution, which amounted to a two-week absence. Most universities had only one astronomer, and he/she had teaching responsibilities, which made two-week travels difficult. By the mid-1950s, coast-to-coast air travel was common, and it became advantageous to have the observatory near a well-served airport.

Aden examined options for the survey, including the design and construction of instruments to measure those parameters of the atmosphere that affect scientific productivity in both stellar and solar astronomy. It was decided that test towers were necessary to get above-ground thermal convection effects that dominate the measurements of the true quality of the site. The Panel decided that Arizona should be given priority because its winter weather was generally better than that of California, and the nation should have a large telescope in weather that is out-of-phase with the weather of either the West or East coasts.[21]

Marjorie and Aden looked over photographs taken from above the atmosphere and found an excellent one taken with remarkably clear weather that showed all of the western states. Aden could easily pick out mountains from which they compiled a list of targets to be explored at closer range. One in particular attracted his attention. There was a significant range west of Tucson, the last mountain range between Tucson and the Gulf of California. A topographic map of that region showed that it was high enough that desert heating should not affect its seeing. It was Kitt Peak. Others that stood out were Hualapai Peak, Summit Peak, and Chevelon Butte.[22]

On March 17, 1955, Aden met with McMath and Goldberg at McMath's home in Ann Arbor. McMath told Aden that he was working for the Panel and that all decisions would made by the Panel. On March 29 the entire NAO Panel, along with NSF representatives and Aden, convened in Ann Arbor to discuss inter-disciplinary relations and NSF's Large-Scale Facilities Program and prepare for a meeting of the National Science Board coming up on May 19, when the status of the National Observatory was on the agenda. The concept of the NAO and the plans for its development were presented to the National Science Board that month, and the NSF Physics, Math, and Astronomy division received approval to proceed with the NAO Panel's plan for an optical astronomy observatory.[23]

Airborne reconnaissance

First on the schedule was air reconnaissance of possible sites in Arizona and New Mexico. Before leaving Yerkes for the Southwest, Aden arranged with Helmut Abt, who was at McDonald, to charter a small airplane and make a preliminary circuit of the sites. He found a pilot named John Casparis in Marfa, Texas, who owned a single-engine Cessna 140. They set out on May 7, 1955, for a three-day aerial reconnaissance to evaluate possible observatory sites in southern New Mexico and Arizona.[24] Abt took notes and photos of 33 mountains, including those west of Tucson that Aden had spotted in the photograph, the Baboquivari Range. He saw that the area on top of Baboquivari Peak was too small and that the range had no flat surfaces. But upon making a circle to the north, he saw Kitt Peak with its almost level summit and proximity to the highway between Tucson and the town of Sells. As it was part of the Papago Indian Reservation, development of city lights was not an issue.[25] The high mountains surrounding Tucson were excluded because of city lights, which would only get worse with time. Farther east, Mt. Graham in the Pinaleno Mountains was too high. Other peaks were ruled out for various reasons—too rugged, too barren, too forested, too smoky from smelting, too high, too steep, no flat area. Abt recorded that Kitt Peak, named by local surveyor George Roskruge for his married sister Philippa Kitt, was a real possibility and should be investigated further.[26]

On a six-day trip beginning May 24, 1955, Abt and Aden then made a ground survey by jeep of Arizona sites, specifically the Signal Peak in the Kofa Mountains, the Hualapai Mountains south of Kingman, Summit Peak south of Williams on the Kaibab Plateau, and Chevelon Butte south of Winslow, Arizona, also on the Kaibab Plateau north of the Mogollon Rim. The purpose of this brief survey was to assess how complicated it would be to obtain ground access to each of those Arizona mountaintops.[27] They would make another reconnaissance trip by jeep in August 1955, visiting Chevelon Butte, Hualapai Peak, and Kitt Peak, among others.[28]

Offices move from Yerkes to Phoenix

Some hardware for the test towers was arriving at Yerkes by the early fall of 1955, but already working out in the open at night was not pleasant when delicate adjustments would be required. Marjorie and Aden decided that as Arizona was the center of attention, they should establish an office in centrally located Phoenix, where tests and operations could commence. Marjorie was concerned about summer heat, having endured several summers in Tucson when her father made solar measurements at the Desert Sanitarium with A. E. Douglass and when Pettit helped Douglass install the 36-inch telescope in Steward Observatory. But air-conditioning now made life more pleasant in the 1950s.[29]

Aden, Abt, and Harold Thompson (a draftsman at the time, but later superintendent of Kitt Peak) drove from Yerkes to Phoenix on October 9–11, 1955, in Abt's Chevrolet convertible and with the Panel's approval rented an office at 221 E. Camelback Road in Phoenix. Aden hired Thompson, a secretary named Margaret Barker, an account secretary, and a purchasing agent. With that, the office was officially established.[30]

Arriving in Phoenix in November 1955, Aden and Marjorie found that escrow on the house they bought had been held up by the Federal Housing Administration (FHA), so they had to find temporary quarters. The budget was too tight to afford a motel for an indefinite period, so they moved into the office on Camelback Road. After a week the home was ready, and they moved into a four-bedroom, ranch-style house in a new subdivision in Scottsdale that had been a grapefruit orchard. It was low and rambling, but still a tight fit for them and their six children (with one more to be born in two years' time).[31]

Slowly but surely the team was growing. Harold Thompson soon arrived from Yerkes to head the field work. Aden gave lectures at the local astronomical society and found a young man who had exceptional optics-processing and machine-shop skills, Don Loomis. Aden hired him to become the observatory optician. Next to join was Joe Wilson, the youngest instrument maker at Yerkes. With the approval of Lick Observatory director Shane, Bill Baustian (chief engineer for the new 120-inch telescope there) agreed to come to Phoenix every other week to be the NAO chief engineer.[32] Even though the observatory site had not been selected, Aden started to design a cost-effective 80-inch telescope.

Aden was still associate director for Yerkes and McDonald Observatories and needed to return to Yerkes for the summer of 1956 when Strömgren traveled to his other observatory in Copenhagen. That was fine with the Meinel family because they would avoid some of the summer heat in Phoenix. They moved back into the Parkhurst House and settled in.[33]

In the fall they were happy to return to their Arizona home. Then as summer approached the family wondered if they should return to Yerkes. There was no need now, since things had smoothed out on the political landscape. Strömgren had left Yerkes and the University of Chicago to take a position at the Princeton Institute for Advanced Studies. Kuiper had replaced Strömgren as the new Yerkes director, and Aden was no longer needed as an associate director. However, he retained his associate professor position but without teaching responsibilities during the summer. So the family decided to remain in Phoenix in the summer of 1957 while Aden concentrated on directing the site survey, doing some astronomical research, and innovating a new design for the 80-inch telescope. He made rapid progress toward the National Observatory, laying out optical and mechanical designs for the two telescopes. The 36-inch would use a novel design for its polar axis that Aden had discussed with Clyde Chivens, who was president and chief engineer at Boller and Chivens, the leading telescope manufacturer of the time. Chivens's son David built a model incorporating some of Aden's innovations in improving the concept.[34] Working models were always essential to show scientists and managers without drafting experience how the telescope worked.

The University of Chicago received an application to the PhD astronomy program from a promising young man with a master's degree in physics who was interested in planetary science. Kuiper asked Aden, in his role as associate professor, to review the application and advise the department.[35] The student was polymath Carl Sagan, who was accepted and completed his doctorate in planetary studies under Kuiper in 1960. Apart from his contributions in astrophysics and astronomy, he would become known to the world as a respected science communicator, especially for hosting the PBS television series Cosmos.

Survey and selection of Kitt Peak

Site survey plan

Once candidate sites were selected based on about 18 criteria in all,[36] each one was visited by jeep, horseback, or on foot to place meteorological equipment for taking quantitative measurements of the atmosphere. The second stage required on-site measurements of microthermal fluctuations, motions of stars, and intensity fluctuations (twinkle) caused by atmospheric turbulence. The third stage required recording astronomical scientific data using 16-inch telescopes to make photoelectric photometry measurements of levels of stellar brightness and reducing the data and then comparing the data and scientific interpretation results to the performance at other known observatory sites. A small solar telescope was used at this stage to assess daytime astronomical seeing for the solar physics observations promoted by McMath.

The next step was to climb each mountain with instruments, measure the seeing, and assess the site further against NAO Panel criteria. It was believed that all of the sites on the short list were on US Government forest service property and easily accessible. However, the site that looked most appealing because of its proximity to a substantial city with a large airport and university was on the Papago Indian Reservation.[37]

Instrumentation

Designing, building, and fielding autonomous instruments to record the physical parameters related to seeing became a priority during 1957–1958. Instruments to measure cloud cover, rain amounts, and wind direction and speed were all part of standard portable autonomous weather stations and available commercially.

However, instruments to measure astronomical seeing, a subjective parameter never before well-defined, required ingenuity. The index of refraction of air depends on temperature, and the wind causes the turbulence that results in turbulent cells of refractive index change moving across the telescope aperture. The size of these cells changes with seeing and typically varies from 10 to 50 cm. With a 10-cm telescope and a 10-cm turbulent cell size, the star is observed to jump around in the field as a single point. For a 1-m aperture with 10-cm turbulent cells, the star image is large, breaks up, and appears to be boiling. For highest angular resolution, astronomers need the turbulent cells that drive the index of refraction variations to be large and few in number across the aperture. Measurements of small temperature variations as a function of time can be used to infer these index of refraction fluctuations.

Seeing monitors had been used at mature observatory sites, Mt. Wilson and Lick for example, but this was the first time that anyone had a large body of data recorded autonomously at a remote site. This breakthrough technology enabled Aden to recommend a new observatory site based on a large quantity of scientific data, which ensured that the best site (within other constraints) out of several surveyed would be selected.

Thermal fluctuations and seeing

Temperature of the air affects seeing, and additional measurements are made as a function of height above the ground. Temperature fluctuations in the atmosphere cause small local changes in the index of refraction, which act like a weak optical lens.[38] The larger the amplitude of the temperature excursions, the stronger is the weak lens. The more rapid the change in temperature, the more quickly the image boils. Aden built an instrument to measure the thermal fluctuations using knowledge gained from his father-in-law, Edison Pettit. In the mid-1930s, Pettit and his colleague Seth Nicholson were the first to measure the temperature of the bright and dark sides of the Moon using a thermocouple built at Mt. Wilson. There are several ways to measure temperature fluctuations in order to assess astronomical seeing.[39] Aden selected fast-response (low thermal heat capacity) thermocouples.

Polaris telescopes

Astronomers had measurements of how astronomical seeing varied with angular distance from the zenith. But there were few quantitative measurements of how astronomical seeing varied over many nights from mountaintop to mountaintop. Aden developed two sets of experimental apparatus to assess the condition of the atmosphere for astronomical seeing. One was direct optical measurement, and the other measured temperature fluctuations in the atmosphere above the ground.

The National Observatory project provided the first comprehensive assessment of astronomical seeing by recording image properties photoelectrically over hundreds of nights. The apparatus needed to be inexpensive. To keep cost to a minimum, the apparatus needed to operate autonomously without operator attention for extended periods of time under a variety of weather conditions at remote locations. Aden conceived, designed, built, and operated several copies of equipment that fulfilled this need for the site survey. Multiple copies were needed to assess several sites simultaneously to meet the schedule for the development of the National Observatory. After all, this was long before inexpensive computer-controlled apparatus and digital storage devices. Data were recorded on analog magnetic tape and displayed on chart recorders co-located on the mountaintop with the optical system and thermal sensors.

The apparatus used Cassegrain catadioptric systems with a 6-inch aperture and a focal length of 216 inches. The telescopes were mounted rigidly pointing at the apparent north celestial pole. The star Polaris is not precisely at the north celestial pole but rather 44 arcminutes away. Aden designed a simple catadioptric imaging system which had about a 7° field of view. The instrument was designed to produce an image of Polaris which moves in a circle 44 arcminutes in diameter just outside and below the spherical primary mirror of the telescope. This image fell onto an annular ring mask or reticle that was nested around the outside of the primary mirror. The reticle was engraved with alternate open and opaque regions to form a ronchi screen (Figure 7.1). As Earth rotated, the image of Polaris orbited in a circle around the outside of the primary mirror, striking the ronchi screen mask and alternately passing and reflecting light. Starlight that passed through the reticle entered a mirrored prism assembly that deviated the beam to the optical axis of the system. The light then passed through a Fabry lens to produce an image of the entrance pupil of the telescope, which filled the photocathode of a 1P21 photomultiplier tube. The spacing of the rulings on the reticle were such that during the night as the seeing changed, the signal to the photomultiplier would change from a square wave to a sinusoid, depending on how bad the seeing was.

To measure astronomical seeing, Aden designed a 6-inch Maksutov (Cassegrain) telescope of just the right focal length that when pointed to the north pole the image of Polaris circled around the outside of the primary mirror, making one full circle in 24 hours. The image of Polaris fell on an annular occulting reticule which was nested around the outside of the primary mirror. As Earth rotated about its axis pointed at the north celestial pole and time passed

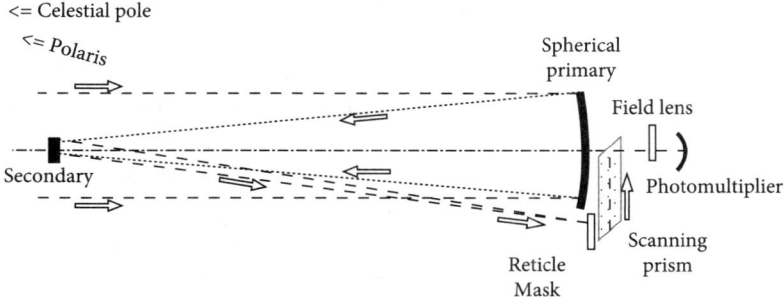

Figure 7.1. Optical layout for the Polaris seeing monitor. Light from Polaris, which is one half degree from the North Celestial Pole, enters from the left, reflects from the spherical primary to the secondary and to the image plane at the reticle mask. The image of the pole star Alpha Canis Minoris falls outside (off) of the primary mirror and is cast onto a fixed reticle pattern. This fixed pattern is an annular ring nested around the edge of the primary mirror. During the night, as time passes, the image of the star Polaris moves around the annular reticle. The reticle was engraved with alternate open and opaque regions to form a ronchi screen.
Drawing by James Breckinridge.

during the night, the image of Polaris would move around the reticle, and the light transmitted through the reticle would flicker on and off as atmospheric turbulence jiggled the image back and forth. Starlight passing through the encoded widths and spacing of lines in the reticule flickered on and off as the seeing moved the image around in the field. The transmitted light passed through a prism assembly (prism periscope) to an extension of the immobile telescope axis and then through a field lens into a 1P21 photomultiplier and electronics which recorded the fluctuations onto an analog tape or chart recorder. This intensity recorded as a function of time, automatically analyzing the image for sharpness and jiggle. The only moving part of the system was the prism periscope, driven by a small clock motor, to track Polaris around the reticule and send the beam to a 1P21 photomultiplier. It was not necessary to track the telescope across the entire sky. Minimal power was required to operate these automatic instruments.[40]

The telescopes were mounted in a 60-foot high tower (Figure 7.2) with two additional concentric wind-shield structures around the core tower to reduce vibration of the telescope mountings which would affect the telescope pointing. Wind-driven changes in the Polaris telescope pointing would cause signal changes at the photomultiplier that might be misinterpreted as contributions to poor seeing. Four towers with Polaris telescopes were built and eventually deployed, one each to Chevelon Butte, Summit Mountain, the Hualapai Mountains, and eventually to Kitt Peak.[41]

Deployment

The upper sections (with the Polaris tracker) of the first test towers were quickly set up in the backyard of the Meinel home in Scottsdale, and the electronics were tested successfully. One of the towers was originally to go to a site in the Sierra Ancha mountains, but it was removed because it had been placed on private land without permission. It would take too long to gain access to that site, and Summit Peak was selected instead. Chevelon and Summit had roads to their summits, so preparing the site for the towers was simple.[42] Roads needed to be constructed for the Hualapai Mountains and Kitt Peak, but the budget was far too small to afford that expense. Local citizens came immediately to help. Aden went to the Kingman Chamber of Commerce, led by Ralph Patey, a Midwesterner who owned a house on Lake Geneva in Wisconsin where Yerkes Observatory is located. Aden may have known him when they were both living there (personal communication from H. Abt to J. Breckinridge). The road up to Hualapai Mountain Park was excellent, but they needed to get to the summit of one of the peaks. They discussed whether the county could grade a tractor trail. A few days later, Patey called to say that the owner of a summit east of the Park agreed to allow the county to blaze the trail.[43]

Figure 7.2. At the summit of Kitt Peak, automatic Polaris seeing monitors were installed in the 10-foot and 60-foot towers with meteorological equipment on the mast. From left, Edwin Carpenter of Steward Observatory, Frank Edmondson, John Duncan (also of Steward), and Aden Meinel.
Courtesy of NOIRLab/NSF/AURA.

16-inch telescopes

Aden wanted two 16-inch telescopes to be used in the site testing. He had a pointed argument with McMath about building these, who said they were unnecessary and too large and expensive for the limited budget. When Aden received bids of $4,000 each for the 16-inch telescopes and their trailers, McMath

206 WITH STARS IN THEIR EYES

saw that they were indeed within budget.[44] Moreover, Harold Johnson, who had joined Aden in Phoenix as his principal aide in the site survey, wanted telescopes large enough that he could carry out meaningful photometry and backed Aden in the request for the two telescopes.[45] McMath relented, and one telescope was sent to Kitt Peak and the other to the Hualapai site. Claude Knuckles (Johnson's night assistant at Lowell Observatory), John Golson, and Leon Salanave joined the growing staff to carry out the fieldwork and the observations at these two main sites.[46]

Access to Kitt Peak to inspect the site was delayed until December 13, 1955, after protracted negotiations described in the following.

Papago permission

Aden's first glimpse of Kitt Peak was a photograph taken by a V-2 rocket launched from White Sands, New Mexico, in 1955. Kitt Peak is in the Quinlan Mountains (just north of the Baboquivari range), which are the highest and farthest west in Arizona and far enough from a city to have a dark sky. But there appeared to be no roads or even a trail to the summit. Aden and Abt consulted Ed Carpenter, a longtime family friend who was the chair of the Department of Astronomy at the University of Arizona. They reported to Carpenter that they had spotted what looked like an ideal mountain for an observatory.[47]

Edison Pettit had helped A. E. Douglass install and align the 36-inch telescope in Steward Observatory in 1931. By 1955 Carpenter was looking for another site for the telescope, far away from the city lights in Tucson, particularly on the high 10-mile long bajada of the Sierrita Mountains close to Tucson (where strip mining for copper ore developed in the 1970s) and not as far as Kitt Peak. A few days later, Carpenter called Aden to say that he had driven out toward Sells and saw Kitt Peak for the first time, hidden behind Coyote Mountain and not visible until one drove past Coyote. He agreed that it looked promising.[48]

The site was on Papago Tribal lands. Prior to 1938 there was no law limiting mining on Native American reservations by non-Natives. Native Americans had seen their reservations essentially looted and naturally were leery of white men and their government. But in 1938 Congress passed the Indian Mineral Leasing Act, which read: "[u]nallotted lands within any Indian reservation," or otherwise under federal jurisdiction, "may, with the approval of the Secretary [of the Interior] . . . , be leased for mining purposes, by authority of the tribal council or other authorized spokesmen for such Indians." It was not until the Indian Mineral Development Act of 1982 that Native Americans were granted full rights to the minerals on their land and were allowed to initiate, change, or terminate leases.[49]

Before Aden approached the Tribal officials, he first drove down to Tucson from Phoenix and had a long discussion with Carpenter regarding potential cultural sensitivities. Ed told Aden of the need for a careful approach with the Papagos and told him of past episodes when less circumspect proposals offended them.[50]

Carpenter sent Aden to talk more about the Papago Tribe with two professors in the Department of Anthropology, Rosamond Spicer and William Kelly. Spicer had lived with the Papagos in 1942–1943 and had compiled a dictionary of their language.[51] Kelly and Spicer emphasized taking everything slow and easy, saying little and listening intently. In a meeting with any Tribal body, they recommended to be there promptly but not to expect it to start on time. The entire Council probably would not be assembled until an hour or two later, and during the meeting the Council would be watching for any mannerisms signaling impatience.[52]

Permission to survey the summit of Kitt Peak

A few weeks later, Aden met with the Bureau of Indian Affairs (BIA) staff in Sells, where he was introduced to the chairman of the Tribal Council. He explained that Kitt Peak looked like a special site for astronomical telescopes and that he and his team would like to climb the mountain to see if it was large enough for the planned National Observatory. The chairman said that permission would have to be granted by the Schuk Toak District Council, which had jurisdiction over the mountain. Approval from the leaders of the village of Pan Tak was also needed. As everyone departed, the chairman said he would bring up the question at the next District Council meeting and communicate the results to Aden. About a month later, Aden received a call from the chairman to say that the District Council wanted Aden to meet with them on the following Sunday at 10 a.m. in the Schuk Toak schoolhouse about a mile from the highway to explain to them why Kitt Peak was important.[53]

Aden arrived at the appointed time and was met by one of the government staff from the Sells BIA named Jones. He cautioned Aden to speak slowly and in simple terms, addressing only the Tribal chairman, who would then translate for the others in attendance. Aden and his team went into the building and sat on the south side of the room on a bench lining that wall of the room. Some women were already beginning a fire outside to cook a pot of beans.[54]

It was noon before the Council was fully assembled and ready to start. The Tribal chairman sat on the same side as Aden and the team. The Council sat on the long bench on the opposite wall. About 20 members were present, some from the District and some from the Tribe. After preliminaries in the Papago language that Aden couldn't understand, the chairman turned to Aden and asked,

"Why are you interested in *Iolkam Duag*?" Jones leaned over toward Aden and said: "They call it "home of the clouds." It's one of their sacred mountains. The other is Baboquivari, which is their chief sacred mountain, the home of *I'itoi* [their mischievous creator god]."[55]

Abt related that the Papagos went to Kitt Peak to hunt and that *I'itoi* had the four winds trapped in a cave on the mountain. If anyone moved the boulder from in front of the cave, the four winds would escape and destroy the reservation. Therefore, everyone on the staff was asked to promise not to move the boulder should they come upon it during a hike (personal communication from H. Abt to J. Breckinridge).

Aden explained to the Tribal Council that their mountain, using the Papago name, was a special mountain, an "island in the desert" from which one might be able to see even to the edge of the universe. He asked permission to place two telescopes there for the astronomers and to climb the mountain and look north, east, south, and west. The chairman then translated what Aden had said, and then Jones leaned toward Aden and said, "That's interesting. He translated telescope into 'long eyes' and you to the 'man with the long-eyes.'"[56]

There was no response from anyone. But Aden had noted that even when the chairman addressed the District Council he did not look at anyone in particular. He simply talked to the opposite wall of the room. The members stared directly ahead to the wall in front of them. No one moved or even changed expressions. Finally, one man spoke up, again not looking at anyone in particular. Then, after a long pause, another. Then, again after a long pause, the chairman made some comments. After about an hour of this low-keyed exchange, the chairman said that they would need to consult with the leader of the village named Pan Tak just east of the Kitt Peak access road. He was an old man and keeper of a calendar stick. The mountain was especially sacred to his village for its petroglyphs that brought good luck in hunting. Villagers would make trips to a cave at the summit to place offerings to the "rain-cloud god" for a wet summer season.[57]

Aden and Abt made two authorized attempts to climb the mountain, first on foot and then by jeep. Both failed, so they requested permission from the Council to climb it on horseback with two Papago guides. A fair price was negotiated, and the Council also requested shoes and hay for the horses. A month later, the old man at Pan Tak gave his permission. They arrived at the summit just before sunset and took photos and movies with time-exposures of the sky glow from Tucson. On the way back down the next morning, Aden's horse slipped on a steep rock, sending him up over the horse. Although the horse did not land on him, his hand was broken. They later found out that the Papagos had not shod the horses, though shoes had been provided for them. A few years later one of the guides was killed when he fell off the same horse; the other guide was also killed in an accident soon thereafter.[58]

The goodwill shown toward the Papagos on the journey and the assurance that the mountain was also special to the astronomers got back to the Tribal Council, so when Aden requested permission to bulldoze a trail to the summit and place a testing tower and telescope there, it was granted without too much delay. Tests revealed that Kitt Peak was an excellent choice, as predicted. The next request was to build an observatory, and Aden promised that the telescope domes and public road would not be visible from Schuk Toak and Sells, not foreseeing the current array of telescopes that now stretch across the summit.[59]

The old man of Pan Tak declined permission, but the younger members of the District and Tribal Councils were inclined otherwise. Decisions were made by a majority vote, but the old man of Pan Tak had veto power, so all seemed lost. Then two events improved the outlook. First, the old man died, removing his veto. Second, Carpenter and Rosamond Spicer invited the Tribal Council to visit Steward Observatory and look through the 36-inch telescope. A date near the end of October 1955 was soon set up, with the Moon at first quarter so that the craters would show especially well at the terminator. One by one the Council members took a turn up on the observing platform. Each made a startled comment as he looked, smiled, and descended the stairs for the next man's turn. Carpenter concluded by saying that the telescopes to be built on their mountain would be much bigger and enable astronomers to see to the edge of the universe.[60]

Site survey campaign results

Site survey instruments measuring wind velocity, microthermal fluctuations, and cloud cover were operational at Hualapai, Summit, and Chevelon Peak from October 1956 through October 1957 and from November 1956 through December 1957 for Kitt Peak. Seeing was automatically recorded using the Polaris telescopes and directly assessed at the sites using the two 16-inch photoelectric photometry telescopes at select intervals and sites during this period. Observations at Kitt Peak were not without problems. The telescope swayed in high winds, even though the tower had a wind screen, an inner wind screen, and an innermost instrument tower. The anemometer was rated for 120 mph winds but blew away in the first winter storm. Aden was concerned that the site was so isolated from other mountains that it might be subject to frequent high winds. Fortunately, they never had winds as high again during the survey days. The total number of criteria for the selection of the observatory was 18. Kitt Peak was selected because it excelled at 16 of them.[61] Aden directed this comprehensive survey, prepared a detailed report,[62] developed the rationale for the decision he recommended to the Scientific Committee of AURA led by Shane, and presented it March 1 and 2, 1958. Kitt Peak was selected.

The formation of AURA

Two weeks after the Russian launch of Sputnik on October 4, 1957, the NAO Panel was replaced by a formal organization incorporated on October 28, 1957. The acceptance of this new organization by NSF removed the Universities of Chicago and Michigan from their business role and enabled Kitt Peak National Observatory to accept funds directly from NSF and truly function in the interests of the entire university community without the perception of a conflict of interest.

Leo Goldberg coined the name Association of Universities for Research in Astronomy (AURA). The original 1954 NAO Panel membership of the NSF group was now in 1957 augmented by one scientific and one business member from each of the seven universities that were asked to form AURA: University of California at Santa Cruz, University of Chicago, Harvard University, Indiana University, University of Michigan, Ohio State University, and University of Wisconsin. Yale University would join the following year. At present there are 47 member institutions. The new AURA "corporation" consisted of a Board of Directors, with McMath as its chairman, and Kitt Peak National Observatory (KPNO), with Aden as director. The Board eventually approved two divisions (Stellar and Solar) and the engineering services and Optical/Machine Shop. A Space Division came later. Missing from the organization was a contracting officer experienced in federal government contracting regulations, rulings, and law to interface with the NSF federal government contracting office. Before AURA was established, this function was performed by financial management divisions within a university administration. Jim Miller joined AURA to be the business manager at its founding. The University of Arizona was excluded from membership for what Aden thought were weak reasons—the astronomy department was not yet up to the same academic standards set by Caltech, University of Chicago, etc. It was only two years old and had yet to graduate a PhD.

This exclusion did not dampen the efforts of President Harvill, the Board of Regents, and Carpenter of Steward Observatory from doing their best to continue close cooperation with the new National Observatory. Carpenter had plans for some years to relocate the Steward Observatory 36-inch to a mountain site far from Tucson city lights. He had explored a site on the high bajada of the Sierrita mountains, between Tucson and Baboquivari. If Kitt Peak were selected, AURA would offer Carpenter a site for his 36-inch telescope, and Aden was instructed to find an area for it just north of where the test tower was located on the University of Arizona section of Kitt Peak.

Two years later, in 1959, AURA Board member Frank Edmondson applied for a senior faculty position in the Department of Astronomy at the university, but negotiations ended over disagreements on salary and benefits.[63]

The KPNO Tucson office

Through eminent domain law, President Harvill arranged for an entire square city block across the street from Steward Observatory to be made available for the Kitt Peak offices. A temporary office was set up in a commercial building close to campus. Aden and Bill Baustian worked with local architect Ed Varney to design an efficient structure for offices, laboratories, and optical/machine shops. The design was for construction in three phases, with one floor of offices and a basement for labs. The first phase was a simple building constructed at a cost of $180,000. Subsequent additions were made for laboratories and a state-of-the-art large optics shop at much higher costs. Offices for the Stellar Division were finished first (about 1961) and are in the front of the building. Offices for the Solar Division are in the middle (about 1962), and the offices for the Space Division, which became the Planetary Science Division, were located at the back of the lot.

Permission to build the observatory

About a month after the Papago visit to the observatory, Aden was asked to come to the District Council meeting to receive their decision. As he drove alone out to Schuk Toak, he wondered about what would soon transpire. He had reason to be optimistic because he had had a good relationship with both Councils. Only a single veto would end the opportunity to put an observatory on Kitt Peak, but the one certain veto was no longer in the picture. He carried with him a map of the site that had been approved by AURA. It showed a 50-acre summit region where the observatory would be built and also a much larger surrounding area of 1,500 acres that would be restricted from development. AURA would pay the tribe an annual lease fee for the summit as well as the reserve area. The crux of the matter came when he told them that the astronomers would want this lease to extend over 50 years. At any time during the 50 years, they could either renegotiate the lease or after that time the mountain and all on it would become theirs.[64]

They looked at the map. A sensitive question arose: What about employment for members of their Tribe? At that time unemployment on the reservation was high. Aden said that they would be given preference for jobs at the observatory, they could sell their handicrafts on site at no cost, and they would be given an annual stipend. President Harvill also had instructed Aden to tell them that the University of Arizona would set up scholarships for their children who wanted to become astronomers or technicians at the observatory. The Papago Tribal Council agreed to negotiate a long-term lease if Kitt Peak proved to be the suitable location for the observatory. AURA was told to come to the schoolhouse on March 5, 1958, for signing a resolution authorizing the lease of Kitt Peak to the National Science Foundation (Figure 7.3).[65]

Then the Tucson law firm of Boyle, Bilby, Thompson and Shoenhair took over the job of drawing up the resolution to authorize the lease agreement and obtaining approvals. On June 11, NSF received clearance from the Bureau of the Budget for the Kitt Peak legislation, introduced in the House of Representatives by Stewart Udall and in the Senate by Barry Goldwater. The bill was signed by President Eisenhower on August 28. Lease-signing by all parties was finally complete on October 31, 1958. Now everything could move forward toward construction and dedication of the Kitt Peak National Observatory, just two years away.[66]

Figure 7.3. A meeting of the Schuck Toak District Council and AURA representatives on March 5, 1958, at which the Council approved a resolution authorizing the lease of Kitt Peak to the National Science Foundation. Seated on left and right at the table are Mark Manuel (chairman of the Papago Tribal Council) and Chester Higman (administrative assistant to the chairman). Harold Thompson, engineer for the site survey, is standing at far left, and Aden Meinel is standing at far right.
Courtesy of NOIRLab/NSF/AURA.

The 36-inch telescope

The original plan called for a 36-inch telescope and an 80-inch. Funds for both were provided by grants from NSF to the University of Michigan in 1955 and 1956. The budget for the 36-inch was about $233,000, with the purchase of the mirror blank accounting for $114,000, almost half of that. Boller and Chivens of Pasadena won the bid to construct the mounting, and Murray J. Shiff Construction of Tucson won the contract for the building, dome, and hydraulic platform.[67] Optics for the 36-inch (0.9 m) arrived in Tucson from Ferson Optical. Don Loomis was now back in Tucson from his Lick Observatory apprenticeship working on the 120-inch. When Loomis and Aden tested the primary, it showed 3 attoseconds (asec) of astigmatism. Ferson said to send it back and he would re-test it, and if it had astigmatism he would refinish it at no cost. Because he had already taken a year longer than was originally agreed for delivery, Loomis and Aden decided to finish it themselves in Tucson on a polishing machine that they promptly assembled. To polish it to remove the astigmatism in the traditional way would have returned it to a spherical shape, requiring aspherizing it all over. But Aden had experience with defective Schmidt plates and knew that the simplest way to solve the problem was to do as he had done at Mt. Wilson when he was a teenager—simply rotate the rigid polishing tool on the mirror. It worked, and with a bit of smoothing with a smaller flexible polisher Loomis finished an excellent 36-inch mirror that remains in use today (Figure 7.4).[68]

The 80-inch telescope

Much of the cost of a large, ground-based telescope lies in the structure needed to cover the telescope during storms and bad weather. On a clear night this structure needs to move so the telescope can follow objects across the sky, complicating the design, construction, and cost of the structure. The distance between the primary mirror and the secondary mirror determines the size of the enclosed structure. The higher the f-number of the primary mirror, the longer the distance from the primary to the secondary mirror, the larger the structure needed to cover the telescope and rotate during the night, and hence more expensive.

Aden created a novel telescope design concept for the 80-inch in order to lower the cost significantly for both the telescope and its enclosure. If he could get the cost low enough, there might be room for another 36-inch telescope within the budget. An effective way to reduce costs was to use a fast focal ratio primary mirror, and Aden's initial engineering concept was for an 80-inch f/1.5. When presented to the astronomers on the NAO Panel, it ignited sparks immediately. Bowen, then director of Mt. Wilson and Palomar Observatories

Figure 7.4. Aden Meinel with the 0.9-m telescope and Cassegrain spectrograph.
Courtesy of NOIRLab/NSF/AURA.

and a founding member of the NAO NSF Panel, was adamant that the f/3.3 of the 200-inch was as fast as one could make a telescope to the quality required for astronomy. Aden argued that he had just completed five 36-inch telescopes at Yerkes Observatory in collaboration with Fred Pearson.[69] Those were f/2.5, and all delivered diffraction-limited images to the focal plane. Bowen relented and approved f/3.0 but no less. Aden also used Don Hendrix's argument that if a null test were developed the optician could make the required f/ratio. The Board supported Aden and reached a compromise: Aden could design the 80-inch for f/2.75 with mirror surface figures to give a Ritchey-Chrétien wide-field performance. The "as-built" telescope is f/2.6, and it performs diffraction limited over a full degree field of view using a corrector designed by Aden.

He could not use his f/1.5 design, and his super compact concept was ruled out. He seriously considered how to make the primary mirror of the 80-inch as light as possible, which would then permit a less massive and thus less expensive mechanical mount. He wrote a technical report outlining an extremely innovative design for a f/1.5 minimum mass telescope, which he called the "Miami" design, for "*Minimum inertia and mass instrument.*"[70]

He proceeded to lay out the mechanical design for the 80-inch that was finally built. Loomis made a scale model for show-and-tell, both to the Board and to the AAS at the next annual meeting. It was a traditional fork design, which called for a more complicated observer's platform than the one at McDonald. It required two degrees of freedom to clear that yoke, not just an up-down motion (Figure 7.5).

The NAO project team had expected Boller and Chivens to win the contract to build the 84-inch mount, but the bid by Willamette Iron and Steel Corporation of Portland came in under the Boller and Chivens estimate and won the contract. Willamette was a shipbuilding company that was failing yet thought that if

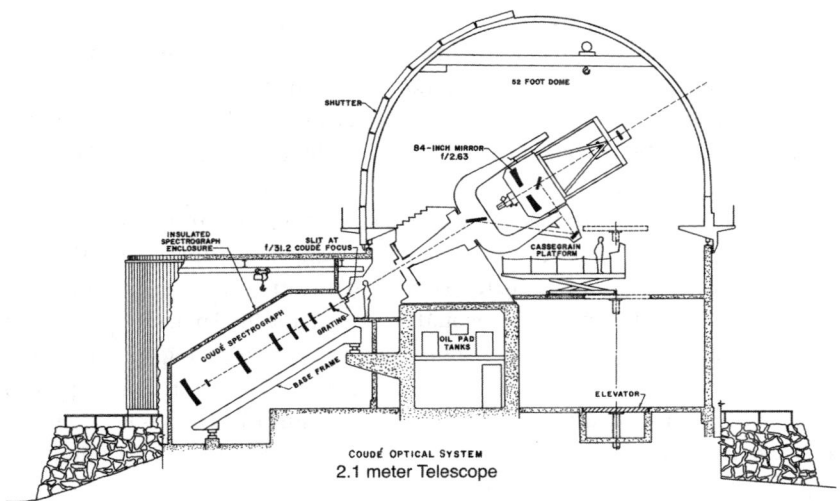

Figure 7.5. A cross section of the KPNO 2.1 meter telescope building shows both the standard telescope operation in Ritchey-Chrétien mode as well as the coudé optical system that feeds the large, stable spectrograph attached to the building. This system has been removed, as there is now a separate coudé feed telescope used with the spectrograph.

Courtesy of NOIRLab/NSF/AURA.

they built a large telescope, obviously a precise engineering job, they would get more business, but that failed. They encountered various difficulties not directly related to the telescope, which caused cost overruns that led to bankruptcy.[71] A few years later, Boller and Chivens did win the contract to build the coudé spectrograph that was intended to be used frequently. The coudé feed telescope was added in the early 1970s that completely separated the coudé spectrograph operations from the Cassegrain operations of the 84-inch telescope.

The mirror

The telescope primary mirror and its mechanical mounting are the largest, most precise subsystem in all astronomical reflecting telescopes. For ground-based telescopes, the direction of the force of gravity changes continuously during the night. The reflecting surface of the mirror needs to be shaped into a hyperbola to a precision of a few nanometers (1 nanometer = 1 billionth of a meter) over its entire aperture, in this case 80 inches (2.1 m). And the motion of this large surface needs to be mechanically constrained to move less than a few nanometers in surface error. In addition, the tilt angle relative to the axis of the hyperbolic secondary mirror cannot move more than 0.01 arcseconds while the telescope scans across the sky, following the motion of the stars during the night.

For the structural design of the primary mirror, Aden proposed a design similar to that for the Palomar 200-inch, more correctly like the 120-inch that was then being polished for the new Lick telescope, and traveled to Corning to discuss its manufacture. He had encountered a new technique there during his Air Force work and felt this new process would make ribbed, lightweight, large mirror blanks without the headaches of the 200-inch casting. He proposed to cast the large, plane-parallel mirror blank as usual but to add a new final step to the process to reheat the glass and slump the assembled mirror over a convex mandrel so that the concave reflecting surface would be close to the final machined, near-net shape needed for the telescope. A significant cost of processing a large, plane-parallel mirror blank is the time required to grind or "hog-out" the deep bowl concave curve shape to achieve small f-numbers. Using this process leaves an excessive mass of glass away from the mirror center, which adds weight and undesired thermal inertia to the mirror but is not needed for structural purposes. He was the first person to use this technique to manufacture a large telescope mirror.

At the time the NAO project team was discussing options with Corning, the company was in the process of developing Pyroceram, a ceramic glass that had almost zero coefficient of expansion, which made it desirable for telescope mirrors. But at that time Corning couldn't cast a piece big enough that could be useful for the telescope, so Pyrex was adopted. (Pyroceram eventually evolved for use as CorningWare because of its resistance to breaking.) Interestingly, the Pyrex glass used for the 80-inch blank was from the same glass melt as that used for the 200-inch, although it turned out to have a larger coefficient of expansion than the larger mirror.

Aden set up a meeting at Corning to discuss his ideas. Corning asked glass physicist George V. McCaulay to come out of retirement to give his thoughts, based on his experience managing the casting of the 200-inch mirror blank and its annealing activities about 25 years earlier. He used his Caltech mechanical engineering training to design the mirror cut-out pattern necessary to light-weight the blank and still maintain the needed optical surface figure. McCauley approved the lightweight design Aden had prepared and believed the re-melting process to be an excellent choice.

Visiting Corning as the mirror was ready for firing, Aden witnessed something spectacular. The mold was piled high with pieces of Pyrex cullet. Corning had filled the furnace with new mix and melted and refined it. After Corning finished the 200-inch mirror blank, they then tore down the furnace and stored the pieces along the banks of the Chemung River just behind the furnace area. Most of it was still there 15 years later when Aden first went to go over plans for the work. Now the selected pieces were ready for re-melting. In the center was the largest piece, surrounded by a single ring of smaller chunks. It looked like a royal crown.

The time came to slump the mirror structure over the convex mold. The glass was heated to just below the temperature where it begins to flow, and all waited for the blank temperature to equilibrate throughout the disk. Then Aden took control of the temperature knob. After all, he had bought the mirror, and it was his to risk the slumping operation. Viewing through a window in the tank he gradually raised the temperature until he could see the 80-inch disk of light-weighted glass begin to move and sag over the convex mandrel. He decided when to lower the temperature, and the mirror blank stopped its sag. After several weeks of cooling, the glass was annealed and the oven opened to reveal an 84-inch diameter concave mirror blank. The light-weighting cores under the surface were intact but slightly distorted by an insignificant amount.

The mirror was now 84 inches in diameter instead of 80 inches, making it larger than the 82-inch telescope at the McDonald. This displaced the Texas telescope to fourth-largest in the world and made the KPNO telescope the third-largest in the world at that time. Aden was concerned about the political sensitivities but proceeded to arrange for the grinding, polishing, and figuring of the mirror. The blank arrived in Tucson in late 1959, and work began by early 1961 to grind and polish.

Grinding and polishing the mirror

Because the planned f/ratio was faster than any previous large telescope mirror, it was not known if some of the standard optical fabrication techniques would work. It was decided to try to achieve this f/ratio on a 16-inch model. The opticians did run into some difficulties, but the solutions they found proved that it would be possible to polish the front and back surfaces of the 84-inch blank successfully.

The observatory had always planned to polish the 80-inch mirror at the Tucson Optical Shop. Aden would teach Loomis using the 36-inch mirror, but more was needed. Don Hendrix was then surface-figuring the 120-inch mirror at Lick, where Shane was then director. Aden asked Shane and Hendrix if Loomis could go to Lick and serve as an apprentice to Hendrix in the figuring of the 120-inch mirror. Young Aden had been Hendrix's assistant on the Mt. Wilson NDRC projects before World War II, and they had mutual respect. Aden asked Hendrix what he thought of the idea and received a positive response. Then Aden approached Shane with a proposal. The KPNO would pay Loomis's salary, and he would work free for Shane and Hendrix at Lick, learning how to grind, polish, and figure large astronomical telescopes mirror during his apprenticeship.

Aden soon got a call back that the arrangement was approved, only Loomis could not have a mountain house. Because Loomis was married and had a family, he would need to live in San Jose and commute each day up to the observatory. Loomis agreed to this arrangement, moved his family to San Jose, and went off to prepare his skills over the next few months as the National Observatory started building its telescopes. He did well, but the daily commute up and down the Mt. Hamilton road was tiring. This came through loud and clear when his Volkswagen turned over on the only straight stretch of road. Although he was not injured, Aden felt it was time for him to return to Phoenix.

Corning delivered the lightweight mirror blank to the KPNO Optical Shop in the fall of 1959, and grinding and polishing began. Aden had taken the precaution of making the outer lip diameter 84 inches so that an optician would have excess glass to enable him to figure the mirror perfectly to the desired 80-inch diameter (Figure 7.6). The extra four inches were to allow for the typical downturned edge. If necessary, the edge could be masked to create a diffraction-limited 80-inch mirror. Loomis did such an excellent job surface-figuring the mirror to its extreme outer edge that the 80-inch telescope became the 84-inch Kitt Peak telescope. No mask was needed.

Most modern telescopes now have hyperbolic shapes that provide wide, coma-free fields of view, as was achieved with the 84-inch. The KPNO Monthly Reports always describe the system as a "coma-free paraboloid." This phrase, now commonly used to describe Ritchey-Chrétien optics, does not appear until

Figure 7.6. Aden Meinel inspects the slumped Pyrex primary mirror for the 2.1-meter telescope that he designed along with William Baustian and others. The first of its kind, its relatively fast focal ratio of f/2.63 allowed a relatively fast Cassegrain focus.

Courtesy of Stephen F. Jacobs.

1962 in a private note by optics engineer Dan Schulte. The camera at focus has a 35-arcminute field of view.

The legacy of the telescope

The 2.1-m telescope was the first telescope optimized to take advantage of changes in seeing during the night. The flip-top secondary mirror enabled a split observing program. For example, Cassegrain focus before midnight and coudé after midnight doubled the efficiency of the telescope. The fast primary, reduced cost of the dome structure, and the slumping process cutting in half the time required to grind, polish, and figure the primary mirror resulted in additional savings. Summarizing, Aden's innovations with respect to the 84-inch were (1) an unusually fast f/2.6 primary and f/8 Cassegrain focus that allowed for a small, cheaper dome, (2) Ritchey-Chrétien coma-free optics, (3) a Pyrex mirror made with the slumping process that was ground and polished in

1.5 years, (4) a flip-top secondary mirror allowing five-minute conversion between Cassegrain and coudé foci, and (5) fast, high-resolution spectrographs.[72] Those features led to the discovery of the Lyman-alpha forest, the first gravitational lens, the first pulsating white dwarf, and the realization that most solar-like stars have companions.[73] Such discoveries were possible because of (1) fast, high-resolution equipment, (2) competitive time scheduling, and (3) relatively long observing runs that allowed for experimentation. Today the telescope is scheduled every clear night and has state-of-the-art, large-array detectors, and an innovative adaptive optics system (Figure 7.7). These features enable the telescope to remain highly productive scientifically almost 60 years after "first light" (the first use of a telescope to take an astronomical image). The 84-inch was an engineering pathfinder for several telescopes: (1) the University of Hawaii 2.2 m (88-inch) on Mauna Kea, (2) the Steward Observatory 2.3 m (90-inch), and (3) the Sahade 2.15 m (85-inch) in Argentina, which was modeled after the Steward 2.3-m design. It is a tribute to the foresight of Aden's thinking in the late 1950s that most of the large modern optical telescopes have primaries that are faster than the f/2, extremely low in weight considering the size of the primary, and Ritchey-Chrétien configurations like the KPNO 2.1-m.[74] More details of the development of the KPNO 84-inch as written by Aden are available in chapter 3 of *Stars and Stellar Systems*, Vol 1: *Telescopes*.[75] Aden's intimate experience with the hands-on fabrication of a large-aperture, primary-secondary mirror pair that produces a Ritchey-Chrétien system would come into use later when he diagnosed the problems 25 years later with the Hubble Space Telescope at NASA/JPL.

By March 21, 1960, McMath had pushed Aden out of AURA and had removed his responsibility for finishing the mirror. Nick Mayall took over the Optical Shop and implemented Aden's optical and telescope design. Aden, in his usual unselfish manner, continued to contribute to the telescope optical engineering when he was director of the Steward Observatory.[76] The telescope was dedicated on September 15, 1964, when Aden was director of Steward Observatory, deeply involved in their 90-inch telescope, and leading the establishment of a center of excellence in optics, the Optical Sciences Center at the University of Arizona.

US Army threatens Kitt Peak

In 1958, shortly after the selection of Kitt Peak for the National Observatory, the US Army Electronics Proving Ground (AEPG) at Fort Huachuca, Arizona, started plans for a new test range for air support electronic equipment. Airplane drones would fly across the 266 miles of southern Arizona from Fort

Figure 7.7. The 2.1-meter telescope at KPNO with the white cylindrical spectrograph at the bottom.
Courtesy of NOIRLab/NSF/AURA.

Huachuca to Yuma, transmitting and receiving electronic communication and radar signals in support of the ground fighter. Aden was informed about this by an architect friend who had heard about it during an annual meeting at Fort Huachuca.[77]

Aden was well aware that astronomical measurements at Mt. Wilson were shut down permanently because of the many television transmission antennas adjacent to the observatory. Radio frequency transmissions interfered with sensor electronics and made sensitive, precision photoelectric photometry impossible for the astronomers. At one point, conducting copper screen wire was placed over the open dome slit and the dome was well grounded, but those extreme measures were insufficient.

Aden telephoned commanding General F. W. Moorman at Fort Huachuca for more information and asked Ralph Patey, Observatory business manager, to write to Colonel George Moynahan, director of Combat Development, AEPG. Aden and Patey met with Colonel Moynahan and members of his staff at Fort Huachuca on January 7, 1959, and learned that Kitt Peak itself would be used as a location for a tracking radar and a microwave relay, and the drone airplanes would pass just north of Kitt Peak and well south of Tucson. During the meeting Colonel Moynahan repeatedly emphasized that his project was related to national security and much more expensive than that of the observatory and thus much more important.[78]

James Edson, Aden's friend and mentor from his days 18 years earlier as a member of the Planet Group (see Chapter 2, section "Youth and education") and Navy rocket project at Caltech (see Chapter 3), was now the assistant secretary for research and development of the Army. Aden telephoned Edson and explained to him the problem he was facing. Aden reported to Shane (president of AURA) that he had talked to Chester Higman, business manager for the Papago Tribe, on the afternoon of January 7, 1959. Higman and the Tribe had no knowledge of the plans that AEPG had for the Papago lands. No one, not even the military, could enter and use parts of the reservation without the consent of the Tribe. Higman was convinced that the Tribe would honor the terms of the Kitt Peak lease limiting the mountain and the area around it to scientific work related to astronomy. On January 9, 1959, AEPG released a news story that said "since much of the land between Ft. H. and Yuma is owned either by the federal government or state governments, only a few acres of private land will need to be acquired." The AEPG officials quoted in the news release did not know that Indian reservation land is tribal land and not federal government land. After discussions with NSF, AEPG at Fort Huachuca quietly withdrew their plans to use Kitt Peak and moved the flight path of the drones 20 miles south.[79]

Edson would come to the assistance of Aden again in 1959 in his role as assistant secretary for research and development, this time in a meeting held at Red Stone Arsenal on August 12, 1959, to discuss the KPNO Space Division.[80] Attendees included Aden, Russ Nidey, Wernher von Braun, Edson, and Gerard Mulders (NSF). Aden presented plans for the 50-inch, remotely controlled telescope and showed how this technology development was applicable to space telescopes for reconnaissance, intelligence, and astronomy.

Astronomy from a space platform

In 1955, Eisenhower's press secretary Jim Hagerty announced that the president had approved plans for going ahead with the launching of small Earth-circling satellites as part of the United States' participation in the IGY.[81] A federal committee was formed to detail the engineering and scientific merit by planning a group of space experiments in support of the IGY. McMath, in his role as president of the AAS, was selected to be a member. This was the same year he was chairman of the NAO committee. His service on this committee should have brought him valuable experience to apply to the AURA vision a few years later for a 50-inch astronomical space telescope.

The launch of Sputnik in October 1957 stimulated several papers on the exciting scientific research possible using astronomical telescopes in space. Aden was one of the first professional astronomers to recognize the potential advantage of astronomical telescopes in space.[82] In a peer-reviewed scientific journal, he presented a detailed description of the scientific discoveries possible based on data from space telescopes. He carefully outlined the spacecraft and optical engineering challenges to developing space telescopes and instruments. His paper was followed by others.[83] Aden presented his work on astronomy with space telescopes at an International Astronomical Union conference in Liege, Belgium, in 1960 and summarized his work in a paper for the proceedings.[84] Earlier, Spitzer had written an appendix to a classified internal technical report for the RAND Corporation in 1946 on astronomical observations from space.[85] But that document was not made available to the public until 1990 when it was reprinted by Pergamon Press.[86]

Lloyd Berkner, chairman of the National Academy of Sciences group on space sciences problems, sent a telegram to McMath for the AURA scientific committee regarding telescopes in space for astronomy. The telegram was read by Shane at a meeting of the AURA science committee held on July 9, 1958, nine days before President Eisenhower authorized a new civilian agency, the National Aeronautics and Space Administration (NASA). Berkner wrote, "I introduced this subject to get the Scientific Committee thinking about it since I feel it is something we cannot ignore, though there appears no way at present for us to contribute to it." Six months later, on January 21, 1959, Aden wrote to Shane, "We are rapidly nearing the completion of creative work on the stellar instruments and my attention has been directed to pending projects in the placement of optical telescopes in space vehicles. This field looks very attractive, inasmuch as my interests lie in the planning and design of new observing techniques."[87] Aden wrote that he hoped that AURA would permit him to devote a major portion of his efforts toward such a program.

Aden proposed the Space Division at Kitt Peak in 1959 with a program to place astronomical instruments aboard rockets and launch them from White Sands, New Mexico. Also included was an effort to develop a remotely controlled telescope, which was a necessary technology for operating telescopes in space. Building telescopes and instruments that could survive launch was expensive, and this cost was in addition to the cost of the expendable rocket needed to lift the telescope into orbit. Unlike ground-based telescopes, which are accessible for adjustments and equipment changes, space telescopes must be launched fully ready to operate at peak performance; they cannot be revisited for new instruments and repairs like the modern Hubble Space Telescope (HST). The enabling legislation for NSF required that its funding be spent on compelling scientific research, not engineering and building space vehicles. That was the responsibility of NASA. It became obvious that about 90 percent of the cost of pursuing astronomy from space was in the launch vehicle, the space telescope, and engineering to assure the survival of the astronomical instruments. The AURA Board soon approved the organization of KPNO into three operating units or divisions: Stellar, Solar, and Space. Keith Pierce, a distinguished and respected solar astronomer, was in charge of the Solar Division. McMath considered himself a solar astronomer and so collaborated with Pierce in the operation of that division. Aden managed the Stellar Division and the Space Division.[88] With his past experience in rocket science, it was natural that Aden would take significant interest in the Space Division.

When Aden was the director of the Observatory, director of the Stellar Division, and manager of the Optical Shop, he pointed out that it might be necessary to add an astronomer to the staff to be in charge of the Stellar Division in order to provide him (Aden) with enough time for this new activity in addition to his duties as Observatory director.[89] The AURA Board never added staff to help Aden but rather added the burden on him to prepare proposals and plans for a major space astronomy effort at KPNO and to "market" those proposals to NSF, NASA, and the Department of Defense. Shane wrote the following to McMath: "After phoning you yesterday I talked to Code about the Space Vehicle and Princeton matter. . . . We have to get into this business or be left behind, and if we can get Princeton to come in and have N.S.F. support we should accomplish great things. *But this must not be at the expense of the projects now under way* [emphasis his]. It may however replace the X-inch [telescope]. . . . The only persons who know about the Space Vehicle and Princeton Business are McMath, Code, Shane, Meinel, Patey, and the latter only because he occupies the same office with Meinel and would have overheard conversations. After January 30 we can increase slightly the number who know about it, i.e. if all goes well."[90]

On January 29, 1959, Shane and Aden, along with then AURA Vice President Edmondson, presented plans for the Observatory's involvement in space astronomy to NSF and Princeton on January 30. The NSF supported the idea, and Princeton joined AURA to facilitate cooperation in a space astronomy program. On March 1, 1959, the Scientific Committee of AURA discussed the proposed space program (which probably included Aden's plans for a 50-inch space telescope) with the entire AURA Board and received a positive response.[91]

A meeting to discuss the AURA space astronomy program with NASA headquarters was held on May 26, 1959, with Nancy Roman (head of the Observational Astronomy Program at NASA), Gerhardt Schilling (chief, NSF Astronomy and Astrophysics Programs), and G. Keller (NSF program director for Astronomy) in attendance. The AURA Board approved a $160,000 proposal to NSF for a conceptual design study of an astronomical space telescope. At the AURA Board meeting on June 2, 1959, Aden reported that AURA's interest in astronomy with space telescopes had been presented to the Army, Advanced Research Projects Agency (ARPA), Army Ballistic Missile Agency (ABMA), and the Air Force. Waterman signed the NSF grant letter on June 9, 1959. Aden reported that there was some degree of reluctance on the part of NASA to the AURA program.[92]

To assess the extent of interest, NASA later convened a study conference on the science and technology readiness for large space telescopes held at Woods Hole Oceanographic Institute in June 1965.[93] Earth's atmosphere absorbs UV radiation below about 300 nm wavelengths and radiation in select regions in the infrared. Observing in those regions of the spectrum requires a space observatory. However, less certain was the scientific requirement for building large space telescopes for astronomical research in the visible range (between 300 and ~5,000 nm). Spitzer chaired the conference, which was attended by Aden, Goldberg, Code, Whipple, Münch, and others. The report called for the construction of a 120-inch space telescope.[94] Roman, often referred to today as the mother of Hubble, was initially and loudly opposed to it but came around after being persuaded by Code.[95]

In 1967, three US military contractors with expertise in the design and construction of cameras, telescopes, and optical systems for Department of Defense applications received contracts from NASA to investigate the feasibility of designing, building, launching, and operating a 120-inch telescope in space for astronomical applications. These were Itek, Perkin-Elmer, and Kodak Federal System Division. Each of these three contractors concluded that astronomical telescopes and instruments on orbiting spacecraft were feasible. They identified several subsystems that needed technology development and reported no insurmountable barriers from their optical and aerospace engineering perspectives for mirror coatings, fabrication of large optical mirrors, detectors, and coating mirrors in space. These three reports constituted the core studies funded by NASA in preparation for the proposal issued by NASA for the Large Space Telescope (LST), renamed Hubble Space Telescope (HST) a few years later.

Aden had described a KPNO space telescope program centered on the development of a 50-inch space telescope.[96] He selected the 50-inch size based purely on his interpretation of the needs of the space astronomy community and his estimate of the telescope/instrument mass and his knowledge of the lift capability of the Saturn V rocket. He used his four years of experience in developing rocket technology at Caltech to develop a credible space systems scenario from launch to observatory. This optical telescope would provide an angular resolution that could not be achieved from a ground observatory even under the best possible seeing conditions. Because it was above Earth's atmosphere, it would record data in the infrared and ultraviolet regions of the spectrum which Earth's atmosphere absorbed.

In 1959, no space telescope of any size had ever been designed, built, and launched. It was not until nine years later, in December 1968, that the first 8-inch telescopes of the Orbiting Astronomical Observatory (OAO) were launched into orbit.[97] To propose a 50-inch space telescope in 1959 appeared absurd to most, especially Roman. About his idea for a large aperture orbital telescope in 1964, she wrote, "Meinel's present attitude reminds me very much of his earlier attitude toward the 50-inch, 24-hour telescope. . . . I cannot help but feel . . . that Meinel is letting his enthusiasm carry him overboard for more projects than he can successfully manage, particularly in view of his past reputation as a very poor administrator."[98] But to Aden this was a long-term creative vision, intended to direct and provide focus for technology-development programs.

In May 1959, Aden wrote to Shane that he had met in Tucson with representatives of the ABMA from Huntsville, Alabama. They discussed technical and political aspects if the 50-inch space telescope were to be launched in a Saturn rocket, comprising seven Jupiter rockets with a combined thrust of 1,500,000 pounds, which was expected to be operational in 1963. The ABMA invited them to come to Huntsville and discuss matters further with the staff, then led by Wernher von Braun. The team from Kitt Peak made that visit on August 12, 1959, with Edmondson representing AURA, Aden and Russell Nidey representing KPNO, Mulders representing NSF, and Edson from the US Army. At the meeting, Aden made a presentation on payload characteristics. The afternoon was spent speaking with von Braun in his office and then watching a static test firing of a Jupiter rocket from the blockhouse. The ABMA staff had prepared a three-foot-tall model of a Saturn rocket with the AURA telescope in the nose cone.[99]

KPNO career sunset

On March 5, 1960, the mirror of the 36-inch telescope was installed on Kitt Peak, and first light was obtained the next night.[100] Kitt Peak had its first major research telescope. Kitt Peak National Observatory was to be dedicated 10 days later, on March 15, 1960. But behind the scenes, rumblings about Aden's administrative decisions had been percolating. The NSF complained about Aden's disregard for his contractual stipulation that no expenditures, purchases, or commitments in excess of $100 were to be made without the approval of the Foundation. John Luton, assistant director for administration of NSF, chastised Aden for paying several hundred dollars for engineering services without permission. He had overspent the FY 1960 budget by almost $100,000.[101]

Resignation

Kuiper became director of Yerkes again when Strömgren left there in 1957 to join the Princeton University Institute for Advanced Study, where he was assigned Einstein's office. Kuiper was director for three years before Strömgren arrived at Yerkes in 1950. At the University of Chicago, department chairs were selected by votes from the tenured faculty. Kuiper's astronomical research centered on the planets and their satellites and the Earth's Moon, but Yerkes faculty members were interested in stellar evolution, cosmology, and theoretical astrophysics. On January 4, 1960, the faculty voted to replace Kuiper with W. W. Morgan. Kuiper resigned the directorship of Yerkes Observatory but remained on the University faculty. In his role as director of Yerkes and McDonald Observatories, Kuiper had been a member of the AURA Board and a strong supporter of Aden and his work for many years. It was he who wanted Aden to do his PhD at the University of Chicago in 1946, and he supported Aden's work on atmospheric optics and auroras when Aden was a research faculty member at Yerkes.

In the early spring of 1960, Aden attended an AURA Board meeting at the University of Chicago. He received a call from Kuiper, who wanted to meet Aden at the University Faculty Club the next evening. Kuiper proposed to Aden that he move his entire research group of about a dozen persons from Yerkes to the University of Arizona. He told Aden that he had research contracts that would make his group completely independent of Carpenter and the Department of Astronomy. He also said that he expected some sizable support from NASA toward a specialized telescope, as well as a contract to build a Lunar and Planetary Laboratory (LPL), some nine months before President Kennedy's inaugural speech that outlined the "race to the Moon." But visionary NASA managers were already planning on future space missions to the Moon and beyond, and they needed not only accurate facts about the Moon for planning, but also the support of a senior reputable astronomer of Kuiper's stature. Aden concluded the meeting by saying that he had always had friendly relations with Harvill and that he would be glad to relay the proposal to him. Kuiper emphasized that Aden should mention NASA's pending contract for the LPL to Harvill. Harvill was excited at the prospect and immediately telephoned Kuiper while Aden was there in his office. He invited Kuiper for an early visit and asked Aden to arrange Kuiper's trip through the University's coordinator of research, David Patrick.[102]

Then it hit. Several AURA Board members were incensed that Aden would talk with Harvill without their approval, and worse yet, leave Carpenter out of the loop. Aden already knew that if he had asked Carpenter to make the arrangements that Carpenter would be unhappy at the prospect of a dynamic person like Kuiper entering the scene through Steward Observatory. Aden responded to the Board that he viewed it as a personal message carried between Kuiper and Harvill and that as Kuiper's friend he was happy to help open the door to him. Aden did not believe that Carpenter would be unduly concerned about the future of Steward Observatory and the Department of Astronomy because Carpenter had already played his trump card, but the AURA Board did not yet know about it.[103]

Why would Carpenter not mind? Bart Bok from Harvard had stopped by Aden's office in Phoenix in early 1957 on his way to Australia. After he was passed over for the directorship of Green Bank Observatory in West Virginia, Bok wanted to get away from the American scene altogether. He had accepted the directorship of Australia's Mt. Stromlo Observatory and was traveling by car through the United States to take a cruise from California across the Pacific. He thought Aden would become the director of Kitt Peak, so he confided in him the day before that Carpenter had asked him (Bok) if he was interested in becoming the new director of Steward Observatory upon Carpenter's retirement in about five years. Harvill extended the offer, and Bok accepted.[104]

By now, Aden had realized that his days as director were numbered, probably in single digits. He had refused to entertain McMath when the latter visited Tucson for AURA Board meetings and left that social burden to another Board member, Jim Miller. Miller had been appointed associate director of administration and, according to Aden, had assumed many of the director's duties and in fact had to approve anything that Aden wanted to do.[105]

So a few minutes before a meeting of the AURA Board, Aden asked Edmondson to come into his office at Kitt Peak. Aden told him that he felt it was in the best interests of the observatory and his own health that he resign as director. He felt that they now needed an administrator with more management skills and social graces.[106] He had academic tenure and would continue as the director of the Stellar and Space Divisions and the Optical Shop.

On March 14, the day before the dedication of KPNO, Aden submitted his letter of resignation to the AURA Board from the Observatory directorship, effective March 31. The Board accepted it and decided not to release the information until after the dedication the next day. However, someone at the meeting leaked the information to the *Tucson Daily Citizen,* which ran the story on the same day as the dedication. Nonetheless, Aden gave a speech on that special day to emphasize the bright future of KPNO. Aden's replacement was Nicholas Mayall, his thesis advisor at Lick Observatory.[107]

For the next 18 months, Aden concentrated his efforts on the development of instruments and technology for scientific measurements and space telescopes and consulting for government and industry. Now free of close Board oversight, Aden threw himself into photoelectric photometry, spectroscopy, and instrument development. Traveling and lecturing, he promoted the need for new, inexpensive telescopes and instruments. The lack of administrative duties enabled him to continue a long collaboration in stellar spectral classification with Abt that was begun when they were both at Yerkes, to work on the optical test and figuring of the new 84-inch telescope, and to help Carpenter plan for the development of University of Arizona telescopes on Kitt Peak. Aden established a program to develop innovative electronic imaging systems to take advantage of the higher quantum efficiency (qe) available using photo-cathodes compared to the qe of the photographic plate.

During this period Aden published several papers and a book chapter summarizing his work on large telescope design and engineering for the KPNO. These included the engineering description of an all-sky camera for motion pictures of auroras, design for a large-aperture space telescope, astronomical seeing measurement and instruments, and fundamental principles for the design of reflecting telescopes. Aden was immersed in the final alignment of the 84-inch telescope at KPNO and recording and analyzing spectra using his spectrograph at the 36-inch telescope. He also helped Morgan prepare *An Atlas of Low-Dispersion Grating Stellar Spectra* using the spectrograph that he had designed for the new 36-inch telescope.[108]

One of the programs Aden started at KPNO was the assessment of electronic imaging devices for use in astronomy. The photographic emulsion had been in use for almost 100 years as the medium of choice for recording images of celestial objects. Further advances in quantum efficiency and photometric accuracy in this technology were not possible. It was clear that new astrophysics applications required electronic imaging systems, not only for ground-based telescopes. Returning film from spacecraft in orbit was an uncertain and expensive technology that could not be borne by limited academic budgets.

Dismissal

During World War II the defense industries developed the photomultiplier (1P21) to measure light to levels well below what can be seen. The qe of the 1P21 was about five times that of the photographic emulsion. By 1960 this detector was in regular use by astronomers to make precision (less than 0.1 percent error) measurements of the brightness of single stars, one at a time. To create a calibrated image necessary for many astrophysics applications required many days. The 1950s saw the rapid development of several electronic imaging devices driven by commercial television. At the time there were two technologies used for electronic imaging. Both required converting the image to electrons, using a photocathode and then reading the image out in some manner.

Many different techniques were developed to store and read out the recorded image. Several groups were conducting research in these areas using different sensor architectures.[109] At the time there were two technologies being pursued to replace photographic film with a high qe recording medium. One was the Return-Beam Vidicon, and the other was called image tubes. At KPNO, Bill Livingston and Aden started work to assess the utility of devices based on commercial TV cameras for astronomy. Aden was particularly interested in the TV camera technology because the image was transmitted electronically along a cable or through space by encoding the image onto radio waves. Transmitting images through space was essential technology if space telescopes were to return astronomical images to the ground for scientific analysis. Commercial industry was developing this technology rapidly, and Aden believed that cost-effective astronomical instrumentation could benefit from industry programs.

By the late summer of 1960, results were beginning to come from the electronic imaging program, and Aden was invited by the Optical Society of America to summarize progress in an oral paper on modern electronic detectors for astronomy to be delivered at their spring meeting March 1961 in Pittsburgh, Pennsylvania.

Aden wrote his paper and distributed a copy to Mayall, the new KPNO director, in the fall of 1960. A few days later, he was called into the director's office to find Mayall and McMath waiting. McMath forbade him from showing the images at Pittsburgh because he felt it was all too speculative and would make KPNO look foolish.[110] But Aden defied McMath and presented the material at the Pittsburgh meeting on March 3, 1961, to show the state-of-the-art technology of electronic transmission of images, with the caveat that it was untested.[111] A crowd gathered around him afterward, wanting to know more about electronic imaging. By this time he was emotionally drained and felt as if he were having a nervous breakdown. His gaze never left the window on the flights to Chicago and then·to Tucson and the shelter of Marjorie and the children, now seven of them with the birth of David in 1957. He seemed to be recovering—until McMath's next visit, when he expected repercussions for giving the talk over McMath's objections.[112]

Aden was called into Mayall's office to find Shane waiting alone, sitting in Mayall's chair. He told Aden that both the Board and Director Mayall requested that he step down from his position as director of the Stellar and Space Divisions and that he be relieved of his duties as the technical leader of the Optical Shop. He was stripped of his opportunities to pursue his passion for astronomical optics, his 84-inch telescope, and the tools for cutting-edge astronomical research. He knew that McMath had pressured Mayall into this as punishment. Rather than be reduced to the rank of astronomer without access to the Optical Shop—even with no reduction in salary—Aden soon submitted his letter of resignation, effective August 31, 1961. This was not what Shane and Mayall had anticipated. In fact, several Board members approached Marjorie to persuade Aden to reconsider and even promised to raise his salary above what he was earning as director, but she did not tell him of the offer until sometime later. She knew that it was time to move on for their professional, physical, and mental well-being. They needed to escape the unrelenting wrath of McMath and look forward to more intellectual and rewarding goals.[113]

Aden used his friendship with University of Arizona president Harvill, established years earlier during the founding of the National Observatory, to begin discussions to move across the street on campus to Steward Observatory. He was fairly certain that Harvill would support his move to the University of Arizona and had begun discussions with the dean of the Graduate College, Herbert Rhodes, to negotiate the conditions of his transfer. He was busy with several ongoing research programs, consulting work for the Air Force and industry, and professional society business and wanted relief from administrative burdens.[114]

The AURA space program continued using instruments mounted on small rockets until it was abolished in 1973. These early rocket launches from White Sands, New Mexico, used captured German V-2 rockets, some retrieved by Aden himself (see Chapter 4, section "Nordhausen/Mittelbau-Dora"). Astronomers preferred to use KPNO rockets to those of NASA because Russ Nidey, with whom Aden worked in the Navy, had joined KPNO as deputy director of the Space Division. Nidey oversaw the entire beginning-to-end process of the development of the scientific payload, along with the rocket launch, operations, and data collection, much more carefully than NASA engineers, who were more focused on spacecraft and new launch vehicles. Congress wanted a single-point contact for all civilian space matters, and in 1969 the Space Division was renamed the Planetary Science Division to reflect the charter of their sponsoring organization, NSF.[115] The space telescope program and NSF's involvement in space astronomy suffered after Aden's departure.

Three months after Aden stepped down from the directorship of Kitt Peak to lead the KPNO Stellar Division, Kuiper joined the Department of Astronomy of the University of Arizona, on July 1, 1960, and started to develop an independent laboratory for the exploration of the Moon, other satellites, and their planets. Ewen Whitaker moved from the UK to Yerkes and then on to UA to work with Kuiper in establishing LPL and creating the most comprehensive, precision lunar atlas of the time for the Apollo astronauts.[116]

Today, the LPL is a leading center of excellence studying our solar system and those of other stars (exoplanets). It employs over 200 scientists and technicians in their own five-story building on campus, situated between the College of Optical Sciences and KPNO offices, within a few hundred feet of each other. Within five years, three major astronomy centers had been created in Tucson: Kitt Peak National Observatory, Kuiper's Lunar and Planetary Laboratory, and Bok's Department of Astronomy. Over the following 10 years, observing stations with their telescopes would be added to mountaintops in southern Arizona. Aden had certainly helped to shape Tucson into the observational astronomy center of the Western world.

Aden was one of the pioneers in the development of electronic imaging. For centuries, all astronomical observations were made by eye. The progressively larger telescopes built by Galileo and Herschel were great refractors designed for visual observations. In 1840, John W. Draper made the first permanent record of an astronomical object: the Moon. It was not until the development of large reflectors such as the 36-inch Crossley at Lick Observatory and Hale's 60- and 100-inch telescopes at Mt. Wilson that photography became an astronomical research tool. For 50 years, astronomers had successfully used photographic emulsions on glass plates to record stars, nebulae, planets, galaxies, and spectra. By the mid-1960s, the sensitivity, uniformity, and stability of photographic emulsions had been refined by industry to the limit of that technology.

Today, after 35 years of development and substantial investment by many astronomers and industry, vacuum-tube imagers have evolved into the solid-state charge-coupled-devices (CCD) responsible for the high-quality images recorded by the HST and used in all astronomical instruments, replacing the old film and glass plates completely with a new highly sensitive and linear recording medium.

Notes

1. Struve, O. 1940. Cooperation in astronomy. *Scientific Monthly* 50: 142–7.
2. Edmondson, F. K. 2005. *AURA and its US National Observatories*. Cambridge University Press, 7, 264; Anonymous. 1951. Report of Committee. *Astronomical Journal* 56: 147–8.

3. Strohbehn, J. W. 1971. III. Optical propagation through the turbulent atmosphere. *Progress in Optics* 9: 73–122.

4. Meinel, A. B. 1958. *Final report on the site selection survey for the National Astronomical Observatory.* Kitt Peak National Observatory Contribution No. 45; Meinel, A. B. 1961a. Astronomical seeing and observatory site selection. In G. P. Kuiper (ed.), *Stars and stellar systems: Telescopes,* 154–75. University of Chicago Press.

5. Stebbins, J., and Brown, F. C. 1907. A determination of the moon's light with a selenium photometer. *Astrophysical Journal* 26: 326–40; Stebbins, J. 1911. The discovery of eclipsing variable stars. *Astrophysical Journal* 34: 105–11.

6. Irwin, J. B. 1952. Optimum locations for a photometric observatory. *Science* 115(2983): 223–6.

7. Anonymous. 1955. Astronomical photoelectric conference, sponsored by the National Science Foundation and Lowell Observatory at Flagstaff, Arizona, August 31 to September 1, 1953. *Astrophysical Journal* 60: 17–18.

8. Letter from L. Goldberg to P. van de Kamp, September 24, 1954, Archives of Kitt Peak National Observatory.

9. Edmondson, *AURA*, 29.

10. Edmondson, *AURA*, 30.

11. Morgan, W. W., Meinel, A. B., and Johnson, H. M. 1954. Spectral classification with exceedingly low dispersion. *Astrophysical Journal* 120: 506–11; Meinel, A. B. 1956. An F/2 [*sic*] Cassegrain camera. *Astrophysical Journal* 124: 652–7.

12. Edmondson, *AURA*, 268.

13. Edmondson, *AURA*, 269.

14. Edmondson, *AURA*, 30–1.

15. Edmondson, *AURA*, 31.

16. Edmondson, *AURA*, 31.

17. Edmondson, *AURA*, 32.

18. Meinel, A. B., and Meinel, M. P. 2002a. *Echoes from a simpler time.* Unpublished.

19. Meinel and Meinel, *Echoes*, 87.

20. Whitford, A. E. 1956. The plan for a new American observatory. *Publications of the Astronomical Society of the Pacific* 68: 115–18.

21. Meinel and Meinel, *Echoes*, 87; Edmondson, *AURA*, 36.

22. Meinel and Meinel, *Echoes*, 87.

23. Edmondson, *AURA*, 36.

24. Interview with Helmut Abt by J. Breckinridge, March 1, 2017, Kitt Peak National Observatory; Edmondson, *AURA*, 37; Abt, H. A. 2020. *A stellar life.* Palmetto Publishing, Charleston, South Carolina, 33–4.

25. Meinel and Meinel, *Echoes*, 87; Edmondson, *AURA*, 37; Abt, *A stellar life*, 39.

26. Meinel and Meinel, *Echoes*, 87; Abt, *A stellar life*, 39.

27. Meinel and Meinel, *Echoes*, 87–8; Edmondson, *AURA*, 38.

28. Edmondson, *AURA*, 41.

29. Meinel and Meinel, *Echoes*, 88; Andrew Ellicott Douglass (1867–1962) founded the astronomy program at the University of Arizona and also the scientific discipline of dendrochronology. He was responsible for the construction of the Steward Observatory and its instruments in the early 1920s.

30. Edmondson, *AURA*, 41; Meinel and Meinel, *Echoes*, 88; Abt, *A stellar life*, 38.

31. Meinel and Meinel, *Echoes*, 88.

32. Meinel and Meinel, *Echoes*, 88.

33. Meinel and Meinel, *Echoes*, 89.

34. Meinel and Meinel, *Echoes*, 89.

35. Abt, *A stellar life*, 28.

36. Abt, *A stellar life*, 34–7.

37. A few years later, astronomers learned that "Papago" was a derogatory nickname given to the tribe by early settlers. It translated loosely to "bean eaters." The name was changed to the Tohono O'Odham Nation in March 1986 and is now the accepted official name for this tribe of Native Americans. This historical narrative precedes 1986, so we are using the Papago name for the tribe for the period of the events described here.

38. Breckinridge, J. B. 1976. Interference in astronomical speckle patterns. *Journal of the Optical Society of America* 66: 1240–2.

39. Skidmore, W., Travouillon, T., and Riddle, R. 2006. Evaluation of sonic anemometers as highly sensitive optical turbulence measuring devices for the Thirty Meter Telescope site testing campaign. *Proceedings SPIE* 6267. https://doi.org/10.1117/12.671518.

40. Meinel and Meinel, *Echoes*, 91.

41. Meinel and Meinel, *Echoes*, 91.

42. Meinel and Meinel, *Echoes*, 91; Abt, *A stellar life*, 38–9.

43. Meinel and Meinel, *Echoes*, 91.

44. Meinel and Meinel, *Echoes*, 94.

45. Harold L. Johnson (1921–1980) was Aden's astronomy classmate at UC Berkeley (see Chapter 5, section "Graduate student"), was awarded the 1956 Helen B. Warner Prize of the AAS, and was elected a member of the National Academy of Sciences.

46. Meinel and Meinel, *Echoes*, 94.

47. Meinel and Meinel, *Echoes*, 99–100.

48. Meinel and Meinel, *Echoes*, 100.

49. Abt, *A stellar life*, 39.

50. Meinel and Meinel, *Echoes*, 100.

51. Spicer, R. B. 1949. *The desert people: A study of the Papago people.* University of Chicago Press.

52. Meinel and Meinel, *Echoes*, 100.

53. Meinel and Meinel, *Echoes*, 100–1.

54. Meinel and Meinel, *Echoes*, 101.

55. Meinel and Meinel, *Echoes*, 101.

56. Meinel and Meinel, *Echoes*, 101.

57. Meinel and Meinel, *Echoes*, 101–2.

58. Meinel and Meinel, *Echoes*, 102–3.

59. Meinel and Meinel, *Echoes*, 104–5.

60. Meinel and Meinel, *Echoes*, 105–6.

61. Meinel, A. B., and Abt, H. A. 1963. *Final report on the site selection survey for the National Astronomical Observatory*. Contributions of the Kitt Peak National Observatory # 45.

62. Meinel, *Final report*.

63. Correspondence between Richard Harvill and Frank Edmondson, October 20, November 5, and November 19, 1959, in the Harvill papers at the University of Arizona, Special collections.

64. Meinel and Meinel, *Echoes*, 106.

65. Edmondson, *AURA*, 88; Meinel and Meinel, *Echoes*, 106; Abt, *A stellar life*, 40.

66. Edmondson, *AURA*, 88–9; Meinel and Meinel, *Echoes*, 106.

67. Edmondson, *AURA*, 96.

68. Meinel and Meinel, *Echoes*, 108.

69. Pearson, chief optician at Yerkes Observatory, was A. A. Michelson's collaborator and optical engineer on his measurement of the speed of light in a vacuum.

70. Meinel, A. B. 1957. Extremely innovative design for a f/1.5 minimum mass telescope. KPNO Technical Report #12. Office of the Executive Secretary. August 15, 1957.

71. Personal communication from H. Abt to J. Breckinridge, June 2020.

72. Abt, H. A. 2009. The Kitt Peak 2.1-meter telescope: An unusually innovative telescope (abstract). *Bulletin of the American Astronomical Society* 41: 187.

73. Burbidge, E. M., Lynds, C. R., and Stockton, A. N. 1968. Further observations of quasi-stellar objects with absorption-line spectra: Ton 1530, PKS 0237-23, and PHL 938*. *Astrophysical Journal* 152: 1077–93.

74. John Glaspey, private communication in article he prepared for the NOAO newsletter titled "The beginnings of the KPNO 2.1-m telescope" by John Glaspey and Sharon Hunt.

75. Kuiper, G. P., and Middlehurst, B. M. 1960. *Stars and stellar systems*. Vol. 1. *Telescopes*. University of Chicago Press.

76. Meinel, Kitt Peak National Observatory, Monthly report for January 1962, available at KPNO, 950 N. Cherry Street, Tucson, AZ 85711.

77. Edmondson, *AURA*, 99.

78. Edmondson, *AURA*, 99.

79. Edmondson, *AURA*, 100.

80. Letter from J. Stock to G. Kuiper, February 26, 1959, Kuiper Archives, Special Collections, University of Arizona.

81. Green, C. M., and Lomask, M. 1970. *Vanguard—a history*. NASA SP-4202.

82. Meinel, A. B. 1959a, b. Astronomical observations from space vehicles. *Publications of the Astronomical Society of the Pacific* 71: 369–80.

83. Spitzer, L., Jr. 1960. Space telescopes and components. *Astrophysical Journal* 65: 242–63; Code, A. D. 1960. Stellar astronomy from a space vehicle. *Astronomical Journal* 65: 278–84; Goldberg, 1959. Astronomy from satellites and space vehicles. *Journal of Geophysical Research* 64: 1765–8.

84. Meinel, A. B. 1961b, c. Design considerations for a large aperture orbital telescope. Report. Les Spectres des Astres dans l'Ultraviolet Lointain; communications

presentees au dixieme colloque International d'Astrophysique tenu a Liege les 11, 12, 13 et 14 Juillet 1960. Institute d'Astrophysique cointe-sclessin, Belgique, 49–59.

85. Zimmerman, R. 2010. *The universe in a mirror: The saga of the Hubble Telescope and the visionaries who built it*. Princeton University Press.

86. Spitzer, L., Jr. 1990. Report to Project RAND: Astronomical advantages of an extra-terrestrial observatory. *Astronomical Quarterly* 7: 131–42.

87. Letter from Aden Meinel to C. D. Shane, Mary Lea Shane Archives of Lick Observatory, filed with the University of California at Santa Cruz, California, Library.

88. Meinel and Meinel, *Echoes*, 114.

89. Edmondson, *AURA*, 109.

90. Edmondson, *AURA*, 109.

91. Edmondson, *AURA*, 109–10.

92. Edmondson, *AURA*, 112.

93. Zimmerman, *Universe in a mirror*, 33–4.

94. Spitzer, L., Jr. 1966. *Space research: Directions for the future*. National Academy of Sciences Space Science Board publication #1403. https://doi.org/10.17226/12410; Zimmerman, *Universe in a mirror*, 33.

95. Zimmerman, *Universe in a mirror*, 29–30.

96. Meinel, Astronomical observations, 369–80; Meinel, Design considerations, 49–59.

97. Code, A. D., Houck, T. E., McNall, J. F., Bless, R. C., and Lillie, C. F. 1970. Ultraviolet photometry from the Orbiting Astronomical Observatory. I. Instrumentation and operation. *Astrophysical Journal* 161: 377–88.

98. Zimmerman, *Universe in a mirror*, 30.

99. Edmondson, *AURA*, 112–13.

100. *Arizona Daily Star*, March 5, 1960.

101. Edmondson, *AURA*, 113–14.

102. Meinel and Meinel, *Echoes*, 113.

103. Meinel and Meinel, *Echoes*, 113–14.

104. Meinel and Meinel, *Echoes*, 114.

105. Meinel and Meinel, *Echoes*, 114–15.

106. Meinel and Meinel, *Echoes*, 114–15.

107. Meinel and Meinel, *Echoes*, 115.

108. Meinel and Meinel, *Echoes*, 115–16.

109. Breckinridge, J. B., Kron, G. E., and Pappiashvili, I. 1964. Transfer efficiency and storage capacity of electronographic image tubes. *Astronomical Journal* 69: 534–5; Kron, G. E., Ables, H. D., and Hewitt, A. V. 1969. A technical description of the con-struction, function, and application of the US Navy electronic camera. *Advances in Electronics and Electron Physics* 28: 1–17; Burns, J., III, Hiltner, W. A., and Miller, R. 1956. Image converters with thin protective foils. *Astronomical Journal* 61: 172; Lallemand, A., Duchesne, M. and Walker, M. F. 1960. The electronic camera, its in-stallation and results obtained with the 120-inch reflector of the Lick Observatory. *Publications of the Astronomical Society of the Pacific* 72: 268–87.

110. Meinel and Meinel, *Echoes*, 117.

111. Meinel, A. B. 1961d. New frontiers of astronomical optics. *Journal of the Optical Society of America* 51: 471.
112. Meinel and Meinel, *Echoes*, 117–18.
113. Meinel and Meinel, *Echoes*, 117–18; Edmondson, *AURA*, 115.
114. Meinel and Meinel, *Echoes*, 117–18.
115. Edmondson, *AURA*, 227–8.
116. G. P. Kuiper archives at the University of Arizona and the National Archives and Records Administration (NARA), US Government Form 171, Personal qualification statement dated June 15, 1976; Whitaker, E. A. 1988. *The University of Arizona's Lunar and Planetary Laboratory: Its founding and early years.* University of Arizona.

8

Opportunity Knocks—Doors Open Wide

As soon as the staff of Yerkes Observatory heard that Aden had submitted his resignation to the AURA Board in the summer of 1961, Director W. W. Morgan sent him a job offer. He and the rest of the faculty wanted Aden and the family to return to Yerkes with as much salary for a six-month period as Aden had received as Kitt Peak director for 12 months. Aden and Marjorie discussed it together with their children, and their unanimous decision was to stay in Tucson. They loved the desert and the University of Arizona (UA) community activities, including sports. And the children would receive free tuition at the university.[1]

Aden wanted more academic freedom and recognition than AURA could offer, and he had been thinking about moving over to UA for several months. University President Richard Harvill had helped Aden to integrate the National Observatory into the academic community in Tucson, and they had great respect for each other. Soon after his resignation as KPNO director and demotion on March 31, 1960, Aden met with Harvill to discuss a potential role for himself at UA. On March 21, 1961, while Aden was still at KPNO, Harvill sent Aden a strong letter of support, thanking him for bringing Gerard Kuiper and his Lunar and Planetary Group from Yerkes to UA six months earlier. The relationship that the two had forged during the founding of the National Observatory was cemented. Harvill had recognized early that this new federal research facility across the street from Steward Observatory was a perfect opportunity for UA to expand its own astronomy department and raise the prestige for all science and technology there.

By the spring of 1961, Aden's situation at KPNO was becoming intolerable to him. Aden met with the dean of the University's Graduate College, Herbert Rhodes. On May 9, 1961, Rhodes wrote a letter to Harvill confirming Aden's interest and his potential value to the university. Negotiations were underway to find funding for Aden to move to UA. Correspondence with the new director of KPNO, Nicholas Mayall, confirmed that the observatory would concur with Aden's desire to move. This was the last political hurdle before Aden received a formal offer of employment from the university.[2]

On August 31, 1961, Aden left the employ of AURA and moved out from under harassment by McMath and the AURA Board to new opportunities at UA.

With Stars in Their Eyes. James B. Breckinridge and Alec M. Pridgeon, Oxford University Press. © Oxford University Press 2022. DOI: 10.1093/oso/9780190915674.003.0008

He was recovering from a nervous breakdown and needed the peace and quiet associated with a respected university professor. The following day, Aden carried his books and files 300 yards across Tucson's KPNO office on Cherry Avenue to his new office at Steward Observatory, directed by Edwin Carpenter.

Before Aden joined UA, Carpenter and Harvill had plans to expand the Department of Astronomy. In 1961, the department had three full professors: Carpenter, Leon Blitzer, and Kuiper. Kuiper would go on to found the Lunar and Planetary Laboratory there and lead the nation's programs in mapping the lunar surface for the Apollo astronauts. There were two associate professors, Walter Fitch and Hugh Johnson, and one assistant professor, Ray Weymann, who nine years later would become the Observatory director. These five faculty members offered 12 formal astronomy classes. During Aden's tenure of six years, the department would expand significantly in course offerings and international reputation.

At UA, Aden was a full professor of astronomy and had no administrative responsibilities. Stressed out and completely drained by his interactions with McMath, Shane, and the AURA Board, Aden looked forward to returning to scientific research, teaching, and consulting. His research continued on the Morgan-Keenan-Kellman (MKK) spectral classification from diffraction grating spectra, infrared astronomy, and planetary atmospheres, along with his work to develop instruments and detectors for astronomical research.[3] Just because his office moved did not mean that he severed his intellectual ties at Kitt Peak. He maintained highly creative professional collaborations with Kitt Peak instrument engineers as well as UA astronomers, atmospheric scientists, engineers, and physicists.

When Aden moved into his new office on the first floor of Steward Observatory, classes had been in session for a week and a half. He had no teaching responsibilities for the 1961–1962 academic year, but he was occupied finishing his observing programs with the KPNO telescopes, reducing his measurements, writing papers, and planning new technologies for astronomical research. In a paper published in the October 20, 1961, issue of *Science*, Aden wrote a six-page essay titled "New Frontiers of Astronomical Technology."[4] The premise was that larger and larger ground-based telescopes offered only asymptotic gains in distance but at higher and higher costs. Only new technologies being developed by the aerospace industry could make a quantum leap in improving resolution and seeing by avoiding the limitations imposed by atmospheric airglow and turbulence. Among these advances might be balloon-borne telescopes orbiting above 80,000 feet where there is no seeing disturbance, orbiting astronomical observatories, and even a manned lunar observatory, all with electronic means of information retrieval such as "ruggedized" image orthicons. The paper was received cautiously by senior members of the astronomy community, who were concerned that this novel concept would drain funds from their more important classical astronomy programs and destabilize the status quo by supporting expensive, high-risk engineering ventures rather than routine astronomical research. Nonetheless, Aden's ideas were prescient in developing satellite technology and data transfer from space, which could be used not only for scientific advancement but also aerial surveillance of targets on Earth. At the onset of the Cold War, the US Air Force was keen to explore some of these possibilities and recruited Aden as a consultant (see Chapter 11).

Before leaving KPNO as its director, Aden had negotiated with UA for a two-acre site for the Steward Observatory's telescopes on Kitt Peak. This was sufficient space for more new telescopes and support facilities. Within a year of his arrival at the University, Aden and Carpenter moved the 36-inch Steward Observatory telescope from the campus to the mountain. Aden started planning and designing a larger telescope project, which is today the Bok telescope located within UA-leased land boundaries on Kitt Peak.

Aden's 1959 paper on design considerations for a large-aperture orbital telescope and a follow-on paper at International Astronomical Union (IAU) conferences described his efforts to build a space observatory for astronomical telescopes.[5] This work attracted the interest of the US Air Force and, with other events described in Chapter 11, led to a second and clandestine career as consultant to both the Air Force and CIA.

Under his salary arrangement, Aden was expected to work full-time for Kuiper, director of the Lunar and Planetary Laboratory (LPL), who asked Aden to complete the design and construction of the 60-inch telescope for Mt. Lemmon where Kuiper was establishing his LPL Observatory. In addition, Carpenter expected Aden to work full-time updating the 36-inch telescope and moving it to its new site on Kitt Peak. Aden enjoyed refurbishing and moving the 36-inch because Marjorie's father had helped A. E. Douglass install that same telescope in the then-new observatory building on campus 40 years earlier.[6] Aden's work for Kuiper had progressed as far as a visit to Corning to receive a newly cast Pyrex mirror blank for the 60-inch telescope, when suddenly everything changed. Carpenter had a heart attack.[7]

During Carpenter's recovery, Aden took charge of affairs for him, including completion of yet another addition to the new Steward Observatory building, this time completing the second floor. He also helped Carpenter draw up a proposal to NSF for a Science Development Grant that would forever change the astronomy program at UA. It would include more astronomy faculty and new class offerings in advanced astrophysics and instrumentation. Aden and Carpenter proposed a science rationale for building both a 60-inch and a 100-inch telescope and gave budgets for both, leaving the final choice up to NSF.[8]

By the end of 1962 Carpenter had recovered, and all was back to normal. Then on the last night of an AURA Board meeting on February 11, 1963, which Carpenter attended, Aden received a call in the middle of the night from Carpenter's wife to say that he had just died in the ambulance on the way to the hospital. Harvill called Aden into his office the next morning to ask if he would take over as director of Steward Observatory. Aden hesitated, saying it was his understanding that Bart Bok in Australia was lined up to be Carpenter's successor. Harvill explained that there were too many important projects in the near future, among them an application for one of the new NSF Science Development Grants for the university, which required Aden's experience and expertise.[9]

Director again

Aden's plan to spend time quietly on astronomical research and technology development was upended after only 16 months. His typical high energy would now be focused on developing a new large-aperture telescope for Steward Observatory, hiring new faculty, consulting for the Air Force, and eventually creating a world-class optical science education and research center.

By the spring of 1963, President Kennedy's initiatives in science and education were moving rapidly through Congress, led by Vice President Lyndon Johnson. It was quickly realized that the major research and teaching universities on both coasts did not have the capacity to train enough high-quality scientists and engineers for the nation's immediate needs. Congressional delegations from states other than California and those in the Northeast recognized their opportunity to bolster their state's science and technology programs. Aden and Harvill were well aware of this and wanted to leverage Arizona's natural resource of clear skies and Tucson's new national observatory into leadership for the state of Arizona in astronomy and planetary science.

Aden learned quickly that it was a pleasure to collaborate with both the dean of the Graduate School and the dean of Liberal Arts. The 1964 UA budget was soon completed and included significant growth for astronomy and planetary science. He was allotted two new staff positions to support the growth of astronomy. The university wanted to see Steward Observatory and the Department of Astronomy expand to meet these new opportunities.

In the meantime, Kuiper's 60-inch telescope at the Lunar and Planetary site on Mt. Lemmon north of Tucson was nearing completion. The work of directing Steward Observatory and heading the Department of Astronomy was growing, and Aden ended his work with Kuiper and LPL. Kuiper was busy with his NASA contracts, expanding LPL into a new building, and leading his lunar-mapping project in support of the NASA Apollo program.

Harvill told Steward Observatory staff and faculty that they were to postpone work on the Science Development Grant proposal in order to submit a grant application to NSF for major equipment with funding for both Steward Observatory and the Department of Physics (principally Professor Stanley Bashkin). If successful, a 60-inch telescope would be added to Steward Observatory's site on Kitt Peak. Bashkin would get a Van de Graaff generator to continue his beam-foil spectroscopy discoveries, and Aden and Bashkin would expand their collaborative work in atomic spectra and spectrometer development for laboratory and observational astrophysics.[10]

Aden was attending a classified meeting in Washington, DC, when he received a message from Marjorie at his hotel. "President Harvill called me today. He had heard back from NSF about our proposal for a new telescope. He reported that NSF said that 60-inches was too small and said to come back with a proposal for a duplicate of the 84-inch telescope. He wants to know how soon you can give him the revised budget." Aden responded: "Tell him I'll have it ready when I step off the plane in Tucson tomorrow evening." The proposal was prepared, submitted on time, and the wait began.[11]

About a month later, Marjorie and Aden stepped off a plane together in Tucson to see a grinning Bashkin standing at the gate and holding up a copy of the *Tucson Daily Citizen* newspaper. The front-page headline read "UA Gets Major NSF Grant." The proposal for the new 84-inch class telescope and the augmented Steward Observatory faculty/staff had been funded. This was to become the 90-inch Bok telescope. Also funded was Bashkin's proposal for a Van de Graaff generator, along with research funds for physics graduate students. Aden and Marjorie were elated and again sent a message to Bok announcing that funding for the new telescope was in hand.[12] Harvill and Aden had been successful in obtaining funds from the Higher Education Facilities Act of 1963.[13]

When the Van de Graaff generator was up and running, it produced moderate energy ion beams for laboratory atomic spectroscopy. One application was to identify electronic transitions in atoms and support the discovery and identification of the composition of stars and nebulae observed using the spectrometers on Kitt Peak telescopes. Aden had worked for several years studying spectra of atomic transitions in stars and needed to know what elements in the periodic table were associated with the spectral lines he observed in stars and nebulae. He collaborated with Bashkin to build a fast spectrograph (f/0.8) capable of recording faint laboratory spectra.[14] Bashkin's Van de Graaff generator was exciting atomic levels of a variety of materials to develop Grotrian diagrams, which are used to calculate wavelengths for spectral lines and reveal both allowed and forbidden unique electronic transitions between energy levels within atoms.[15]

Ion-beam polishing

At Steward Observatory, Aden was experimenting with vacuum deposition of thin-film optical coatings, trying to find the optical coating prescription that would have maximum transmittance in the UV with maximum absorption in the visible. This was difficult and led some astronomers to use liquid filters which were complicated to fabricate and often had short life spans.[16] Aden searched for a better solution using ion-beam deposition. Part of the processing technique to make these mirror coatings is to view the source and the substrate to be coated through a window in the vacuum chamber.

Aden built the spectrograph for Bashkin to observe and measure atomic spectra from his beam foil experiment (Figure 8.1). Aden also built another spectrometer to measure the optical bandwidth of filters as they were being coated inside the chamber. Aden noticed that the window in the vacuum chamber was becoming distorted during the evaporation process. Don Loomis had so carefully polished the windows flat and now they were showing small optical distortions. Aden could see the ion source through the vacuum window and

Figure 8.1. Stanley Bashkin (left) and Aden Meinel shown in the ion-beam spectroscopy laboratory at the University of Arizona ca. 1964.
Courtesy of the University of Arizona.

concluded that the ions emitted by the source were impinging on the chamber window, causing the flat optical figure to change to a curve. He hypothesized correctly that ions were bombarding the glass, removing small amounts of material in a controlled fashion.

He then wondered whether ion beams could be used to remove glass from selective areas on optical surfaces to create a surface that will correct wavefront aberrations and improve image quality in optical systems. In figuring out what physical processes could lead to that phenomenon, he discovered what we now call ion-beam polishing or ion-beam milling. This process is today used in the commercial optics industry to remove small amounts of substrate material in a precise, deterministic manner.

Aden realized that these ion beams could also be used for controlled polishing of optical surfaces for astronomy.[17] He and his team developed hardware and performed experiments to demonstrate the practicality of the process.[18] In 1965 the University of Arizona applied for a patent in the interests of Aden and Marjorie, Bashkin, and Loomis for the invention of ion polishing; one was awarded (Meinel, Bashkin, Loomis. US Patent # 3548189, December 15, 1970). In the early 1970s, Perkin-Elmer paid the University $48,000 for exclusive rights, reimbursed to them by the US Air Force. Aden and Marjorie received 40 percent of the amount, Bashkin and Loomis 20 percent, and the university the balance.[19]

Neither Perkin-Elmer nor UA commercialized ion-beam polishing, and by 1990 the patent was allowed to expire. Today ion-polishing of optical surfaces is routinely used in industry for cost-effective, large-aperture, lightweight telescope mirrors.[20] Aden had recognized that these same beams of charged particles could be used to figure optical surfaces more accurately than the classic approach of contact-rubbing glass surfaces with abrasives or polishing materials. Ion-beam polishing removes microscopic amounts of glass without warping the substrate and thus is ideal for mirrors that have low stiffness and large area, such as lightweight astronomical primary mirrors, and keep telescope system cost to a minimum.

The final proof of the commercial importance of ion-figuring came over 10 years later, when Caltech and UC Berkeley Keck telescope engineers discovered they needed to warp each mirror segment mechanically after they cut the hexagonal segments from a circular disk of glass. Touch-up was essential to meet the specifications for the 10-m, segmented-mirror, Keck telescopes on Mauna Kea in Hawaii. Ion-beam figuring was suggested. The Keck project members talked it over with Itek, which did the original mechanical polishing of the Schott ZeroDur glass ceramic segments, and also with Kodak, which had the ion machine. The first tests showed promising results. It proved so effective that finally all 72 segments for both Keck-I and Keck-II were finished by ion-beam polishing.

Hyderabad telescope

While quietly sitting at his desk at Steward Observatory one day in 1963, Aden received a phone call from Jason Nassau of the Warner and Swasey Observatory at Case Institute of Technology, Cleveland, Ohio. Nassau was a friend and colleague of Marjorie's parents. He had been asked by the State Department to assist the international aid program CARE in the construction a telescope for Osmania University at Hyderabad, India.[21] Funding was supplied by the US Wheat Loan to India as part of the Food for Peace Act of 1966.[22] Academic research and teaching programs in India were to receive, among other equipment, a 48-inch (1.3-m) telescope designed and built in the United States. The law stipulated that interest payments would be translated into scientific apparatus for Indian universities, the equipment to be built in the United States. Hyderabad University wanted to have the largest telescope in India, so all agreed upon a 48-inch telescope. It was nearing completion at the manufacturer near Pittsburgh, Pennsylvania. Nassau had made several trips to India to develop the requirements for the telescope, identify its location, and commission a contractor in the United States to build the telescope. On the telephone call with Aden, Nassau announced that he was dying from cancer and asked Aden to finish it. As was his custom, Aden agreed.[23]

An astronomical observatory in India was important to astronomers. To measure time-variable celestial objects and those invisible from the Northern Hemisphere required telescopes spread around the earth at different longitudes and latitudes. The great Melbourne reflector in Australia, which had first light in 1868, is an example. Rapidly changing astronomical events, such as solar eclipses, comets, novae, variable stars, and meteor showers, need to be observed during the night on a 24-hour basis, not just when the object is above the horizon at a particular observatory site. Before the advent of orbiting telescopes, such global coverage required that astronomical telescopes be located at different longitudes around the earth as well as different latitudes. Astronomical observatories located in Europe and the United States were limited in their coverage, so an observatory in India (68–97 degrees E) would be ideal to be able to cover astronomical events around the clock. In addition, astronomy is truly a global science and plays an important role in the development of a globally educated workforce.

The telescope was under construction at J. W. Fecker Optical in Pittsburgh. By 1963, construction of the observatory building, dome, and observing platform were finished at the observing site in India, ready for the telescope.[24] Seeing an opportunity to help an astronomy community in need as well as "see the world," Aden immediately accepted the paid consulting job.

He traveled from Tucson to Pittsburgh to inspect the progress on the telescope, directed the optical testing of the 1.3-m primary mirror, and made several recommendations on how the clock-drive should be engineered. Unfortunately, the project ran out of funds at the company in Pittsburgh, and the telescope was shipped—unfinished—to India with the expectation of completion there. The remaining work consisted of precision machining and adjustment of the drive mechanism that moves the telescope to point at objects in the sky and track them precisely as Earth turns.

Aden and Marjorie made five trips to the Osmania University Observatory in Hyderabad, India, over the next few years. Security was always an issue in the mid-1960s when they traveled in India. There was much political unrest and misunderstanding about why India was paying for astronomy when citizens were starving. Of course, the citizens of India were not paying for it. It was part of a US aid package, but the average uneducated citizen could not understand the big picture. The period covered the leadership transition from Jawaharlal Nehru to Indira Gandhi, when unemployment and inflation were high. There was also local civil unrest across the entire state of Hyderabad.

The telescope's drive mechanism suffered backlash and was not able to track stars smoothly as they moved across the sky, reducing its usefulness. Aden worked with the engineers in India to machine the parts, but a lack of skilled machinists and proper machining equipment there limited what could be built. The mechanical backlash problem plagued the telescope its entire useful life as a research instrument, and it could not be used to record good, long-exposure, high-resolution images.[25] The basic design, not the fault of Indian scientists and engineers, was flawed. Unfortunately, neither the financial nor engineering resources were sufficient to make the telescope track properly for the length of time needed to record the faint objects essential for cutting-edge research.

But the telescope was successful in some of its functions, and its presence was much appreciated by the astronomy community of India. P. Vivekananda Rao, retired professor of astronomy, Osmania University, Hyderabad, India, recently wrote:

The 48-inch telescope performed very well for 25 years (1968–1994) without any major breakdowns. During these fruitful years it was used extensively for making photometric and spectroscopic observations of various astronomical sources. It has produced about 80 research publications and 14 doctoral theses. Graduate students who received training on this telescope have worked and are still working in leading astronomical institutes within and outside India. I was fortunate to have been able to use this telescope extensively (maximum user) for my research which helped in furthering my career. Apart from this, I have trained many graduate students in observational astronomy, many of whom went on to work for their doctoral degrees in this field. Presently (2018), the telescope is not being used for any regular observations. However, it is being used to train graduate students in observational astronomy and for public outreach activities. The staff and students of the Department of Astronomy, Osmania University, are grateful to Dr. A. B. Meinel, Director of Optical Sciences Center, University of Arizona, Tucson, who had been the consulting astronomer for this telescope right from its installation.[26]

Steward Observatory's 90-inch telescope

Aden had successfully built the 82-inch at Kitt Peak at lower cost than anticipated, and this new Steward Observatory telescope was almost a copy. In 1963, NSF awarded a contract to UA, with Aden Meinel as principal investigator, for $1,300,000 to build the telescope and basic scientific instruments. Aden laid out an aggressive 36-month schedule so that work was underway by the fall of 1965, several months before Bart Bok was to arrive as the new director.

Back in 1958, working with Corning glass manufacturers, Aden had designed and supervised the casting of the honeycomb Pyrex mirror blank for the KPNO 84-inch (2.1-m) telescope. To reduce the mirror-machining time and thus cost, Aden risked the thermal slumping of the mirror at Corning to the near-net final shape. He supervised the optical figuring and test of this mirror in the optical shop at the Kitt Peak facilities in Tucson. The telescope was less expensive than the comparable 82-inch at McDonald Observatory. He understood that new technology was needed to reduce the mass and the thus the cost of the telescope. At that time (1963) structural computational models (static and dynamics) and material physics understanding of fused silica and Pyrex at these sizes was immature, and full-scale construction was necessary to assess the structural engineering details of different mirror architectures and materials.

In addition to being the Steward Observatory director, Aden was also the director of its Optical Shop. He completed the detailed optical design and the assembly view mechanical designs for the 90-inch (2.3-m) telescope to provide both Cassegrain and coudé focus, as well as the concept for the structure of the building and dome to give extra space for instruments that might be invented in the future. The primary and secondary mirrors for both the coudé and Cassegrain focus were designed by Aden, as well as the optical test procedures for individual mirrors and the entire end-to-end optical system. These were fabricated in the optical shop by Loomis and his team.

The 7,000-pound blank for the primary mirror of the 90-inch was cast by General Electric's Lamp Glass Division by thermally (1,650°C) fusing together more than 250 hexagonal ingots of fused silica and grinding the edge round. Aden chose fused silica (quartz) because the material is stable with a low coefficient of expansion, and the mirror surface will not warp to blur images during nighttime temperature changes, as experienced in other ground-based telescopes, including the Pyrex 84-inch mirror. The material is much more difficult to work with than standard borosilicate glasses. But Aden led his team to develop the grinding, polishing, and figuring processes that made the fabrication of this hard surface mirror material economically feasible.

There was no room at the observatory facilities on campus to grind and polish the mirror, so Aden rented a storefront in a strip mall on nearby Park Avenue. Aden and Loomis equipped the room with a turntable capable of grinding, polishing, and figuring the 90-inch mirror. The ceiling of the room was low and the center of curvature inaccessible, so the mirror surface could not be tested and verified from the center of curvature. Aden designed innovative optical tests for Loomis to use to figure the polished hyperbolic surface onto the mirror surface, such that when used with an aspheric secondary, the primary/secondary mirror system would create the required Ritchey-Chrétien (R/C) coma-corrected telescope. The process took three years, and Aden remained the project director of the telescope long after the arrival of Bok in 1966 and into 1968 when the telescope was finished and dedicated.

Years later, Perkin-Elmer used a far more sophisticated test system with an expensive test tower and null corrector at the center of curvature of the 2.4-m (94-inch) Hubble Space Telescope, resulting in the wrong R/C optical figure in that mirror.

Lunar exploration

During the NASA manned lunar program, the cadre of astronauts destined for the Moon regularly passed through Tucson to learn lunar geography from Kuiper and atmospheric physics from Aden. A course in lunar geography was needed to understand the landing topography and what they might expect upon arriving at the Moon. Maps of the lunar surface at the same quality as those used by airline pilots for Earth travel were vital. Astronauts also needed to understand how the return capsule would interact with Earth's upper atmosphere as atmospheric drag slowed re-entry for a safe return home; the capsule would have to return to Earth without reflecting off the atmosphere and bouncing uncontrolled into space. Based on the research that Aden performed at Yerkes, he was a world expert on the chemistry and physics of Earth's upper atmosphere. Both Kuiper and Aden made fast friends among the astronauts (Figure 8.2). Kuiper and his staff at LPL provided detailed lunar maps to the astronauts, who carried some of them aboard the Apollo missions.

Figure 8.2. Signed photograph of the Group 3 NASA astronauts, dedicated to Aden Meinel. Back row (left to right): Michael Collins, Walter Cunningham, Donn Eisele, Theodore (Ted) Freeman, Richard (Dick) Gordon, Russell (Rusty) Schweickart, David Scott, Clifton Williams. Front row (left to right): Edwin (Buzz) Aldrin, William (Bill) Anders, Charles Bassett, II, Alan Bean, Eugene (Gene) Cernan, Roger Chaffee. All served on Apollo missions except for Bassett, Chaffee, Freeman, and Williams, who died in training accidents before their missions. Aldrin, Cernan, Collins, Gordon, and Scott each flew on a Gemini mission, and Bean on Skylab. Aldrin, Bean, Cernan, and Scott walked on the Moon.
Courtesy of Ed Meinel.

National needs in optics

Soon another national initiative would consume Aden and Marjorie. In 1959, while Aden was busy directing and building KPNO, a few members of the Optical Society of America (OSA) urged OSA leadership to investigate training of the optics workforce in the United States. A booklet was prepared by members under the leadership of Robert E. Hopkins of the Institute of Optics, Rochester, New York. The OSA submitted a proposal to NSF to fund the publication and distribution of the booklet. A few weeks after Aden moved from KPNO over to

UA in 1961, the Optical Society of America began publishing and distributing the booklet, titled *Careers in Optics*.[27] It had a profound effect on an entire generation of young engineers and scientists.

Early in World War II, the US defense establishment became painfully aware of deficiencies in optical science, engineering, and technology in the United States. There was little or no optics design and manufacturing industry in this country. Germany was the main source of binoculars, range finders, camera lenses, and small telescopes, all critical to the war effort. This supply was cut off just as the United States needed to expand its military in 1939.[28] Allied troops were almost blind.

Technicians from the two or three astronomy departments across the nation who were trained in optics fabrication and testing of astronomical optics were pressed into service. For example, Mt. Wilson Observatory astronomers and their staff at a facility in Pasadena manufactured prisms and lenses for binoculars, range-finders, and cameras for the war effort. The Schmidt camera, known for its high acuity and wide field, was developed in the optics shop of Mt. Wilson Observatory by young Aden Meinel, under the guidance of Roger Hayward and Don Hendrix. Twelve years after the end of World War II, the supremacy of Soviet technology was demonstrated by the surprise launch of Sputnik 1 in October 1957. American optical surveillance technology had failed. In the following years, funding to support science and technology education in all technical areas and in particular those related to defense grew exponentially.

In October 1959, two years after the launch of Sputnik 1, Aden published an essay that was the first (by one month) serious publication in a respected journal by a respected astronomer discussing both the engineering challenges and astrophysical science measurement objectives for telescopes in space.[29] He outlined the astrophysical phenomena expected to be visible from space with a telescope and predicted the structural, thermal, mechanical, scattered light, materials science, and radiation-field challenges we face in our large space-telescope imaging systems today. Like his 1961 paper titled "New Frontiers of Astronomical Technology," and others, this paper garnered great interest from the US Air Force and opened important doors for Aden.[30] It may have been these papers, or Aden's many talks promoting space telescopes, that prompted his invitation to brief Allen Dulles, then CIA director, on the reality and technical challenges of space telescopes.[31]

The invention of the optical laser in 1960[32] and its many rapidly developing applications by the mid-1960s, along with the expanding commercial (entertainment industry) and military interest in imaging systems, drove urgent needs for optical science and engineering training in the United States.

Figure 5.2. Aurora image captured at midnight on April 10, 2015, in Delta Junction, Alaska. The storm was triggered by a coronal mass ejection that erupted from the sun on April 6, according to NOAA's Space Weather Prediction Center. The yellow-green portion is caused by electrons from the magnetosphere of space colliding with oxygen molecules in the atmosphere, whereas the blue-violet portion results from electrons striking nitrogen molecules.

Image courtesy of Sebastian Saarloos. NASA Aurora Image Gallery.

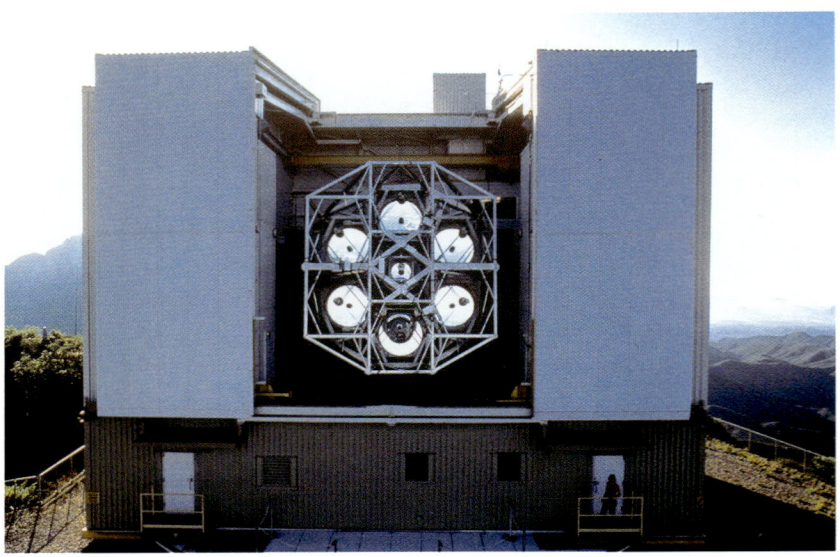

Figure 9.5. Completed in 1979, the MMT incorporated several innovations, including the entire building that could rotate, not simply the dome. Note the worker standing in the lower right doorway for scale.

Figure 11.3. GAMBIT 1 (KH-7) reconnaissance satellite in the Space Gallery at the National Museum of the U.S. Air Force.
Public domain photograph: US Air Force.

Figure 11.4. The Lockheed A-12 in flight.

Photo: US Air Force. Courtesy of Lockheed Martin Corporation via Smithsonian National Air and Space Museum (NASM 7A28869).

Figure 11.5. Certificate of appreciation presented to Aden Meinel from the U.S. Air Force for services rendered.

Courtesy of David Drach-Meinel.

(a)

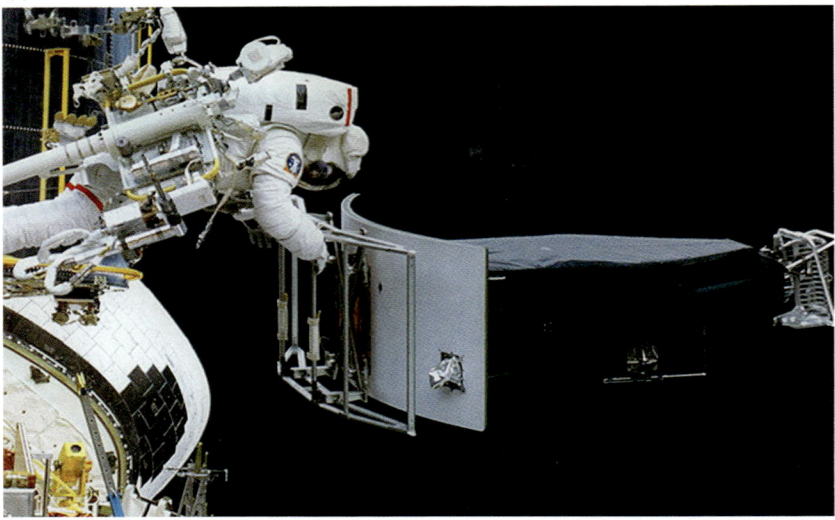

(b)

Figure 12.5a,b Replacement of the camera on the Hubble Space Telescope during the first servicing mission in December 1993. a: Astronaut Jeffrey Hoffman removes the WF/PC-1. b: Astronauts Jeffrey Hoffman and Story Musgrave install the WF/PC-2.

Photos courtesy of NASA/STS-16.

Figure 13.2. University of Arizona campus looking southwest photographed from a drone over the intersection of Campbell and Speedway roads showing the buildings built or influenced by Aden Meinel. Building 2 is the 157,000 sq. ft. Meinel building which houses the James C. Wyant College of Optical Sciences. Building 4 is the National Optical Astronomy Observatory headquarters building, designed by Aden Meinel. This was formerly called the Kitt Peak National Observatory Offices. Building 5 is the Steward Observatory, where Aden, as observatory director, was responsible for a major addition. The section of the stadium labeled as 1 at the far left of the image is the location of the Richard Caris Mirror Laboratory where the seven 8-meter diameter mirrors for the Giant Magellan Telescope are being fabricated under the leadership of Roger Angel, a faculty member hired and supported by Aden Meinel. Building 3 is the Lunar and Planetary Laboratory created by Gerard Kuiper who was recruited by Aden Meinel to join UA in 1960. Downtown Tucson is seen in the distant upper right.
Photo by James B. Breckinridge.

Figure 13.3. Recent aerial photograph of Kitt Peak National Observatory showing the telescopes and observatory infrastructure built during Aden's directorship. Center foreground is the McMath-Pierce Solar Telescope complex below the two water tanks. The dome of the 84-inch telescope is shown at 1, and the one for the 50-inch remotely controlled telescope is shown at 2. Inset: Left, the 36-inch telescope that Aden moved from Tucson; right, the 90-inch Bok Telescope for which Aden designed the optics, supervised the detail mechanical structure, and invented the dome structure to meet severe cost limitations.

Photo courtesy of the National Optical and Infrared Observatory.

Optical Society of America exposes
a vulnerability

Optical Society of America leadership, composed of businesses and government senior technical personnel, recognized the severe shortage of properly trained optical engineers and scientists to staff the needs of the astronomy, space surveillance, and laser communities. Between 1961 and 1964, the OSA established two committees, the first to investigate and define the problem, called "National Needs in Optics," and the second, called "Optics—An Action Program," to recommend a solution.

The OSA Board of Directors commissioned the National Needs in Optics committee in 1961. Defining the full scope of optics was one of the first tasks that the committee undertook. Van Zandt Williams, president of OSA, was asked by industry leaders and the OSA Board of Directors to lead a government, industry, and academia team to formulate a plan for how the United States was going to create a globally competitive workforce in optical science and engineering.

Under the leadership of President David MacAdam of the University of Rochester in 1962, the OSA began discussions of how to respond to the pleas of its members and optics users for more optical science and engineering education in the United States to satisfy evolving academic, military, and commercial needs. Commercial needs for training in optics were growing fast, driven by the entertainment business. In 1965, one of us (JBB) was working at Zenith Radio in Chicago in an industry that was mass-producing color TV tubes and image-converter tubes for the new low-dose X-ray machines. Engineers needed to understand image quality and electronic bandwidth quantitatively for transmitting quality images over the relatively narrow radio frequency bandwidth airways from electronic cameras (orthicons and vidicons). Military brass had an urgent need for "see-in-the-dark" infrared cameras, but there were no basic educational opportunities in optics and the image-formation science required to understand how to design and build competitive optical system devices.

Van Zandt Williams, a chemist and an officer of OSA, was skeptical of the need for optics education, and in that vein he wrote on November 19, 1962: "In so far as Government agencies are concerned, 'Optics' represents a fuzzy and variable area of scientific endeavor. It is certainly not a recognized, integral field of scientific pursuit as is the field of physics, chemistry, or biology, nor a specific subfield as low energy nuclear physics, polymer chemistry, or genetics. No agency has sufficient recognition of 'Optics' as an entity to get overly concerned about its problems or its needs. . . ." He further noted that ". . . the basic question is . . . What is Optics, and is research in Optics in a dangerously low state? I have not yet found many critical people who think so."[33] It was necessary to circumscribe what was included in the concept of optics so that the Optical Society could determine its own role and how to obtain funding. The OSA Committee on Needs in Optics was officially formed on March 13, 1962, at the OSA annual meeting in Washington, DC. Its members, appointed by OSA president MacAdam on October 2, 1962, comprised S. S. Ballard of the University of Florida, W. R. Brode of the National Bureau of Standards, OSA executive secretary Mary Warga, David MacAdam of Eastman Kodak, and Harold Stewart of the University of Rochester.[34] Dudley Williams, an infrared spectroscopist, in collaboration with Bill Wolfe of the infrared lab at the University of Michigan, drew up a list of research areas that they believed constituted the field of optics.[35] The areas included were (1) radiometry (infrared, visible, and ultraviolet); (2) geometrical optics, including element design and fabrication; (3) physical optics; (4) physiological optics; (5) spectroscopy (molecular, atomic, and solid state); (6) quantum optics; (7) systems of instrument design (as opposed to component design); and (8) astronomical and space optics.

By the annual meeting in 1963, "Optics—An Action Program" was presented to the OSA membership. The Needs in Optics Committee Report to the OSA read:

> During the last few years, there has come a seeming contradiction in the field of optics. Simultaneously, there are indications of a de-emphasis of optics in academic teaching and research together with a tremendous increase in the application of optics with a consequent shortage of personnel with training in optics. Optics people have shared this concern, and at the Fall 1961 meeting of the OSA two proposals for optics programs were submitted to the Board of the Society. At the Spring 1962 meeting of the OSA, the Board formed an ad hoc committee with the following charge: ". . . to consider the problem of current needs in optics teaching and research and present a definite plan of action to the Board at its meeting in October, 1962."[36]

The committee was specifically charged to consider the two proposals that had been submitted to the OSA Board of Directors on October 16, 1961. These two proposals were:

1. The Boston Proposal. A description of a proposed National Institute of Modern Optics (NIMO) and a program to determine its need and feasibility. Ballard summarized the Boston Proposal as follows.[37]

   It was proposed that NIMO should be primarily concerned with fundamental or theoretical optics. NIMO would be operated by a consortium of Boston area universities, with professors serving both the research institute and their home university. The model used funding by the National

Science Foundation, much like KPNO is funded or like DOE funds the nuclear physics laboratories. The plan seemed to make a division between theoretical and applied optics and lacked a strong organization and individual leader for the effort.

2. The Rochester Proposal. Recommendations were that the Optical Society conduct a survey to determine the extent and the nature of the national demand for persons trained in optics and the nation's present capability to meet this demand and that, when the preliminary results of the survey are available, a session of the Society be convened in Rochester, New York, for the presentation and discussion of the survey results.[38]

This committee submitted its report to the OSA Board on October 2, 1962. This report was coauthored by Lucian Biberman of the Institute for Defense Analysis (IDA), a federally funded think tank chartered to advise the defense department. The report, with its recommendations, was accepted by the OSA Board, and the committee was discharged.[39]

In early 1964 a survey was sent to all 4,238 OSA members asking: (1) What is your field of interest in optics? and (2) Where do you see the shortfall of knowledge? About 20 percent of the members responded, and their responses were summarized.[40]

Table 8.1 provides a summary of this 1964 survey and lists the 14 critical elements to an education in optical science, engineering, and technology. This survey clearly showed the many dimensions of optical science and engineering for the evolving national security, commercial industries, and academia.

The chair of the "Optics—An Action Program" Committee was OSA president Van Zandt Williams, who was also director of the American Institute of

Table 8.1. Summary of the 1964 OSA's membership survey that defined the broad scope of optical science and engineering.

Category	
Optical properties of materials	Physiological optics (vision and psychology)
Optical components	Photometry and radiometry
Detectors	Spectroscopy
Instruments and systems	Atmospheres (planets and stars)
Geometrical optics	Lasers
Physical optics (E & M + quantum)	Colorimetry
Information theory (correlation)	Education methods

Physics. Other members included Rod Scott of Perkin-Elmer, Bob Hopkins of the University of Rochester, W. Lewis Hyde of American Optical, and F. Dow Smith of Itek.

At that time the Institute of Optics at the University of Rochester was the US center of excellence in optics, located not far from Eastman Kodak, Bausch and Lomb, and several smaller optics industries, all concentrated in Rochester, New York. The committee advised that the United States needed more than one center of excellence in optical science and engineering but had not yet created a plan when Williams suddenly died on May 13, 1966, while on a trip to London to consult with the Royal Society.[41]

Several earlier efforts beginning in 1962 to obtain national recognition for optical science and engineering education had failed when a source of funds could not be identified. In the early 1960s a group of distinguished scientists in Boston proposed a National Institute for Modern Optics (NIMO) with the charter to perform fundamental or theoretical optics and secondarily with applied research application.[42] Funding was proposed to come from NSF in a manner similar to the national labs for research into nuclear science, such as Brookhaven and the National Center for Atmospheric Research (NCAR). Some support would come from the defense department. NIMO lacked a hard-driving, scientifically respected leader who had the energy and the personal connections to identify the clear need, develop a compelling plan to integrate the resources of government/academia/industry, and harvest funding from multiple agencies. But the landscape changed by 1966. With the death of Van Zandt Williams, the committee had no chair to put the "action" into "Optics—An Action Program."

With the concurrence of committee members, OSA executive director Mary Warga asked Aden to chair the committee. Bart Bok had arrived at Steward Observatory in the spring, and Aden was no longer chairman of the Department of Astronomy and director of Steward Observatory. However, he was deeply involved in optical surveillance projects, his astronomy research, the construction of what became the Bok telescope, and the new optical telescope in Hyderabad, India. Nonetheless, he accepted in June 1966 without hesitation, knowing that he could make a significant global impact. But he also knew that he had to convince sponsors and his peers across the nation that Arizona was the place for another center of excellence in optics and engineering.

He assembled administrators at the University of Arizona, senior officers from the US Air Force, and leaders in the academic and industrial optics communities. Two factors at the time drove the broader national need for science and engineering graduates with formal education in optical science and engineering: the invention of the laser in 1961 and the national security need for optical surveillance imaging. The Cold War was raging. Would the Soviets soon successfully weaponize high-energy lasers before the US could understand the threat? The basic physics was understood, but the optical systems that used lasers remained elusive, as was an educated workforce to lead efforts for research, engineering, and development to build hardware for the field. In addition to military applications, the scientific community wanted to use lasers as research tools to understand the interaction of light and matter, and commercial industry recognized the laser's potential as a precision cutting and forming tool. Defense, commercial, and scientific applications drove the need for academic training in optics. Today laser technology permeates commercial products in medicine, microcircuits, communication, construction, the automotive industry, and on and on. But in 1966, all these applications were still unimagined. In 1965 the US Air Force needed officers to manage optics technology and industry hardware development contracts for systems. And industry needed them to successfully bid the contracts and accomplish the defense work. A knowledgeable workforce was simply necessary to face Cold War perils.

The intelligence community urgently needed reliable overhead imaging systems to assess threats to our national security. NASA required optical imaging and spectrometer systems for space-based scientific research in Earth resource management, solar/stellar astrophysics, and the exploration of the solar system. The commercial telecommunications industry was starved for basic research and technology tools to convey TV images over cables and free-space efficiently. Today, this basic research into the physics of image formation and development of imaging systems has led to high-quality images for the entertainment industry, medicine, national security, defense, data visualization, and scientific inquiry.

OSA committee members recognized that the National Astronomical Observatory in Tucson was evidence of NSF's strong commitments to technology for astronomical optical instruments, making UA a strong candidate for a new center of optical science and engineering research and education. Aden supported this enthusiastically, as did UA president Harvill, who committed resources of the university for facilities and faculty positions to make UA the premier academic center of excellence in optics in the world. The Arizona Board of Regents, recognizing the commercial economic impact to the state and worldwide prestige, enthusiastically endorsed Harvill's initiatives.

With the arrival of Bart Bok from Mt. Stromlo Observatory in Australia on June 30, 1966, to assume leadership of the Steward Observatory and the Department of Astronomy, Aden could move on to his next adventure: the development of a 90-inch telescope for Steward Observatory and then creation of optical sciences as a research and academic teaching center.

The concept of the national astronomical observatory found expression several years earlier in meetings of distinguished scientists and engineers who clearly stated a national need for new astronomical facilities accessible to all qualified researchers, based on merit. When Congress established NSF in 1951, a source of funds became available to the university community to establish a plan for a national astronomical observatory. Aden spearheaded implementation of that plan successfully as the Kitt Peak National Observatory. Others worked the funding issues within the government, and Aden's role was that of a charismatic leader who inspired respect from all who worked for him and who had a reputation for getting the job done.

In a similar manner 15 years later, two independent national committees, one comprising leaders in the defense community and the other members of the academic and commercial industry communities, were formed to investigate the national need for formal training in optical sciences and engineering. After some discussion, both committees independently recommended establishing a center of excellence in optical sciences for teaching and research at UA.

Debates about the need often centered around what the optical sciences and engineering are, and why a new center was needed, leading to this description of the scope of both:[43]

Optical science is the study of the generation, propagation, imaging, and sensing of light and from the ultraviolet through the visible and into the far infrared.

Optical engineering is the application of optical science to the hardware and software development of devices, methods, and processes for commercial, academic, and government applications.

Optical science is interdisciplinary and requires knowledge of physics, chemistry, mechanical, structural and thermal engineering, material science, nano- and micro-structures, waveguides, spectroscopy, polarimetry, vision (psychology and physiology), detectors, solid state physics, electronics, and atmospheric science. The theoretical physics tools necessary for a comprehensive education in optical science include geometric ray-trace (trigonometry), scalar and vector electromagnetism and diffraction (calculus), linear algebra (polarization), radiometry (thermal radiative transfer), statistical optics, image processing, information theory, computer science, and quantum theory. Little or no interdisciplinary contact was rewarded or encouraged. In the classic academic institutions of 1966, these topics were taught and research was performed in separate, independent departments, often spread out across a campus and well insulated from each other by university administrators and facilities. No classic academic program had the resources or the vision to bring it all together under one roof and create the program the nation needed.

The report of Aden's committee did recommend the establishment of a new Center of Excellence in optics. They also went further. As Tucson was fast becoming the astronomical center of the United States, and because making large optics seemed to be something that would be around for quite some time in the future, the committee agreed that Tucson was the ideal site for the Center. Harvill thought that made good sense, and he was prepared to ask for state support to start a graduate study program. The big problem was where to house such a large group and where to find sustaining funding from contracts and grants. Optics were spread among several different directorates at NSF. The science communities (astronomy, biology, and medicine, for example) used optics as a tool for research, and not as a science and technology discipline unto itself. Harvill released some monies to begin a study, but significant support was to come from a different source—the US Air Force.

The genius of Aden was that he was able to cut seamlessly across barriers and fiefdoms among long entrenched academic departments to create a Center which over time has evolved to become what is today an internationally recognized research and academic program granting degrees at the bachelor's, master's, and PhD levels in a 160,000-square-foot research facility with a faculty (including three Nobel Prize winners) that today (2021) numbers over 100 and a total number of graduates that exceeds 2,500.

Federal agencies respond

On May 7, 1964, President Johnson, in an address to students at Ohio University and in another to students at the University of Michigan on May 22, 1964, outlined his goals for higher education under his Great Society program. Congress supported education with the Higher Education Facilities Act of 1963, which was signed into law by Johnson a month after becoming president upon the assassination of John F. Kennedy. This act authorized several times more college aid within a five-year period than had been appropriated under the Land Grant College federal funding bills in an entire century. It provided for college libraries, 10–20 new graduate centers, and several new technical institutes. A major piece of legislation followed with the Higher Education Act of 1965. These increased federal funds were awarded to universities to create scholarships and provide low-interest student loans.

The NSF responded by setting up a program of Science Development Grants, better known as the Centers of Excellence program.[44] Spurred on by pressure from the executive branch, the agency wanted to increase the number of institutions of recognized excellence in research and education in the sciences. Criticism had been heavy for some time for the agency to redistribute science funding. By deliberately excluding the top 20 elite universities and concentrating its funding on second-tier institutions, and by emphasizing geographical dispersion, the Centers of Excellence program not only responded to outside criticism, but reflected the philosophy of the Great Society. Relatively large awards were made to hire new faculty, support graduate students, and construct new facilities.

Guidelines for proposing to these programs required that the CEO, in this case President Harvill, be the principal investigator. In mid-1964, Harvill and Aden submitted a proposal from UA to NSF for a science development program grant, which was funded later that same year and was used to hire faculty for education in optics and astronomy; funding of "bricks and mortar" facilities would come later. The first optics faculty appointments were Bob Noble and Roland Shack from Perkin-Elmer, hired in 1964, and Stephen Jacobs from TRG (Technical Research Group), who joined in 1965.[45] Funds for these faculty appointments came from an NSF Science Development Program grant. Each new faculty member had several students and research associates to work on grants. So, three faculty members might add up to 12 persons who needed offices and lab space. The first faculty were on board but were sharing space with two other large programs that were rapidly expanding—the Lunar and Planetary Laboratory and Steward Observatory. It was clear that new facilities were needed if UA were to sustain an active academic program in the growing interdisciplinary field of optical science and engineering.

In October 1964, at the annual meeting of the Optical Society of America, Aden delivered an address, including the results of the OSA survey and summarizing the recommendations of his "Optics—An Action Program" Committee. Aden emphasized his concern that American universities has become "fossilized" in their approach to technical and scientific education.

There were few or no opportunities for interdisciplinary studies. A graduate physics degree typically accepted only a series of narrowly focused physics classes toward graduation, and an electrical engineering degree typically accepted only a series of narrowly focused electrical engineering classes toward graduation. These narrow curricula were needed to foster competitive PhD research dissertations. They focused on self-preservation of the disciplines and provided society with few highly trained generalists who could synthesize complex ideas across multiple domains for the future of mankind. Industry and government needed MS and PhD graduates and leaders with cross-disciplinary training. More and more research was requiring broad training in addition to specific focused skills. Aden waxed eloquent in his address titled "Some Aspects of Graduate Study in the Optical Sciences" that justified the need for graduate study in the optical sciences:[46]

Optical science in the university community today faces a dilemma. It is particularly critical at the graduate level, and I would like to focus your attention on the problem.

For two years now, the Optical Society of America has had the Needs in Optics Committee examining the status of optics in the United States and the possible programs for its improvement. Interesting facts have been noted. One that I would particularly like to mention is the role played by industry in optical education.

Quite clearly, industry has acquired—through neglect by the university—the principal role in training its own specialists at all levels. Industry has played this role well. By starting with scientists and engineers with solid backgrounds in other fields, industry has created a remarkably productive and creative national team.

I hardly need refer to the statistics again, but with only 34 of the PhD theses in the physical sciences in the field of optics since 1957 (that amounts to about 1%), the magnitude of the role filled by industry is shown by the American Institute of Physics scientific and technical roster figure of 13.5% who list optical sciences as their principal field of endeavor.

Let me spotlight the remarkable change that the last decade has seen in the relative decline of the importance of the university in research. By any dollar figure, universities across the country have more research than ever. At one time the university was the principal source for basic research. Industry was concerned with development and only occasionally did it humor certain individuals with the privilege of pure research. Today we see industry in all fields of science engaged in massive programs of research and development. With a few notable exceptions, a percentage comparison shows a decline in the percentage of the national research and development effort that is carried on by the university. Although it is hard to separate the budgetary figures into separate research and separate development dollars, it is clearly evident that industry is displacing the university as the home of the researcher in many fields of science. I even see the tendency for industry to view with alarm attempts by the university to compete in seeking Federal contracts for large-scale basic research efforts.

In the light of this fact, I would like to raise four basic questions: (1) Does the university have a role in the production of specialists in optics at the MS and PhD level? (2) Is not the national need met in the present role of the university in producing persons competently trained in certain broad and basic academic skills? (3) Can the university train a person as rapidly and at as great a depth as an industry training program? And finally, (4) Can a university program remain abreast of the many fields in the optical sciences?

To answer these four questions will take longer than my allotted time. I, moreover, doubt that the answers are clearly defined, although I am personally biased to a strong university program at PhD level in the optical sciences. We are, in fact, attempting to start a new graduate program at the University of

Arizona, but as I look across the nation I am not at all confident that we will succeed. Let me explain this remark.

The university structure on a nationwide basis faces an unprecedented influx of students in the next decade. Universities are, in fact, feeling the pinch on facilities that has swelled in a flood tide up through the grade school and the high school in the last decade. Universities, therefore, have a real challenge in the burgeoning capital construction and operating budget needs just to meet the basic education programs. To expect the university to outlay from its own sources of revenue the millions needed to initiate a strong new program in a scientific and technical area so competently handled by industry is not realistic.

Where does the source of money come if the university is to successfully enter a new field, such as optics? Quite clearly the answer must be from the Federal sector. When I look about I see a curious orthogonality to this answer. Federal support for our few pioneering graduate programs in optics is drying up at an alarming rate, yet contracts for rather basic support are going to industry at an accelerated rate. Clearly we must insist that the Federal Government face this problem and assume its responsibility as a partner with the university for seeing programs of national importance kept strong.

In this context I would like to draw your attention to a recent statement that would appear to give support to this assertion. [Robert] McNamara recently was quoted as saying, "We consider it extremely important that our vital contacts with the creative research people in these institutions be continued. These are the people who in the past have been responsible for some of the most important technological improvements in the equipment used by our military forces, and we should not deprive our national defense of the benefits of their creativity." We hope that optics will be so considered.

One cannot expect a university program to be strong if its sole role is training of persons who in due course will pass through the doors of industry. This role harks back to the teacher's or normal college philosophy that for years in the past held such a check-rein over the destinies of many of our emerging universities. The leaven that makes the difference is the stimulus of research.

The support of a strong graduate program in optics needs the support of research. We cannot expect to attract the competence nor keep the faculty abreast of the field *unless* there is ample support and contact with basic research. This research must be free of the "end product," a current philosophy in evidence. Some academic people absolutely love the challenge posed by a need or sought end product or capability. I am one. Others simply want to explore some curious corner of theirs without the encumbrance of deadline or need. Some, a few, simply like to quietly think and distill the rolling tide of research to its ultimate in truth and simplicity. We must encourage these rare qualities within the community of the university.

I would like to outline how I would expect a successful new graduate program to be developed.

The initial step is to assemble a core faculty selected carefully from industry. This step is not an expensive one, but it does place a premium upon the vision and faith of those first persons.

The second is to obtain a commitment for a basic support budget, both Federal and local, for a multi-year period at a liberal growth rate.

The third, and biggest hurdle, is the money for capital construction of the physical plant required by the program. This is not a small sum.

The fourth is the funding of a number of research projects of specific objective, which, combined with the institutional type funds from the basic support, provide the means of engaging in active research at advanced level.

In addition to the four questions that I have enumerated and the four steps to achieve a new program, there are four reasonable expectations. It is reasonable to expect that a major university could and should assume the responsibility for the academic program and tenure for the senior staff in due time. It is reasonable to expect the Federal Government to assure the availability of graduate traineeships and the new type institutional grants for basic support. It is reasonable for industry to expect the competition by university staff with industry for basic research contracts in specific areas. It is likewise reasonable for industry, on the other hand, to expect that the university will not directly compete with them in applied research or product production, either directly or through thinly disguised peripheral companies.

To make a success of the task that we of the Optical Society of America have undertaken, we must create a three-cornered team: The University, The Industry, and The Federal Government. Each one must accept its responsibility, and we will achieve our national goal.

"Optics—An Action Program"

These early discussions resulted in a three-year Advanced Research Projects Agency (ARPA) contract totaling $600,000 and an allocation of $180,000 in state funds for the construction of an office and laboratory adjacent to Steward Observatory.

Following his address to the membership, there was a coffee break on the hotel parapet just outside the meeting room. A number of young scientists talked to Aden about his vision for the Optical Sciences Center (OSC) and life in Tucson. One of those was physicist Stephen Jacobs who worked for Gordon Gould, one of the early (1959) inventors of the laser, at the company Technical Research Group (TRG) in New York City. Aden convinced Jacobs to join him to lead the academic laser group at the UA, and he arrived in the summer of 1965. A few years later, Jacobs's work attracted Marlan Scully from Yale, who in turn recruited physics Nobel Prize winners Willis Lamb and Nicolaas Bloembergen to join the optical sciences faculty with teaching and research appointments.[47] Over the years, this process was repeated for other optical science areas: geometrical optics, optical fabrication and test, radiometry, detectors, image processing and vector-waves (polarization), spectroscopy, and so on, until 1973, when Aden stepped down to pursue solar thermal power.

The Air Force response

The plan that Aden laid out in 1964 for a strong academic program in optics found support both in the civilian scientific research community (NSF) and the military. In May 1960, U-2 pilot Francis Gary Powers was shot down over the Soviet Union, and future spy-plane reconnaissance missions over the USSR were no longer an option because they could be tracked at will. Eisenhower needed another way to verify the claims made by Khrushchev that the Soviets had missile superiority over the United States. Downward-looking space telescopes became an option as international treaties did not forbid them, and there were no known defenses against imaging from space. But US Air Force officers had little or no training in how to design and build the size of telescopes needed for space-based imaging systems that would resolve strategic military details on Earth's surface.

The invention of the laser in 1960 and its further rapid development caused Washington to believe that it might be possible to weaponize the technology. The Soviets had already demonstrated levels of technical competence in lasers similar to those of the United States. The 1964 Nobel Prize in Physics was shared by Charles Hard Townes (US) and the USSR team of Nikolay G. Basov and Alexander M. Prokhorov. The importance of the invention of the laser to science and society is emphasized by the incredibly short, four-year time span between its discovery and the award of the Nobel Prize.

Optical science and technology became essential to US Air Force strategic needs in both surveillance and lasers. Both the laser programs and the surveillance programs needed advanced optics technology. The Office of the Secretary of Defense recognized the importance of optical science and engineering to the future of the nation's security.

Training of Air Force officers at the MS and PhD levels to make knowledgeable engineering and acquisition decisions was essential. In 1964 the US Air Force started negotiation with the University of Rochester Institute of Optics about sending 5–10 officers per year to obtain graduate degrees in optical science and engineering. The University of Rochester faculty balked and would accept only two students per year.[48] This was insufficient to satisfy the urgent needs of the nation's defense programs. In the meantime, another Air Force office began discussions with the Rochester Institute of Technology (RIT) about accepting Air Force officers into their graduate program in optics technology. The Institute specialized in imaging science and technology and had little to offer in laser research and development or in optical engineering, fabrication, and test of imaging systems.

About 1963, an Air Force Colonel from Wright-Patterson Air Force Base in Dayton, Ohio, was negotiating with the University of Rochester and RIT to put a center of excellence in lasers and surveillance optics in Rochester, when yet another Air Force Colonel and Aden's friend, Lew Allen, heard about it. He called Aden, and together they created a plan that trumped the Wright-Patterson offer.[49]

The academic program that Aden was putting together, under his 1964 funding from NSF, followed the recommendations laid out in OSA's "Optics— An Action Program" and included geometrical optics, diffraction, opto-mechanical engineering, fabrication, test, laser physics, physical optics, statistical optics (coherence theory), interferometry, and instrumentation, along with image science and visual perception. Included in these plans were faculty and facilities in computer-aided design (CAD) of optical systems, detectors, radiometry, and optical material science. Psychology of vision and spectroscopy (atomic and molecular) were integrated into the program through Aden's personal contacts in the Departments of Psychology and Physics on the UA campus. Image processing was added later through the auspices of signal processing, probability theory, and statistical analysis instruction in the Departments of Mathematics and Electrical Engineering. Later this image-processing capability would find significant application in the medical sciences.

Then-Lieutenant Jim Mayo at Wright-Patterson AFB asked Aden how many officers UA was prepared to educate. His response was that UA would support 5–15 per year, provided the candidates could pass admission criteria and succeed in academic classes.[50] In an age when science and technology graduate students flowed exclusively from undergraduate programs, this was a non-traditional approach. Aden himself was an unconventional graduate student. He had been away from college for four years before he returned, motivated to finish his BA and PhD in three years on the GI Bill and no longer distracted by the war. Aden was sensitive to the needs of the student who returned to school after real-world engineering experience because his own academic life had followed a non-traditional approach. These veteran students had an understanding and were anxious to obtain the skills only graduate training can provide.

The nascent UA program in optical sciences and engineering reflected a new area for academic studies, and many of its first students arrived with undergraduate degrees in engineering or math. Others came with several years of experience working in the field of optical engineering, knowing what they still did not know. Some of these quit their well-paying engineering jobs and uprooted families to move to Tucson to join the new OSC for a formal education in optical engineering. Two agencies in the Department of Defense moved to fund optical science and engineering—ARPA and the Air Force.

Advanced Research Projects Agency

In the summer of 1963, Aden was invited by his Air Force contacts at Kirtland Air Force Base in Albuquerque, New Mexico, to consult on telescope design and instruments for meter-class telescope systems to observe satellites from the ground.

Not long after President Kennedy's assassination on November 22, 1963, the phone rang in Aden's office in Steward Observatory.[51]

Ralph Zirkind, deputy director of ARPA, was calling to invite him to attend a classified meeting at the RAND Corporation in Santa Monica, California. His experience with astronomical seeing was invaluable (see Chapter 11). He also offered ARPA support to try new techniques of polishing and testing mirrors for three years at $150,000/year. Aden hastily submitted the proposal, which was funded even more quickly.

With that support, Aden drove to Kitt Peak and talked with his close friend and skilled optician, Don Loomis. He knew that Loomis was thinking of leaving KPNO and asked him to head the Steward Observatory Optical Shop and work on their 84-inch telescope under the ARPA grant to innovate optical processing. Loomis then joined the growing optics team at Steward Observatory.[52]

At the spring 1964 meeting of the American Physical Society, held that year in Tucson, Zirkind found Aden in his office at Steward Observatory. During this meeting, Aden outlined his vision for a program in optical sciences at UA. Aden's energy and infectious enthusiasm captured Zirkind's interest, and the university was able to start its first stages of a comprehensive program in the optical sciences.

At Steward Observatory, things were happening. Kuiper's team was growing so rapidly that he soon forgot about Aden's sudden departure from LPL matters. The Steward Observatory copy of the KPNO 84-inch telescope, which became the Bok 90-inch, was now Aden's big challenge. Aden made several improvements, learning from the KNPO 84-inch. The new mirror would be fused silica, not Pyrex. General Electric (GE) had just developed a method of fusing hexagonal boules of fused quartz into large mirrors. They had made one attempt to make a 4-m disk for KPNO's new 4-m telescope, but it had failed dramatically. When the 4-m furnace was opened, the mirror in front of them was reduced to perhaps a million small chunks of fused quartz.[53]

What had happened? The furnace overheated, causing the surrounding ring of fire brick to fuse. When it cooled, the mirror fractured from the compressive force imposed. Now GE needed to test a revised process and wanted to fabricate a fused silica mirror to a smaller, less risky size. The Steward Observatory's planned mirror (today the Bok telescope) fitted GE's needs, and UA received an attractive price and delivery time. When it arrived, it was not an 84-inch blank; it had finished up at 90 inches. No one complained. But Aden checked with Boller & Chivens (B&C), the contractor that was awarded the bid to build the telescope, to be sure that they could enlarge the cell to accommodate the larger mirror blank at no increase in cost. Clyde Chivens, president of B&C, told Aden that it was not a problem.[54]

Department of Astronomy

The Department of Astronomy started to grow in 1964 with the arrival of a new faculty member, Ronald Hilliard, from the University of Saskatchewan. Aden had completed the move of the 36-inch telescope to Kitt Peak and its reassembly, and Hilliard was hired to finish building new instruments for the telescope, including a coudé room for high-resolution stellar spectroscopy. Hilliard recalled that Aden would show up every once in a while and sit in his office, telling the staff about what he had learned on his many trips. One time he came back all excited about holography, reporting that it was "the wave of the future." None of the staff had any idea what he was talking about. At this time holography was a technical solution looking for a problem.[55]

By 1965, the number of astronomy graduate students and professors had more than doubled in one year. Three strong personalities—Bok, Kuiper, and Aden—were competing for office, teaching, and research space. New hires were arriving monthly. Bok accelerated his date of arrival in Tucson by a few months in view of the rate at which things were happening. In the spring of 1966 he moved into Aden's office, and Aden moved back into his old office. Aden continued to consult on the Steward Observatory's 90-inch project until it was ready for dedication in June 1969.[56]

Bok was dedicated to teaching and took that role seriously. In the fall of 1966, Meinel taught his class in stellar spectroscopy, but he was distracted by consulting for the Air Force, developing his optical science program, and wrapping up his work on the telescope in India. He missed several classes. Bok called Aden into his office and shouted his displeasure at Aden's teaching habits. Everyone in the building heard the loud dressing down.[57] Afterward, Aden sought help from Kuiper and moved the optical sciences faculty, staff, and laboratories out of Steward Observatory over to LPL.

When Aden took over the astronomy department in the spring of 1963, Carpenter was in the process of moving the 36-inch telescope from campus to Kitt Peak. At that time there were five astronomy professors, including Carpenter and Kuiper, offering 10 courses to a graduate student body numbering less than 6.[58] By 1966, when Aden turned the department over to Bok, the department listed 12 faculty members teaching over 20 courses to a student body of over 20. Aden had recruited new faculty, found funding for a second-floor addition to Steward Observatory, and created a first-class optics shop. Space astronomy, astrophysics, physics of the solar system, and three classes in optical instrumentation and optical sciences had been added. The 36-inch telescope was up and running at the University's site on Kitt Peak and the construction of the Steward Observatory 90-inch telescope was well underway. On campus, the square footage devoted to astronomy had more than doubled. Kuiper's lunar work was expanding rapidly in response to the needs of the Apollo program. In late 1964, Kuiper submitted a proposal to NASA to construct a new five-story building on campus with an underground optics shop for large mirrors. The new LPL building, with 51,600 square feet, was finished in mid-1966. Three new astronomy faculty (in reality, optical scientists)—Roland Shack, Bob Noble, and Steve Jacobs—plus their graduate students moved into temporary offices in LPL. Jacobs's student laser lab remained at Steward because of lack of space at LPL.

Coincident with completion of the LPL building, Aden prepared a proposal to the Air Force for OSC, with encouragement from Brockway McMillan, who was assistant secretary for research and then undersecretary of the Air Force and later the second director of the National Reconnaissance Office (NRO). McMillan and Alexander Flax, succeeding McMillan as director of the NRO, assigned Dan Anderson to guide the Optical Sciences proposal through the Pentagon. Harry Davis, deputy assistant secretary of the Air Force (Special Programs), was instrumental in procuring 3–5 years of Air Force funding for the Center and had to sign off on all procurements related to Air Force research and development, including intelligence and reconnaissance. When Bob Shannon decided to move to OSC from Itek in 1969, there were two jokes. One was "Oh, you're joining the HDMI" (Harry Davis Memorial Institute). The other was "I see you're joining the ABM program" (Aden Baker Meinel, not anti-ballistic missile).[59]

The optical sciences academic program first grew within the Department of Astronomy. As word spread, more and more students were arriving each year. The optical engineering industry had fallen into a brief recession, and young engineers returned to college wanting formal training in their profession to become more competitive when the economy turned around. The 1966–1967 catalog of classes at UA included six optical science classes within the Department of Astronomy: Introduction to Optical Sciences taught by Robert Noble; Optical Design by Aden; Optical Instrumentation team-taught by Noble and Aden; two courses in Interference, Diffraction, and Image Formation by Shack; and Classification of Stellar Spectra by Aden. In addition, 16 astronomy courses and two in solar system astronomy were taught by faculty from LPL.[60]

In November 1965, Harvill used state funds to authorize the architectural design for the OSC Building as part of the proposal the Air Force requested. The contract stipulated operation and delivery by August 1967. It was, in fact, dedicated on January 22, 1968, with John Lucas, undersecretary of the Air Force, delivering an address at the formal gathering on the mall in front of the building. A year earlier, the Arizona Board of Regents had granted full academic status to OSC and had approved the degrees of MS and PhD in optical sciences, making UA one of only two universities in the United States to offer degrees in that discipline.[61]

All the cards were stacked in the right direction. Aden wrapped himself in the prestige of NSF research and education and had worked the national crisis in optics education and the surveillance/laser education needs of the Air Force to create OSC, originally part of the Department of Astronomy and then a separate center of excellence managed as an independent unit—the College of Optical Sciences—reporting to the Provost with its own dean and administrative status equal to that of the College of Medicine. Aden had promised the Air Force that OSC would admit officers who met UA academic standards and then graduate MS and PhD officers. From 1968 through 1976, 15 to 20 percent of the graduate students were Air Force officers. The Air Force paid their living allowance, salary, and tuition in exchange for adding three years to their enlistments. Often the research for the dissertations was finished at their home base (using Air Force equipment and super computers) either in Dayton, Ohio, or Albuquerque, New Mexico. In many cases, faculty advisors were consultants to the Air Force and would travel to air bases to supervise research. In some cases, the students would return to Tucson to complete a residency requirement, defend their thesis or dissertation, and attend graduation. This approach to PhD education, supported by Harvill and all UA administrators, was completely nontraditional. But then nothing that Aden did professionally was traditional.

Notes

1. Meinel, A. B., and Meinel, M. P. 2002a. *Echoes from a simpler time*, unpublished, 120.
2. Correspondence from Mayall to Harvill in 1961 microfiche archives for President Harvill, University of Arizona, Special Collections.

3. Meinel, A. B. 1962. Infrared astronomy (abstract). *Astronomical Journal* 67: 118; Meinel, A. B., and Hoxie, D. T. 1962. On the spectrum of lightning in the Venus atmosphere. *Publications of the Astronomical Society of the Pacific* 74: 329–30; Meinel, A. B., Bashkin, S., and Loomis, D. A. 1965. Controlled figuring of optical surfaces by energetic ionic beams (abstract). *Applied Optics* 4: 1674.

4. Meinel, A. B. 1961e. New frontiers of astronomical technology. *Science* 134: 1165–71 (Reprinted as *New frontiers in astronomical instrumentation*. Kitt Peak National Observatory, Contributions no. 11, 1961).

5. Meinel, A. B. 1959. Astronomical observations from space vehicles. *Publications of the Astronomical Society of the Pacific* 71: 369–80 (Reprinted as Kitt Peak National Observatory, Contributions no. 1, 1959).

6. "Steward Observatory formally dedicated with appropriate ceremonies at the U. of A. *Tucson Daily Citizen*, April 24, 1923.

7. Meinel and Meinel, *Echoes*, 123.

8. Meinel and Meinel, *Echoes*, 123–4.

9. Meinel and Meinel, *Echoes*, 124–5.

10. Meinel and Meinel, *Echoes*, 125.

11. Meinel and Meinel, *Echoes*, 128.

12. Email from A. B. Meinel to W. Patrick McCray, November 9, 1998.

13. Facilities Act of December 16, 1963 (Higher Education Facilities Act of 1963), Public Law 88-204, 77 STAT 363, "to authorize assistance to public and other nonprofit institutions of higher education in financing the construction, rehabilitation, or improvement of needed academic and related facilities in undergraduate and graduate institutions." 16 (NWCTB-11-LAWS-PI159E6-PL88(204)).

14. Meinel, A. B., and Bashkin, S. 1964. Optical spectra from fast ion beams (abstract). *Astronomical Journal* 69: 552–3; Bashkin, S., Fink, D., Malmberg, P. R., Meinel, A. B., and Tilford, S. G. 1966. Collisional excitation atomic spectra in accelerated beams of light elements. *Journal of the Optical Society of America* 56: 1064–75; Bashkin, S. 1968. Beam foil spectroscopy. *Applied Optics* 7: 2341–50.

15. Grotrian, W. 1928. *Graphische Darstellung der Spektren von Atomen und Ionen mit ein, zwei und drei Valenzelektronen: Zweiter Teil. Struktur der Materie in Einzeldarstellungen* 7. Springer-Verlag, Berlin.

16. Kron, G. E. 1967. A simple design for liquid filters. *Publications of the Astronomical Society of the Pacific* 79: 76–7.

17. Noble, R. H., Statham, R. B., and Meinel, A. B. 1965. Effect of surface conductance on ionic polishing (abstract). *Journal of the Optical Society of America* 55: 1580.

18. Meinel, A. B., Bashkin, S., Loomis, D. A. 1965. Controlled figuring of optical surfaces by energetic ionic beams. *Applied Optics* 4: 1674.

19. Meinel and Meinel, *Echoes*, 129.

20. Wang, Y., Dai, C., Li, W., Meng, X., Dong, H., and Wang. P. 2016. Polishing an off-axis aspheric mirror by ion beam figuring. *Proc. SPIE* 9683. https://doi.org/10.1117/12.2242718; Li, Y., Zhang, Q., Wang, J., Xu, Q., and Ye, H. 2017. Precision grinding, lapping, polishing and post-processing of optical glass. In *Comprehensive materials finishing*, vol. 1, ed. Hashmi, S., 154–70. Elsevier Science, Amsterdam.

21. Meinel and Meinel, *Echoes*, 143.

22. Ahlberg, K. L. 2008. *Transplanting the great society: Lyndon Johnson and food for peace*. University of Missouri Press, Columbia.

23. Meinel and Meinel, *Echoes*, 143.

24. Nassau, J. J. 1962. Case Western Reserve University Archives, Shelf 19IN2, Box 14.

25. Srinivasan, S. R. 1986. Astronomy at Osmania University, India. *Journal of the British Astronomical Association* 96: 339–41.

26. Vivekananda Rao, P. 2017. History of the Japal-Rangapur Observatory (JRO) at Hyderabad, India: Its construction and research program. University of Arizona, Meinel Archives.

27. Howard, J. N. 2006. Early OSA efforts in optics education. *Optics News* 17: 16–19.

28. Hartmann, P. Jedamzik, R., Reichel, S., and Schreder, B. 2010. Optical glass and glass ceramic historical aspects and recent developments: A Schott view. *Applied Optics* 49: D157–76.

29. Meinel, A. B. 1959. Astronomical observations from space vehicles. *Publications of the Astronomical Society of the Pacific* 71: 369–80 (Reprinted as Kitt Peak National Observatory, Contributions no. 1, 1959).

30. Meinel, A. B. 1961c. Design considerations for a large aperture orbital telescope: Report. *Mémoires de la Société royale des sciences de Liège*, Sér. 5, tome 4: 49–59 (Reprinted as Kitt Peak National Observatory, Contributions no. 7, 1961); Meinel, A. B. 1961d. New frontiers of astronomical optics. *Journal of the Optical Society of America* 51: 471.

31. Personal conversation with J. Breckinridge in 2005.

32. Schawlow, A. L., and Townes, C. H., US patent number 2,929,922 for the optical maser, now called a laser, dated March 22, 1960, Bell Labs.

33. Memo to the Executive Committee of the OSA, MIT Archives, Richard C. Lord Papers, Collection MC178 box 18, file on "Needs in Optics Committee."

34. Memo to the Executive Committee of the OSA, Folder: Needs in Optics Committee, Richard C. Lord Papers, Collection MC178, Box 18, MIT Archives.

35. Memo to the Executive Committee of the OSA, Folder: Needs in Optics Committee, Richard C. Lord Papers, Collection MC178, Box 18, MIT Archives.

36. Personal correspondence dated Feb 6, 2012 from Stephen R. Wilk of Saugus, Massachusetts, to author JBB. Meinel archives, College of Optical Science, University of Arizona.

37. Ballard, S. 1962. Optical activities in the universities. *Applied Optics* 1: 96.

38. Memo to the Executive Committee of the OSA, Folder: Needs in Optics Committee, Richard C. Lord Papers, Collection MC178, Box 18, MIT Archives.

39. Reprinted from *Journal of the Optical Society of America* 53: 772–4.

40. Biberman, L. M., and Zandt Williams, V. 1964. *Optics—An Action Program*. Task III: Basic long range research programs in optics. OSA Member Survey. *Applied Optics* 4: 205–7.

41. Scott, R. M. 1966. Van Zandt Williams 1916–1966. *Journal of the Optical Society of America* 56: 1149–51.

42. Ballard, S. 1962. Optical activities in the universities. *Applied Optics* 1: 96.

43. Class notes from a lecture given in 1968 by Roland Shack from J. Breckinridge.

44. https://nsf.gov/about/history/nsf50/nsf8816.jsp.

45. Meinel, A. B., Patrick, D. L., and Harvill, R. A. 1966. A proposal to the United States Air Force for a grant in support of an optical science laboratory at the University of Arizona, Tucson, Arizona.

46. Meinel, A. B. 1964. Some aspects of graduate study in the optical sciences (abstract). *Journal of the Optical Society of America* 54: 1385.

47. Interview with Stephen Jacobs by J. Breckinridge, September 29, 2016, in Tucson, Arizona.

48. Interview with Jim Mayo by J. Breckinridge on February 14, 2017, in Tucson, Arizona.

49. Interview with Jim Mayo by J. Breckinridge on February 14, 2017, in Tucson, Arizona.

50. Interview with Jim Mayo by J. Breckinridge on February 14, 2017, in Tucson, Arizona.

51. Meinel and Meinel, *Echoes*, 130.

52. Meinel and Meinel, *Echoes*, 130.

53. Meinel and Meinel, *Echoes*, 130.

54. Meinel and Meinel, *Echoes*, 130–1.

55. Interview with Ronald Hilliard by J. Breckinridge on May 24, 2017, in Tucson, Arizona.

56. Meinel and Meinel, *Echoes*, 131.

57. Interview with John Glasby by J. Breckinridge on February 28, 2017, in Tucson, Arizona.

58. University of Arizona Graduate Record of course offerings and academic department descriptions, 1962–1963.

59. Interview with Bob Shannon by J. Breckinridge on February 27, 2019, in Tucson, Arizona.

60. University of Arizona Record Graduate Catalog for 1965–1966 and 1966–1967.

61. Harvill Papers microfiche, University of Arizona Library Special Collections and Archives.

9

Inventing Modern Optical Sciences

Bart Bok arrived in 1966 from Mt. Stromlo Observatory to become director of the Steward Observatory and chairman of the Department of Astronomy at the University of Arizona (UA). This enabled Aden to concentrate on developing a sustainable Optical Sciences Center (OSC) to continue his research on astronomical telescopes and instruments and to teach. It also gave him more time to consult on Air Force projects and develop the Air Force as a reliable funding source for the new Center. Thereafter until 1973, Aden provided initiative and leadership as director of the OSC to establish the foundation for today's James C. Wyant College of Optical Sciences at UA. When the Center became self-sufficient by 1973, he returned to his tenured faculty position to continue research in optical system science and engineering for large telescopes and innovative astronomical instruments. From 1973 to 1984, Aden and Marjorie collaborated closely to establish the technical and scientific foundation for today's commercially viable generation of solar thermal power across the globe. This phase of their careers was so comprehensive that Chapter 10 (written by Donald E. Osborn) is devoted to it.

As discussed at length in the previous chapter, Aden's report on the need for a new optical center was well received, but no sponsor was apparent. Optics was a field outside the sphere of activity of NSF, which was designed for classic academic science departments and not interdisciplinary studies. Funding from mission agencies and/or private sources was considered, but proprietary research within a university environment in the United States is awkward and was soon ruled out. Roderick Scott and Robert Hopkins (both members of the OSA Optics Action Committee) suggested that the Department of Defense might be the proper sponsor. In the meantime, Aden had become involved in the rapid growth of interest in space optics developments (see Chapter 11). New avenues were opened up by his associations with the secretary of the Air Force and with James Eyer, who chaired the Optics Panel of the Scientific Advisory Board of the CIA and later served on the faculty of the OSC. It was apparent that a new center was timely, but the problem was how to create it, particularly after Washington turned its back on new brick-and-mortar projects paid for by the Department of Defense.[1]

With Stars in Their Eyes. James B. Breckinridge and Alec M. Pridgeon, Oxford University Press. © Oxford University Press 2022. DOI: 10.1093/oso/9780190915674.003.0009

Faculty

Faculty for a new academic program requires years of development. Aden's work on two committees of the Optical Society of America (OSA), as well as President Harvill's commitment to the program, inspired Aden to begin his search for optics faculty as early as 1964. His roles as director of Steward Observatory and chairman of the Department of Astronomy provided him with the clout to hire faculty who shared his vision for an optics institution. At the same time, his work on national advisory committees provided awareness of cutting-edge research and what was needed.

Aden began to hire faculty on "soft money" (non-tenure track) with strong optics backgrounds while he was department chairman from February 1963 to June 1966. The problem was where to find them. During the early 1960s, optics was not a recognized field of engineering and was taught only in physics departments as a one-semester undergraduate course covering both geometrical and physical optics. In the United States, physics departments emphasized particle- or high-energy physics to investigate the fundamental nature of matter. Many senior physics faculty had worked on the Manhattan Project and continued to teach the subject most comfortable to them—high-energy physics. In 1960, only three colleges or universities in the West could reasonably claim to have comprehensive optics graduate programs: (1) Institute of Optics at the University of Rochester in New York; (2) Imperial College in London; and (3) École Supérieure d'optique in Paris. The Soviets had optics programs and facilities in Leningrad, but at the height of the Cold War there was little interaction with the West.

Faculty were needed in Tucson with demonstrated expertise in geometric ray-trace, scalar and vector wave diffraction, radiometry (radiative transfer), statistical optics, image processing, instrument design and development, and quantum mechanics. Candidates had to be able to create compelling research topics in these areas, write and manage proposals, and create self-sustaining research groups for graduate students and the university.

As a result of the Soviet launch of Sputnik 1 in 1957, academic departments were flooded with students in the classic academic areas of physics, mathematics, electrical engineering, mechanical engineering, and chemistry. By the late 1960s, there were more graduates than academic jobs in these fields. Some had completed research in quantum optics, optical materials, and thin-film/micro-circuits and were looking for exciting work. Many graduates of English-speaking institutions had well-paying jobs in industry, so it was difficult to recruit the best and brightest. Commercial industry (e.g., Eastman Kodak) had generous pay-packages that universities could not match. Although federal government contracts employed most of these graduates, it was an unstable work environment because the government was required to offer contracts to companies submitting the lowest bids.

Aden was able to attract faculty candidates who wanted a more stable work environment and home life for their families. But at the same time, candidates were taking a risk by coming to Tucson to teach in a non-traditional department. Would the program succeed? Would faculty be tenured? Would there be job security?

Of course, no single person can be an expert in all of these areas, and Aden needed to assemble a faculty capable of teaching and research in each of these and manage that faculty to work well together toward a common goal. By 1970, Aden had his core faculty, and the full academic program offered over 25 lecture semester classes and 4 laboratory classes in 1970. Six of these classes were cross-listed with other departments, such as physics, chemistry, psychology, mathematics, and electrical engineering. The PhD program required 120 class units. If a class met for 50 minutes three times a week, it was considered a 3-unit class. Units were awarded for dissertation (PhD) and/or thesis (MS) research, so the number of required classes was about 30, which included four courses for a minor.[2]

As early as 1961, Aden started to develop a network of optical scientists who would become his core faculty, joining him at UA from 1964 to 1973, using his reputation in applied optics and astrophysics, along with contacts in the defense optics industry and at military/research contracting agencies, to select highly qualified candidates. For example, Aden was consulting with Perkin-Elmer (P-E) on advanced space optical systems. Dick Perkin, who cofounded the company, admired Aden for his ability to visualize an optical system end-to-end, identify optical engineering challenges, and create solutions that worked technically and economically. Aden served on national panels reviewing technical proposals on the East Coast. Government contractors relied on his advice. Through his connections with the Air Force and CIA, he was able to know who was doing what, not only at P-E but also its competitors (principally Itek and Eastman Kodak). This enabled Aden to select the "best of the best" for his faculty.

In the fall of 1963, a few months after he was appointed director of Steward Observatory, Aden attended the annual OSA meeting in Rochester, New York, where he spoke with Robert Noble. Aden had been introduced to Noble by Perkin during one of Aden's Air Force program reviews. Noble had taught at the University of Michigan prior to joining P-E. Perkin knew that Noble wanted to return to a university career and had relayed this information to Aden in 1963, which made it easier to recruit Noble for the Center in the spring of 1964.[3] Noble's expertise was instrument design and development; he was also a capable lecturer on a wide variety of topics in optics. At UA, Noble taught classes in instruments and radiometry and later assumed responsibility for coordination of the admission of MS and PhD students and shepherding their careers at the Center.

The next three faculty members to be hired by Aden were close friends before they were hired: Roland Shack, Stephen Jacobs, and Nick Stavroudis. Aden's consulting work for P-E and his respect for the Shack's depth of understanding of optics was key to finding the initial faculty. Shack would brief Aden on optics issues that were pending at the company. In the fall of 1955, Shack had graduated from the University of Maryland with a BS in physics and a BA in fine arts and was working half-time at the National Bureau of Standards with Stavroudis. At the same time, he was working as an artist and set designer for a live production theater in Washington, DC, where he met his future wife Pamela, an optical designer. In 1957, Pam and Roland married, and Roland accepted a new job at P-E, but they lived in New York City, where they met Steve Jacobs.

Aden told Perkin of his great admiration for Shack's intuitive understanding of optical engineering. Perkin asked Aden where Shack should go to graduate school to learn more about optics, and Aden suggested Imperial College, London. At the recommendation of Aden, Perkin as president of P-E paid for Roland to complete his PhD in optics at Imperial College from 1961 to 1963. Pam earned a graduate diploma in technical optics in 1962 and was hired for the following year as a teaching assistant at Imperial College. Robert Shannon was one of many visitors to Imperial College during that time. The Shacks returned to the United States in 1963, and Roland resumed work at P-E. He had been offered a teaching position at the University of Rochester but had not accepted it when Aden approached him about coming to Arizona. Although Shack had not completed his doctorate, he was hired as a lecturer in the Department of Astronomy with the proviso that he would finish his dissertation. He did finish in 1965, and Aden sent him to conferences in Paris, Stockholm, and London. Shack was instrumental in the hiring of Jacobs and Stavroudis and taught a broad range of classes, including first-order optical design, optical testing, aberration theory, image formation and interference, coherence, and diffraction, until the faculty expanded after 1970 when teaching loads were rebalanced. Shack invented the Shack-Hartmann interferometer commonly used today in many instruments.[4]

The Air Force wanted their officers to learn details of laser physics, and Aden needed a respected laser scientist who could teach. Jacobs graduated from Johns Hopkins in 1956 in physics and went to work at P-E on satellite-tracking cameras. Four years later, Jacobs left P-E and found work with Gordon Gould at the Technical Research Group (TRG) on Long Island. Gould is usually credited as the inventor of the laser in November 1957; Jacobs contributed to making the laser practical. After a few years, Control Data Corporation bought out TRG, and the laser research work stopped.

Jacobs met Shack out on the parapet of the Hotel New Yorker at the Optical Society meeting in October 1964. That was the same meeting where Aden gave his address titled "Some Aspects of Graduate Study in the Optical Sciences." Jacobs thought that Shack had gone to the ends of the earth when he took a position in Arizona, but Shack invited him to come look it over. He did, and Aden, who needed a laser specialist, liked him. Lasers were brand new, so nobody knew anything about them. Jacobs arrived in Tucson in the summer of 1965 to teach his first course on lasers in the fall.[5]

Stavroudis received his bachelor's and master's degrees from Columbia University and earned a doctorate at Imperial College in 1959. An expert in the mathematics of geometrical and physical optics, he joined OSC in 1967 as one of its earliest professors. He advised eight MS and PhD students during his tenure, which extended to 1988, when he retired.

The fourth employee Aden hired was Phil Slater from Armour Research Labs in Chicago, Illinois, now called the Illinois Institute of Technology Research Institute (IITRI). He received his PhD from Imperial College and specialized in airborne imaging systems, atmospheric optics, photointerpretation, and instruments.

Aden recruited several faculty members from industry, among them William (Bill) Wolfe, Jr., and Bob Shannon. Neither had a PhD, but they had ample industrial experience and loved the idea of dealing with students and teaching. The fact that there were several industrial people on the faculty earned the respect of students for the applied field of optical engineering. Aden quickly recognized that industry experience in the design, building, and testing of practical optical systems was essential and just as important as academic credentials. As Jacobs recounted, the students said, "Hey, these are guys who've been there. They've done these things, and they're worth listening to." That's what helped make the Center's program so successful.[6]

An important aspect of understanding the principles behind the design of surveillance telescopes is an in-depth appreciation for how images are analyzed. By looking at an image, sometimes blurred and out of focus, how does one decide what the objects in the field of view are? Aden needed an image analyst who could work with the optical engineers to develop requirements for systems that could be built. James Eyer, a graduate of Rochester Institute of Technology (RIT) who had several years of experience working as an image analyst-interpreter for the CIA, joined the faculty. His hands-on experience in the interpretation of images provided important feedback for engineers teaching the design of optical systems, detectors, and radiometry.

At OSC, Aden established research groups in optical materials (lenses, filters, mirrors) and coatings and found financial support for graduate students and laboratory facilities. In 1968, he recruited and hired Clarence Babcock, retired chief glass scientist and Associate Director of Research at Owens-Illinois glass company. Babcock was a major contributor to the invention of the low-expansion glass ceramic, Cer-Vit, used at the time in large-aperture telescope mirrors. He taught classes in glass science. Arthur Francis Turner, retired head of the physics department at Bausch + Lomb, joined OSC in 1971. His research focused on formulation of glass compositions for evaporated thin-film waveguides, precision lens development for integrated optics, and multilayer thin-film waveguide systems. He earned the Scientific and Engineering Award from the Academy of Motion Picture Arts and Sciences, served at OSA president in 1968, and was awarded OSA's Frederic Ives Medal in 1971. Aden recognized the great demand for practical thin-film theory and practice, and so he had Turner and John Poulos (Jacobs's associate at TRG) set up a thin-film lab course in an ordinary classroom in Steward Observatory in the fall of 1967. The class was a huge success for several academic years. Turner was active until 1979, when he retired from UA as professor emeritus. Bernhard Seraphin served in the German army in World War II and received his doctorate from Humboldt University of Berlin in 1951. He immigrated to the United States in 1959 and worked at NOTS at China Lake, where he became a leader in modular spectroscopy. Aden hired him for OSC in 1970, and together they worked on developing efficient solar energy technology, such as spectrally selective surfaces for high-temperature photothermal solar energy conversion. His other research focused on chemical vapor deposition of semiconductor and metal films of optical quality, stabilization of thin-metal films against agglomeration at high temperatures, optical properties of thin refractory-metal films, and thermo-reflectance of amorphous semiconductor compounds. Structural engineers Ralph Richard and Allan Malvick also supported students and faculty.

During the spring of 1967, OSC became a degree-granting entity within UA at the March meeting of the Arizona Board of Regents.[7] In 1968 and 1969, the OSC faculty grew from 20 (four of them joint appointments with other departments) to over 65. There was an obvious trend in Aden's pattern for recruiting and hiring faculty. They all had basic education in optics or optics-related fields and had worked in industry on "real-world" optics problems for a few years, gaining experience. Most faculty were hired before 1970, and not one of them came directly from academia.

After Aden hired the first faculty members, he became a "hands-off" manager, neither directing, reviewing course material, nor demanding that specific topics be taught.[8] As long as the general topics, such as lasers, ray-trace design, interferometry, optical testing, aberration theory, electricity and magnetism for optics, and quantum optics were being taught, Aden left the academic program in the hands of the faculty. He was an active promoter of the research program that provided the fabric of support for the whole Center.

Teaching faculty were paid 15 to 20 percent of their salary from State of Arizona funds, the remaining 80 to 85 percent from research grants and contract awards. Almost all of the faculty were on soft money and expected to submit proposals to NSF, mission-oriented agencies, and private foundations for funding. Graduate students were supported half-time on research grants to their major professor, with 20 percent funding awarded to the best students from State funds.[9]

By the time Aden stepped down as director in May 1973, there was strong academic teaching and research in several critical areas: computer-aided design (CAD), ray-trace optical design and engineering, precision structures for optical systems, optical materials, physical optics, radiometry, detectors (IR and visible), optical testing, quantum optics, and lasers.

Table 9.1 lists the names and the specialties of the 21 founding faculty members who were recruited and hired during the nine-year period that Aden was the director of OSC. Column 1 gives the person's surname, column 2 the starting date at UA, column 3 the graduating college or university, column 4 the associated industry or government laboratory, and column 5 the technical/scientific field of expertise. Three faculty from other departments held joint appointments with optics. These were structural engineers Malvick and Richards and psychologist Wheeler, a specialist in the psychology of vision. This table illustrates the diverse technical and scientific backgrounds of the faculty and the depth and breadth of experience each brought to the Center. This unique faculty enabled graduate students to be exposed to a broad set of theoretical and applied science/ technology disciplines in optics and also fulfilled the recommendations of the Optical Society of America.

Bricks and mortar

By the fall of 1965, nine months before Bart Bok arrived to take over Aden's role as director of Steward Observatory and chairman of the Department of Astronomy, it was obvious that the Air Force was serious about helping to establish a university-level education and research center in optics at UA. Aden, along with D. L. Patrick (UA coordinator of research) and UA president Harvill, prepared a 101-page proposal containing architectural drawings and floor plans for a five-floor, 80,000-square-foot OSC and Laboratory.[10]

The Department of Astronomy was rapidly growing, and Bok needed Observatory space for incoming astronomy students and laboratories. Bok and Aden had several disagreements about the allocation of space.[11] Until the new building could be completed, the optics program that Aden had created occupied two floors of Kuiper's Lunar and Planetary Laboratory (LPL), as

Table 9.1 Founding Faculty Members of the Optical Sciences Center

Name	Year	College	Industry/ Government	Subjects
Babcock	1968	Purdue University	Owens-Illinois Glass	Glass, ceramics, optical materials
Bartels	1970	Göttingen	Leitz	Biomedical imaging
Eyer	1965	Optics, University of Rochester	CIA	Photointerpretation
Frieden	1966	Optics, University of Rochester	IBM/US Navy	Image restoration & aberration theory
Gaskill	1968	Stanford University	USAF & Stanford electronics lab	Fourier & statistical optics
Marathay	1969	Boston University	Technical Operations Research	Diffraction, polarization, partial coherence
Noble	1964	Ohio State University	Perkin-Elmer	Optical instrument design & engineering
Nudelman	1973	University of Maryland	Naval Ordinance Lab	Photoelectronic imaging devices, diagnostic radiology, radiometry
Seraphin	1970	Humbolt University, Berlin	Michelson Laboratory, US Navy, China Lake	Optical properties of thin-metal films
Shack	1964	Applied Optics, Imperial College, London	NBS & Perkin-Elmer	Image formation & scattered light
Shannon	1969	University of Rochester	Itek Corporation	Digital modeling and optimization of optical systems, synthetic aperture optical systems
Slater	1966	Applied Optics, Imperial College, London	IIT research institute	Design and evaluation of orbital remote sensing systems, atmospheric optics & photographic science
Stavroudis	1967	Applied Optics, Imperial College, London	Lockheed Electronics	Optical design and engineering theory
Swindell	1968	University of Sheffield, England	English Electric Company	Analog transaxial tomography, Pioneer-Jupiter image processing

Table 9.1 *Continued*

Name	Year	College	Industry/ Government	Subjects
Turner	1965	University of Berlin	Bausch & Lomb	Evaporated thin films, integrated optics, glass composition
Wolfe	1969	EE & Physics, University of Michigan	Honeywell Radiation Center	Optical properties of IR materials, radiometry
Jacobs	1965	Physics, Johns Hopkins	TRG, Inc. & Perkin-Elmer	Laser physics & interferometry, optical heterodyne
Hopf	1970	Physics, Yale University	USAF Cambridge Labs	Theory, free-electron lasers, quantum electrodynamics
Sargent	1969	Theoretical Physics, Yale	Perkin-Elmer and Bell Labs	Optical bistability & phase conjugation; quantum modeling of molecular absorption spectra
Shoemaker	1972	Physical Chemistry, University of Illinois.	IBM research Labs	Optical coherent transient effects & microwave spectroscopy
Scully	1970	Physics, Yale University		Lasers and quantum mechanics

Cross-listed from other academic departments

Malvick	1965	Mechanical Engineering, Notre Dame		Mechanical structures, mirror support systems
Richard	1972	Structural Mechanics, Purdue University	McDonnell Aircraft Corporation	Active figure control of large telescope mirrors
Wheeler	1969	Psychology, Indiana University	National Physics Laboratory, UK	Psychophysical and electroencephalographic evaluation of image quality

well as facilities off campus. The LPL proposal for 51,600 square feet at a cost of $1,200,000 had been funded by NASA primarily to support lunar mapping and studies for missions to the Moon. Ground was broken in 1965 and was complete in the fall of 1966.[12]

The 1965 plan reflected the needs of the Air Force, framed within the comprehensive program prioritized for membership of the OSA by the "Needs in Optics" committee, which was followed by the "Optics—An Action Program" committee report. Laboratories at the Center were dedicated to: (1) precision large optics for the design and fabrication of telescope mirrors up to 4 meters in size; (2) precision aspheric optics fabrication and testing; (3) electronographic and image tube systems; (4) lasers and coherence systems; (5) thin-film coating; (6) glass technology; (7) psychophysics; and (8) photoreconnaissance.

A crisis arose in early 1967, when there was waning interest in federal funding of science and technology. Funds for the construction of facilities disappeared. More importantly, at that time laws were changing and disallowed direct funding of construction on university campuses by the military. But the patience and skill of the Deputy Undersecretary of the Air Force, Harry Davis, and the determination of the UA Foundation, saved the project.[13] Plans for the funding of OSC resulted from two meetings at P-E with Richard Perkin and Rod Scott, a visit to Tucson by Robert Hopkins (professor at University of Rochester Institute of Optics), and a meeting in the office of F. Dow Smith (corporate scientist and vice president) at Itek. Aden then had key discussions with Brockway McMillan (Undersecretary of the Air Force and later director of the National Reconnaissance Office), Harry Davis, and Eugene Fubini (Assistant Secretary of Defense) about making a proposal to the Air Force for the OSC. Harry Davis was enthusiastic and agreed that his office would endorse the proposal and find a way to implement it.

At the end of the first decade of the OSC, Aden wrote his personal recollections on raising the funds for the building:

> With the general idea established, practical problems then came into focus. It was first determined that, indeed, no building funds could be obtained from the Air Force budget. The first of two meetings between Harry Davis and Dan Anderson for the USAF and key Congressional leaders, meeting in the late Senator [Carl] Hayden's office with Barry Goldwater, an enthusiastic booster from the start, brought forth that Congress would look favorably on a maximum effort to find an avenue to allow the USAF to sponsor the Center, At this meeting President Richard Harvill suggested that the University, with the cooperation of the University of Arizona Foundation, might find a solution to the bricks and mortar problem if the Secretary of the Air Force would indicate an intent to support the Center through a lease fee in the basic contract. The last action by McMillan before leaving as Under Secretary was to endorse this plan of action.

> The ensuing phase, with the full support of Alexander Flax, the Assistant Secretary of the Air Force for Research and Development, was largely played

between Harry Davis's office, the Air Force and DOD's legal offices, and Congressional staff. In Arizona the same scenario was being played between the University and the University of Arizona Foundation. SAMSO [Space and Missile Systems Organization] was then asked to sponsor further preliminaries including firming the construction planning. The result of these combined efforts was the drafting of the baseline contract with SAMSO, now concluded.

The concluding event in Arizona was headed by Gil Bradley, now President of the Valley National Bank. He organized a consortium of Arizona banks to provide a loan to the Foundation to build the Center. The state legislature appropriated funds requested by the Board of Regents to equip the Center and establish an academic program. The Regents further selected a site on campus and transferred title to the land to the banks until the day the loan was paid, a date passed in March 1975. On 25 November 1967 I received a much appreciated birthday present—the banner headline in the Arizona Daily Star: "5.25 Million Contract Awarded UA by USAF."[14]

The original plans for the building called for the top floor of the structure to be secured for classified Air Force and Department of Defense work. This never materialized because of the timing of student protests across the nation against the Vietnam War. Protesters believed that funding from the Defense Advanced Research Projects Agency (DARPA) and the USAF made UA part of the military-industrial complex, and therefore on-campus classified research never became an active part of OSC research programs.

Curriculum and textbooks

Aden was both the director of OSC and the chair of the academic committee responsible for the definition of course content and syllabi. From 1967 until 1989, OSC offered only MS and PhD degrees. Most students completed their undergraduate work in applied physics, mathematics, electrical engineering, or mechanical engineering, with a few arriving from undergraduate programs in astronomy or optical engineering (University of Rochester). The Air Force officer students arrived with degrees in engineering, mathematics, or physics from the Air Force Academy and from other universities. The Arizona Board of Regents granted the University authority to award MS and PhD degrees in optical science in March 1967.[15]

In the fall of 1965, the academic program in optical sciences commenced with five faculty, four graduate students, and a curriculum of seven courses. Each class earned the student three units and nominally met for 120 minutes of total lecture time per week. The seven classes offered were: (1) Introduction to the optical sciences; (2) Optical design; (3) Spectroscopy; (4) Optical instrumentation; (5) Interference, diffraction and image formation I; (6) Interference, diffraction and image formation II; and (7) Laser physics. As chairman of the department, Aden's role was to coordinate the syllabi across each of the classes. Each faculty member was responsible for the specific content of each class, as well as the students' success on written and oral exams before faculty committees.

From its inception, the PhD program at UA has required students to take approximately 54 units of coursework (roughly 18 three-unit courses), complete a satisfactory dissertation, and pass two major exams. The first exam is the comprehensive examination (called the preliminary examination in 1967) and is typically taken after two to three years of graduate study. This written exam is closely tied to the core curriculum. The second exam is the traditional oral dissertation defense before a faculty committee upon completion of the dissertation and all coursework.

At the beginning of the academic program there were not enough optics classes to fill out a full set of 18 three-unit courses in optics, and so a policy was initiated that students could apply courses taken, at the appropriate level, in other departments such as physics, mathematics, engineering, and astronomy to satisfy the University's requirement. While the majority of the students took all of their graduate coursework in optical sciences, OSC always had a liberal policy in allowing courses from other departments to be counted as part of students' PhD coursework as long as those courses were relevant to their degree program in some substantial way. For example, Aden arranged for the OSC academic program to accept a class in the psychology of vision. An important part of the optical design and engineering of visual optical systems is the psychological and physiological interpretation of image content. This open policy fits well with the interdisciplinary nature of applied and theoretical optics research. New faculty were arriving frequently, and students would petition to repeat the same course title and number, but with a different professor, to obtain another unique perspective on the same or similar subject.

One requirement for all PhD students is that they take at least two optics laboratory courses. This policy was adopted early in OSC's history and was intended to ensure that no student would graduate without having had hands-on experience in an optics laboratory. Laboratory classes in computer-aided optical design, optical testing/aberration theory and analysis, laser physics, optical fabrication shop, and radiometry and detectors were unknown in 1965. Laboratory syllabi were developed by the new faculty. Aden taught the optical fabrication shop class, Shannon the design lab, Shack the optical testing lab, Jacobs the laser lab, and Wolfe radiometry and detectors. Aden published his laboratory notebooks on topics such as glass grinding rates, optical figuring, subsurface damage, scratch/dig requirements, edging, and optical testing/alignment.

When Aden stepped down as OSC department chair in 1973, the core classes required of all PhD students were: (1) Fundamentals of optics (essentially an electromagnetic waves course); (2) Linear theory in optics; (3) Geometrical optics; (4) Interference and diffraction; and (5) Interaction of radiation with matter.

Over the past 50 years, the field of optical science and engineering has expanded rapidly, and today the academic department is partitioned into applied optics (optical engineering), image science, photonics, and optical physics. Since its implementation, this flexible core curriculum has been successful. A major strength is that it can be adapted to the changing needs of students and faculty by adding, eliminating, or modifying the courses within it. Adapting the curriculum to meet both the needs of the students and the faculty examiners was an idea originated by Aden and continues at this writing.

In the beginning, almost no textbooks on how to design and build cameras and telescope optics were available. There were none that dealt with the important areas of radiometry (how bright is the image?) and image quality (how sharp is the image?). There were also no books on computer-aided ray-trace design, only computer and software operator manuals, because the field was changing so fast. In the rapidly evolving field of lasers and quantum optics, original journal papers were used in classrooms, and professors' notes were available.

Students in the early classes at OSC would tape-record each lecture at the same time they were taking in-class notes. In addition to classes, homework, and research, they would transcribe in longhand their recordings, discuss the content among themselves, and merge these with their in-class notes to create loose-leaf "textbooks" for the courses. In some cases, groups of students would study together and create chapters or sections and then make photocopies of those books for incoming students. Later students had the benefit of these comprehensive notes for each course. Eventually, professors and their students found time to publish textbooks along with critically important problem sets, often reflecting real-world problems. Today there is a large collection of textbook-quality materials for all students of optics.

Setting the research direction

Aden set the research direction of the Center based on funded proposals from government mission agencies, such as the Department of Defense (DOD), Air Force, Office of Naval Research (ONR), NSF, and others. In the beginning, almost all of the optical sciences faculty received their support from direct contracts for theoretical and applied research. Deliverables on these contracts were technical reports, facility in-person teaching lectures, and concept hardware devices and systems. Routine manufacturing jobs were not accepted, but technology transfer to industry and small businesses was common.

The Air Force Project THEMIS began in October 1968 and ran to August 31, 1971, with a total of $800,000.[16] This four-year contract from the Air Force to

UA was led by principal investigator Aden as a member of the Fubini committee, which advised the Secretary of the Air Force on scientific matters. Aden was also president-elect of the OSA for the 1971–1972 term. Among his many credentials, he had been a regular consultant to the Deputy Secretary of the Air Force, HQ SAMSO, and P-E. These activities provided him with unique insight into the scientific and technical challenges facing the modern Air Force during the intense days of the Cold War. Project THEMIS was a DOD program to create new academic centers of excellence by broadly distributing defense research funds to universities. It succumbed to budget cuts in 1971, when there were student protests against the funding of university research by defense entities. Secretary of Defense Robert McNamara founded the program in 1967 as a "program to strengthen academic institutions" in response to congressional demands for wider distribution of federal funds and specifically to President Johnson's 1965 directive instructing federal agencies to find new ways to help universities trying to establish themselves as research institutions.

Sustaining funds

Finding sustaining funds to operate OSC was always a challenge because the funding source was primarily soft or contract money intended to deliver a fixed product to a fixed schedule. Contracts seldom ran for more than three years. Faculty were principal investigators or co-investigators on research and engineering proposals. Students were expected to be productive employees when they were fresh arrivals with undergraduate degrees. Students with physics, mathematics, or engineering backgrounds arrived with no optical engineering experience, but in their first semester they were expected to be productive on research contracts. Faculty members found themselves laboring to teach optics to the new students as well as keep up with the contract work. This worked well, however, because the technical skill set at OSC was unique in the nation; in many cases, the customer had nowhere else to turn in order to accomplish applied and basic research. In contrast, the University of Rochester's Institute of Optics discouraged its faculty from accepting government and private industry contract work.

The laser program

Aden turned the responsibility of developing the laser academic and research program to Steve Jacobs. He immediately developed a two-semester teaching syllabus for quantum optics (lasers) and a parallel series of innovative, hands-on laboratory experiments. These were targeted for science/engineering graduate students whose undergraduate background was not necessarily physics. Laboratory equipment was purchased under the Air Force contract, and by the fall of 1968 there were over 10 students enrolled in the two-semester class. The laboratory was at Steward Observatory, and Jacobs's office was across the street at LPL.

One of Jacobs's first PhD students, Jack Hanlon, introduced him to a young quantum physicist, Marlan Scully, of MIT in 1968. Scully and Murray Sargent were both students of Nobelist Willis Lamb and had started to write a book on laser theory. Sargent recommended that Lamb take on Scully as a graduate student. Both Sargent and Scully visited Tucson in December 1968, interviewed with Aden, and received an enthusiastic response from both Jacobs and Aden, who wanted them to come to OSC. Scully suggested that Sargent go to the UA's new OSC to write their book on quantum theory of the laser; Sargent was hired by Aden and arrived in Tucson in June 1969. The book, coauthored by Scully, Lamb, and Sargent, was published in 1974.[17] Jacobs taught practical laser engineering, while laser theory and quantum physics were taught by Sargent and Scully for many years. Sargent also taught in the computer science department. Scully was influential in bringing Nobelists Lamb and Nicolaas Bloembergen to OSC faculty.

Two of the hallmarks of OSC quantum optics efforts were the summer schools and winter conferences devoted to physics of quantum electronics. They were intended to bring together students, young faculty, and the founders from the new field of lasers who came from diverse backgrounds in government, academia, and industry. These were held annually for several years. Attendees were graduate students, professors, and industry scientists who wanted or needed to learn lasers and quantum optics to support their careers. It was perfect for the time, and Aden encouraged Jacobs's and Scully's leadership of the school/conference, ably assisted by Jacobs's wife Kathy. The *Proceedings* of these annual summer schools and conferences were written up in a series of books, revealing clearly how the academic field of quantum optics evolved over the years to become the photonics industries of today.

Aden recognized the importance of clear communication among faculty, students, and sponsors, in addition to the usual scientific publications. A department based on contracts must maintain communication channels with current and future sponsors. Research results needed to be interpreted clearly to sponsors, who in general had technical backgrounds but often little direct background in the new fields of optics, lasers, imaging systems, and optical information theory. Aden remembered how effective Marjorie had been as a technical writer and editor for the Caltech defense research programs, first as his fiancée and then as his wife. Throughout their lives together, she remained his guiding light on technical writing and editing issues for scientific papers, technical reports, and proposals. At OSC, Aden hired Martha Stockton to be the Center's chief technical editor and writer. He established several in-house technical publications, including *OSC Technical Reports* (bound in orange covers), a monthly newsletter titled *OSCillations*, and weekly newsletters. These publications provided important cross-faculty technical communications and also a spirit of cohesiveness among the disparate technical disciplines that comprise optics. New hires and major milestones achieved by the Center's employees were often highlighted. By the mid-1980s these publications were discontinued.

Optical imaging systems

Between 1965 and 1975, optical imaging systems such as cameras, telescopes, television, movies, and digital display systems were playing a major role and becoming ubiquitous in all areas of society. These systems collect and manipulate light to create images for our eyes to see and our brain to record, interpret, and entertain.

Mission agencies needed optical science and technology. The Air Force had a unique need for optical scientists and engineers to support the gathering of intelligence critical to national defense. The National Aeronautics and Space Administration (NASA) needed to record and transmit images across the vast distances of interplanetary space. The National Institutes of Health (NIH) needed high-quality images for medical research. Astronomers wanted to image the skies through the turbulence of the earth's atmosphere and look farther and farther into the cosmos with space telescopes. Commercial television and the movie industry required continuous improvement in image quality from their optical systems for their growing customer base. The capabilities of photographic film were stretched to their limits, and the technologies for digital image recording, transmission, and storage were emerging. Images transmitted over wires or the airwaves early in the development of television required large communication bandwidths and high electrical power, limiting the quality of pictures transmitted from space and television towers into homes. The OSC contributed to the scientific understanding of image formation by showing how to reduce the bandwidth and power needed to transmit images from space.

From experience with space telescopes at KPNO and the DOD, Aden knew that the key to space-imaging was communicating image signals electronically over wires or through space by radio. Until television became popular in the 1950s, images were always communicated on sheets of film or photographic prints. It was impossible to image from deep space without converting images to electronic signals. Astronomers could not afford to drop film from orbit. Ralph

Baker and Richard Cromwell at OSC both made significant contributions to the recording and communication of images to analysis centers.

Three critical advances were needed: (1) evolve the design, fabrication, and testing of optical systems and devices such as mirrors, lenses, and filters from art to science to routine engineering; (2) develop and implement a quantitative understanding of image formation, transmission, and display for interpretation; and (3) conduct research into optical materials needed to understand the fundamental nature of the interaction of light and matter. It is the interaction of light and matter that enables us to manipulate light to form images and convert light to electronic signals for transmission and display.

Telescopes for astronomy

The period from 1966 to 1973 saw a burst in the construction of ground-based astronomical telescopes. New technologies in glass and ceramic science enabled lightweight mirrors. At the same time, in optical-surface fabrication pioneered by industry for the Air Force, space optics programs led to large telescope apertures, both of which reduced telescope costs. Aden facilitated the transfer of technology from industry to academia; as a result, cutting-edge, ground-based astronomical telescopes were being built around the world.

But an astronomical telescope needs more than just mirrors. It needs a large, precision mechanical structure to separate the primary from the secondary mirrors and maintain that separation at the level of nanometers (nm). The telescope needs to point to an accuracy of a few milliarcseconds and track an object in the sky over periods of hours with milliarcsecond precision. A milliarcsecond is about the size of a dime atop the Eiffel Tower as seen from New York City. In addition, the separations between optical surfaces—for example, the primary to the secondary and then to the instrument—must be constant and stable to less than a 1/10 wavelength of light or typically 20 nm. Mechanical and structural engineers meet these requirements by using stiff structures and mirrors. Historically this requisite stiffness was achieved by using massive components, but in the mid-1960s the development of increasingly more massive telescopes led to prohibitive cost increases. Fear arose in the scientific community that the inability to build larger telescopes would prevent further deep-exploration and discovery. But new ways to manufacture mirrors, design telescopes, and record/ analyze images developed at the OSC, along with other work on adaptive optics and detectors, precipitated a continuing revolution in the astrophysical sciences.

Space telescopes require innovative structures that survive the harsh vibration environment of launch yet retain the nanometer-precision necessary for the optical system to deliver high-resolution, angular images at the focal plane.

Ground-based telescopes use stiff structures to maintain precise alignment while the telescope pointing swings through the sky in the presence of the ever-changing direction of the gravitational field of Earth warping the structure.

In the early 1980s, flexible mirror technology was developed, which enabled adaptive, real-time control of optical surfaces.[18] These new technologies enabled lightweight telescopes to be built but with the physical properties of massive telescopes. New technologies in optical design (decreasing the f/ratio of the telescope), deformable mirrors (material science), servo optimization and adaptive structures (mathematics), high-speed electronics (micro-circuit industry), detectors, and also understanding image formation (system engineering) enabled new, large-aperture telescopes at acceptable costs.

Resignation as director

President Harvill retired in 1971, and in the fall of 1972 Aden announced that he would resign from the directorship of the OSC to devote more time to personal research. He was replaced by Peter Franken of the University of Michigan, who had been deputy director of DARPA. Upon Franken's arrival in May 1973, Aden continued his research in astronomical telescopes and solar energy development.

Faculty and students soon felt Aden's departure. Jacobs described the difference in leadership style between Franken and Aden: "Aden was always encouraging. He gave you the best 60 seconds of consulting possible before he wanted to move on. Peter called himself a '*nichtsager*' [naysayer]. He was rarely encouraging, but pointed out (if possible) why it wouldn't work. On Peter's door was a poster showing a young lioness and the words: 'It's only my opinion, but it's absolutely correct.'"[19]

Franken quickly realized that many students were enjoying themselves working as paid employees of the faculty and continuing to gain optical science and engineering knowledge but not graduating. This blocked junior students from moving up and graduating. He firmly believed that the purpose of graduate school was to graduate students, and so he formed an "exit committee" of faculty to move students up and out. The department's student-academic productivity increased noticeably as Aden returned to his solar thermal power research and the design and engineering of astronomical telescopes and instruments.

International science and astronomy

Aden's empathy with international scientists grew out of his direct experiences as a young 22-year-old officer near the end of World War II when he spent six months in Europe interviewing impoverished scientists and investigating science and technology for optics, astronomical, and defense needs. He carried this personality trait forward throughout his entire life, as shown in his collaboration with astronomers in India when he was director of Steward Observatory. As director of OSC and later during his and Marjorie's promotion of solar energy to developing countries, he was always generous with his time and energy toward international science and education. He collaborated equally with astronomers in both Taiwan (Republic of China) and China (People's Republic of China).

Aden was an active leader in the International Astronomical Union (IAU), first in his roles as director of KPNO and Steward Observatory and then as chairman of IAU Commission #9: Telescopes and Instruments. The predecessor of the IAU, the International Union for Cooperation in Solar Research, was founded by George Ellery Hale in 1905. At its establishment 14 years later, the IAU prohibited scientists from those countries defeated in World War I from joining; Hale objected and withdrew his support. By the end of World War II, however, astronomers from all countries were invited to join, and astronomy blossomed across the world.

Aden and Marjorie became popular speakers on the international scene, teaching and lecturing both on solar energy and astronomical telescopes and instruments. Their trips were subsidized by the US State Department's International Communications Agency (ICA), a subset of the US Information Agency (USIA). In 1979, Aden and Marjorie began a long trip around the world trip to Taiwan, China, India, Saudi Arabia, and Europe for the ICA.

Professor C. Y. Fan of the Department of Physics at UA, an old friend of Aden from their days together at the University of Chicago 25 years earlier, became aware that the Chinese government was planning to build a large astronomical telescope. In 1979 Aden and Marjorie were invited to make their first trip to mainland China as guests of Academia Sinica (Chinese Academy of Science) to discuss the design of a large, ground-based, astronomical telescope for Chinese astronomers. The OSC in Tucson was hosting Wu Shutung from the Shanghai Optics and Fine Instruments Factory at the time, and they traveled to the Purple Mountain Observatory as guests of Y. C. Chang, its director. On their way they stopped off in Taiwan to visit the National Central University, Chung-li, to check on the progress of a 60-cm telescope under construction by Boller and Chivens for P-E. They gave some lectures and then traveled to Nanjing through Beijing and stopped off in Shanghai to visit the Shanghai Optics and Fine Instruments Factory as guests of Director Wang Zhijang.

During this visit Aden collaborated with his Chinese colleagues—Hu Ningsheng, Hu Qiqian, and Pan Chunua—to develop a detailed design for a minimum cost 4-m telescope.[20] The primary was a monolithic 4-m mirror, and it is obvious that this telescope shares many of the design elements that Aden hoped he could have used in NSF's KPNO 4-m telescope (now called the Mayall telescope) that he was not able to work on after his resignation as director of KPNO.

The most striking differences between the 4-m aperture telescope he designed for the observatory in China and the 4-m aperture Mayall telescope were the f/ ratio, mass, and cost of the telescope, and the cost and size of the dome structure. Aden's 4-m aperture telescope for the Purple Mountain Observatory had an aperture of f/1.5, a primary mirror with a mass less than 7,000 kg, and an innovative mechanical structure to separate the primary from the secondary, along with an unconventional location for the elevation axis of the telescope.

This 1979 trip was the first of three trips to China to teach optics and consult on the design and construction of astronomical research facilities. One of the authors of this book (JBB) accompanied them to the Beijing Observatory, Nanjing Astronomical Instruments Factory, and the Shanghai Glass Factory over three weeks in the spring of 1986. Several days in Beijing were spent lecturing on ground-based telescope design and engineering, as well as astronomical spectroscopy and interferometry at the Institute for Astronomy and Astrophysics. They toured several historical sites during the academic interchange.

At the Nanjing Astronomical Instruments factory, Aden and JBB lectured and collaborated with scientists and engineers on the innovative design of the LAMOST (Large Sky Area Multi-Object Fiber Spectroscopic Telescope) and discussed observing programs for the new facility (Figure 9.1). They also explored the design and engineering of low-expansion glass for the primary mirrors of astronomical telescopes at the Shanghai Glass Factory. At the time, modern telescopes built in the West used either fused silica or a glass-ceramic called Cer-Vit (developed by OSC's Clarence Babcock when he was in industry). Fused silica melts and pours at a high temperature and is difficult to use. Cer-Vit, although capable of being formed at a lower temperature, was expensive without much lower expansion than Corning's Pyrex. The Chinese VO-2 glass was a vitreous ceramic similar in physical properties to the Cer-Vit glass. The LPL Mt. Lemmon Observatory north of Tucson uses two 1.5-m mirrors fabricated using Cer-Vit.

By 1980 Aden was consulting with Chinese astronomers and telescope designers on the optical design and plans for inexpensive, large-aperture, astronomical telescopes. One of the outcomes of this consulting work, which began in 1975 and continued for 25 years, is the LAMOST.[21]

On the way to Taiwan, Aden suffered coronary difficulties. While hospitalized there, he wrote lectures that Marjorie typed and delivered until it became urgent for him to return to Tucson for a heart operation. Undetected internal bleeding led to some heart damage, but he recovered well.[22] After his recovery, solar energy had ceased to be worth the effort, and they returned to optics and

Figure 9.1. Marjorie and Aden Meinel and Hu Ningsheng at the Nanjing Astronomical Instruments factory in March 1986 examining an astronomical mirror blank manufactured of vitreous ceramic VO-2.
Photograph by James Breckinridge.

astronomy. A few years later, they retired from UA and on to a second career at the Jet Propulsion Laboratory.

The Multiple Mirror Telescope (MMT)

Astronomical optics and optical engineering were always central in Aden's life. He firmly believed in the excitement of making discoveries and the ease with which discoveries were made by building new larger telescopes and optical instruments. In the early 1950s, he recognized that the cost of giant telescopes was limiting the number of astronomical discoveries by limiting the number of astronomers who could afford the cost of a giant telescope. There were less than 10 astronomers in the entire world who had access to the then-new Palomar 5-m telescope. He was increasingly aware of the importance to maximize telescope aperture for minimum cost. He had successfully built the Bok 90-inch telescope to fit within the $1,300,000 grant that UA had received from the Higher Education Facilities Act of 1963.

Aden knew that by 1970 Corning had developed lightweight, low coefficient-of-expansion mirrors and had demonstrated the manufacture of several large mirrors for a classified Air Force program. The program was canceled, and the mirrors were about to be destroyed when Aden set out to save them for ground-based astronomy. To show the Air Force how astronomers would use the mirrors in an astronomical telescope, he quickly completed a conceptual optical design and made several sketches showing how astronomers would use the mirrors in an inexpensive system to create the world's largest-aperture, ground-based telescope. One of these drawings presents a perspective view of Aden's COLT or six-shooter telescope concept (Figure 9.2). In his youth, he learned to communicate his ideas using sketches and perspective views in classes in high school and from his friend, engineer-artist Russell Porter, who drew the assembly views of the Palomar 200-inch telescope for George Ellery Hale and others at Caltech. Aden convinced many sponsors to support his programs over the years using sketches similar to that shown in Figure 9.2. Examples of Aden's more finished artwork are at the James C. Wyant College of Optical Sciences at UA.

Ever since he was at Yerkes in the early 1950s, Aden believed he would someday build the world's largest telescope. In 1939, as a 17-year-old student-optician at Mt. Wilson Observatory, he worked next to the team that was building the world's largest telescope at that time: 200 inches, or 5 m. Even at that young age, he became infected with the excitement of building large telescopes and the opportunity to make great discoveries.

In 1970, while he was director of OSC, an opportunity arose to build a telescope with the angular resolving power of a 6.8-m telescope at a small fraction of the cost of traditional construction, even larger than the resolving power of the 200-inch (5-m) telescope. It was to be a joint venture between UA and the Smithsonian Institution.

Three factors were key to this opportunity. One was the sudden availability of the seven nearly identical 1.8-meter, lightweight, fused-silica mirror blanks manufactured by Corning that were surplus by the cancellation of the Manned Orbiting Laboratory (MOL) program in June 1969. The second was Aden's innovative discovery that large optical telescope apertures could be combined to "synthesize" the imaging performance of a large telescope but using smaller, less expensive telescopes. Radio astronomers, who observed much longer wavelengths, were using multiple telescopes (antennas) to record signals that could be used to reconstruct images. In the 1920s, Mt. Wilson astronomers had recorded optical interferometer fringes to reconstruct images of simple objects. But no one had succeeded in creating images of complex objects in the optical/IR (0.4–10.0 microns wavelength) region of the spectrum. Up until that time (1970) all optical systems for astronomy used a continuous-surface collecting aperture. Aden, however, recognized that the large aperture provides two functions—angular resolution, determined by the outermost diameter of the aperture (mirror

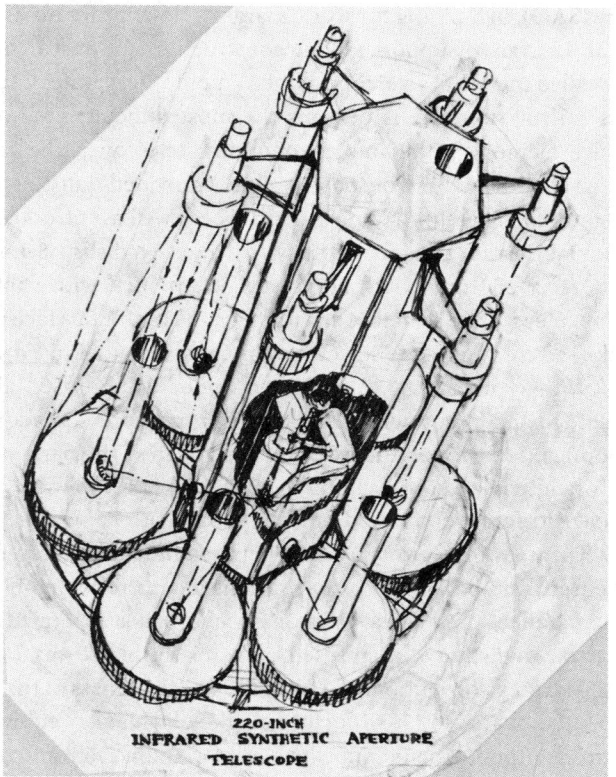

220-INCH
INFRARED SYNTHETIC APERTURE
TELESCOPE

Figure 9.2. Sketch of a large, sparse-aperture astronomical telescope made by Aden Meinel to show how astronomers would use six large telescope mirrors. Aden named this telescope COLT, after the Western six-shooter. This was the first telescope of this design to be proposed and was used with additional design/ engineering materials to convince the Air Force not to destroy the technology demonstration mirrors at Corning Glass. This design concept led to construction of the Multiple Mirror Telescope at the Whipple Observatory in Arizona at a small fraction of the cost of other, less-innovative, large-aperture telescopes.
Drawing by Aden Meinel.

or lens), and also the light-collecting surface area, which determines the brightness of the image. He had studied image quality from sparse aperture telescopes and concluded that important optical astrophysics data could be obtained with high angular resolution, sparse-aperture systems.[23] Frank Low of LPL showed that this was particularly true for astronomical science measurement objectives across the IR (1–10 microns) region of the spectrum. The third key factor was the interest on the part of Fred Whipple, director of the Smithsonian Astronomical

Observatory (SAO), in building the world's largest telescope for his astronomers. The SAO and UA raised support for the construction and research staffing of what is now called the MMT telescope of the Whipple Observatory.

The glass mirror substrate is perhaps the most difficult, time-consuming, and expensive element within an astronomical telescope. The large, light-weight mirror blanks that had become available provided a distinct advantage for the design of the telescope support structure, as well as introducing unique challenges for fabrication and mirror support. The seven disks 1.8-mm in diameter that OSC received from the Air Force were 30 cm thick, with a hollow center filled by an open egg-crate structure separating two continuous face plates, each 2.5 cm thick. The core egg-crate structure was constructed of fused silica plates 1.0 cm thick with 7-cm spacing. A circular cylinder wall of fused silica 0.6 cm thick surrounded the circumference of the mirror to provide additional stability and connection between the front and back plates. The parts were assembled by high-temperature spot-welding the fused silica components into place. This assembled structure composed of fused silica material was then heated to about 3,000°F in an oven to provide fusion between the internal egg-crate structure and the front and back plates. At this point the mirror assembly consisted of two plane parallel, fused silica disks or circular plates separated by a lightweight structure. The mass of each fused silica mirror blank was 545 kg, compared to 1,820 kg for a conventional solid blank. The total mass of the six mirrors is thus 3,270 kg compared to 35,000 kg for a single solid blank of equivalent light-collecting area. Without Aden's innovative mirror and wavefront control technology, a single solid blank would need to be thicker to maintain the stiffness to achieve the required telescope system wavefront performance. The savings in mass exceeds a factor of 10, and the entire telescope structure is much less expensive than would be a telescope with a classical monolithic mirror.

There are at least three ways to combine light from six mirror substrates to create a single, high-performance, large-aperture astronomical telescope (Figure 9.3). The top portion of the figure shows how the mirrors within each of the three candidate telescope architectures project onto the sky. The lower portion shows how the optical paths from each of the six mirrors are combined to form one image. The configuration shown in A requires tilting each mirror with respect to each other and grinding to remove material from each mirror blank. The configuration in B requires that the circular mirror blanks be cut into hexagons so that the flat sides nest with each other to form a continuous surface. This was not possible because of the structural design of the fused silica core structure of the disk. Also, the telescope aperture is reduced, making the overall telescope smaller and thus decreasing the optical system's angular resolution. Architecture C retains the high angular resolution of A and was thermally well baffled for higher IR signal-to-noise ratio in the 1970s compared to that in architecture A. An expansion of

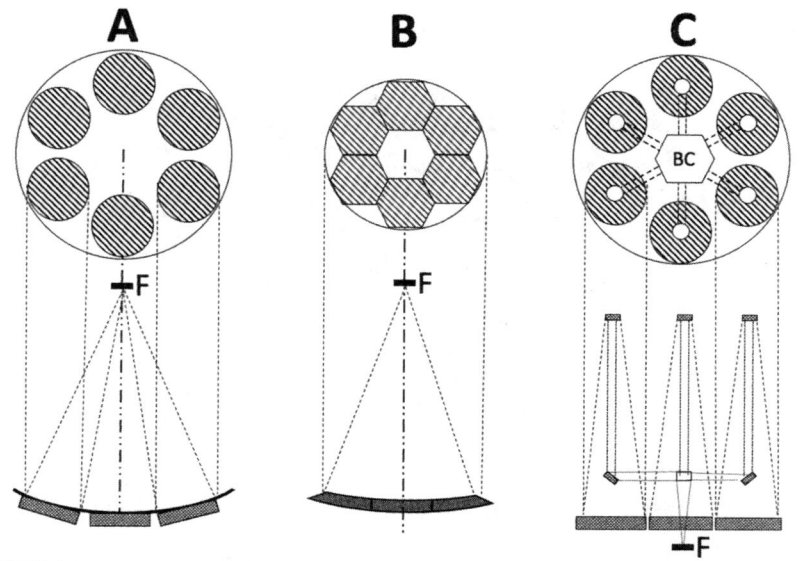

Figure 9.3. Three candidate topologies considered for the MMT are shown here as A, B, and C. Individual mirrors are shown as hashed circles. Light comes from above into each of the candidate telescope configurations. F = focal point, BC = beam combiner.

Drawing by James Breckinridge.

the segmented mirror architecture shown in B is used today in the 10-m aperture Keck telescopes and the NASA James Webb Space Telescope.

The merits of using one large, monolithic aperture versus several smaller ones for optical and IR astronomy were raised by H. L. Johnson and W. L. Richards.[24] Low concluded that within the limits of the available IR detectors of that period, the optical architecture proposed by Aden, which employed six independent, co-mounted, and bore-sighted telescopes, was best (Figure 9.4). Later, Aden's student, James Breckinridge, made a detailed quantitative investigation of sparse-aperture image quality to show the advantages and disadvantages of sparse-aperture telescopes with modern detectors.[25]

In 1967, Low visited Meudon, France, where Pierre Connes had constructed a mosaic mirror telescope.[26] Low realized that such a telescope would have problems with its background thermal radiation if it were to be used at a 10-micron wavelength. But these problems would be minimized if each mirror of the mosaic were part of its own telescope and able to be baffled separately, as one can ultimately achieve with the MMT architecture. He discussed the idea with Aden

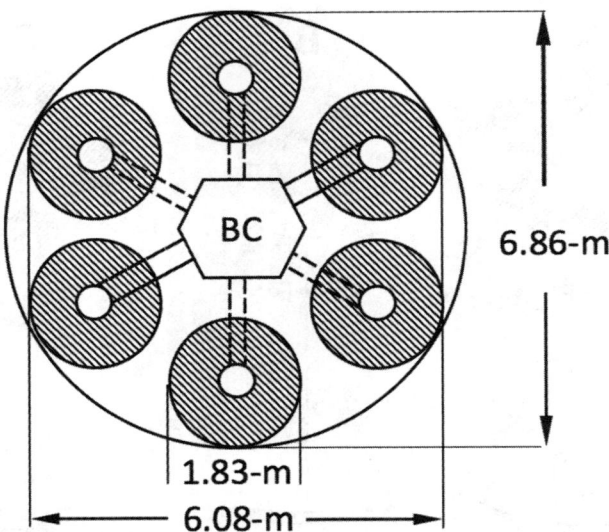

Figure 9.4. Mirror configuration selected by Aden Meinel and Frank Low for the Multiple Mirror Telescope (MMT). Since the six mirrors do not fill the area of the outside circle, the aperture is said to be sparse. Each of the six 1.8-m mirrors produces a beam of light, and these six beams are combined at the beam combiner, shown as BC in this figure. When these six beams are superposed coherently at IR wavelengths, the telescope's angular resolution is equivalent to that for a 6.86-m telescope. This exceeds the size of the Palomar 5-m telescope. However, the light-collecting area is equivalent to a 4.57-m filled aperture.
Drawing by James Breckinridge.

and Kuiper, and this information was folded into selection of the primary mirror architecture for the MMT.

The MMT project started about 1970, following a meeting on synthetic optical apertures held at OSC. Aden proposed the idea of taking the six 1.8-m mirrors and combining them into a single large telescope.[27] At the same time, Low was suggesting a large, multiple-aperture, IR system for radiation (light)-collecting and spectroscopy, along with an optical interferometer. Concurrently, Whipple saw the need for a new type of telescope that would use new technologies developed over the past few years and wanted the SAO to play a major role in ground-based astronomy.

Funding and siting

By 1970, Aden and Low had fleshed out the first-order optical design concept or architecture for the MMT.[28] They began to look for a source of funds to build the telescope when Whipple called. Aden, Low, and Kuiper were sitting in Aden's office when the phone rang, and discussions began about creating a joint venture between the SAO and UA to build the MMT. Aden, Whipple, and Low knew that no single state government, funding agency, or private foundation had enough money to build a large telescope capable of cutting-edge research. But by combining resources, this new telescope could be built to open new windows into the universe. As time went on, it became clear that the 1969 Mansfield Amendment prevented the DOD from funding it.[29]

The SAO required Congress to sign off on the MMT project, and so Whipple and his staff drafted a proposal for $1.5 million, spanning three years, to the Office of Management and Budget (OMB) with Aden's sketch of the design on the cover (Figure 9.2) and a photo of his model on the frontispiece. The NSF was loath to support Smithsonian projects, believing that it, not the Smithsonian, controlled ground-based astronomy in the United States. The NSF also believed that the Smithsonian, as a quasi-federal agency with its own access to congressional funds, should use that funding line rather than NSF's. What began as a concept soon evolved into negotiations between UA and SAO, ultimately hinging on the site for the telescope, which would determine how much each of the partners would contribute and what their roles would be. The SAO was a research division of Harvard University, which did not have the best relations with UA, a public university. Would it be built on Mt. Lemmon (which UA, Low, and Kuiper wanted for its better infrastructure) or Mt. Hopkins (where SAO already had an observing station under darker skies)? The Forest Service on Mt. Lemmon was under pressure from the local recreation industry to open up the mountain for more skiing and hiking by the citizens of rapidly expanding Tucson, and it appeared that more telescopes on Mt. Lemmon would be an uphill battle. After extended discussions between UA and SAO at the Lunar and Planetary Laboratory on the UA campus in January 1972, Whipple called for a vote. Surprising everyone, Aden voted for Mt. Hopkins and thus settled the issue.[30] Aden did not want to waste his energy during what would obviously be an extended period of confrontation for little return to OSC. There was much more to do to sort out new funding for OSC and maintain a competitive academic program. As usual, Aden left the detailed implementation to others and moved on to his next creative venture.

The team introduced several new innovations that reduced the cost of large-aperture, ground-based telescopes and significantly expanded access to the skies for a large number of astronomers: (1) lightweight mirrors using a welded core structure that was enabled by the selection of fused silica; (2) the first use of a segmented or multiple mirror primary and early developments in co-phasing segmented mirrors using adaptive optics; (3) the first use of an exoskeletal (box) structure to hold the primary and secondary mirrors in precise separation; (4) co-rotation of the floor of the observatory, offices, and telescope control center rooms with the telescope to reduce the physical size and thus cost compared to the cost of the classic hemispherical dome structure; (5) laser metrology with feedback to hold the structure stiff during the time the telescope guides across the celestial sphere to track objects; (6) active thermal control; (7) with the exception of the Russian Bolshoi Telescop Azimutalnyi (BTA-6) and William Herschel's 40-foot telescope, built in 1787, major optical telescopes prior to the MMT used equatorial mountings. Several technologies pioneered by the MMT contributed to the success of the subsequent generation of large telescopes. These included high dynamic-range servos for the altitude-azimuth mount; highly accurate pointing that eliminated the need for sky charts; improvements to optical performance by attention to the thermal environment of the facility; contributions to vacuum coatings deposition, optics cleaning, and maintenance; early experiments in co-phased adaptive optics; and co-alignment and co-phasing of multiple telescopes.[31]

When commissioned in 1979, this was the largest telescope in the world in terms of angular resolution and remained so for the next 14 years until the 10-m Keck telescopes, also of multiple mirror architecture, came online in 1993. The cost of the MMT was less than 20% that of the Keck telescopes. Upon completion it was the heaviest and most expensive dome after the Palomar 200-inch (Figure 9.5; consult the color insert for this figure).[32]

In 1978, seven years after he conceived of the optical configuration for the MMT, Aden identified the major cost elements that comprise a typical large-aperture telescope, and in 1979 he published his first detailed analysis of the cost-scaling laws that govern large optical telescopes.[33] The objective of his work was to identify specific high-cost factors used in specific telescope architectures in anticipation that further or alternative technology development might lead to reduced cost and higher telescope productivity. By 1982 he had completed a study that showed that fast primary mirrors of f/1 offered the possibility of using endostructure designs rather than the conventional exostructure.[34] He applied the results of this study to his work for the University of Texas (UT) on the 300-inch telescope for McDonald Observatory. Aden's 50-year interest in reducing the cost of large astronomical telescopes is summarized in a 2004 paper where he and Marjorie collaborated with van Belle to show that ground-based telescopes costs increase as the 2.5 power of the telescope aperture diameter and those of space-based telescopes increase as the 2.0 power of telescope aperture diameter.[35]

By the mid-1970s, Aden had turned the detailed design and engineering over to others. The project was managed by Steward Observatory, and Aden became consumed with the OSC business of academic department operations and raising funds. He continued to consult on the implementation of the MMT, but daily construction and operation became the responsibility of a joint-venture team of UA and Smithsonian astronomers and engineers. Although successful, the MMT had one fundamental drawback in its design—the six mirrors were not optimized for the full span of the yoke. Roger Angel of Steward Observatory suggested casting one filled-aperture, 6.5-m, Pyrex spin-cast mirror using technology initially supported at OSC. His idea was so successful that the MMT was decommissioned after only 19 years in 1998, and the six individual mirrors were stored in the basement.[36]

The Eye of Texas

Aden and Marjorie returned to Tucson in late November 1979 after a two-month trip to Taiwan and mainland China, where they consulted on the design and construction of large aperture astronomical telescopes and presented several lectures on optical system design and engineering.[37] Bypass surgery to correct a problem that arose while on the trip and his recovery occupied them until about January 1980.

Their work to identify telescope design approaches for the UC Berkeley 10-m telescope was finished until that university could find funds to continue the development of what eventually became the Keck telescopes. Harlan Smith, director of McDonald Observatory in Texas, called them to say that he was trying to convince UT to build a big telescope and that he needed Aden's help to design a credible 300-inch telescope for their observatory.

Smith announced that he saw a way to raise funds to build a 300-inch (~8-m) telescope quickly, before the envisioned but unfunded 10-m telescope of UC Berkeley could be built. This would give UT the world's largest-aperture telescope for at least a few years. He was able to convince the university administration to provide some funding to get started. Smith asked if Aden would come to Austin on a regular basis, two weeks per month, to work with Texan optical scientists and engineers on the design and costing of the telescope. Aden agreed and helped to prepare perspective drawings of the concepts so that UT administrators and funding sources could easily visualize concepts.

The Texas State Legislature solidly supported authorizing funds for the telescope in a House Rules and Regulations entry for 1983 (HCR 266), commending UT on its construction of a 300-inch telescope, signed by Democratic governor Mark White.

Smith had expected the cost to be about $30 million. The Aerospace Corporation estimated that it would soar to about $65 million. The work culminated when Aden gave a briefing to the Regents at a meeting held at McDonald Observatory near Fort Davis in western Texas, where Aden had served as associate director during his Yerkes days 30 years earlier. Marjorie detected a coolness in the response from the attendees. A final report was prepared to be submitted to the Regents. For some reason, all of the work and drawings that Aden had so laboriously prepared were not included in this report. Only some old drawings of a concept that had apparently been discarded were in the presentation package. Aden was upset, and they departed for Tucson, never to return. Smith apologized for the snub, but Aden and Marjorie had new projects in mind.

The decision for funding was expected from the Texas State Legislature during the 1984 session. That was the year the oil industry collapsed in Texas. The bill perhaps did not even get out of committee for a vote. At the time there was a perfect storm of (1) dwindling oil reserves, (2) a catastrophic fall in crude prices from reduced demand in Europe and North America, and (3) OPEC price cuts. As a result, banks in Texas failed and unemployment in oil-related fields was rampant. The problem was that Texas was essentially a one-industry state in the oil boom of the early 1980s before it all collapsed. An analogy is the potato famine in Ireland in the mid-19th century—a one-crop country devastated by the water mold, *Phytophthora infestans*, that attacked the leaves and roots of potato plants.

A few weeks later, Aden and Marjorie learned that the Regents had rejected the 300-inch telescope, and it died. About a year later, Smith called Aden asking for approval to send copies of Aden's drawings and reports on the 300-inch to the Japanese to build a new 8-m telescope on Mauna Kea, which became the Subaru Telescope of today.[38] Aden, always selfless and thinking of being helpful to others, agreed, and a copy of the entire package was shipped to Tokyo. How they used the information is unknown.

Aden and Marjorie sold their Austin house and moved back to Tucson. The children had gone their separate ways, and so they decided to downsize and buy a condominium in a new development closer to Tucson. They were not there long before opportunities at Caltech/JPL in Pasadena arose.

At OSC they moved into a small office on the ground floor and wondered what would happen next. Aden had been away from academic affairs of UA for several years. He taught no courses and had no students. He was promoted to University Regents professor and commented that this appointment seemed to be a barrier to performing regular department duties. His solar energy contracts had expired long ago, and now with the collapse of the 300-inch project he felt left out. There was a strong engineering team at UC Berkeley to build the Keck telescopes, but there were no funds to move the project forward and hire Aden.[39]

Aden and Marjorie observed that there were no textbooks on the subject of stellar spectral classification and the design of spectrographs for astronomers. In 1980 Aden corresponded with his old friend W. W. Morgan at Yerkes Observatory and Yerkes director Lewis Hobbs, asking if there was interest in writing such a book. If funding could be found, he and Marjorie were willing to spend a year back at Yerkes writing it. They prepared a brief proposal seeking travel and subsistence funds to NSF, which was not funded. Aden wrote to Morgan to ask if the University of Chicago/Yerkes could support it, but that, too, never materialized.[40] The mainline interests of astronomers had moved on from stellar research to galaxies and cosmology, and Aden's reputation was now in telescopes and optical sciences.

Aden and Marjorie's adventures always seemed to originate with telephone calls. In 1984 an invitation arrived from James Breckinridge for the Meinels to join a NASA/JPL project definition team to identify the next-generation space telescope. The Hubble Space Telescope project was well underway, and NASA scientists/technologists wanted to begin to plan the follow-on to the Hubble, which is now known as the James Webb Space Telescope. They traveled to the Jet Propulsion Laboratory (JPL) to present space telescope concepts. Having retired from UA after 22 years, they accepted jobs at JPL, which is owned by NASA and managed by Caltech. They had come full circle and had returned to Caltech after a 40-year absence.

Notes

1. Late in 1969, an amendment to the Military Authorization Act, introduced by Senator Mike Mansfield (D-Montana), confused and alarmed both the defense and civilian research enterprises. The amendment barred the Department of Defense from using its funds "to carry out any research project or study unless such project or study has a direct and apparent relationship to a specific military function."

2. UA academic catalog 1970–1971 and *Optical Sciences Center Newsletter*, January 1970; Meinel, A. B., Eyer, J. A, Noble, R. H., and Slater, P. N. 1970. The Optical Sciences Center: Its history, organization, and relation to government and industry. *Applied Optics* 10: 243–7.

3. Letter from A. B. Meinel dated September 9, 2002, to Roland Shack upon his retirement in 2002, given by Pam Shack to J. Breckinridge, April 2019.

4. Breckinridge, J. B. 2012. *Basic optics for the astronomical sciences*. SPIE Press, Bellingham, Washington, p. 293.

5. Interview of Stephen Jacobs by J. Breckinridge, September 29, 2016, in Tucson, Arizona.

6. Interview of Stephen Jacobs by J. Breckinridge, September 29, 2016, in Tucson, Arizona.

7. Eyer, J. A. 1970. USAF contract report F33657-C-0803 section: Foreword.

8. Interview of Stephen Jacobs by J. Breckinridge, September 29, 2016, in Tucson, Arizona.

9. Personal communication from Robert Shannon to J. Breckinridge.

10. Meinel, A. B., Patrick, D. L, and Harvill, R. A. 1965. *A plan for an Optical Sciences Center*. Optical Sciences Center Technical Report No. 0.

11. Interview of John Glasby by J. Breckinridge on February 28, 2017, in Tucson, Arizona.

12. Whitaker, E. A. 1985. *The University of Arizona's Lunar and Planetary Laboratory: Its founding and early years*. University of Arizona.

13. From 1961 to 1969, Harry Davis was Assistant Secretary of the Air Force (R&D), and from 1969 until he retired in 1973 he was Deputy Assistant Secretary (Special Programs), specializing in intelligence, reconnaissance, navigation, etc.

14. Meinel, A. B. 2014. Personal reflections of the first decade. Appendix 1, James C. Wyant. Optical Sciences Center/College of Optical Sciences: 50 years of excellence. *Proceedings SPIE* 9186, 25.

15. Papers of University President Richard B. Harvill, Special Collections, University of Arizona Library.

16. Meinel, A. B., et al. 1971. Third annual report: project THEMIS—Evolutionary studies of precision optical systems, Contract# F44620-69-C-0024, September 1, 1971. University of Arizona, Optical Sciences Center. OSC Library QC350/O85/AFOSR/F44620.

17. Sargent, M., III, Scully, M. O., and Lamb, W. E., Jr. 1974. *Laser physics*. Addison-Wesley Publishing Co., Reading, Massachusetts; Jacobs, S. F., Sargent, M., III, and Scully, M. O. (eds.). 1974. *High energy lasers and their applications*. Addison-Wesley Publishing Co., Reading, Massachusetts.

18. Benedict, R., Jr., Breckinridge, J. B., and Fried, D. L. 1994. Introduction: Atmospheric compensation technology. *Journal of the Optical Society of America* A11: 257–62.

19. Personal communication from S. Jacobs to J. Breckinridge, April 21, 2017.

20. Meinel, A. B., Meinel, M. P., Hu, N., Hu, Q., and Pan, C. 1980. Minimum-cost 4-meter telescope developed at 4 October 1979 Nanjing study of telescope design and construction. *Applied Optics* 19: 2670–9.

21. Smith, M. C. 2010. Progress and plans for Chinese surveys. GREAT Plenary, Brussels, June 23, 2010; Cui, X.-Q., Zhao, Y.-H., Chu, Y.-Q., Li, G.-P., Li, Q., Zhang, L.-P., Su, H.-J., Yao, Z.-Q., Wang, Y.-N., and Xing, X.-Z. 2012. The Large Sky Area Multi-Object Fiber Spectroscopic Telescope (LAMOST). *Research in Astronomy and Astrophysics* 12: 1197–242; Wang, S.-G., Su, D.-Q., and Chu, Y.-Q. 1996. Special configuration of a very large Schmidt telescope for extensive astronomical spectroscopic observation. *Applied Optics* 35: 5155–64.

22. Letter from A. B. Meinel to W. W. Morgan, dated December 18, 1980.

23. Meinel, A. B., Meinel, M. P., and Woolf, N. J. 1983. Multiple aperture telescope diffraction images. *Applied Optics and Optical Engineering*, vol. IX (ed. R. R. Shannon and J. C. Wyant), 150–201. Academic Press, New York.

24. Johnson, H. L., and Richards, H. L. 1970. Optimum size of infrared photometric telescopes. *Astrophysical Journal* 160: L111–16.

25. Breckinridge, J., Bryant, N., and Lorre, J. 2008. Innovative pupil topographies for sparse aperture telescopes and SNR. *Proceedings SPIE* 7013–3E.

26. Chevillard, J.-P., Connes, P., Cuisenier, M., Friteau, J., and Marlot, C. 1977. Near infrared astronomical light collector. *Applied Optics* 16: 1817–33.

27. For a detailed account of the declassified transport of the mirrors to Tucson, see Chapter 11, MOL/DORIAN (KH-10) and KENNEN (KH-11).

28. Meinel, A. B. 1970. Aperture synthesis using independent telescopes. *Applied Optics* 9: 2501–4; Meinel, A. B., Shannon, R. R., Whipple, F. L., and Low, F. J. 1972. A large multiple mirror telescope (MMT) project. *Optical Engineering* 11: 33–7.

29. DeVorkin, D. H. 2018. *Fred Whipple's empire: The Smithsonian Astrophysical Observatory, 1955–1973*. Smithsonian Institution Scholarly Press, Washington, DC, 210.

30. DeVorkin, *Fred Whipple's Empire*, 211–13, 216, 222–4.

31. Hege, E. K., Beckers, J. M., Strittmatter, P. A., and McCarthy, D. W. 1985. Multiple mirror telescope as a phased array telescope. *Applied Optics* 24: 2565–76; Hebden, J. C., Hege, E. K., and Beckers, J. M. 1986. Use of the coherent MMT for diffraction limited imaging. *Proceedings SPIE* 628. https://doi.org/10.1117/12.963510.

32. Meinel, A. B., and Meinel, M. P. 2002a. *Echoes from a simpler time*. Unpublished, 340.

33. Meinel, A. B. 1978. An overview of the technological possibilities of future telescopes. In *ESO Conference on Optical Telescopes of the Future* (ed. E. F. Pacini, W. Richter, and R. N. Wilson), 13–26. ESO, Geneva; Meinel, A. B. 1979. Cost-scaling laws applicable to very large optical telescopes. *Optical Engineering* 18: 645–7.

34. Meinel, A. B. 1982. Cost relationships for nonconventional telescope structural configurations. *Journal of the Optical Society of America* 72: 14–20.

35. Van Belle, G. T., Meinel, A. B., and Meinel, M. P. 2004. The scaling relationship between telescope cost and aperture size for very large telescopes. *Proceedings SPIE* 5489, Ground-based Telescopes (September 28, 2004); https://doi.org/10.1117/12.552181.

36. Meinel and Meinel, *Echoes*, 341.

37. Meinel, A. B., and Meinel, M. P. 1980. Aden and Marjorie Meinel's China trip October–November 1979. *Applied Optics* 19: 2666–9.

38. Personal communication between Tom Barnes of McDonald Observatory and J. Breckinridge, January 2019.

39. Meinel and Meinel, *Echoes*, 313.

40. Letter from A. B. Meinel to W. W. Morgan, December 18, 1980; letter from W. W. Morgan to L. M. Hobbs, December 24, 1980; letters from A. B. Meinel and M. Meinel to L. M. Hobbs, July 27, 1981, and October 5, 1981; letter from A. B. Meinel and M. Meinel to W. W. Morgan, February 1, 1982; letter from W. W. Morgan to A. B. Meinel, February 12, 1982, all in the W. W. Morgan Papers, Box 35. Folder 14, University of Chicago Library.

10

Power for the People: A Solar Odyssey

Donald E. Osborn

Beyond Aden Meinel's seminal work in astronomy and optical sciences, he was also a transformative figure in the development of large-scale applications of solar energy. His work fundamentally changed the way we looked at how solar should be deployed and helped lead the way to the "Solar Century" that we now see unfolding.

Solar dawn: Early exposure

While this interest did not express itself until Aden was in his late forties, both Aden and Marjorie grew up not only steeped in science and technology, but also with strong exposure and connections to early developments in the solar energy field. In late 1919 and early 1920, Marjorie's father, Edison Pettit, a noted solar astronomer at Mt. Wilson, became acquainted with Robert Goddard of rocketry fame and Charles Greenly Abbot, who both were interested in solar energy. Abbot was a renowned American astrophysicist and director of the Smithsonian Astrophysical Observatory where he had started as an aide in 1895 and later served as secretary of the Smithsonian Institution from 1928 until 1944. As an astrophysicist, he was a leading expert on solar radiation, determining the solar constant at the highest reaches of Earth's atmosphere by deploying high-altitude balloons with pyrheliometers. While at the Smithsonian, "Dr. Abbot decided to place observatories at as high an altitude as was practical in as clear a climate as possible. Following this aim he looked carefully at mountains in the Southwest, in particular in Arizona, much like Aden would do half a century later in finding the best site for the National Astronomical Observatory."[1]

Abbot was also one of the leading researchers in the applied use of solar energy. His research led him to build improved solar cookers, solar boilers, solar stills, and some 15 patented solar energy inventions, including his last patent in 1972 when he was 101 years old. Abbot devoted much of his long life (103 years) to solar energy research. His obituary succinctly summed up his career: "Nothing of immediate practical value emerged from his [solar] experiments, but Dr. Abbot possessed remarkable foresight, in recognizing the importance of

With Stars in Their Eyes. James B. Breckinridge and Alec M. Pridgeon, Oxford University Press. © Oxford University Press 2022. DOI: 10.1093/oso/9780190915674.003.0010

finding sources of energy that did not depend on fossil fuels."[2] In his name, the American Solar Energy Society's instituted the Charles Greeley Abbot Award to honor the person who has made a significant contribution to the Society or the field of solar energy. The solar oven that Abbot built at Mt. Wilson Observatory would play a role in Aden's solar experience.

Aden's first experience with solar energy came early. In 1925, when Aden was about three years old, his father added a solar water heater to their home in Pasadena.[3] At the time many "Day & Night" solar water heaters were being sold throughout the Southwest. Later that decade, as natural gas was becoming common in the area, Aden's father replaced it with a new "Day & Night" gas water heater. Both Aden and Marjorie remember as children in the 1920s seeing the Eneas Solar Water Pump that had been installed at the Pasadena Ostrich Farm.[4] In the 1930s, Marjorie's father had a solar furnace at home which Marjorie, Aden, and fellow students would use to cook hot dogs.[5] Marjorie would say that her first date with Aden was to solar-cook hot dogs.

In 1940, while Aden was working at Mt. Wilson as an apprentice optician, he found Abbot's old, abandoned solar oven (Figure 10.1) on the hillside.[6] Years later, after Aden became involved in solar energy research, Aden found that the solar oven on Mt. Wilson had been dismantled and the parts sold to a junk dealer. Little did he know that Abbot would pop up later in his career.[7]

Four years later, John Burnham, a wealthy landowner with property at the Salton Sea, asked young students Aden and Marjorie if solar energy could be used to keep the level of the Salton Sea constant—or even lower it a bit—to recover "lost land" from rising water levels.[8] Their results, presented in a 150-page report, "were disappointing." Using solar energy looked too formidable and the "cost enormous." They noted that "Marjorie's father said 'cost had always been a problem with solar energy applications,' so we looked at this side [cost] of the equation." They ended up recommending either water sprays to enhance evaporation of the seawater or building a pipeline to pump excess water from the Sea to the Gulf of California. They were paid $150, "a princely sum for two months' work" of two quasi-students at the time.[9] They shortly thereafter got married. As they were later to say, "Solar energy then faded from our view over next few years."[10]

Picking up the solar gauntlet

In 1970, Aden was 48 and the director of the UA Optical Science Center. After many years of intensive academic and research endeavors in astronomy and optical sciences, Aden and Marjorie decided it was time to take a sabbatical. They decided they "wanted to do something completely different from telescopes and

Figure 10.1. The Abbot solar cooker.
Drawing by Aden Meinel.

optical science, something they could do together now that the family had grown up. Solar energy!"[11] They had been in Phoenix in 1955 setting up the working office in the search for a site that was to become Kitt Peak National Observatory, at the same time the 1955 World Solar Symposium and Solar Exposition was being held in Phoenix and the corresponding technical sessions in Tucson. They wanted to understand why, after the surge of interest after the 1955 World Solar Symposium, there was so little solar research. They wanted answers to basic questions: "What about those millions of solar homes that were predicted to be everywhere by 1975? Where were they? Why did the 1970 report of the National Academy of Sciences . . . dismiss solar energy 'as holding little promise for meeting future energy needs of the United States'?"[12] As they would put it later in 1972, "Our involvement with solar energy came strictly from curiosity. Neither of us are members of the solar energy establishment." They simply wondered whatever happened to the bright predictions made for solar energy in the decade of the 1950s.[13]

While most people who take a sabbatical go someplace else, Aden and Marjorie stayed in Tucson. Both of them threw all of their energy into the question of why solar hadn't gone further and what it would take to make a real difference in our energy needs. "Getting re-acquainted with solar energy was an exciting prospect. We started in two arenas, at the University of Arizona, and doing solar projects at home. . . . The start was to take a sabbatical so that we could give our full attention to this goal: where has solar energy been? And where was it going?"[14] Their intense study of the history of solar invention and development showed them that it was characterized by "brief periods of intense interest followed by silence, then a renewal as new travelers [researchers] appear on the road."[15]

In studying solar, they came to the conclusion that the problem with solar was twofold. First, solar applications were all aimed at small uses and primarily in developing countries. Aden and Marjorie felt strongly that if a developed country couldn't afford a technology, it wasn't going to be cost-effective for underdeveloped countries with more dire economies. When reviewing the solar energy literature, they discovered that "there had been nothing new to move the field past the temporary stimulus of the [1955] World Symposium. And the OPEC oil embargo still lay several years in the future."[16] Secondly, in their minds the whole purpose of a technology like solar energy was to make a quantum leap for mankind, not just nibble around the edges. So they were talking about gigawatts, not kilowatts or thousands of BTUs. Aden was also a technologist and felt that we had to apply our best and newest technology to help solve the problem. "The Meinels had no patience with the notion that solar was only for undeveloped peoples who cannot afford 'sophisticated' fuels. They also dismissed the 'rooftop' concept because while such projects were often cleverly engineered by ingenious enthusiasts, they were on far too microscopic a scale ever to succeed economically."[17] Emphasis on the individual household was important to reduce demand on fossil fuels, but too few households had gone solar to make a significant difference.[18] "In the Meinels' eyes, land-based [large scale] solar [high temp thermal] power was the answer. . . ."[19]

Aden also took a look at the emerging area of photovoltaics (PV, or solar cells), which at that time had been used for satellites and some remote applications. But the commonly accepted wisdom, which was justified at the time, was that the cost of PV would have to be reduced by nearly a thousand-fold to make economic sense. When they crunched all the numbers, small-scale solar thermal would need to lower the cost by only 25 percent. But with high-temperature, large-scale solar thermal, they saw a pathway in which it was reasonable to assume that the cost curve could be lowered.

Photovoltaics at the time was not at all considered something that was going to be practical on a large scale. In 1970, the price of solar cells was about $100 to $300 per watt. Photovoltaics, or "direct conversion," did not seem a viable option, as Aden stated. "If we look at direct conversion by means of silicon and cadmium sulfide cells, we see a rather large industrial effort, stimulated by the space programs. The cost today [circa 1972] of silicon cells with battery storage for night operation is $300,000 per kW [$300/W], a thousand times more expensive than commercial power plants, which cost about $300 per kW. There is no obvious breakthrough in sight that can reduce the cost of silicon solar cells by this large a factor. Direct conversion, therefore, does not seem to be an option at this time."[20] Indeed, stimulated by the space program and remote power applications, the price of PV in the 1970s was far too high to be considered for grid-connected rooftop or large power plant uses. In 1975, the cost of the PV modules was down to about $100 per watt (http://solarsouthwest.co.uk/solar-panel-cost/) and by 1977 would still about $76 per watt.[21] A full PV system would be 2–3 times that. The target for cost-effective PV modules in the grid-connected power sector would be somewhere around $1/W. So PV was some 100 times too high for cost-competitive solar energy via solar cells. This would have made Aden's dismissal of PV realistic at that time. The technical/cost state of batteries for storage of PV-generated electricity was a non-starter as well. Indeed, in late 1972, Joseph J. Loferski, a well-known pioneer in the development of modern photovoltaics, would note that "the current cost of silicon cell arrays . . . is $7000/ m^2 ($650/ft^2). While it is possible that cost reduction by a factor of 100 might be obtained by making currently conceivable changes to the manufacturing process, it is not clear at present whether silicon systems based on current concepts can ever reach competitive cost levels."[22] It was simply the case that in the 1970s PV was not a viable alternative to conventional power plants, and it was far more expensive than solar thermal plants being proposed.

The other development that caught their attention at the same time was Peter Glaser's concept of the power satellite to collect solar energy in space and beam it back to Earth via microwaves. What Aden liked about Glaser's concept was that it was looking at how to deploy solar at a large scale for large-scale power production using advanced technology. What he didn't like about it was that to gain a just a fourfold increase in the solar resource, the idea of having to launch all this technology into orbit did not make any sense to him whatsoever. But what he loved about it was that Glaser was one of the first to start thinking about a new direction for solar.

Glaser was talking about the first gigawatt solar farms. This idea of a thousand-megawatt (gigawatt) solar power plant was beyond the pale at the time but central to their concepts. During Aden's sabbatical in 1970, he and Marjorie developed a detailed, complete concept of a National Solar Power Facility (NSPF) based on high-temperature, large-scale, solar thermal power plants utilizing advanced thin-film selective optical coatings to increase solar power plant efficiency and cost-effectiveness substantially. In that year, they not only thoroughly researched past work in the field and developed the bold, large-scale, high-tech approach of the solar farms, but they also identified and solved many of the technical challenges.

Aden had the rare gift of being able to see the big picture while completely filling in the details. In December of that year, their vision for how solar could contribute to the large-scale energy needs of the United States and the world had its public unveiling with a major, full-page spread in the *Tucson Daily Citizen*[23] about their solar work and the proposed NSPF. In that article, Aden and Marjorie played out the key elements of an expansive vision for building thousand-gigawatt solar energy stations over 100 years in a 5,000-acre corridor stretching from Yuma to Las Vegas, straddling the Colorado River; the stations would have high-temperature, planer solar collectors utilizing optical thin films with liquid sodium for the heat transfer fluid and molten salts for the thermal storage. Aden and Marjorie suggested that they could have a prototype completed in four years and "within the next decade start building 5 to 10 solar stations per year each occupying less than two square miles of solar collection area." When asked who would pay for it, Aden said he would look to the federal government for funding, as well as from a consortium of utility corporations. He pointed out that "The cost of any kind of mass energy production is tremendously high. For example, the new Navajo plant in Northern Arizona is costing $516 million. Our estimate of the cost per 1,000-megawatt solar energy station is about $700 million. But it doesn't use up your customary fuels—or any at all."[24] This was the most expansive and detailed solar power plan ever proposed, with the possible exception of Glaser's space-based solar power concept. It was not talking kilowatts or even megawatts, but starting off with gigawatts and planning for a thousand gigawatts. At the time, the electric system load of Tucson Electric Power was about 542 MW peak.[25]

Power for the People

Only a month later, it was clear that the Meinel solar power proposal had caught the imaginations of many. In an article in the *Tucson Daily Citizen*, Aden had had a "tremendous response" to the plan. The article stated that "[two] of the nation's largest energy-related corporations are new entrants talking tentatively but ardently to the Tucsonan [Meinel] in terms that may lead in substantial financial backing for the projects. . . . Both US Sen. Barry Goldwater [R-Arizona] and Rep. Morris K. Udall [D-Arizona] wrote that they were excited." In a throwback to their earlier work as young researchers at the Salton Sea, this article went on to describe the Meinels' plan for taking water from the Pacific to Yuma and deploying a desalination plant powered by the proposed solar power stations.[26]

With a small $5,000 grant from the UA Foundation and NSF in early 1970, the Meinels had developed a rich and detailed plan by early 1971 for the high-tech, massive deployment of photothermal conversion based on their solar farms concept. They generated enough interest with their plan to get an additional $65,000 from the RANN (Research Applied to National Needs) program of NSF.

The *Tucson Daily Citizen* articles caught the attention of a young freshman engineering/physics student, Don Osborn (author of this chapter). Osborn had been an organizer for the First Earth Day at his high school in April 1970 before starting at UA that summer. He wanted to combine his twin passions of science and environment. The article on Aden's proposed solar farms caught his imagination and led him to seek out Aden early in the spring semester of 1971. Although the Optical Sciences Center (OSC) was purely a graduate program in 1971, Osborn walked in to the director's office, announced that he wanted to study solar energy, and asked Dr. Meinel if he would teach such a course.

Aden listened patiently and then told Osborn, "Well, tell you what. If you can get the administration to authorize offering a solar course, I'll be happy to teach it. You'll have to work on the petition, run the paperwork, and make sure there's the necessary number of students to take it."

Most likely Aden thought that would be the end of it. Osborn headed to the Administration Building and found out that, with the proper signatures, a special-purpose seminar class could be established as early as for the fall semester. Encouraged, Osborn returned to OSC and presented the paperwork to Aden. But since fall was "so far off," Osborn also asked if something could also be held that summer. Aden and Marjorie had recently completed a draft of a book that brought together all the threads they had developed during Aden's sabbatical. He thought that it would be helpful to have a group of students review a draft of *Power for the People: Solar Energy* and be a sounding board for concepts developed including the NSPF proposal. Aden agreed to take on a few undergraduate and graduate students for a summer seminar under the Independent Studies program. Later that spring, Osborn had rounded up six other students—in such diverse fields as engineering, physics, government, and urban planning—to sign up for a seminar that summer. As Aden later put it, "Our sabbatical over and *Power for the People* in hand, we started a solar energy class, undergraduate students also welcomed. We would explore solar energy together. It would provide a foil for scrutinizing each facet of both what was old and what could be added that is new."[27]

Aden would use the students to review, test, and refine the concepts that he and Marjorie had developed, as well as putting the concepts in the history of solar context that they had uncovered using the draft of *Power for the People* as their text and guide, telling the students: "You are the next generation from the surge of the 1950s. There is already a new surge beginning, but keep this past history in mind as we see this surge develop. The question we will address is: will it finally be successful and solar maintain and grow in application to the needs of us and the rest of the world? Or will it fail once more?"[28]

The class would explore the various sources of energy, established and emerging, along with their promises, problems, and impacts. This focus on Areas of Concern included:

1. Large-scale use—Power supply for US needs in years 2025 and 2525;
2. Practicable;
3. Economic—between present cost of power and twice that by 2025;
4. Long life and non-use of critical resources;
5. Clean, little destruction to environment;
6. Non-use of Earth's resources—sustainable to 2525.[29]

As part of the question of sustainability, Aden had the class research the Hubbert Curve showing the typical extraction history of finite resources such as fossil fuels and put them into the context of long-term historical thinking.[30] This mix of students would be guided by Aden in topics such as high-temperature thermal conversion using thin-film coatings to show how ideas get started and pushed forward. They would work out and analyze solar thermal power plant designs. Aden was truly in his element in teaching, guiding, and challenging these students, some of them still early in their academic careers. The students in turn would challenge the ideas presented, help flesh out critical parts of the NSPF proposal, and provide that sounding board Aden wanted. Aden also provided the draft manuscript for a paper being developed for the *Bulletin of Atomic Scientists* and included input from the summer class in what would be his significant paper, published in October 1971,[31] as well as in *Power for the People*.

The solar class expanded in the fall 1971 semester as OPTI 298a, meeting twice a week, with 15 students signed up and several others just "showing up." *Power for the People*, revised after the summer seminar, was printed by the McDonnell-Douglas Corporation and used as the main textbook for the solar class. Given Aden's other academic and OSC responsibilities, he would run the class one day a week, and the other would have a guest lecturer. As part of the deal Aden made with him to agree to teach the class, Osborn (now a sophomore) would act as the teaching assistant, an unusual position for a lower-division undergraduate in a graduate class at OSC.

The solar class ran for four semesters. The topics covered included:[32]

- The history of solar energy development and applications, including a detailed historical and optical analysis of the Battle of Syracuse, 212 BC.
- Optical thin films for high-temperature conversion applications;

- Selective absorber panel design and analysis;
- Material problems at high temperatures;
- Electric utility system design and operations (with field trips);
- Water impacts of power production;
- Optical concentration design and analysis;
- Resource utilization and limits;
- The wide range of other energy alternatives;
- Environmental impacts and sustainability;
- Engineering and mass production of high-tech components;
- Economic analysis;
- The solar resource.

And always the class would come back to tackle key elements of the NSPF concept, components, and system design, helping Aden refine and strengthen the concept.

The solar energy classes held by Aden were:

- Summer 1971: Opti Sci 199/299. Summer Seminar;
- Fall 1971: Opti Sci 298a. Solar Seminar;
- Spring 1971: Opti Sci 299. Solar Studies;
- Fall 1972: Opti Sci 298a. Solar Seminar.

Not only did these solar classes provide Aden with the sounding board he had hoped would provide valuable feedback, but the students also helped test and flesh out many key aspects of the NSPF concept and further refine the material that made up *Power for the People*. It also created an ongoing passion for solar that would live on in many of the students he would teach and mentor, none more so than young undergrad Osborn.

Unveiling the technical concept

At a May 1971 solar conference at NASA's Goddard Space Flight center, Aden and Marjorie presented an invited paper on their NSPF concept, which had a mixed reception. Aden and Marjorie started their paper with provocative comments: "We realize that a number of you here today have spent many years working in this field, and we, being newcomers, hope that you will excuse us if we inadvertently fail to give credit where credit is due. We hope that a view of the subject of 'large-scale uses of solar energy' from persons outside the traditional field may contribute something towards this goal." They continued, "To condense 16 years of work is difficult (since the Phoenix Symposium), but it seems that aside from silicon and related solar cells the main emphasis has been on small-scale solar energy applications. . . . We wondered whether the philosophy of the 'individual rooftop' system and the 'under-developed country' themes had not conspired to limit the range of ideas. A thesis which lays emphasis upon simple devices to match the limited skills and limited requirements of the users is definitely not compatible with the aim of a national power system for an *advanced* [their emphasis] nation—our topic today."[33] Aden would later say that their remarks caused a rift with the "old hands" at solar energy uses there, which was perhaps an understatement.[34]

After the paper's presentation, Aden and Marjorie ran into Charles Abbot at the banquet where Abbot was the guest of honor at age 101 ("and still sharp"). Abbot asked about Marjorie's parents (who had passed a few years earlier) and sister and commented, "I see the torch has been passed on. I read the title of your paper but couldn't hear it. I wish you both much success."[35] Little did Aden and Marjorie know that this was not to be the last solar connection, nor the last torch to be passed between Abbot and the Meinels.

By mid-1971, momentum was building, and optimism was high. An *Arizona Daily Star* article quoted the Meinels predicting that a 100 MW solar field could be up by 1976 and the buildout of enough solar plants along the Colorado Valley to "produce a major portion of the country's necessary electrical and desalt seawater a year for 120 million people" could be in place by 2076.[36] In the article Aden called for a task force at the Arizona Power Authority to "insure that the study progresses and to help finance a one megawatt pilot plant by 1974." The article concluded: "the Meinels believe they have answers for the six main problems encountered in the harnessing of solar energy so far: (1) high initial capital facilities, (2) maintenance of energy collectors, (3) low conversion efficiency of sunlight to heat, (4) heat losses within the system, (5) damage to sun-collecting mirrors by dirt and time and (6) storing enough energy for night and cloudy day operations." In a follow-up article a few days later, it was reported that Aden announced to Yuma audience, "We expect an announcement from Washington any day now on the first monies for this project," adding that the "money would be an estimated $3 million for the one-megawatt plant, which would be the start of a century-long project."[37]

A three-acre area at a university lab near the Tucson International Airport had been identified as the site for a 1 MW demonstration solar farm.[38] The *Tucson Daily Citizen* article also went on to say that a 100 MW plant could be up and running by 1976. At the time, Tucson Electric Power had a peak load of about 600 MW. The federal government was identified as the "only feasible funding source because of the magnitude of the concept," with Meinel lamenting that the government did not "yet have a solar (or other new source) power development body comparable to the Atomic Energy Commission."

The first full-fledged publication of the Meinel concept came out in October 1971 with their paper in *Bulletin of the Atomic Scientists*.[39] They had worked on it since early 1970 during their sabbatical and ended up with a well-thought-out, well-researched idea from engineering, deployment, and economic perspectives, complete with drawings and paintings by artist Don Cowan. For the 1970s and even into the 1980s, their thinking about how to deploy solar was prescient. In fact, starting in the mid-1980s, hundreds of megawatts of solar thermal power plants of the Meinels' basic design have been built in southern California deserts.

By December 1971, interest had mushroomed. Senator Mike Gravel (D-Alaska) called the Meinel Solar Plan "the most innovative and exhaustively researched idea for the large-scale conversion of solar radiation to electricity."[40] A "Citizens Group for Solar Energy" had been formed in Tucson to spread the word nationwide to conservation groups. The magazine *Innovation* "extolled the Meinel plan." The Atomic Energy Commission also expressed interest, and Senator Gravel had entered a detailed description of the Meinel Solar Plan into the *Congressional Record*. The article mentioned that "the Meinels have made more than 20 lectures by invitation from a variety of organizations and universities outside Arizona." It also reported that "Helio Associates Inc. has been formed in Arizona, looking ahead to becoming the parent of engineering and manufacturing of related equipment should the UA team's concept be propelled into regional or national usage. Helio would carry on those aspects, including plant construction, that would lie outside the University of Arizona field. The UA, however, is one of the shareholders of Helio." A few weeks later, Aden pointed out that Helio Associates was "one example of an industrial spin-off that the university can stimulate for the area."[41] As many other researchers had done (and still do), Aden and his team had set up a spin-off company to follow up on the more commercial aspects of their research, doing so through and in conjunction with the university, and expectations were running high as 1971 closed out.

Forging a solar pathway

Aden and his research team at OSC would leverage the initial $65,000 RANN grant to obtain a follow-on grant through the RANN Program for $173,000, still well short of the expected $3 million, covering June 1971 to January 1973, for "Research Applied to Solar-Thermal Power Conversion." This research grant covered the "laboratory fabrication and measurement of solar-absorbing coatings of high selectivity, (a/e), and their evaluation for use in electrical power production by means of thermodynamic cycles."[42] The project team consisted of Aden B. Meinel (principal investigator), Bernhard O. Seraphin, Dean B. McKenney,

William T. Beauchamp, Victor A. Wells, Francis Turner, and Fernando J. Lopez-Lopez. This research project would look at the key technical aspects of applying advanced thin-film technology to create solar absorbers that would efficiently capture the irradiance (be highly absorbing in the wavelengths of sunlight, high "a" or very "black") but would retain the thermal energy by being a poor emitter in the infrared (a low "e" or emittance). The measure of how effective that absorber coating is in capturing and retaining the solar radiation is the unit "a/e." The higher the a/e, the higher the selectivity, the more efficient the absorber, and the better able to generate high temperatures with less or no optical concentration (less need for mirrors or lenses).

The NSF (RANN) grant was used to fabricate and measure high selectivity (high a/e) coatings and evaluate their impact on improving the performance of high-temperature thermal cycles for electric power production. Two selective coating types were produced and evaluated: (1) interference optical thin-film stacks, and (2) intrinsic optical absorption of chemical vapor-deposited CVD silicon. The thin-film interference stacks yielded a/e of up to 30 at 300°C, "an improvement in the state of the art by a factor of 3."[43] The CVD stacks yielded a/e of over 20 at 500°C. McKenney, Beauchamp, and Turner had extensive expertise and facilities at OSC in the area of interference thin films, while Seraphin and Wells focused on CVD coatings. Numerous coatings of each type were developed and evaluated for optical performance and for durability. Using the basis of high a/e coatings, the Meinels focused on design details and performance of linear collectors with the coatings (Figure 10.2). They also studied the actual mix of direct to total irradiance (sun plus sky) in Tucson, showing that even in "sunny" Tucson, the amount of direct sunshine was much less than would be expected from US Weather Service records of "percent of sunshine," indicating that lower-concentration (or no-concentration) systems would have an advantage due to real-world weather patterns—an important point in the days before much-improved solar radiation records were available. This research project would form the technical basis of the proposed solar farm proposal.

In concert with the RANN grant, a solar thermal test bed to test the optical and thermal performance of the high a/e absorbers was funded through a $25,000 contract to the newly formed Helio Associates from four Western utility companies: Tucson Electric Company, Arizona Public Service Company, Southern California Edison, and the Salt River Project. Helio was formed as a spin-off from the OSC research group in late 1971 at the suggestion of James Bailey. Helio Associates was formed with the Meinels, Bailey, Dean McKenney, and Pat Beauchamp.[44] The Board of Directors included two Tucsonians: retired Major League baseball star Hal Warnoc (UA alumnus) and UA Foundation president Leicester Sherrill.

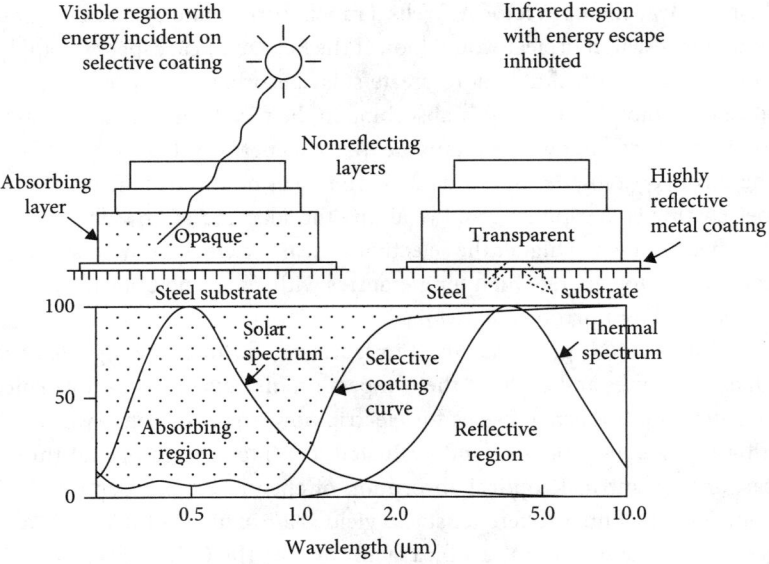

Figure 10.2. Selectivity of absorber showing absorptivity over solar spectrum and emissivity over the infrared at operating temperatures.

Drawing by Aden Meinel appearing in *Optical Sciences Newsletter* 6(3), December 1972.

Full steam ahead

As 1972 began, Aden said, "Solar energy looks like the hottest new business prospect in Tucson" and that 1972 might "very well prove to be the turning point for his idea."[45] In January, the Meinels would tell the American Physical Society that they believed that "they had solved a problem whose answer had long eluded man, an efficient way of tapping the boundless cost-free energy of the sun . . . with a demonstration plant within five years."[46] They announced that the thermal test bed for the advanced solar thermal collector, or "credibility model," would be ready by late spring and called for a 25 MW, $100-million demonstration plant to be built within five years. In February they published a paper titled "Physics Looks at Solar Energy" in *Physics Today*, based on the presentation they had made the previous month at the American Physical Society in San Francisco.[47] That paper detailed the fully thought-out application of solar thermal power for the large-scale application of solar in a "technologically advanced nation, utilizing all the arts of mass manufacturing on a large scale. Hundreds or thousands of megawatts per power plant begins [*sic*] to make sense when we want to obtain power at today's low electrical energy rates."[48]

They threw down the gauntlet to start the paper by stating, "The idea of using the sun as a source of energy has had a long history, but so far it has been a history of bright hope and dismal failure."[49] It was a statement that would not endear them to other solar researchers. After dismissing efforts to develop solar for underdeveloped countries and for rooftop applications, they discussed the various solar conversion methods for large-scale electricity generation. They pointed out, for example, that photovoltaic "prospects are clouded for the cost-reduction of a thousand-fold that would be required to get the cost close to a few hundred dollars per *kilowatt* [their emphasis] typical of bulk electrical power systems."[50] They concluded that high-temperature (high-efficiency) solar thermal conversion using high "a/e" selective coatings and modest optical concentration ("X") would permit solar power stations that could compete with and replace modern fossil fuel steam power plants. A complete system based on an a/e of 10 and X of 10 permitted an operating temperature of 500°C with a liquid sodium working fluid.[51] Based on this solar collector module design, the Meinels presented a complete solar power plant or solar energy farm that incorporates a thermal storage system based on either salt eutectics or liquid sodium for operation at night and cloudy days. With an operating temperature of 500°C, conventional steam turbines would then be used to generate electricity, as in a conventional power plant.

In May, Aden and Marjorie gave the keynote address at the Regional Environmental Education, Research and Improvement Organization Conference at New Mexico State University, proposing a 1 MW, $10 million demonstration solar farm to be in operation near Yuma by 1976. The Meinels told the audience that "the earth's deserts may be God's greatest gift to this world in terms of natural resources."[52]

By summer of 1972, Aden and the OSC/Helio team has built a solar thermal test bed (Figure 10.3) on the roof of the OSC building designed "to test the performance of the thin films" being developed as high "a/e" selective coatings and "to demonstrate at full scale all of the essential features of the solar collector system."[53] While NSF funding supported the theoretical and engineering design development for the solar farms concept, the modest joint utility funding to Helio made possible the construction of the test bed. As described in the December 1972 *OSC Newsletter*:

The model itself was built by Helio Associates, Inc., an organization recently formed as a spin-off from the Optical Sciences research group, for the purpose of carrying the development of solar energy utilization beyond the point appropriate within the traditional role of the University. The test bed was funded through a contract to Helio from four Western utility companies: Tucson Gas and Electric Company, Arizona Public Service Company, Southern California

Figure 10.3. Marjorie and Aden Meinel with the first thermal test bed for the advanced solar thermal collector on the roof of the Optical Sciences Center building, 1972.
Courtesy of Ed Meinel.

Edison, and the Salt River Project. The purpose of the test bed is twofold: It will be used to establish through experimentation that the concept will work as predicted, and it will be used to obtain baseline engineering data on the collecting and heat transfer subsystems.[54]

As mentioned earlier, Dean McKenney transitioned from OSC to full-time at Helio to work on the project and, with William (Pat) Beauchamp, did much of the experimental work on the test bed. "Two absorber pipes were built for this test bed, both made to the same size and physical description. The first pipe is chemically coated stainless steel; the second pipe is coated with multilayer thin-film coatings having a selectivity ratio (absorptance vs emissivity) of approximately 14. . . . Both pipes were cut into several sections to facilitate coating, as no coaters large enough to prepare a 10-ft section were available at any reasonable cost; the short (30-in.) sections" were coated and then coupled via special machined ends that would permit a high vacuum seal. The test bed was then used to verify, through scaled testing, the predicated performance of larger-scale high-temperature collector systems. At the end of 1972, Aden and the solar team were expecting "that a 1-megawatt experimental solar power farm can be operational near Yuma, Arizona, by 1976."[55]

Aden and Marjorie's promotion of the solar farm concept and their efforts to move forward with its development was prodigious. In just the second half of 1972, they gave some 22 talks on solar energy across the entire United States, including to United Aircraft Research Laboratories, Stanford University, MIT, and NASA.[56]

In December, James Bailey, now vice president of Helio Associates, announced that Helio and Meinel's research group were seeking about $4.2 million to develop a 1.8-acre demonstration plant in Tucson from a consortium of utilities, government agencies, and industry groups.[57] The demonstration solar farm would generate enough to serve the electric needs of about 30 homes. Reporting that utility company reactions had been "very positive," Bailey indicated that they hoped for "a 'positive' result from current negotiations within a few months." Bailey observed that "at least $400,000 is needed to buy basic equipment to get the farm operating on a limited basis. If not enough money is provided, then the project would remain small."[58]

So by December 1972, Aden's team had (1) secured about $263,000 in funding, out of the millions that they had sought and clearly stated were needed, (2) fully fleshed out the full concept and details for an expansive development program for large-scale solar farms, (3) established Helio Associates, Inc., as a university spin-off corporation to be the focal point for solar-farm engineering and development, (4) begun detailed research in key selective film development for the solar absorbers, (5) built the thermal test bed for the advanced solar thermal collectors central for the concept, and (6) successfully popularized the solar farm proposal across the nation and worldwide. It was clear that their expectations were that major funding for the demonstration solar power plant was around the corner, and plans were made to have that as the next step in an aggressive developmental program resulting in major commercial deployment by the late 1970s.

At the end of 1972, it was announced that Aden would be stepping down as the director of OSC in order to take a year's leave of absence. During his leave Aden planned to continue work on the solar energy project that he Marjorie began in 1970 and to promote the growth of Helio Associates. After his leave, he planned to return to OSC as professor of optical sciences. The search began for a new OSC director.[59] He explained that his solar research and development activities and developing Helio Associates were taking up almost all his time. "I'm away from the UA almost all the time now giving talks and explaining the [solar] project to other universities and national labs interested in developing it. . . . There will be a much larger effort in solar science next year. That's why I'm retiring as director."[60] An appreciative article in the *Tucson Daily Citizen* a few days later noted that "[his] efforts to convert solar rays into energy have won him international ac-claim."[61] In that same article it was publicly announced that Gerard Kuiper was planning to step down as the director of the Lunar and Planetary Laboratory as well to "spend more time on research and teaching."

Uphill slog with a Tahiti interlude

As the new year began, the Meinels were already feeling headwinds in their quest to develop large-scale solar farms as indicated by the headline in the *Tucson Daily Citizen*: "Solar plan funds delayed."[62] Despite funding delays, Aden and his team were still upbeat on the long-term prospects. At the multi-week UA Colloquium on Technology, organized and co-chaired by Aden's undergrad student Osborn, Aden remarked on a panel discussing energy futures that "[though] production of solar energy now faces a 'cost problem,' research has been negligible com-pared to other forms of energy."[63] While extolling the promise of solar energy to the Southwest and the country at large, Aden explained how "research on solar energy is handicapped at present by comparatively small grants—$3.8 million spread among 20 universities and 5 industrial contracts," adding that "though 'the problem has been to raise money for (solar energy) programs,' the signs for the future are 'all encouraging.'"[64] At the Tucson Press Club, Aden said that he was "hoping for the day when the confidence level" would permit greater funding for solar research.[65]

While the search for adequate funding was going slow, interest was growing from an unexpected source. Late in 1972, Marjorie picked up the phone just be-fore dinner. "This is Hawaii calling," as Aden got on the other phone. "This is Marlon Brando. I have an island in the Pacific, and I'd like to have solar power for it." As Marjorie reported it, she thought, " 'This is the way they sell real estate,' " so I hung up, but thankfully Aden stayed on the line." Marjorie added, "I guess I'm the only woman who has hung up on Marlon Brando." Brando owned and lived part-time on an atoll in the Tahitian island group Tetiaroa that he wanted to de-velop in an ecologically and sustainable manner and had plans for an ecotourism resort and research center. He wanted a solar system to supply the power needs for about 500 people. This would be about the size of the $4.3 million demonstra-tion solar farm that Aden was trying to get funded. Aden said that Brando's "in-volvement might give the solar energy field that initial boost it needs to really get started,"[66] hoping that the planned Tucson demonstration system could be built in Tucson, tested, and then moved to the South Pacific. Brando quietly traveled to Tucson in the spring of 1973 to meet with Aden, Marjorie, and a few other se-lect UA environmental researchers to discuss the proposal.[67]

Excitement quickly grew among Aden's research group and students, with many hoping to be among the ones selected to go to Tetiaroa for an on-site field inspection trip, and some sported leis in the weeks that followed. Artist Don Cowan, who produced many paintings, drawings, and illustrations for Aden and OSC, painted a large and beautiful watercolor of what such a tropical island solar installation might look like.

While unfortunately the "Solar in Paradise" project was not to be, Brando did institute some of the sustainable practices at his island and started the development of the ecotourism facilities. The Brando resort on Tetiaroa, completed in 2014, 10 years after his death, is today fully solar-powered with a large photovoltaic and battery storage system, along with a host of other sustainable applications.

In May 1973 it was announced that after a national search, Peter Franken from the University of Michigan would replace Aden as the director of OSC.[68] At that same time, Representative Morris K. Udall (D, Arizona), as chairman of the Environment Subcommittee of the House Interior Committee, was pushing for immediate federal funding of $20–30 million to support the Meinel solar farm demonstration project. However, Nixon's Secretary of the Interior Rogers C. B. Morton responded, "It is a dangerous thing to say we are ready to move. You can't just throw dollars at a problem and expect solutions." To this Udall responded, "But you can't throw press releases and speeches and expect results either."[69]

An article in the British magazine *New Scientist* in May 1973 titled "The Sunshine Spreaders" nicely summed up the Meinel's solar journey to that point: "One of the more extraordinary visions called up as a solution to the long-term energy crisis is the idea of covering thousands of square miles of the deserts of Arizona and New Mexico with 'solar power farms.' The credit for making this look [like] a credible solution to at least some of America's energy problems must go to a larger-than-life husband and wife team, Aden and Marjorie Meinel. Starting from scratch less than four years ago they have put solar power on the map in the USA, and if this leaves some rather disgruntled unpublicised workers who have been plodding away in the same direction for years—well, that's science." The article pointed out that "they hope to raise money for a much bigger project. But their main achievement to date has been to put solar power on the agenda as a subject for serious discussion."[70]

At this time, Aden and his team had more closely analyzed actual weather patterns in the "sunny" Southwest and found that even the long stretches of "sunny" weather had a surprising number of hours when the sun's face was obscured. Because concentration systems required direct irradiance from the unobscured solar disk, they were gravitating from development of a high-temperature (high-efficiency), fairly high-concentration system to a flat plate or low-concentration system to extend the effective operating hours while also lowering system costs.

The largest international conference on solar to that date, the UNESCO/ISES "Sun in the Service of Mankind" Congress, was held in Paris in July 1973. It attracted some 600 solar scientists from 70 nations, including Aden and Marjorie, as well as Bernhard (Bernie) Seraphin of Aden's team at OSC. Two divergent views dominated the congress, that of "low-tech" applications such as solar water and space heating, agricultural crop-drying, and the like, meant for developing countries, and "high-tech" solar electric generation systems represented by Aden's approach. Here Aden referred to the new direction of lower-concentration systems, stating, "We are not after maximum efficiency. We are after the maximum number of kilowatt hours per dollar. You may have the option of operating at 200°C at about 12 percent conversion efficiency. If you go to a higher temperature you may have 25 percent efficiency. But you can only spend twice as many dollars if you are to have equal payoff. If the higher temperature and efficiency mean going to concentrating systems that work only half the time [due to weather factors], you cannot afford to spend any more money than you would spend operating at 12 per cent efficiency. It took us a long time to learn that." Aden added that his team was now "trending to concentrate on inexpensive collector systems operating at low[er] temperatures."[71] In that same article, reporting on Meinel's involvement with the solar congress, it was reported that President Nixon had proposed a large increase in energy resource development funding, suggesting that, "It must have been great news to the Meinels. Hopefully, they would get a large enough share to proceed with their solar energy demonstration at all due speed."

Meanwhile, in 1972–1973 a rival group at Honeywell and the University of Minnesota received about "$1 million from NSF to study central electric power generation from 'sunlight.' Honeywell selected 'thermal collection of sunlight via concentrators—similar to Meinels' 'Solar Farm.'"[72] Indeed, the proposed work of their NSF grant, "Research applied to solar-thermal power systems," closely followed what Aden's team had been developing but was funded at a considerably higher level than the OSC/Helio team. In the same report, Roger Schmidt of Honeywell reported that the "essence of the idea is to concentrate 'sunlight' onto a selective-absorber-coated heat pipe, using a trough mirror" with heat pipes rather than "flowing liquids" to run a generation system in the 300–500°C range.[73] Schmidt noted that Honeywell was working with the Meinels on coatings. He concluded that "Honeywell economic analysis (corroborates Aerospace's more complete analysis)—'solar field' would produce power at about twice current costs for fossil/nuclear plants."

In September of that year, Aden and Marjorie stated at a seminar sponsored by the Arizona Council of Engineering and Scientific Associations in Tucson that "a one-acre demonstration solar power 'farm' could be built for $4 million."[74] In the article, Aden "noted that the cost is far beyond what the Federal government has budgeted for solar power research," but that the solar farm concept "could be commercially available in 10 to 15 years" and that "solar power is a logical thing for the Southwest."

The Arab oil embargo hit in October 1973, and the nation's focus was riveted more than ever on energy issues, with much of it looking for quick fixes. While many were calling for a Cabinet-level "Department of Energy," Aden was calling for a "NASA-type energy agency" with a "Manhattan Project" or Moon-shot scale program for solar development, saying that solar's potential as a viable source of power was no longer "a matter of economic debate. It is for real."[75] He argued that "the technology already exists, especially for early application of solar energy to full climate control (heating and cooling)" and that "the principal question of 'How do we industrialize it?' needs an answer." "How are we going to go about making solar energy use possible at reasonable cost?" He related that Marjorie had received a call from San Antonio Light & Power "seeking a solar power plant to substitute for curtailed natural gas supplies" and responded, "There is no answer to these inquiries . . . except to say there is nothing on the market today and that it will be at least summer of 1974 before there can be any effective industrialization, even with a major effort. The technology is there; what is needed is money and industrialization."[76]

Meanwhile, in an article a couple of weeks earlier, it was announced that George Lof of Colorado State University and John Yellott of Arizona State University had both received relatively large grants from NSF to design, build, and test solar heating and cooling systems.[77] "Perhaps the West's most ardent solar energy advocates, astronomers Aden and Marjorie Meinel of Tucson, assert that the sun can be a moderately economical source of warmth, cooling and hot water within three to six years. The Meinels' timetable includes solar-generated electricity within 15 years, fuel and chemical processes within 20."[78] The article revealed that "[she] and her husband have a standing offer to build a prototype $4 million substation to provide solar energy for a small community. So far, there have been no takers."

At the end of 1973, Aden and Marjorie had succeeded in putting their fully fleshed-out vision of large-scale solar farms at the forefront of the national and international energy debate, and Aden had stepped down as director of OSC to devote more time to his twin passions of solar and the Multiple Mirror Telescope (MMT). Meanwhile, however, the modestly substantial funding to demonstrate and further develop the solar farm approach was still elusive. Helio Associates was selling $5 design packets for a low-cost, low-tech, do-it-yourself solar air heater designed by Aden and son Wally Meinel for people to add to their homes as supplemental heating.[79]

Cloudy weather

As the new year broke, Aden was meeting with state legislative leaders to discuss funding for the larger demonstration solar project with the expectation that $4 million would be forthcoming from two Arizona utilities "to build a demonstration solar energy project."[80] A follow-up article a week later referred to the $4 million demonstration proposal, with Aden remarking, "Our problem is that funding is lagging far behind the progress we've been achieving. . . . I hope the legislature will consider what it can do to encourage these developments."[81] If anything, that was an understatement. Clearly the proposed program had demonstrated its readiness for substantially greater funding to move forward but was essentially stalled because of a profound lack of funding despite ever-growing support from the public and elected officials.

In February, the disconnect between the competing visions of a solar future came into stark relief. Much of the traditional "solar establishment" had for decades focused on small-scale uses of solar, such as for heating and cooling, as opposed to the large-scale power generation focus of the Meinel proposal. At a congressional hearing before the US Senate Science and Astronautics Committee, both sides pushed their cases for how the proposed $204 million federal solar program should proceed; the Meinels had been invited to testify by Senator Barry Goldwater (R-Arizona), an influential member of the committee.[82] James C. Fletcher, NASA administrator, and H. Guyford Stever, director of NSF, emphasized that demonstrations of rooftop solar for heating and cooling of homes and businesses should be the focus of these new funds under the direction of their departments, with a goal of demonstrating the usefulness of these systems by 1979. Aden and Marjorie, however, "counseled congressmen not to bank on getting Americans to heat and cool their homes with sunshine." They argued that "conversion of the sun's rays to electricity is a far more promising line of research" for which their approach was one avenue, which conflicted with the testimony of Fletcher and Stever. While the federal program administrators proposed that the US Department of Housing and Urban Development (HUD) be the lead federal agency for solar development and "emphasized development of suncatching panels for rooftops as the key to solar energy, Marjorie Meinel, always the more blunt of the two, warned, 'Don't be fooled—the panels are only a third of the cost of an installation.'" This exchange brought out into the open the clear fault lines within the government and within the community of solar researchers over big solar electric, utility-scale solar development versus the emphasis on smaller applications for heating and cooling and related building-oriented applications.

A column a few days later titled "Solar Energy: Are We Locked in Old Paths?"[83] began by stating, "Occasionally the contrast between the bureaucratic mind and the unshackled thinker can be demonstrated with shocking vividness, and it just happened recently at a congressional hearing on solar energy." The column recounted Stever's and Fletcher's testimony and stated that the US government "is locked in on a theory of employing solar energy to heat and cool buildings." The columnist pointed out that "[this] idea, as Dr. Stever himself observed, has been around since the early 1940s" and that Stever recommended that USHUD be the lead agency in solar "since architecture and construction of future homes will be so thoroughly involved." Solar cells (PV) were also mentioned by both Stever and Fletcher as being 100 times too costly then for general use. The column continued that "Marjorie Meinel, in her turn, enlightened the senators with the simple truth," stating why people would not turn to solar for home heating and cooling but that the "best avenue of research, she suggested, is the use of solar energy to generate electricity." Stever had been asked by the senators about the efficacy of more funds for solar, to which he replied that more funds would result in more demonstration units but was "unlikely to result in swifter technology development." The column concluded, "There is always the possibility, of course, that a lay mind can miss some subtlety known only to government scientists and management experts. But it certainly seems that $204 million spent on . . . solar energy panels on rooftops [i.e., for heating and cooling] would be a tragic waste when so much more might be achieved in learning how to use the sun's heat to run power plants."

A few days later, the *Arizona Daily Star* weighed in with an editorial titled "Listen to the Meinels."[84] The editorial opened with "Tucson's nationally recognized experts in the field of solar energy, Dr. Aden Meinel and his wife Marjorie, have told congressmen that $204 million federal solar research dollars might be directed the wrong way" and "puts them in conflict with NASA and the National Science Foundation." The editorial concluded that "[the] contract between the dreamworld approach of NASA and NSF and the practicality of the Meinels is marked. Congress should listen to them, lest development of a valuable new source of energy be needlessly delayed."

At this time, the main funding for the Meinel solar team was through NSF. The public argument between these two prominent scientists and the heads of the key funding agencies and the resulting public condemnation of the federal approach would not endear the Meinels to these key agency heads, nor to much of the rest of the established solar research community. Indeed, Bernie Seraphin, Aden's key collaborator at OSC, was moved to rebut the *Star*'s editorial that heaped praise on the Meinels, apparently fearing that their funding prospects were being jeopardized.[85] "As a co-worker of Professor Meinel on the solar project, I was shocked by your editorial, 'Listen to the Meinels' in which you construe a contrast 'between the dreamworld approach of NASA and NSF and the practicality of the Meinels' approach.' Unfortunately, the editorial's style, lack of accuracy, and tendency for oversimplification does more harm to the public image of solar energy conversion." Seraphin observed that NSF funds were supporting their solar work. Aden and Marjorie were also moved to respond with a letter to the editor[86] in which they pointed out that the *Star*'s editorial "gave a good summary of testimony to the U.S. Senate given by NSF, NASA and by us" and that "Dr. Seraphin's letter (March 12) taking you to task, on the other hand, gave a good summary of the important NSF support we have received." Knowing it was not a good idea to being seen as biting the hand that was feeding you—even in an inadequate way—the Meinels wrote, "We pointed out our concerns about the wisdom of this emphasis on small-scale uses of solar energy." They recalled that their research into why the solar field had stalled showed that "the field of home applications of solar energy is especially susceptible to disappointment." They said that solar would be more successful in larger apartment and commercial buildings, which would have more operation and maintenance support, and that "solar energy may ultimately have its greatest impact and benefit for this country" as "electrical power production."

The schism within the solar community between the smaller, rooftop solar applications and the large-scale approach for solar generation of electrical power proposed by the Meinels was on display in an April solar exhibition and an informal hearing on the state of solar art in Arizona at Arizona State University in Tempe; the exhibition was attended by Senator Paul Fannin (R-Arizona), the ranking minority member of the Senate Committee on Interior Affairs.[87] Key solar researchers such as John Yellott and Jeff Cook of Arizona State University and John Peck of the University of Arizona touted the benefits and progress in solar heating and cooling for homes and other buildings. Yellott said that solar was on "the verge of coming out of the lab and onto the rooftops" to which Marjorie countered, "Solar heating and cooling is a great idea. . . . But will there be public acceptance?" and added that "solar collectors just simply are unsightly on homes" and advocated solar delivered via utilities from solar farms rather than on rooftops.[88] Arizona State University architect Jeff Cook "acknowledged that solar power needs science to make it usable, and art to make it used" and did not feel that it posed any significant problems in architectural design.

It seemed that the gulf between the Meinels' team and most of the rest of the solar community was widening, not closing. In addition, the Meinels were starting to express doubt that even large-scale, solar-generated electricity would ever become competitive, saying, "As traditional fuels rise, the cost of materials and labor needed for solar collectors will also rise." Aden was still holding out hope for the development of substantial funding to further his solar work in a statement to Congress submitted by Senator Fannin in which Aden called for $120 million per year for five years "to answer whether solar power in an option or illusion."[89] Meanwhile the administration, under the energy leadership of the

Atomic Energy Commission's Chair Dixie Lee Ray, was proposing just $200 million over five years "divided among solar heating and cooling projects, windmill power, ocean thermal systems and electricity production," compared to some $4 billion over the five years for nuclear fission.[90] In response, Paul Rappaport of RCA Laboratories and the Meinels agreed "that solar energy could produce relatively inexpensive electricity within 10 years if adequate research funding were available." Rappaport, who would later lead the future Solar Energy Research Institute (the predecessor to the National Renewable Energy Laboratory), was pinning hope on dramatic cost reductions in photovoltaics, while the Meinels were pushing their photothermal, solar farms approach saying, with funding, they could "eventually compete with oil- and natural gas-fired plants."

The sad state of federal energy research and development (R&D) funding had real-world impacts on the Meinel solar project and relationships between it and other segments of the solar R&D community. The increasing clashes between the Meinels' large-scale, utility-based solar power approach and the cadre of solar researchers focused on smaller-scale applications (mainly heating and cooling) had perhaps less to do with a philosophical disagreement over the appropriate scale of solar applications and much more with a real competition for the extremely limited R&D funds needed to pursue these approaches. Federal solar R&D funding was almost nonexistent, with a paltry $32 million allocated in 1974–1975 for all renewables (solar, geothermal, biomass) and energy-efficiency efforts, less than 1% of the total federal energy R&D budget of some $4 billion.[91] It is important to emphasize that federal funding levels were so low that it actually was a "zero sum game." While the funding needed to progress to the pilot demonstration stage was multiples of the entire budget available to solar, the small research grants needed to keep even low-level efforts going were difficult to obtain; when one small effort was funded, another was of necessity left out in the cold.

Against this backdrop, Aden was having to spend much more time trying to secure enough funds to just keep his solar project team together and alive than in making actual progress in the work. Meanwhile, during this entire period Aden was also hard at work shepherding though the development of a new generation of telescopes, the MMT, an effort for which funding was much more abundant and available. Either of these research efforts would have amounted to a full-time commitment for most people, but Aden would typically have several major pursuits running concurrently.

In October 1974 the federal Solar Research, Development, and Demonstration Act was signed into law, calling for the establishment of a national Solar Energy Research Institute (SERI). As a staffer for Senator Fannin stated a few months later in January, "Everybody's jumping into solar energy—everybody and anybody wants to support it and help it out."[92] Arizona's "legislative leaders, the congressional delegation, trade associations and academic leaders" were making what Arizona state Fuel and Energy Office director C. W. Myers described as "an all-out effort to get the institute" to Arizona. Meinel, part of the Arizona team to land SERI, summarized the NSF solar funds "flowing into the state for solar research" as "about $100,000 a year to Meinel, $150,000 a year to his firm Helio Associates, $150,000 to Arizona State University and perhaps another $100,000 in small academic grants scattered around the state." Meinel added, "We'll be disappointed if the amount of money coming into the state doesn't at least double within the year." One of the projects featured was the "Decade 80" solar demonstration home that was being built by the Copper Development Association in the Tucson foothills, featuring a number of solar-related features, including a photovoltaic roof, despite photovoltaic cells being described as "fabulously expensive." Also in January 1975, Marjorie was one of nine finalists nominated as "Woman of the Year" in the field of education by the *Ladies' Home Journal*.[93]

Despite favorable headlines, the funding needed to move the Meinel solar project forward was still elusive.[94] Helping Arizona's effort to win SERI and statewide enthusiasm for that effort seemed to re-energize Aden's optimism about solar and prospects for finally getting the serious funding needed to move the solar project forward to the pilot demonstration stage. Arizona's congressional delegation was formidable, influential, and determined that "sunny Arizona," with its solar and research resources, would be the home for the new SERI.

By April some progress appeared to have been made. Helio Associates, with E-Systems of Dallas, announced that the three major Arizona utilities had given tentative approval to build a prototype of a new solar power design that Aden had developed, the "Solar Power Bowl," based on a large telescope design he had first proposed in his early Kitt Peak days [see Chapter 6, section University of Chicago 500-inch telescope].[95] The Power Bowl was a fixed spherical reflecting "bowl" with a tracking receiver boom, designed for high-temperature solar thermal power production. This design had the advantage of needing only the receiver to track rather than the much more massive reflecting surfaces. Osborn, then still an undergrad and working at Helio, performed the optical analysis for this project. Helio looked at this design as an additional avenue of development for solar power generation. Dean McKenney, speaking for Helio, said that he expected "the prototype collector to be placed in the Tucson valley in late summer or early fall" of that year. Funding for a full-size, 400-foot (122 m)-diameter power bowl was pending from the national Electric Power Research Institute. Aden and Helio proposed such a system to E-Systems of Dallas and provided a technical analysis and a scale model. When some funding finally did materialize for the Power Bowl concept, the bulk of it went not to the concept developer, but to E-Systems and Texas Tech University. The funding resulted in a 20-m

Stationary Reflector/Tracking Absorber (SRTA) in Crosbyton, Texas, in 1980, which generated steam at 540°C.[96]

During a May meeting on energy and water requirements for arid lands held at UA, Aden declared, "We have the necessary technology to develop solar energy. The question we must answer is if we can do it cheaply enough."[97] He indicated large-scale deployment of solar for power generation as an alternative "to the idea that each family should buy a solar collector to put on the roof for heating, cooling and operating appliances." He also argued that the "national solar institute should be located in Arizona because many of the specialists already are here and there in a positive attitude toward the development of solar power."

As part of Arizona's effort to lure SERI there, Governor Raul Castro and the state legislature created the Arizona Solar Energy Research Commission to coordinate the state's efforts. Among the 15 commissioners appointed was Marjorie Meinel.[98] Keith Tyrley, president of the state's largest utility, Arizona Public Service Company, was appointed chairman of the Commission. The Commission, however, was budgeted "a paltry $25,000 a year" by the legislature for a three-year period.[99] This was in contrast to other states, such as Colorado, New Mexico, and California, which offered up to $80 million in bonding authority to attract SERI.[100] *Arizona Highways* (a popular, photo-rich magazine published by the Arizona Department of Highways) devoted the August 1975 issue to solar energy in Arizona, proclaiming, "Now is the time, Arizona is the place!" with a lead article titled "Solar Energy Research Institute" by Senator Fannin. It also featured major descriptions of the Meinel solar farm concept, as well as ongoing work in Arizona on photovoltaics, passive solar and other solar applications, the history of solar in Arizona (including the 1955 First World Conference on Solar Energy), the Decade 80 Solar House, and the McMath Solar Telescope at Kitt Peak.[101]

The Meinels continued to promote both large-scale solar and Arizona as the ideal place for that development. In August, they would tell a joint meeting in San Francisco of astronomical societies that the United States "must convert to solar energy sooner or later" and that the technology "was already available to convert whole cities to the use of electricity from the sun."[102] That same month, Helio Associates, with McKenney as president, was acquired by Mountain States Mineral Enterprises. While it seemed strange that a mining engineering consulting firm would be a good match for Helio's solar focus, it was the only option that was available at the time to keep Helio solvent. In the announcement it was said that the two Tucson-based firms hoped to "pool the technology of both companies to come up with sun-powered systems that can be supplied to the mining industry as complete packages."[103] Osborn, working on optical analysis of various solar systems ranging from the Solar Power Bowl to air-inflated solar crop dryers, joined the Helio staff in its move to Vail, Arizona, on the southeast outskirts of the Tucson. Any synergy imagined never did seem to emerge, but it did permit Helio to continue to limp along while the hunt for substantial funding continued. As far as Helio staff could see, there was never any real interaction with Mountain States staff or programs.

In October, the mayor of Tucson, the Meinels, and the UA Foundation announced that the City of Tucson would put up $1.5 million to pay for half the cost of a potential site near UA for SERI if Tucson were selected, and the state would put up $1.5 million as a match for any site if Arizona were selected.[104] The city's plan, largely crafted by Aden, proposed using the university's power of eminent domain to acquire the proposed SERI site. Local pushback on that aspect caused the university administration to disavow the plan in part less than a week later with the UA vice president for research, Richard Kassander, announcing, "It's illegal, immoral, unethical and probably fatting to use the U of A's power of eminent domain to acquire property to convey to anyone else."[105] Kassander added that UA "does not want to be put in the position of saying that the institute will be better put in Tucson than anywhere else in Arizona." Despite this ruckus, the Commission and Meinel continued to refine and promote the Arizona bid for SERI.[106]

Into the wilderness as the dream fades

After six years of concentrated effort and traveling around the world to evangelize large-scale solar power, having the funding needed to advance and prove their concept always seemed a mirage—appearing to be near but always just out of reach. Early in 1976 the solar textbook that the Meinels had written, *Applied Solar Energy: An Introduction*,[107] was published without serious solar funding on the horizon. Solar water- and pool-heating systems had become established in limited markets. Passive solar homes and climate-tempered homes had been demonstrated and built in several locales. Solar cells (PV) were being used in space and remote applications, and various other solar applications were being worked on. But almost no funding had materialized for the type of large-scale solar power-generation systems that the Meinels had devoted the past years to developing and promoting. Despite a nearly continuous stream of promises and expectations of significant funding to truly demonstrate the viability of the Meinel concept and make real progress, such funding had never materialized, nor was it likely to materialize anytime soon. Solar was becoming more of a grind without tangible results than the adventure they first saw. "Keeping Helio alive was no small task. It seemed like all of our time during the working day was writing new proposals for Federal support grants or contracts. Writing up required reports on the progress was managed after-hours and on weekends."[108] As discussed in an overview paper on solar thermal power generation, the Meinels still had a questioning eye on the possibilities for the fledgling area of PV, stating, "It will be interesting to see what results the present intensive effort to develop inexpensive solar cells [PV] may yield in a few years ahead."[109]

The Meinels began to take a decidedly skeptical view of solar as a major contributor to our energy needs. In an interview with *The Arizona Republic*, Aden questioned if solar power was really "the answer to the country's future needs, as millions are being led to believe."[110] Aden was quoted as saying, "The dream of many is that the sun will become a major energy resource, eventually to supplant fossil fuels entirely. This dream has been frustrated in the past and could be once more as the perpetual barrier of economics is faced. The sun is yet to be domesticated." He also said that smaller solar heating and cooling systems for households faced the same cost barrier, as they were 3–10 times more expensive than electric or natural gas systems. The article added that "despite their concerns about the economic feasibility of using the sun as a major source of energy, the Meinels believe Arizona should do all it can to obtain the National Solar Research Institute and the 10-megawatt solar power plant." Aden warned that "[if] the power station doesn't work out, solar energy will be in real trouble."

Aden and Marjorie delivered a paper in Dublin, Ireland, which was covered in an editorial in *The Arizona Republic*: "Dr. Aden Meinel and his wife Marjorie have concluded that domesticating the sun for dependable, cheap energy is not realistic. Solar energy still must be considered an exotic energy source. . . . The only long-range energy hope for the world, the Meinels told colleagues, is the nuclear breeder reactor." The editorial castigated nuclear opponents and promoted nuclear and coal as the way forward.[111]

Suddenly, the Meinels were being embraced by pro-nuclear and utility groups and were facing backlash by the fledgling solar industry and environmentalists. An article in the *Arizona Daily Star*[112] featured many customers of solar pool-heating expressing great satisfaction with their solar pool heaters, only to be followed at the end with a comment by Aden that "solar heating plants do not raise the temperature of swimming pool water enough to be very useful"; he mentioned his disappointment with the solar pool heater that he built for his own pool. This was followed a few days later by a letter to the editor in *The Arizona Republic* in which a local engineer promoted nuclear and used the earlier remarks by the Meinels to support his view.[113] Five days later, a Tucson solar pool-heating company took out a half page "open letter" that defended the effectiveness of solar pool-heating and the large number of satisfied customers, saying that "if Dr. Meinel's heater flunked the test he should try another."[114]

In June 1976, a welcomed break took place with the wedding of Aden's secretary Cynthia Sims to student Donald Osborn, who was then working at Helio Associates on several solar projects. Aden and Marjorie were honored guests.

That summer, Osborn submitted a paper on solar crop-drying, coauthored by Aden and Beauchamp. The paper was to be presented to the International Solar Energy Society meeting in Winnipeg, Canada, in August. It was so typical of Aden that he encouraged and supported his young student to travel to the conference and present the paper as the lead author. About a year later, as Osborn prepared to leave to take a position as associate director of the Arizona Solar Energy Commission in Phoenix, the Meinels would become godparents to the Osborns' first child, Laura.

With the state's largest utility push for building and opening a large nuclear plant outside of Phoenix, energy was becoming a bitterly contested topic. Starting in the summer of 1976, Arizona Public Service (APS) published a series of full-page ads in *The Arizona Republic* featuring the Meinels with the headline message, "When we get solar power the energy situation will look brighter." The ads touted that APS and others were researching solar but that "a lot more money and efforts will have to be invested to mass-produce electricity from the sun." They quoting the Meinels, saying, "The sun is not yet domesticated." Below a photo of the Meinels, the ad read, "They are right. The truth is solar power will not be domesticated soon enough to prevent an Arizona energy shortage in the 1980s. Two important energy sources *must* be developed, coal and nuclear."[115] The ad did fail to note that APS and the other Arizona utilities had failed to deliver on the funding that had been "pending" for several years to build the solar power demonstration plants that Aden had been promoting.

In various speeches, talks, and papers, Aden and Marjorie continued to downplay the viability of solar. They told reporters and utility officials at a seminar in New Mexico in late September that "[people] may be expecting too much from sunlight as a solution to the nation's energy problems." Asked why he had apparently lost his enthusiasm for solar, Aden replied, "I guess the closer you get to the practical problems of exotic fuels the more you realize why they are exotic."[116] These doubts were further elaborated in an article in the *Arizona Daily Star* reporting on Aden's utility seminar speech and quoting Aden: "We're still enthusiastic, but we're becoming very realistic." However, in hindsight, the most telling Aden quote was, "We've spent a decade arguing. There is very little support for actually building things."[117]

Aden and Marjorie had been spending so much time and effort researching, developing, and promoting their proposed solar electric power system, only to see the promised funding evaporate, that they had begun to sour on the concept and to overemphasize the barriers. This pessimism was, at one and the same time, promoted by anti-solar groups and ill-received by the solar community. Such solar notables as Yellott began to push back in a letter to the editor of *The Arizona Republic*: "Fortunately, despite the statements made by Dr. Meinel, there are new ideas in the world of solar energy, and many of them are making use of materials not available to our Victorian forbears. Concentrating collectors, roof-top heating and cooling systems. And other examples of ingenuity are proving to be both effective and economical."[118] An ongoing round of debates in con-ferences, speeches, and in the press started over solar's role in the nation's (and world's) energy future and of the "soft vs. hard path" of energy development. Aden, who had been the most prominent proponent of large-scale solar power, now was increasingly the portrayer of solar as "not ready for prime time" and uneconomic.

Meanwhile, on top of all this controversy, by the end of 1976 Helio Associates was failing to thrive, and Aden had to step in as president, replacing McKenney, who returned to OSC.[119]

In March 1977, after two years of a strongly contested and highly competi-tive bidding, the announcement was made that Golden, Colorado, was chosen as the site for SERI (now the National Renewable Energy Laboratory, NREL).[120] Despite high hopes, considerable effort, and strong congressional delegation support, the Arizona effort to which Aden had contributed so strongly had failed, and the new wave of solar support funding would be directed to and through SERI in Colorado, rather than in Arizona. Paul Rappaport from the Davis Sarnoff Laboratory at Princeton was to be the new SERI director.

Later that month, the Meinels published a paper titled "Soft Energy Paths: Reality and Illusion," in the utility magazine *Electric Perspectives*.[121] That was the first in a series of papers, talks, and congressional testimony over the next three years in which the disillusionment with solar was shared broadly by the Meinels and strongly encouraged by utilities, nuclear proponents, and others who were pleased to see one of the bright lights of solar casting doubt on its vi-ability. Their conclusion was starkly summarized by Marjorie's testimony when the Meinels made a joint presentation to Congress in September 1977: "Solar energy is widely acclaimed as the ideal energy option. Some enthusiasts even urge abandonment of our nuclear capability because of the 'low cost' and im-minence of solar energy. We once dreamed the same dreams of a nonnuclear future via solar energy, but reality has dawned as the magnitude of the eco-nomic barrier has become clear. We have come to the same conclusion reached by the French Government in the 1890's: That solar energy is expensive and is undependable."[122]

In the mid-1970s the energy futures debate took on the form of the debates be-tween supporters of the "Soft Energy Path" and the "Hard Energy Path," launched in a 1976 article by physicist Amory Lovins in *Foreign Affairs* titled "Energy Strategy: The Road Not Taken?"[123] Lovins argued that the United States had to choose between continued use of fossil fuels and nuclear fission. The "hard path," which promised a future of steadily increasing reliance on coal, oil and gas, and nuclear fission, had serious environmental risks and emphasized the short term.

Renewable energy sources (solar, wind), "the soft path," offered long-term sustainability. In what became known as part of the "hard path vs. soft path" energy debates, some of the pronouncements and locales representative of this new and decidedly negative view on solar's role by the Meinels included:

- "Economic Aspects of Solar Energy." First International Scientific Forum on an Acceptable World Energy Future, November 7, 1977.
- "Hard Realities of the Soft-Path to a Solar Future." First International Scientific Forum on an Acceptable World Energy Future, November 7, 1977.
- Congressional Briefing, House Subcommittee on Advanced Energy Technologies and Energy Conservation, Research, Development, and Demonstration, February 7–21, 1978.
- "The Solar Path: How Soon?" Public forum debate with Amory Lovins, April 27, 1978.
- A. B. Meinel and M. P. Meinel, "Solar Energy: Promises and Problems." Seminar on the hard and soft energy path: Do we have a choice? South Dakota School of Mines and Technology, August 30, 1978.
- "Solar Energy, Fuel of the Future?" Viewpoints, Christian Perspectives on Social Concerns, Augsburg Publishing House, 1978.
- "Cloudy Days Ahead for Solar Energy." *The Sciences*, February 1979.
- A. B. Meinel and M. P. Meinel. Federal DOE Advanced Energy Technologies Hearings, Congressional Briefing, House Subcommittee on Energy Development, February 28, 1979.

In March 1978, Aden and son Walter (Wally) Meinel filed a patent assigned to Motorola for a non-imaging trough collector that would improve the efficiency of solar energy collection.[124] In addition, Aden and Wally also researched how to convert troublesome tumbleweeds (also known as Russian thistle), which are prevalent in Arizona, into compressed fireplace logs called "Tumblelogs." Wally's company, Arizona Scientific Research, was able to produce a number of demonstration samples. However, Tumblelogs (Figure 10.4) never "caught fire" as it were because the fuel density was too low.[125]

During this period, Aden and Marjorie made numerous international trips and lectures for both their observatory and solar work. Ironically, much of the focus of their lectures abroad ended up being on small solar applications in developing nations, such as the heating and cooling devices that they had disparaged:

- 1978 pilgrimage to Syracuse, where they saw for themselves the site at the Bay where Archimedes was said to have repulsed the Roman fleet using concentrated solar energy, a story they had researched and written up;[126]
- Sabbaticals in Italy and Sicily;

Figure 10.4. Aden and Marjorie Meinel with tumbleweed and Tumblelog.
Courtesy of David Drach-Meinel.

- Meetings in Saudi Arabia;
- Talks in Ghana, India, France, Taiwan, China, England, Mexico, Ireland, and Yugoslavia;
- Second trip to India for the National Solar Energy Convention in Bhaningar, where they were given a copy of Walter Adams's 100-year-old solar energy book.

By 1979, when serious funding for solar had finally begun to flow, with some $1.36 billion being allocated to large-scale solar projects, Aden had largely withdrawn from solar R&D activities. Then in 1981, at the start of the Reagan era, federal energy R&D budgets were slashed, and large-scale solar efforts were largely terminated. Meanwhile, the other major and celebrated advance in astronomy that Aden had also spent much of the 1970s developing and promoting, the MMT, was dedicated in May 1979.

Disillusionment to triumph

In 1980 Aden was awarded the Optical Society Association's Frederick Ives Medal, one of the most prestigious awards in the optical sciences. The tribute was "for his contributions to thermal solar energy, analysis of the principles of coherently combined, independent telescopes, and the leadership he has given to several major optical and astronomical research centers."[127]

By this time, Aden had completely withdrawn from solar activities. It would be another two decades before solar power on the scale he was contemplating really started to occur. In 1982 Aden and Marjorie left Arizona and solar pursuits and moved to NASA/JPL in Pasadena, California, where they had started their "solar odyssey" in their youth. Many years after the Meinels' interaction with Charles Abbot at the 1971 Goddard Space Flight Center Solar Conference, there was to be one more solar connection linking the Meinels, Abbot, and solar. In 2000, Aden's student, Donald Osborn, was awarded the Charles Greeley Abbot Award, the most prestigious honor given by the American Solar Energy Society (ASES) for his "persistent and effective contributions to the many fields of solar energy." Osborn had gone on from working for Helio and as a research engineer at the OSC to become the associate director of the Arizona Solar Energy Commission (1977–1981) and then director of the UA College of Engineering's Solar & Energy Research Facility (1981–1991). In 1991, he left UA to develop and manage what would become the first, large-scale commercialization of grid-connected photovoltaics as superintendent and manager of the Sacramento Municipal Utility District (SMUD) Solar Program.

Under Osborn's direction, the SMUD Solar Program was responsible for the installation of over 3,000 SDHW systems and the operation of the world's largest (as of mid-2002) distributed generation PV system—over 10 MW including more than 1,000 rooftop "PV Pioneers." At that time, SMUD was responsible for over half of all grid PV in the United States. For his efforts at SMUD, Osborn has been called "the Father of U.S. Grid-connected PV Commercialization."[128] The successful deployment of solar at SMUD, in both central utility solar plants and as distributed generation on customer's homes and businesses, directly led to the California Solar Initiative Program and today's booming solar market. It was largely due to Aden's influence and mentoring that Osborn pursued his career in solar, which has lasted to this day.

There are now some 900 MW of Meinel-type, concentrating parabolic trough, solar energy-generation stations operating in the California and Arizona deserts, with another 400 MW of central receiver (power tower) power stations in California. But Aden's dream of large-scale solar deployment for serving the United States' and the world's power needs has finally become a reality primarily due to the extraordinary decline in PV power costs.

Today (2021) there are now something like 82 gigawatts of solar in the United States and about 650 gigawatts worldwide.[129] Almost all of this is photovoltaics, with about 60 percent of it as large utility-scale solar (PV) farms and 40 percent as smaller distributed generation on homes, commercial buildings, and agricultural sites. One major reason for this rapid growth and the beginning of the solar century can be attributed to Aden and Marjorie Meinel. Even though they were focused on solar thermal, they were driven even more by a more general concept. Solar thermal was their technology focus, but the Meinel concept was to use advanced technology to deploy solar on a broad scale in a technologically advanced country like the United States to do something environmentally sound and sustainable to meet our energy needs and support our civilization. If we could not do it in the advanced countries, they reasoned, there's no hope for the developing countries. So we had to deploy our best technology and get that out there on a large scale. That's what we ended up doing, first in the United States, then in Germany and Japan, and now in China and worldwide. That is how, some 40 years later, the Meinel solar dream has become a reality and key to a sustainable future and making the solar century a reality.

While the Meinel solar vision may not have played out precisely as they had thought, many key elements of it have helped guide solar's development and have led to where we are today. These include:

- Using large-scale high technology for developed countries first as the key path to practical, high-impact deployment of solar and reducing costs to enable its reach worldwide;
- Grazing under solar arrays to control vegetation growth;
- Solar for tropical islands (Brando's island finally did become a solar-powered paradise);
- Large-scale PV plants in the Southwest deserts of the United States and worldwide using the basic designs developed by Aden;
- Large-scale PV plants along with the now common, distributed PV system on a broad scale;
- Giving scientific and technical credence to the concept that solar could and would be a significant contributor to the world's power needs in a sustainable and environmentally acceptable way.

By envisioning and promoting these key components, Aden and Marjorie truly helped lead us into the solar century.

A solar coda

In 2006, with Marjorie in failing health from Alzheimer's disease, the Meinels moved to Henderson, Nevada, to be near their son, David Drach-Meinel, who was (and still is) co-pastor with his wife Diane at Christ the Servant Lutheran Church. In February 2011, Pastor Dave was approached by a Boy Scout leader and church member, who also happened to work for Bombard Electric in Las Vegas, and was told that a government grant was offering solar panels at no cost on a first-come, first-served basis to not-for-profit organizations. Dave jumped on it, signing the papers for the church's solar system, winning approval from the Church Council after the fact. Aden and Dave would go out to lunch almost every day so that Aden could see the installation process and the snowcapped Spring Mountains behind. After some silence, Aden, now in his late eighties, commented, "I never thought I'd live to see the day."

On May 22, 2011, a presentation of the NV Energy Solar Grant presentation at Christ the Servant Lutheran Church was held, with Aden as the guest of honor. With Aden looking on, his son, holding a copy of his parents' *Applied Solar Energy* textbook, accepted the grant for the solar system. At the presentation, Aden said, "After all these years, from extensive research in the early 1970s to advances in today's technology, I am now so happy to see my dream come true" (Figure 10.5).

Figure 10.5. NV Energy Solar Grant presentation at Christ the Servant Lutheran Church in Henderson, Nevada. Left to right: Aden Meinel; his son, Pastor David Drach-Meinel, holding a copy of *Applied Solar Energy*, written by Aden and Marjorie; Church Council President Jason Hood; representatives from Bombard Electric and NV Energy.
Courtesy of David Drach-Meinel.

Notes

1. Meinel, A. B., Meinel, M. P., and Meinel, B. 2003. *The solar odyssey: Adventures along the way*. Unpublished, 133.
2. Menzel, D. H. 1977. Obituaries: Charles Greeley Abbot. *Quarterly Journal of the Royal Astronomical Society* 18: 136–9. http://articles.adsabs.harvard.edu/cgi-bin/nph-iarticle_query?1977QJRAS..18..136M&defaultprint=YES&filetype=.pdf
3. Meinel et al., *Solar odyssey*, 141.
4. Meinel et al., *Solar odyssey*, 141.
5. Meinel et al., *Solar odyssey*, 142.
6. Meinel et al., *Solar odyssey*, 137.
7. Meinel et al., *Solar odyssey*, 139.
8. Meinel et al., *Solar odyssey*, 142–3.
9. Meinel et al., *Solar odyssey*, 143–5.
10. Meinel et al., *Solar odyssey*, 145.
11. Meinel et al., *Solar odyssey*, 147–8.
12. Meinel et al., *Solar odyssey*, 147.
13. Meinel, A. B., and Meinel, M. P. 1972a. A harvest of solar energy. *Optical Sciences Newsletter* 6: 68–75.
14. Meinel et al., *Solar odyssey*, 155.
15. Meinel et al., *Solar odyssey*, 73.
16. Meinel et al., *Solar odyssey*, 147.
17. Halacy, D. S. 1973. *The coming age of solar energy*. Avon Books, New York, 207.
18. Meinel et al., *Solar odyssey*, 155.
19. Halacy, *Coming age*, 207.
20. Meinel, A. B. and Meinel M. P., Harvest of solar energy. *Optical Sciences Newsletter* 6: 68–75;
21. https://commons.wikimedia.org/wiki/File:Price_history_of_silicon_PV_cells_since_1977.svg.
22. Loferski, J. J. 1972. Some problems associated with large scale production of electrical power from solar energy via the photovoltaic effect. Paper 72-WA/Sol-4, ASME Winter Annual Meeting, New York, November 1972.
23. Sun can power U.S. forever: US's Dr. Meinel outlines project. *Tucson Daily Citizen*, December 2, 1970.
24. *Tucson Daily Citizen*, December 2, 1970, 30.
25. Electric power statistics: production of energy and capacity of plants, fuel . . . , by United States. Federal Power Commission, September 1970.
26. UA man offers fresh water problem solution—solar energy desalination on a huge scale. *Tucson Daily Citizen*, January 21, 1971, 25.
27. Meinel et al., *Solar odyssey*, 161–2.
28. Meinel et al., *Solar odyssey*, 162.
29. Class notes taken by D. Osborn.
30. Class notes taken by D. Osborn.
31. Meinel, A. B., and Meinel, M. P. 1971. Is it time for a new look at solar energy? *Bulletin of the Atomic Scientists* 27: 32–7, 61.

32. Class notes taken by D. Osborn.
33. Meinel et al., *Solar odyssey*, 158–9.
34. Meinel et al., *Solar odyssey*, 159–61.
35. Meinel et al., *Solar odyssey*, 160.
36. UA scientists propose solar energy use. *Arizona Daily Star*, June 12, 1971.
37. Demonstration of solar power set in Tucson. *Arizona Daily Star*, June 17, 1971.
38. Solar power farm site here picked. *Tucson Daily Citizen*, June 30, 1971.
39. Meinel, A. B., and Meinel, M. P. 1971. Is it time for a new look at solar energy? *Bulletin of the Atomic Scientists* 27: 32–7.
40. The nation is warming to Meinel's solar energy plan. *Tucson Daily Citizen*, December 22, 1971, 31.
41. Solar energy theory to be tested in '72. *Arizona Daily Star*, January 14, 1972.
42. Meinel, A. B., et al. Research applied to solar-thermal power conversion. NSF (RANN) SE Grant GI-30022, Final Report, January 31, 1973.
43. Meinel, A. B., et al. Research applied to solar-thermal power conversion.
44. *Optical Sciences Newsletter* 6(3), December 1972.
45. Solar energy theory to be tested in '72. *Arizona Daily Star*, January 16, 1972.
46. Arizona couple unveils device to convert sunlight to power. *The SUN*, Flagstaff, Arizona, February 3, 1972, 3.
47. Meinel, A. B., and Meinel, M. P. 1972b. Physics looks at solar energy. *Physics Today* 25: 44–50.
48. Meinel and Meinel, Physics looks at solar energy, 44.
49. Meinel and Meinel, Physics looks at solar energy, 44.
50. Meinel and Meinel, Physics looks at solar energy, 46.
51. Meinel and Meinel, Physics looks at solar energy, 47.
52. Scientists offer solar energy plan. *Tucson Daily Citizen*, May 5, 1972.
53. UA couple tests solar theory. *Arizona Daily Star*, July 6, 1972; *Optical Sciences Newsletter* 6(3), December 1972, 65.
54. *Optical Sciences Newsletter* 6(3), December 1972, 65.
55. *Optical Sciences Newsletter* 6(3), December 1972, 67.
56. *Optical Sciences Newsletter* 6(3), December 1972, 81.
57. Solar energy "farm" building funds sought. *Arizona Daily Star*, December 16, 1972; Solar plan for energy seeks funds. *The Arizona Republic*, December 16, 1972.
58. Solar Energy 'Farm' Building Funds Sought. *Arizona Daily Star*, December 16, 1972, B-1.
59. *Optical Sciences Newsletter* 6(3), December 1972.
60. Meinel giving up UA office. *Tucson Daily Citizen*, December 15, 1972.
61. Key scientists step aside at UA. *Tucson Daily Citizen*, December 22, 1972.
62. Solar plan funds delayed. *Tucson Daily Citizen*, February 8, 1972, 31.
63. Natural gas cutbacks traced to price rules. *Arizona Daily Star*, April 6, 1973.
64. State seen as source of energy. *Arizona Daily Star*, April 12, 1973.
65. Solar power called an alternative. *Arizona Daily Star*, April 14, 1973.
66. *Tucson Daily Citizen*, April 27, 1973, 41.

67. Marlon Brando offering islands for research, U of A scientists say. *The Arizona Republic*, April 28, 1973.
68. Nationally known physicist succeeding Meinel at UA. *Tucson Daily Citizen*, May 16, 1973.
69. Udall asks Meinel solar aid. *Tucson Daily Citizen*, May 17, 1973.
70. Johnson, T. 1973. The sunshine spreaders. *New Scientist* 58: 337–9.
71. Harvesting the sun. *Arizona Daily Star*, July 4, 1973, editorial comment page.
72. Terrastar, NASA CR-129012, September 1973, B-25.
73. Terrastar, NASA CR-129012, September 1973, B-27.
74. Uphill climb foreseen in easing energy crisis. *Arizona Daily Star*, September 30, 1973.
75. Directive agency advocated for energy shortage. *Arizona Daily Star*, November 29, 1973.
76. Directive agency advocated for energy shortage. *Arizona Daily Star*, November 29, 1973.
77. Solar energy promising but years away. *Tucson Daily Citizen*, November 10, 1973.
78. Solar energy promising but years away. *Tucson Daily Citizen*, November 10, 1973.
79. Conserve that fuel! Build a solar heater. *Tucson Daily Citizen*, December 6, 1973.
80. Legislators, Meinel set solar power talk. *Tucson Daily Citizen*, January 19, 1974, 10.
81. Bills would push solar energy use. *Tucson Daily Citizen*, January 26, 1974, 10.
82. Experts cool to solar heating. *The Arizona Republic*, February 27, 1974.
83. Cole, B. Solar energy: Are we locked in old paths? *The Arizona Republic*, March 2, 1974, editorial section.
84. Listen to the Meinels. *Arizona Daily Star*, March 5, 1974, A-8.
85. Seraphin, B. O. Oversimplifying. Letter to the editor, *Arizona Daily Star*, March 12, 1974.
86. Meinel, A. B., and Meinel, M. P. Uses of solar energy. Letter to the editor, *The Arizona Daily Star*, March 20, 1974, Letters to the editor.
87. Solar art shines at ASU symposium. *The Arizona Republic*, April 18, 1974.
88. Scientists focus on the sun for future survival. *Casa Grande Dispatch*, April 29, 1974.
89. Sun-power support, Fannin says. *The Arizona Republic*, June 28, 1974, A-20.
90. New power sources vital. *Tucson Daily Citizen*, August 7, 1974.
91. U.S. federal investments in energy R&D: 1961–2008, PNNL-17952, USDOE, Pacific Northwest National Laboratory, October 2008 DE-AC05-76RL01830.
92. Arizona pushing to be nation's solar-heating center. *Arizona Daily Star*, January 26, 1975, A-10.
93. UA researcher named for magazine's honor. *Arizona Daily Star*, January 30, 1975.
94. Meinels' solar concept catching on. *Tucson Daily Citizen*, October 15, 1974; Sun-power support needed, Fannin says. *The Arizona Republic*, June 28, 1974; Solar energy to become necessary, scientists tell NAU audience. *Arizona Daily Sun* (Flagstaff, Arizona), October 11, 1974; The rays of the sun provide ray of light for Yuma future. *Yuma Daily Sun*, November 10, 1974.
95. Solar collector gets OK, prototype planned here. *The Tucson Citizen*, April 16, 1975.
96. Dish systems for CSP. J. Coventry and C. Andraka, SAN2017-3433J, 2017.

97. State could power all of U.S., Prof says. *The Arizona Republic*, May 10, 1975, B-1.
98. Governor selects policy commission on solar energy. *Arizona Daily Star*, May 29, 1975.
99. The new solar push: A need for speed. *Arizona Daily Star*, October 14, 1975.
100. Mayor asks lure for solar center. *Tucson Daily Citizen*, October 25, 1975.
101. *Arizona Highways*, Arizona Department of Highways, August 1975.
102. Sun power could run U.S., 2 say. *The Arizona Republic*, August 10, 1975, A-4.
103. Tucson firms merge solar energy efforts. *Tucson Daily Citizen*, August 11, 1975, 32.
104. Mayor asks lure for solar center. *Tucson Daily Citizen*, October 25, 1975.
105. Solar-unit land lure is attacked. *The Arizona Republic*, November 1, 1975, C-5.
106. Arizona is in thick of "poker game" for solar institute. *Arizona Daily Star*, November 29, 1975, B-1.
107. Meinel, A. B., and Meinel, M. P. 1976a. *Applied solar energy: An introduction*. Addison-Wesley, Reading, Massachusetts.
108. Meinel et al. *Solar odyssey*, 193.
109. Meinel, A. B., and Meinel, M. P. 1976b. Solar photothermal power generation. *Environmental Conservation* 3: 15–21.
110. Cost might make solar energy use unfeasible, 2 say. *The Arizona Republic*, April 25, 1976, B-1.
111. Reality of energy. *The Arizona Republic*, April 28, 1976, A-8.
112. Jury still out on solar heating. *Arizona Daily Star*, May 9, 1976.
113. Efforts to cripple America. *The Arizona Republic*, May 11, 1976.
114. An open letter to pool owners concerning solar heating. *Arizona Daily Star*, May 16, 1976.
115. "When we get solar power the energy situation will look brighter," ad by Arizona Public Service, *The Arizona Republic*, July 20, 1976, B-24.
116. "Too much expected from solar energy, U of A Prof Fears" *The Arizona Republic*, Oct 1, 1976, A-21.
117. Solar expectations dim, expert says. *Arizona Daily Star*, October 5, 1976, B-1.
118. The Arizona Republic.
119. UA astronomer takes over solar firm. *Tucson Daily Citizen*, November 5, 1976.
120. Buck, A. L. 1982. A history of the Energy Research and Development Administration. US Department of Energy, Washington, DC.
121. Meinel, A. B., and Meinel, M. P. 1977. "Soft" energy paths: reality and illusion. *Electric Perspectives* 77: 24–7.
122. Testimony by Marjorie and Aden Meinel. Hearings before a Subcommittee of the House Committee on Governmental Operations, House of Representatives, September 20–22, 1977, 1404.
123. Lovins, A. B. 1976. Energy strategy: the road not taken? *Foreign Affairs* 55: 65–96.
124. U.S. Patent 4,131,485. They also received Patent 4,337,758 for a solar collector and converter.
125. Personal communication from Ed Meinel, September 23, 2020.
126. Meinel et al., *Solar odyssey*, 38–54.

127. https://www.osa.org/en-us/get_involved/awards_and_honors/awards/award_d
escriptions/ivesquinn/.

128. "Member spotlight." *Solar Today Magazine*, Summer 2016, 66.

129. U.S. solar market insight. Solar Energy Industries Association, June 2020. https://
www.seia.org/us-solar-market-insight.

11

A View to National Security

On October 4, 1957, an R-7 rocket modified as an SS-6 ICBM lifted off from Tyuratam Cosmodrome in the Kazakh Soviet Socialist Republic, carrying into orbit a metal sphere just under two feet across and weighing 184 pounds. Sputnik 1 was little more than a beeping radio transmitter, but it was still a propaganda coup. The Eisenhower administration had announced in July 1955 that the United States would launch earth-orbiting satellites in the Vanguard series during the International Geophysical Year (July 1, 1957 through December 31, 1958), which was designed to promote scientific collaboration. The Soviets, however, wanted to be the first in space and thereby demonstrate their scientific superiority using rocket technology and German personnel they had captured at Peenemünde in May 1945. They had already tested the first atomic bomb on August 29, 1949, and their first hydrogen bomb on August 12, 1953. There was evidence that they were producing long-range missiles and bombers to deliver those bombs across the Atlantic. And now they had preempted or "scooped" America on satellites.[1]

Public response to the Sputnik satellites in the United States was swift and largely partisan, pointing fingers at the Eisenhower administration and Republicans for allowing the Soviets to put Americans at a scientific and strategic disadvantage. Memories of the surprise attack at Pearl Harbor on December 7, 1941, fueled the fear of a nuclear attack with little or no warning. On November 7, 1957, the Office of Defense Mobilization released the Gaither Report titled "Deterrence and Survival in the Nuclear Age," which documented America's vulnerability to attack by Soviet ICBMs. Eisenhower welcomed the Report "like a skunk at a picnic" and chose not to respond.[2] When he did not act on it, members of the Gaither Committee leaked key elements to the press, which led to construction of public fallout shelters and publication of DIY manuals for home shelters.[3] Democrats quickly began to draft legislation that would become the National Defense Education Act, aimed at producing more scientists and engineers. The media questioned President Eisenhower's priorities and his ability to defend the nation as an elder statesman.[4] Five days after the launch of Sputnik 1, President Eisenhower held a press conference to quell fears, saying, "Now, so far as far as the satellite itself is concerned, that doesn't raise my apprehensions, not one iota. . . . The mere fact that this thing orbits involves no new discovery to science. They knew it could be done—at least they say so, and they have for a long

With Stars in Their Eyes. James B. Breckinridge and Alec M. Pridgeon, Oxford University Press. © Oxford University Press 2022. DOI: 10.1093/oso/9780190915674.003.0011

time—so that is no new discovery, so in itself it imposes no additional threat to the United States."[5] At that same press conference and following the content of a draft by Secretary of State John Foster Dulles, Eisenhower congratulated Soviet scientists on placing a peaceful satellite into orbit ("one small ball in the air") and announced America's contribution to the International Geophysical Year (IGY)—a scientific satellite program of its own known as Vanguard, all of this intended to calm and reassure Americans.[6] But it was an issue and embarrassment for Eisenhower that would not go away, particularly after the Soviets launched Sputnik 2 on November 3, with radio transmitters, scientific instrumentation, and a dog named Laika that died a few orbits later (inspiring the satellite moniker Muttnik). Senator Lyndon Johnson chaired hearings to investigate why America's missile program was suffering.[7]

The onset of Project Vanguard can be traced to September 1955, when the Department of Defense Committee on Special Capabilities selected the Naval Research Laboratory's proposal (one of three) for the IGY program on the basis of its superior tracking system and satellite instrumentation to be launched on Viking three-stage rockets.[8] Failures of the first Vanguard rocket launch in December 1957 ("Kaputnik," "Sputternik") and its backup in February 1958, before the third, successful one in March 1958, only heightened the anxiety in America.[9]

Through it all, Eisenhower could not reveal top secret operations that his administration had begun in the previous five years and were presently in development to strengthen defenses and enable surveillance of Soviet bombers and missiles. The administration, in fact, knew that the Sputnik launch was being prepared as early as June 1957 when a U-2 spy plane photographed the SS-6 rocket on the launch pad at Tyuratam. There was a "missile gap," as *Washington Post* columnist Joseph Alsop dubbed it; Senators Stuart Symington and Lyndon Johnson were relentless in politicizing it and used it to expand Democratic majorities in the House and Senate in elections for the 86th Congress.[10] John F. Kennedy campaigned for president on the issue. There was in fact a missile gap, but it was one in favor of the United States, as U-2 photographs of the entire Soviet territory clearly showed. Although the Soviets monitored every one of the U-2 overflights and raised diplomatic objections, they did not want to publicize it, nor did Eisenhower want to humiliate them publicly and thereby provoke a war.[11]

The national defense proposals and operations that led to this point and afterward are summarized here to describe the environment that called upon Aden for the breadth and depth of his expertise, beginning in 1960, as well as to introduce the CIA and corporate and military personnel who later recruited him as a special scientific consultant—at high clearance levels—for aerial surveillance operations from 1962 to at least the early 1970s. Much of what he accomplished is still classified even after 50 years, and the CIA "can neither confirm nor deny the existence or nonexistence of records responsive to a [Freedom of Information Act] request concerning his activities. The fact of the existence or nonexistence of such records is itself currently and properly classified. . . ." Consequently, our Freedom of Information Act request for documents related to Aden's involvement in several projects and events was denied.[12] However, there is sufficient information in his autobiography and correspondence, as well as declassified documents, to allow a secure if limited glimpse into the scope of his participation in top-secret CIA and Air Force projects, even as he was working at the Steward Observatory, founding the College of Optical Sciences, and establishing the basis for solar energy applications with Marjorie.

Postwar catch-up

As mentioned in Chapter 3, General Henry "Hap" Arnold enlisted Caltech's Theodore von Kármán to organize other scientists in Washington, DC, to project needs for aerial warfare for the next 20–50 years. In March 1945, General Arnold asked Kármán (with William Bollay and Tsien Hsue-shen) to fly to Germany as part of Operation LUSTY. After his surrender just inside the Austrian border on May 2, 1945,[13] Wernher von Braun agreed to write a detailed paper that dealt with rockets traveling between Germany and the United States in 40 minutes and satellites circling the globe in one and a half hours (see Chapter 3). His report led the US Navy to propose on March 7, 1946, a joint Navy-Army venture to explore satellite surveillance. The Joint Army-Navy Aeronautical Board of Research and Development consulted Major General Curtis LeMay, director of research and development, who then assigned Project RAND (Research ANd Development) of the Douglas Aircraft Company to prepare a brief on the feasibility of satellite surveillance, which was delivered on May 12 as "Preliminary Design of an Experimental World-Circling Spaceship." RAND had been formed in 1945 as an Air Force think tank based in Santa Monica, California, to propose solutions to long-range problems.[14] Despite some interest and more elaborate designs a year later involving Schmidt telescopes and television equipment, the Navy shelved the idea because at that time there were no missiles to launch the satellites into orbit; in December 1947 future satellite studies were assigned to the US Air Force.[15] That same year, Congress passed the National Security Act, which established the National Security Council and Central Intelligence Agency (the "Agency"). The Air Force created its own Scientific Advisory Board in 1947, which accomplished relatively little other than balloon reconnaissance with the RAND Corporation (which had formally separated from Douglas Aircraft in 1948) as the disastrous Project GOPHER in 1950.[16]

Interest in satellites re-emerged in April 1951 when RAND presented a paper titled "The Utility of a Satellite Vehicle for Reconnaissance" to the Air Force. The Air Force then contracted with RAND to take the next steps into making satellites a reality as Project FEED BACK [sic]. At about the same time, Lt. Colonel Richard Leghorn was recalled to active duty from Eastman Kodak to head up the Reconnaissance Systems Branch at Wright-Patterson Air Force Base just northeast of Dayton, Ohio. He was later selected to serve as the Air Force Liaison for the Beacon Hill Study Group, commissioned to explore new approaches to aerial reconnaissance of the Soviet Union, thereby excluding intelligence-gathering by secret agents, decoding of radio messages, methods of detecting atomic explosions, and battlefield strategies. It was headquartered above a secretarial school on Beacon Hill in Boston, from which it took its code name. Among the Study Group's other notable members were physicist and electrical engineer Carl Overhage of Kodak; astronomer James Baker and physicist Edward Purcell of Harvard University; Edwin "Din" Land, founder of the Polaroid Corporation and with whom Aden would collaborate closely; and Richard Perkin, cofounder of the Perkin-Elmer Corporation, where Aden was later hired as a special consultant. For this study, Baker dealt with photographic surveillance, Purcell with electronic intelligence (e.g., radio and radar), and Land with photo reconnaissance. Their report, released on June 15, 1952, emphasized the immediate need to place cameras on rockets and supersonic planes flying at 70,000 feet to overfly the Soviet Union as long as the aircraft could avoid detection and interception. To fly at supersonic speed at such high altitudes and also house all the equipment for collecting intelligence, it was critical to limit the weight of the aircraft. With foresight, the Study Group also realized the critical need for trained personnel and some automation to interpret the many thousands of photos gathered by the cameras. Because there were no members of RAND on the Study Group, the report ruled out further balloon reconnaissance and did not envision satellites as an option for consideration until at least 1960.[17]

In January 1953, two months after the election of Dwight D. Eisenhower as president, the Beacon Hill Group had become discouraged by the lack of progress and persuaded the Air Force Scientific Advisory Board that a new panel on intelligence should be formed. The first meeting of the Air Force's new Intelligence Systems Panel on August 3, 1953, was attended by members Overhage, Land, Baker, Allen Donovan of Cornell Aeronautical Laboratory, Stewart Miller of Bell Telephone, Duncan McDonald of Boston University, and the CIA's Edward Allen of the Office of Research and Reports (ORR) and Phillip Strong of the Office of Scientific Intelligence (OSI). Baker then toured facilities producing a variety of reconnaissance platforms and realized the opportunities and advantages afforded by satellites.[18]

Several events shook the Air Force out of this perceived complacency. One was the first Soviet test of a hydrogen bomb on August 12, 1953. Another was the observation of a Myasishchev-4 intercontinental bomber ("Bison") at an airfield south of Moscow by a military attaché that same year.[19] A third was Eisenhower's invitation to Trevor Gardner, the impulsive founder of a reconnaissance camera company in Pasadena named Hycon, to become special assistant to Harold Talbott, Secretary of the Air Force. Gardner shook up the Air Force brass and eased requirements for producing Atlas ICBMs to counter the Soviet threat. He organized the Strategic Missiles Evaluation Committee, headed by computer expert John von Neumann and including Hendrik Bode of Bell Telephone, George Kistiakowsky of Harvard (later Eisenhower's science advisor), and Clark Millikan of Caltech. They first convened on November 9, 1953, and urged haste in ICBM development.[20] A critical member of the team joined the CIA in 1953—Arthur Lundahl. Lundahl (1915–1992) had trained photointerpreters during World War II and remained with the Navy until the Agency recruited him in 1953 to direct the Photographic Intelligence Division (PID) and then the National Photographic Interpretation Center (NPIC) beginning in 1961. He was to remain the leader in photointerpretation for two decades in the top four floors of the Steuart building at Fifth and K Streets, NW, in Washington, DC, code-named HT/AUTOMAT, which opened in January 1956. Lundahl solicited input from specialists in fields as diverse as photogrammetry, graphic arts, and data processing to analyze reconnaissance photos, which would ultimately de-bunk the "bomber gap" and "missile gap" as Soviet propaganda. From the outset, Eisenhower relied on "his" scientists for impartial advice, regardless of whether it complemented or contradicted recommendations from the military.[21]

Edwin Land (1909–1991) numbered among those favored scientists, not just for Eisenhower but also for succeeding presidents. He acquired the nick-name "Din" when his sister Helen, older by five years, could not pronounce his name.[22] Although his chief fame lies in his invention of instant photography and founding the Polaroid Corporation, he spearheaded projects such as U-2 op-tics, surveillance satellites, and the Manned Orbiting Laboratory (MOL). Land owned 535 patents in all, including his system of instant cameras and polarizing filters. Like Aden, Land was a visionary and a polymath, who viewed optics as a "mother science" because it led to ideas in other areas of physics.[23] Land's phi-losophy was that discoveries are made by those who have developed an uncon-ventional way of approaching concepts and methods and "mastered the art of a fresh, clean look at the old, old knowledge."[24] Aden had this same multidiscipli-nary philosophy as Land and Lundahl in founding OSC and College of Optical Sciences. By the end of his career, Land had received 16 medals, including the National Medal of Science, Presidential Medal of Freedom, Frederic Ives Medal, and National Medal of Technology. He had been elected as president of the

American Academy of Arts and Sciences and was inducted into the National Inventors Hall of Fame and the United Kingdom's Royal Society.

Trevor Gardner read a report by the RAND Corporation that a surprise attack by the Soviets would wipe out 85 percent of the Strategic Air Command bombers, and so he approached Lee DuBridge, who was president of Caltech after Robert Millikan and also chairman of the Scientific Advisory Committee of the Office of Defense Mobilization (ODM). On the Committee's recommendation, Eisenhower asked MIT president James Killian on July 26, 1954, to recruit experts to determine how technology could address the nation's current problems. The Technological Capabilities Panel (TCP), as it became known, had separate committees to deal with three projects: Continental Defense, Striking Power, and Intelligence Capabilities. The chairman of the Intelligence project was Land, joined by James Baker, Edward Purcell, Joseph Kennedy of Washington University, John Tukey of Princeton University, Allan Latham, Jr. (formerly of Polaroid), and Allen Donovan. About two weeks after Eisenhower's directive, Land and Baker flew to Washington and interviewed generals, admirals, and the CIA, only to have their worst fears confirmed—intelligence resources were inadequate and needed immediate innovations by scientists. While there, however, Land was approached with Kelly Johnson's drawing of the Lockheed CL-282, a high-altitude plane that would ultimately evolve into the U-2.[25]

Clarence "Kelly" Johnson (1910–1990) was one of the most gifted aeronautical engineers since the Wright brothers and their mechanic/inventor Charles Taylor; at the same time, he was also one of the most fearsome and respected forces of nature in aeronautics. According to his chief engineer, Ben Rich, who later succeeded him as director of the Skunk Works, Johnson resembled W. C. Fields—without the sense of humor.[26] He joined Lockheed at the Skunk Works in Burbank, California, in 1933. At that time their secret, makeshift facility was adjacent to a plastics factory that produced a horrible stench, which was likened to the "Skonk Works" in the Li'l Abner comic strip by Al Capp. After one of the engineers answered the phone with "Skonk Works," the name stuck, although it had to be changed to Skunk Works by copyright law. With a crew of engineers and shop mechanics, Johnson designed and built the P-38 Lightning, P-80 Shooting Star, and F-104 Starfighter fighter planes during the war and later designed the L-188 Electra and Constellation civil airliners there.[27]

In July 1953, the Air Force initiated Project BALD EAGLE to develop a high-flying, reconnaissance aircraft and invited proposals from Bell Aircraft Corporation, Fairchild Engine and Airplane Corporation, and Glenn L. Martin Company. Undeterred by its exclusion, Lockheed submitted unsolicited plans in April 1954 for its single-engine CL-282 flying at 73,000 feet with a range of 1,400 nautical miles on glider-like wings and landing on skids instead of wheels. This concept for a jet-propelled glider was turned down by the Air Force when General

Curtis LeMay, commander of the Strategic Air Command, refused to countenance "a plane with no wheels and no guns." However, Hycon founder Trevor Gardner, who was in the audience at the presentation, was intrigued enough to meet with Philip Strong, chief of operations in the OSI of the CIA. Strong was the person who showed Land the drawing of the CL-282. Land discussed it with the other Project Three members and invited camera specialists Henry Yutzy of Eastman Kodak and Richard Perkin (president of Perkin-Elmer) to optimize the platform in every way. Plans were complete by October 1954. In a letter with memo dated November 5, 1954, Land and his committee recommended to Allen Dulles that a civilian organization, the CIA, should be responsible for unarmed, unmarked overflights of the Soviet Union; otherwise, if the military had oversight, a war might ensue. Eisenhower strongly supported this reasoning for the aircraft. The plane would fly at 70,000 feet to minimize radar detection by the Soviets, with a speed of 500 knots and a range of 3,000 nautical miles. The lone pilot would be in a heated, pressurized suit. In one flight it could photograph territory 200 miles wide and 2,500 miles long. The next month Eisenhower met with Land and Killian, found their arguments persuasive, and signed off on Project AQUATONE on November 26. He placed the Director of Central Intelligence (DCI), Allen Dulles, in charge, with the stipulation that the Air Force would assist as needed. Dulles delegated responsibility for the top-secret project to Richard Bissell, Jr., who had been hired in early 1954 as the special assistant for planning and coordination for the DCI and then CIA Deputy Director for Plans (DDP). He would continue to supervise all aerial reconnaissance until 1962.[28]

Richard Bissell (1909–1994) received his doctorate in economics from Yale University in 1939, then worked as an economist in the Department of Commerce and elsewhere before being hired as the economic advisor to James F. Byrnes, Director of War Mobilization and Reconversion in 1945–1946. After the war, he was professor of economics at MIT until 1952 and chief administrator of the Marshall Plan (European Recovery Program) to rebuild Western European economies. He worked closely with Kelly Johnson on AQUATONE and also spearheaded the CORONA programs before becoming co-director (with Joseph Charyk) of the National Reconnaissance Office (NRO).[29]

The CL-282 was called the "Angel" by Lockheed and designated the U-2, the U for Utility, and -2 because there was already a U-1, the De Havilland Otter utility transport. It was almost 50 feet long with a wingspan of 80 feet and powered by a Pratt & Whitney (P&W) J57 turbojet engine. Production of 20 planes for less than $1 million apiece got underway at the Skunk Works on December 27, 1954. Baker and Perkin adapted a Hycon K-38 camera and lenses as the A-1 camera system with two, 24-inch framing cameras, one mounted vertically to cover a spread of 17.2 degrees below the plane and the other in a rocking mount (designed by Aden's friend, Rod Scott, chief project engineer of Perkin-Elmer and later vice president) so that it could turn obliquely left and then obliquely right to 36.5 degrees. In addition, there was a Perkin-Elmer tracking camera with a 3-inch lens for photographing areas underneath the plane from horizon to horizon. Baker later made improvements for the A-2 camera system using 24-inch f/8.0 lenses with several aspheric surfaces so that they could resolve 60 lines per millimeter. In the search for even better resolution, Baker, Scott, and Hycon's William McFadden continued their research and produced the B camera system, which used one 36-inch lens instead of two 24-inch lenses and increased the oblique coverage to 73.5 degrees and a vertical spread of 21.5 degrees, achieving a resolution of 100 lines per millimeter and spanning 40–50 miles on each side. From an altitude of 60,000 feet, the lens could resolve an object 2.5 feet in diameter. A C camera system with a 180-inch f/13.85 lens was designed by Baker and built by Hycon; it had higher resolution but less coverage and was also sensitive to vibrations. It was the B camera that was primarily used in the U-2.[30] While Aden was still working at the Phoenix headquarters of the KPNO project, occasionally Rod Scott would stop by to talk telescopes. One night over dinner, Aden asked him what took him to Los Angeles so often. Scott replied cryptically that he could not tell them, but if it was successful they would never hear about it, and if it failed they would certainly hear about it. The whole world heard about it when U-2 pilot ("driver") Francis Gary Powers was shot down over Sverdlovsk, USSR, on May 1, 1960.[31]

After the Soviets detonated the first atomic bomb on August 29, 1949, fear of a nuclear surprise attack on the United States galvanized Congress to pass the Federal Civil Defense Act on January 12, 1951. Soon public school students were diving under their desks or out into the hallways to "duck and cover" in surprise drills. Fear of a nuclear attack of magnitudes greater than Hiroshima and Nagasaki gripped the nation. Films such as The Day the Earth Stood Still (1951) exploited that fear, warning of what would happen to planet Earth without worldwide disarmament. On April 3, 1954, eight months after the Soviets tested their first hydrogen bomb, the New York Times published a map showing the range of projected damage if a hydrogen bomb were dropped on lower Manhattan. Everything within an 8-mile diameter would be vaporized, with severe damage 16 miles out, moderate damage 28 miles out, partial damage at 32 miles, and little or no immediate damage 50 miles away (except for radioactive fallout). All five boroughs of New York City and much of New Jersey and Long Island would be devastated.[32] By 1955, half of America believed that they would die in a thermonuclear war instead of diseases associated with old age.[33] Public and private fallout shelters were designated or constructed throughout not only the United States but also Europe and the Soviet Union. In an effort to calm fears and promote transparency, Eisenhower announced his Open Skies proposal at the Geneva summit on July 21, 1955, whereby the United States and the Soviet Union would allow reciprocal aerial photography of airfields in supervised overflights. Nikita Khrushchev, who at that time was secretary of the Communist Party and de facto premier, rejected it out of hand, condemning it as a ploy to locate strategic targets in the Soviet Union.[34] As Eisenhower would later learn, however, the Soviets simply did not want the world to know that they did not have the ICBM inventory that they proclaimed.

Turf battles between the Air Force and the Agency soon arose because of the original, vague assignment of the responsibilities for the U-2 and CORONA. It was a feud that lasted well into the 1960s and beyond, over control of not only spy planes but also satellite reconnaissance in particular. Aden later became embroiled in one jurisdictional dispute and was forced to choose which organization to support. In this case, the animated discussions between Air Force Chief of Staff Nathan Twining and DCI Allen Dulles ended amicably, but only after Eisenhower intervened to reiterate that he wanted AQUATONE to be a totally civilian (i.e., CIA) operation. On August 2, 1955, Twining and Dulles jointly signed an agreement named Project OILSTONE that made the Air Force responsible for providing pilots (drivers) and training them, as well as for weather data, mission plotting, and operational support. The CIA was put in charge of cameras, security, contracting, film processing, foreign bases, and also allowed input into pilot selection.[35]

The Agency recruited only the most experienced pilots from the Air Force, those with at least 1,500 hours of flying time (900 of those consisting of first pilot/instructor time), superlative records, and recommendations from the wing commander. They were subjected to lie detector tests, a battery of physical and psychological tests, background checks and security clearances, and interviews that could last up to two weeks; no more than half of those considered were selected. They would be paid $1,500/month while training in the United States and $2,500/month overseas for 18 months, more than captains of commercial airlines received. Although they would need to resign their Air Force commissions and become civilians to work for the Agency, they could be reinstated in the Air Force without loss of time or grade. The reassignment became known as "sheep-dipping," referring to the process of disinfecting sheep of parasites and cleaning the wool before shearing.[36]

The first U-2 was ready for testing just nine months after approval on July 25, 1955, at Groom Lake next to the Nevada Proving Ground (later the Nevada Test Site and currently the Nevada National Security Site), which was divided into numbered grids. The grid with Groom Lake, a hardpan lake bed, was in Area 51, then commonly referred to as Watertown, Paradise Ranch, or simply the Ranch. The U-2 was disassembled, shipped to Groom Lake in a C-124 cargo plane, reassembled, and had its first official flight on August 8, 1955, piloted by Tony LeVier.[37] After months of test overflights across the United States, the

CIA showed Eisenhower photos of his Gettysburg farm and cities such as San Diego from 70,000 feet; automobiles and lines in parking lot spaces were clearly visible.[38] In light of these detailed photos and also Khrushchev's repudiation of Open Skies, Eisenhower authorized the first U-2 flight across the Soviet Union on June 21, 1956. On July 4, pilot Hervey Stockman flew from Wiesbaden over Belorussia to Leningrad's shipyards and over military airfields to assess the fleet of Bison bombers before returning to Germany over the Baltic states.[39] The next day, pilot Carmine Vito again flew over sites where Bison bombers were being built and tested and this time directly over Moscow itself. The Soviets tracked those flights, as well as several others, by radar until July 10 and protested the incursions as violations of international law through diplomatic channels.[40]

Two major consequences emerged from these first overflights. One was the discovery that there was indeed a "bomber gap," but one in favor of the United States. The Soviets had none of the 100 Bison bombers that the Air Force had estimated, and therefore the Air Force was denied the additional B-52 bombers that it had requisitioned for parity. The second was recognition of the need to reduce the radar cross section of the U-2 by applying radar-absorbing materials or using other methods. The Scientific Engineering Institute in Cambridge and also Edgerton, Germeshausen & Grier (EG&G) undertook research to make the U-2 more invisible beginning in November 1956 under the code name Project RAINBOW. The first attempts involved gluing radar-absorbing material to the fuselage, nose, and tail ("Wallpaper") and also attaching wires with ferrite beads to introduce interference reflections ("Trapeze"). Kelly Johnson hated the contraptions attached to "his" planes and called them "dirty birds." Trapeze introduced enough weight and drag to lower the operating ceiling by 1,500 feet. Worse, Wallpaper caused the engine to overheat, and in April 1957 one of the U-2s (Article 341) overheated and flamed out. The pilot, Robert Sieker, ejected, only to be struck and killed midair by the tailplane (horizontal stabilizer). Nine "dirty bird" missions were flown before the project was canceled in May 1958. One in particular, on August 5, 1957, identified a missile installment in Kazakhstan that came to be known as Tyuratam. A few weeks later, the Soviet news agency TASS announced the successful launch of an ICBM and their ability to send another anywhere in the world.[41] Reducing the radar cross section in other ways was one of main goals of those working on the successor to the U-2 (including Aden), the Lockheed A-12 (Project OXCART).

Overflights of the Soviet Union continued off and on into 1960 as Eisenhower weighed the value of the intelligence gathered against the political and military consequences, knowing that the Soviets were following every flight and concerned that they would close access to Berlin in retaliation—or worse. Despite the risk, he wanted data on the status of Soviet missiles to determine whether

Khrushchev's claims of missile superiority were facts or propaganda. Eisenhower examined all proposed flight plans, even altering some, before approving them.[42]

Then Sputnik 1 hit the headlines on October 4, 1957, and Sputnik 2 a month later, which along with the ICBM launch and Soviet propaganda compounded the fear that Americans already felt from years earlier. On October 15, Eisenhower convened the Scientific Advisory Committee of the ODM, chaired by Columbia physicist Isidor Rabi with Land and Killian present. Rabi won the Nobel Prize in Physics in 1944 for discovering the magnetic properties of atomic nuclei and served as a consultant in the Manhattan Project.[43] From that hour-long meeting resulted the creation of a new post, the Special Assistant to the President for Science and Technology, and the President's Science Advisory Committee (PSAC) that reported directly to the president in an urgent effort to introduce scientific reform into the Defense Department and NATO and more science education in schools and to reassure the American people. Eisenhower offered the advisor post to Killian, who accepted. Killian was not a scientist by training but a journalist and an editor, who was promoted to president of MIT in 1949 largely because of his administrative skills, applied to the wartime efforts of the university.[44] Killian's job description was detailed in a confidential letter from Eisenhower in December 1957. In part he was to advise on scientific and technological matters to the president, the Cabinet, the director of the ODM, the Special Assistant to the President for National Security. He would have "full access to all plans, programs, and activities involving science and technology in the Government, including the Department of Defense, Atomic Energy Commission, and CIA."[45] Eisenhower and PSAC members developed a mutual admiration and respect over time. From a bed in Walter Reed Hospital a few months before his death in 1969, Eisenhower told Killian, "This bunch of scientists was one of the few groups that I encountered in Washington who seemed to be there to help the country and not help themselves."[46]

In Eisenhower's decisive State of the Union address on January 9, 1958, he acknowledged the threat of communist imperialism and announced the production of Thor and Jupiter ballistic missiles and submarine-launched Polaris missiles. He pledged to end the bureaucracy and "jurisdictional disputes" among the branches of the armed forces and advocated more scientific and economic cooperation with America's allies. The 1959 budget, he announced, included increased funding for missiles, nuclear submarines, atomic energy, military pay increases and incentives, and a one-billion-dollar investment in science education and research over four years, including doubling the appropriations for the National Science Foundation. Four months later, Eisenhower saw the need for a civilian agency devoted to space exploration and, on July 29, 1958, signed into law the National Aeronautics and Space Act, creating the National Aeronautics and Space Administration (NASA).

A month later, on February 7, 1958, Killian and Land brought to Eisenhower a proposal for a reconnaissance satellite with photographic equipment in a recoverable capsule. The satellite would circle the earth three times and eject the capsule with film for retrieval by the Navy within an area of 10 square miles. It was cast as a joint project between the Navy and the Air Force but to be overseen by the CIA.[47] Submitted for Eisenhower's approval on April 16, 1958, it would come to be known as CORONA and would spark Aden's progressively deeper involvement within the Agency and the Air Force, beginning about 1960 after he had resigned as director of KPNO.

The origins and evolution of CORONA

As mentioned earlier, the Air Force had contracted with Project RAND to study the feasibility of satellite reconnaissance. RAND submitted its two-volume report (R-262), titled Project FEED BACK [sic], on March 1, 1954, edited by J. E. Lipp and Robert Salter. Among the consultants were astronomers Fred Whipple of the Smithsonian Astrophysical Laboratory (with whom Aden would later collaborate on the Multiple Mirror Telescope) and Lyman Spitzer, Jr., of Princeton University. RAND engineers had been working on a satellite that could orbit at 300 nautical miles altitude and send back photos of strategic sites for television by subcontracting with the Radio Corporation of America (RCA). Using television would obviate the need for problematic recovery of film from the vehicle, which, as described in volume two of the report, would house an auxiliary powerplant unit, ascent and orbital guidance, television camera system, data storage and communications system, and climatization.[48]

The Air Force used the RAND study to invite industry to submit proposals for a reconnaissance satellite that was stable on three axes with high aiming accuracy for its cameras and could communicate with ground controllers. Three bids were received: RCA, Martin, and Lockheed Missile Systems Division in Sunnyvale, California. Lockheed proposed a system like the one RAND detailed, with the exposed film processed on board, scanned electronically, and the signal transmitted to ground stations as the satellite passed over them. It also proposed one in which capsules with exposed film would return to Earth for retrieval. In late October 1956, Lockheed was awarded a contract for both the nonrecoverable and recoverable programs as Weapons System 117-L, or simply WS-117L. RAND was also developing its own recoverable satellite and submitted it to the Air Force in March 1956 but soon withdrew it from consideration, apparently preferring the real-time reconnaissance by television for more timely warning of a surprise attack. However, the concept of a nonrecoverable satellite sending back images in real time would have to wait until the invention of the charged-coupled device that was used in KENNEN (CRYSTAL), as discussed later in the chapter.[49]

As the next several months passed, support for a satellite program (including Vanguard) lagged for lack of focus, administration support, and government funding. On November 12, 1957, nine days after the launch of Sputnik 2, Merton Davies and Amrom Katz of RAND issued a study of a recoverable reconnaissance capsule system titled "A Family of Recoverable Satellites," which was based on a June 1956 RAND Research Memorandum by John Huntzicker and Hans Lieske titled "Physical Recovery of Satellite Payloads: A Preliminary Investigation." The Davies and Katz report recommended three systems in a spin-stabilized satellite to the Air Force: (1) a 12-inch camera with 500 feet of 5-inch-wide film and 60-foot ground resolution; (2) a 36-inch camera with a resolution of 20 feet; and (3) a 120-inch camera with even higher-quality photos. The entire payload or just the film would be recoverable.[50] Certainly precipitated by the hysteria and bad press over the Sputnik satellites and the so-called missile gap, Eisenhower met with Neil McElroy (Secretary of Defense), Donald Quarles (Undersecretary of Defense), Allen Dulles, Killian, and Land on February 6, 1958, to discuss the Air Force proposal for a Thor-boosted satellite with recoverable film. The next day, Land and Killian met with Eisenhower again to explain that the satellite would orbit the earth three times, photographing the Sino-Soviet bloc, and then deorbit for retrieval of the film capsule. Eisenhower placed the program under the Advanced Projects Research Agency (ARPA), which had been approved that same day under the Office of the Secretary of Defense to centralize and oversee all space projects (including missile defense), and deferred to Land to develop the program further. Land then went to Bissell's office and informed him that he was to separate the film return system (Lockheed's Program II-A) from WS-117L and direct it like the U-2 program, with security and reconnaissance aspects managed by the CIA and launch, communications, and vehicle recovery by the Air Force.[51] Bissell knew little about satellites and so convened his U-2 colleague Air Force Brigadier General Osmond "Ozzie" Ritland and a few contractors. All they had to go on were the RAND studies, a few drawings, an untested Thor booster, and only a conceptual upper stage. In a CIA meeting led by Bissell later, someone asked what the project should be called. Another looked down at his typewriter and reportedly said, "Let's call it CORONA." A different account credits the branded paper ring around a cigar for the name.[52]

News of the satellite recovery Program II-A had leaked, and so ARPA publicly terminated it on February 28, 1958, without warning, much to the dismay of the Air Force. However, the WS-117L program for developing the film on board the satellite, scanning the image, and then sending it by radio to ground stations remained under the name SENTRY and then became SAMOS (Satellite and Missile Observation System). The CIA invited Lockheed, Fairchild Camera, General Electric, and the Air Force Ballistic Missile Division into the design phase of the top-secret CORONA program. A cover story for CORONA released to the press was that a new series of satellite flights with a polar orbit and re-covery named DISCOVERY/DISCOVERER would test hardware, the impact of environmental conditions (e.g., radiation) on humans in flight, early-warning devices, tracking stations, and procedures for recovery of a satellite payload after flight.[53] Eisenhower approved CORONA in April 1958.[54]

Bissell had decided to use the Fairchild camera system with a lens focal length of 6 inches and a ground resolution of 50–100 feet; however, the camera in the spiraling satellite would activate only when the camera window was pointed toward the earth. Nevertheless, Fairchild had significant support from the Air Force, RAND, and the CIA. Then Bissell received an unsolicited bid from the Itek Corporation, founded in 1957 by Richard Leghorn (with substantial start-up capital from Laurance Rockefeller) as a manufacturer of reconnaissance cameras. Itek promised a spin-stabilized, panoramic camera with an improved focal length of 24 inches and ground resolution as little as 20 feet; horizon sensors and gas jets would keep the camera continuously pointed downward to Earth, a much more complicated and expensive approach than Fairchild's. The Itek pro-posal was initially accepted only as a backup to the Fairchild system; initially, the first cameras designated C and C' (C prime) were designed by Itek and assem-bled by Fairchild. After the first few missions, Itek both designed and manufac-tured improved cameras for CORONA.[55]

To launch the satellites into a near-polar Earth orbit, the first stage was a Thor rocket built by Douglas Aircraft Company that would burn out at an altitude of 70 nautical miles. Then an Agena upper-stage built by Lockheed would take over to supply the additional velocity to go into orbit; once there, Agena would also control altitude, supply battery power, and shield the film capsule from overheating during re-entry. As the mission was ending, the Agena would go into a downward attitude, and the payload capsule would separate from it fol-lowing a microwave command signal to descend within a designated wide area near Hawaii. After the capsule had fired its retrorocket to reduce its velocity to 1,300 feet per second and had reached an altitude of 50,000 feet, a parachute would deploy and cut the descent to 30 feet per second. Air Force C-119s out of Hawaii would then catch it with a long loop in mid-air; failing that, ships and helicopters would fish it out of the sea.[56]

That was the plan, anyway. The launch of DISCOVERER 1 (CORONA 1) out of Vandenberg Air Force Base on February 28, 1959, failed to orbit and disappeared forever. Successive launches were also failures—insufficient fuel to achieve orbit, no separation or retrieval of the recovery vehicle, malfunction of the camera, guidance, stabilization system, or retrorocket, etc.

Figure 11.1. Capsule of DISCOVERER XIII, the first successful CORONA/
DISCOVERER mission, presented to President Eisenhower at the White House on
August 15, 1960. Left to right: Secretary of the Air Force Dudley Sharp, Secretary of
Defense Thomas Gates, Eisenhower, Air Force Chief of Staff General Thomas White,
Col. Charles G. "Moose" Mathison.
Courtesy of the National Park Service and Dwight D. Eisenhower Presidential Library.

Finally, on August 10, 1960, after 12 consecutive failures, DISCOVERER 13
(CORONA 13) entered orbit. After 17 orbits, the capsule ("bucket") was successfully recovered at sea some 300 miles from the intended location and transported
to Andrews Air Force Base for a photo session with Eisenhower on August 15
(Figure 11.1). Although it carried no film, CORONA 13 showed that it was possible to send an object into space and recover it at sea, which would prove invaluable for manned and unmanned NASA missions. DISCOVERER 14 (CORONA
14), launched on August 18, was even more successful. It was the first mid-air recovery of the bucket in a designated area southwest of Honolulu (Figure 11.2). The
20-pound payload of exposed film—3,000 feet of it—covered 1,650,000 square
miles of the Soviet Union. Resolution was not as high as that in U-2 photos, only 35
feet, but the images revealed sites and installations previously unknown.[57]
Less than a week later, with the U-2 flights suspended and its successor the
Lockheed A-12 still in development, Eisenhower approved security for the

Figure 11.2. U.S. Air Force C-119J recovers a CORONA capsule returned from space on August 19, 1960. This aircraft made the world's first midair recovery of a capsule returning from orbit when it snagged the parachute lowering the DISCOVERER XIV satellite at 8,000 feet altitude and 360 miles southwest of Honolulu, Hawaii.

Public domain photograph: U.S. Air Force.

satellite program under the TALENT-KEYHOLE system. The TALENT Security System was intended to limit the number of personnel with knowledge of over-flight photography by aircraft (the U-2) in the 1950s, and TALENT-KEYHOLE was now extended to satellites. As of August 24, 1960, there were 1,370 certifications, mainly among the CIA (including the Photographic Intelligence Center), Army, and Air Force.[58] Satellites were designated KH followed by a number. KH-1 through KH-4, KH-4A, and KH-4B were CORONA programs, succeeded by KH-5 (ARGON), KH-6 (LANYARD), KH-7 (GAMBIT-1), KH-8 (GAMBIT-3), KH-9 (HEXAGON), KH-10 (DORIAN, the Manned Orbiting Laboratory), and KH-11 (KENNEN, later known as CRYSTAL).

Through the CORONA series (including ARGON and LANYARD), cameras were continually being improved. The C camera used in KH-1 was replaced by the Itek/Fairchild C' in CORONA/DISCOVERER 16 on October 26, 1960, as KH-2, which continued for just under a year. The C' added variable image motion compensation allowing for different orbits and had a resolution of 25 feet. In February 1961 the KH-5 (ARGON) was launched under the CORONA program as an independent Army mapping project with a 3-inch focal length lens and a disappointing resolution of about 460 feet; half of the 12 ARGON missions were failures (not all of them, as reported by Day, 1998).[59] Itek alone produced the C''' (C triple prime) camera with a rotating Petzval f/3.5 lens for KH-3, which first launched on August 30, 1961. The C''' improved on prior iterations in several ways: (1) the larger aperture lens allowed more light and thus films with an ASA as low as 2; (2) resolution was maximized at 180–200 lines per millimeter, down to 12 feet; (3) the film payload was 39 pounds, almost twice that of KH-1; and (4) maintaining focus was more reliable.[60]

There were three versions of KH-4 for the M (MURAL) cameras and KH-4A and KH-4B for the J-camera systems. The M or MURAL camera system comprised two Itek C''' cameras overlapping from front to back, creating stereoscopic imagery; when the two film strips were aligned, the effect was three-dimensional. First launched as DISCOVERER 38 on February 27, 1962, it was also the last DISCOVERER but not the last CORONA, which would continue until 1972. There would be 15 more KH-4 launches in 1962. The KH-4 was called the "washing machine" because the camera rotated continuously; 30 percent of the cycle was spent moving over a long strip of film, and the remaining time moving the film to ready for the next passage of the camera arm. The "two-bucket system" of KH-4A with the J-1 camera (J for Janus, the Roman god with two faces) was introduced as the 58th launch of the CORONA series on August 24, 1963, and returned an overwhelming amount of data for the photointerpreters. Whereas KH-1 through KH-4 produced 285,472 feet of film, KH-4A produced 1,293,025 feet over 52 missions up through September 1969. The final version of KH-4, the KH-4B, comprised two panoramic cameras constantly rotating through 360 degrees, two horizon cameras, and a Dual Improved Stellar Index Camera (DISIC). It not only combined the stereoscopic capabilities of the MURAL camera system and the two-bucket system of the J-1, but also increased the flexibility of control of camera operations. Resolution was now down to 5–7 feet, and flight ceilings could also drop down to 80 nautical miles. On CORONA flight 138 over Moscow, a line of tourists waiting to enter Lenin's tomb in Red Square was clearly visible from orbit. There were 16 KH-4B missions between September 15, 1967, and May, 25, 1972.[61]

The other CORONA program was the short-lived KH-6, or LANYARD. In 1962, Secretary of Defense Robert McNamara saw the need for satellites to investigate KH-4 photos of suspected ABM installations in Tallinn, Estonia, but with higher resolution. The program known as GAMBIT-1 (KH-7) was not fully developed by that time, and so the Agency's Office of Special Activities and the Air Force agreed to retrieve SAMOS E-5 cameras from storage and transport them to the Lockheed Advanced Projects Facility. With the camera's 66-inch focal length lens, they hoped for a resolution of two feet. Three launches that began on March 18, 1963, and ended on July 31 that same year were all failures, either because the Agena-D vehicle never reached orbit, the camera was never activated, or it failed.[62]

Over the course of 145 CORONA missions through 1972, not only did they pioneer satellite intelligence, recovery of objects from orbit, and stereoscopic photography, they also imaged all 25 Soviet ICBM facilities and about 750 million square nautical miles of Earth's surface. For the first time, it was possible to photograph any area on Earth. In doing so, they helped to dispel the myth of Soviet superiority in ICBMs and therefore prevented unwarranted expansion of US defense budgets by billions of dollars. They facilitated the resolution of the Cuban Missile Crisis of 1962 and negotiation in the Strategic Arms Limitation Talks (SALT I) a decade later by virtue of the ability to verify compliance with agreements between the United States and the Soviet Union. CORONA also played a major role in monitoring the events of the Middle East war in 1967 and the progress of Chinese nuclear programs, as well as increasing our knowledge of geography, weather patterns, natural resources, and natural and technological disasters worldwide.[63]

Aden worked on the satellite imaging systems for CORONA KH-4 through KH-6, as well as GAMBIT-1 (KH-7), GAMBIT-3 (KH-8), and "a bit" up to KH-10 and KH-11, although details are still classified.[64] While he was consulting on CORONA he was also involved in Projects FULCRUM, OXCART, and others, as discussed elsewhere in this chapter. He was a regular visitor to both the Pentagon and Agency headquarters. How an astronomer came to be respected and trusted over four presidential administrations during the Cold War—and even plucked from among horses in a remote Colorado pasture by a helicopter and flown to Washington, DC, during the Cuban Missile Crisis—began with an invitation to attend a top-secret meeting in Santa Monica, California.

Down the rabbit hole

Following Aden's resignation as director of Kitt Peak National Observatory in 1960, word leaked out almost immediately within the astronomy community, which at that time was such a tightly knit group that job applications were often unnecessary. The director at an institution or observatory would call or correspond with department chairs at universities and ask about new students. Because astronomy students graduated at the rate of only four or five each year in that era, there were negotiations behind the scenes.[65] Among the requests Aden received was one from the University of Michigan to visit Maui and discuss progress on what was then known as the ARPA Midcourse Optical Station (AMOS) on the summit of Haleakalā. When founded in 1961, its mission there was research and development in the space sciences and also measurement and imaging of re-entry vehicles and satellites (both US and Soviet) in the infrared and optical spectra. Aden accepted the invitation and helped to solve problems with vibrations in the first domes. The Itek Corporation developed optics that could correct for atmospheric blurring there.[66]

Not long after that, he was called for a top-secret meeting with scientists, engineers, and Air Force officers at the RAND Corporation in Santa Monica, California. They knew that as an astronomer, observatory site surveyor, and director of KPNO he had studied the effect of looking up at stars through the atmosphere. Now they wanted to know what resolution could expected through a camera lens looking *down* through the atmosphere, that is, how small an object could be resolved when imaged from orbit.[67] To that point most scientists and engineers assumed that the angular resolution would be the same for both—no better than 1 arcsecond—because of atmospheric turbulence. However, looking up into the phase screen (turbulence) of the atmosphere is in the near-field of a telescope, whereas looking down through the atmosphere is in the far field near the object. For example, if you view an object held some distance away from a pane of ground glass or a glass block used in homes for privacy, it will be a blur. But if you place the object directly in contact with the glass (here the functional phase screen), details will be more easily discernible.

Recall that the resolution achieved with DISCOVERER 14 (CORONA 14) in August 1960 was 35 feet; the J-3 cameras (KH-4B) brought it down to 5 feet by 1967. Aden showed on a blackboard his "back-of–an-envelope" equation developed during the Kitt Peak site survey. Imagine the shock wave in the room when he showed that the theoretical limit to angular resolution from orbit would be a *fraction* of one arcsecond, in this case, about two inches.[68] At that moment, Aden's double life began in earnest at the top-secret level of clearance.

After John F. Kennedy was inaugurated as president, Aden was invited to a classified meeting sponsored by the National Academy of Sciences at Woods Hole, Massachusetts. The site was the turn-of-the-century Marshall Mansion on Quissett Harbor, which had been used for meetings not just by the National Academy of Sciences, but also by the Department of Defense and the Agency. Marjorie did not have clearance, so she walked on to see the Nobska Point Lighthouse. The topic of the meeting was whether or not lasers could be concentrated and aimed with a telescope to destroy a bomber or missile. Blackboards were filled with equations and calculations. Demonstrations outdoors showed that a laser could indeed track and strike a drone aircraft, but it could clearly not destroy an incoming bomber or missile. This was the harbinger of the Strategic Defense Initiative ("Star Wars") program under Ronald Reagan 20 years later, to which Aden and Marjorie returned when they were employed at JPL.[69]

To test the two-inch resolution claim, Aden met at the Itek Corporation with a young Colonel Lew Allen, other Air Force officers, and engineers. They decided to launch a large balloon carrying a 36-inch camera from an Air Force base near Phoenix on a day with calm upper atmosphere winds. It would remain aloft over an industrial center, taking photos for two days and one night, a secret to everyone but the Federal Aviation Administration (FAA). However, it was visible at twilight and came to be known in the newspapers as the "UFO over Phoenix." The astounding results confirmed Aden's prediction of two-inch resolution, which helped to advance Lew Allen's eventual promotion to the rank of general.[70]

One day in 1962, Aden received a telephone call from James Eyer, then at the Institute of Optics at the University of Rochester and also involved with the Orbiting Astronomical Observatory (OAO) project of NASA. He also served on an advisory panel to the Agency under the new DCI, John McCone (1902–1991), who was appointed in November 1961 by President Kennedy to replace Allen Dulles in that position after the disastrous Bay of Pigs invasion. McCone was trained as an engineer and had served as Undersecretary of the Air Force (1950–1951) and chairman of the Atomic Energy Commission (1958–1960). Eyer asked Aden to meet him at the Key Bridge Hotel across the Potomac River in Rosslyn, from where they would be picked up and taken to an undisclosed location for a long briefing. They were driven through the woods upriver to a newly completed, large, white building complex in Langley, Virginia. Inside the lobby was an impressive seal on the floor that read "United States Central Intelligence Agency." They were cleared before passing through several high-security doors to be greeted by Bud Wheelon, the Deputy Director of Science and Technology (DDS&T).[71]

Albert "Bud" Wheelon (1929–2013) graduated from Stanford University with a bachelor's degree in engineering, then took his doctorate in physics from MIT. In the next few years he helped to develop guided missiles at Douglas Aircraft and then the first ICBM at Ramo-Woodridge (later Thompson-Ramo-Woodridge, or simply TRW). His contributions as director of the Radio Physics Laboratory at TRW led to intelligence work deciphering telemetry from Soviet missiles and satellites. At age 33, he was recruited by Herbert "Pete" Scoville (Deputy Director for Research, or DDR) to become director of the OSI. Scoville resigned over disagreements with McCone and the director of the NRO, Joseph Charyk (who had succeeded Richard Bissell), over whether the Agency or the Air Force would control overhead reconnaissance. McCone persuaded Wheelon to take Scoville's place as DDS& T.[72] After all, it was the Agency that had successfully produced invaluable photointelligence with the U-2 and CORONA, whereas many Air Force projects had failed, and McCone wanted both the CIA and Air Force to cooperate and share their resources. However, Wheelon was no diplomat, and neither was his nemesis, Brockway "Brock" McMillan, who would serve as the Undersecretary of the Air Force and then second director of the NRO (DNRO). Claiming that McMillan and the NRO wanted to assume complete control over aerial surveillance, McCone even threatened to "liquidate the NRO if necessary."[73] Aden would be caught in the middle of the feud and later would be forced to choose between them.

Wheelon told all assembled in the room that the president had directed him to seek out optical experts who could find ways to watch Soviet missile developments from space now that U-2 flights over the Soviet Union were no longer an option after Francis Gary Powers had been shot down over Sverdlovsk on May 1, 1960. Wheelon wanted them to examine photos from earlier CORONA missions and make improvements. This was the beginning of Aden's work on satellites and government-paid weekly trips to Washington, DC. One summer trip was especially memorable.

Aden and Marjorie had taken their seven children to a resort in Colorado named "Ah, Wilderness!" to get away from it all, a trip they would make many times in future summers. The resort, now demolished, was then accessible only on foot, horse, or a steam-engine train traveling on a narrow-gauge railroad that shuttled passengers on a few round trips a day from Durango to Silverton, a trip of about 45 miles. It was located in the San Juan National Forest, along the Animas River between Tacoma and Tank Creek stations, about halfway between the terminus points. The cabins lacked creature comforts such as television, radios, and telephones. Guests occupied themselves with hiking, horseback-riding, trout-fishing in streams, and whitewater rafting on the river; skiing and ice-skating dominated the activity board in the winter. One day the only telephone in the resort rang in the owner's office; the caller wanted to speak to Aden. The following morning a helicopter landed in a nearby horse pasture, scattering the horses, and flew Aden back to the small Durango airport, where he boarded a four-engine Air Force Lockheed JetStar. Because the short runway was not certified for jets, the takeoff and ascent were seemingly straight up, a ride that Aden said everyone should experience—but only once. As soon as he arrived in Washington in his cowboy hat and boots, he was taken to a still classified destination, most likely CIA headquarters, the White House, or the NPIC. Later Aden returned by JetStar to Durango and made his way back to the resort. This was not to be Aden's last summons from Ah, Wilderness! to the east coast.[74]

The Cuban Missile Crisis

After the dismal failure of the Bay of Pigs invasion by Cuban exiles in April 1961, sponsored by the United States and approved by Allen Dulles and President Kennedy to bring down the Fidel Castro regime, the Soviets approached Castro and offered a five-year deployment of nuclear missiles there. Khrushchev believed that a nuclear base about 90 miles from south Florida would help to defend and preserve the communist regime in Cuba and at the same time counter American Jupiter missiles in Turkey, Italy, and Great Britain, within striking range of the Soviet bloc.[75] Castro jumped at the offer. The continuing U-2 overflights of Cuba detected Soviet arms deliveries there in early August 1962. A mission on August 29 showed there were at least eight surface-to-air missiles (SAMs) in western Cuba, followed by others, as well as a MiG-21 fighter in September, raising fears of another U-2 being shot down. Nonetheless, on October 14, film from a U-2 overflight of the island by the Strategic Air Command went straight to Art Lundahl at the NPIC and showed (1) medium-range ballistic missile (MRBM) sites around San Cristóbal; (2) 12 MRBM launchers and 12 intermediate-range ballistic missile (IRBM) launch pads under construction; (3) four MRBM sites with 16 fully operational launchers; (4) 24 SAM bases, 40 MiG fighters, and 208 IL-28 nuclear bombers; and (5) 43 Soviet ships in Cuban ports or headed there. Kennedy was briefed on October 16. The number of U-2 overflights then increased to more than one a day, supplemented by Navy and Air Force planes. Satellites were not useful because their orbits put them over Cuba at the wrong angle and wrong time of day to gather enough useful intelligence. On October 18, U-2 photos clearly showed IRBMs with a range of 2,000 miles, twice that of MRBMs.[76]

As a special consultant serving the Agency's Bud Wheelon, Aden was told that he was urgently needed in Washington, DC. Aden was in Wheelon's CIA office on October 27 while Bud was listening to radio communications with a U-2 flying over Cuba in real time. The pilot had taken revealing photos of the missiles but was having engine trouble, so Bud scrambled a plane to intercept the U-2 and accompany it back to Florida.[77] Robert McNamara reported the incident to President Kennedy on October 27.[78] That same day, McNamara also had to relay the message that Rudolph Anderson had been killed when his U-2 was downed over the northern coast near Banes by a Soviet SA-2 missile like the one that had shot down Francis Gary Powers.[79] Wheelon showed Aden some of the recent U-2 images, and Aden agreed that they showed Soviet MRBMs and that all 24 missile bases in Cuba were fully functional, which concerned Kennedy and brought the world closer to nuclear war than it had ever been. U-2 flights continued their flyovers to monitor the removal of the missiles and launch equipment after Kennedy secretly agreed to remove American missiles from Turkey,

Italy, and Great Britain as a quid pro quo. All Cuban missiles and launchers were dismantled by November 8.[80]

GAMBIT-1, GAMBIT-3, and sea monsters

The downing of the U-2 piloted by Francis Gary Powers and immediate suspension of overflights of the Soviet Union by Eisenhower had galvanized the intelligence community in the United States to adopt a three-pronged approach to aerial surveillance: (1) the Lockheed A-12 of Project OXCART, successor to the U-2 (discussed at length to end this chapter); (2) the SAMOS E-6 satellite intended to replace CORONA; and (3) the GAMBIT satellite series, intended to monitor Sino-Soviet bloc targets broadly identified by CORONA but at a higher resolution than the minimum 10–15-feet range that CORONA could then provide. By the end of May, Eisenhower asked George Kistiakowsky (who had become science advisor upon the resignation of James Killian) to oversee a task force commissioned to revamp the satellite program. The group included Land, Killian, Bissell, Jr., Overhage (then with Lincoln Laboratories), and Air Force Undersecretary Joseph Charyk. Charyk, who became the first director of the newly established National Reconnaissance Office in September 1961, succeeded in casting the new program under the control of the Air Force rather than the CIA.[81]

In March 1960 Eastman Kodak had submitted a proposal to the CIA for a 77-inch focal-length camera (a proposal called "Sunset Strip," after the then-popular television show "77 Sunset Strip"), followed in June by a 36-inch camera system called "Blanket" that would provide stereo coverage in satellite reconnaissance. Land then took the "Sunset Strip" proposal to Charyk, who favored it for the long focal-length camera and film-recovery procedure like that of CORONA. The RAND Corporation was also developing a spin-stabilized satellite, and in July personnel from RAND and Space Technology Laboratories suggested that a satellite could be concealed in the re-entry body of a ballistic missile. If the satellite were pointed downward from latitude 55°N, it could image all areas between latitudes 40°N and 70°N. In the end it was the "Sunset Strip" or E-6 that would evolve as the covert program named GAMBIT.[82]

GAMBIT 1 was 15 feet long, 5 feet in diameter, and had a Maksutov f/4.0 lens with a 77-inch focal length (Figure 11.3; consult the color insert for this figure). At 95 nautical miles in altitude it could produce a ground resolution of 2–3 feet. Three thousand feet of 9.5-inch diameter film moved through a strip camera and could image an area of the earth 10.6 nm wide. On completion of the mission, the recovery vehicle would be de-orbited with parachute deployment at 55,000 feet.[83] In its January 20, 1962, report to President Kennedy, the

President's Foreign Intelligence Advisory Board (PFIAB) strongly advocated pushing forward the OXCART and GAMBIT systems because the Samos E-5 program had failed, and the CORONA program was also having difficulties.[84] Kennedy approved the recommendation and assigned to it the highest national priority and secrecy.[85] Every new satellite program required top-secret clearance, and entire nondescript buildings were designated for research and development. Security was so tight that Aden usually had to fly to a different city, rent a car, and drive to the site.[86] By December 1963 he was formally listed as a member of the Optics Panel, chaired by James Eyer, of the Scientific Advisory Board to the DCI (John McCone).[87]

The first GAMBIT-1 flight was launched on an Atlas D booster and Agena D upper stage from Vandenberg Air Force Base on July 12, 1963, and returned with 198 feet of film; average ground resolution was 10 feet, although as remarkable as 3.5 feet on some stretches of film. On the second flight, on September 6, 1963, the recovery vehicle completed 34 orbits and retrieved resolutions previously seen only by reconnaissance aircraft. The third, on October 25, 1963, clearly showed a football game in Great Falls, Montana. In one of the photos a place kicker was shown putting the football into place, and in a subsequent photo the players had lined up for the kickoff—all of this from 90 nautical miles in altitude. After the sixth mission, however, resolution was much poorer, and most failed for various reasons. The last GAMBIT-1 mission (KH-7) of 38 in total flew on June 4, 1967, even though resolution had finally improved to 2 feet, up to five times better than anything that CORONA had produced.[88]

In an effort to improve resolution even more to one foot or less on a side, camera designers from Kodak presented a new system to DNRO Brockway McMillan and General Robert Greer, one with the pointing accuracy of GAMBIT-1 and an increased focal length (which is still classified). Two other GAMBIT systems had been proposed in 1962–1963, namely GAMBIT-2 and GAMBIT-4. GAMBIT-2 was not a significant improvement over GAMBIT-1, and GAMBIT-4 had financial and production drawbacks. That left GAMBIT-3 (KH-8), sometimes called GAMBIT Cubed or G-Cube, which incorporated several improvements over the original, including (1) increased focal length; (2) a thin-base, high-resolution film with a higher exposure index of 6.0, allowing three times as many photographs per mission than GAMBIT-1; and (3) the use of invar, an iron-nickel alloy with a low coefficient of thermal expansion, for the optical barrel and other assemblies. Invar is a shortened form of "invariable," referring to its relative lack of expansion and contraction with changes in temperature. Toward the end of 1963, McMillan authorized Greer to green-light GAMBIT-3, launched not with the Atlas-Agena D combination but with the new Titan-III booster to accommodate the higher payload weight; formal approval came on January 4, 1964.[89] The following August, McMillan called Aden and asked him to fly to Boston for a meeting with him, Land, and others at Land's Polaroid office in Cambridge. Aden replied that he was leaving for Durango the next day, a Sunday, with the family for another stay at the Ah, Wilderness! resort. McMillan persisted and said that he would send a Lockheed JetStar to pick up Aden in Durango on Sunday evening. Aden took a cab to the Durango airport from the train station and saw another Presidential JetStar with "United States of America" emblazoned on the fuselage waiting for him. After the rapid ascent from the short runway, the plane refueled in Texas before arriving in Boston at 1 a.m. on Monday. Aden was amused that the steward was forced to charge him $6.00 for his dinner when it must have cost the government several thousand dollars to fly him back and forth to Boston. Aden met the committee at 8 a.m. the next morning in Land's office and heard about GAMBIT-3 and Kodak's plan to use invar for the primary mirrors. He strongly advised against metal mirrors after his experiences with them in the KPNO Space Division, and the other committee members concurred. After the meeting, he arrived at the Boston airport at 4 p.m. for a jet to Denver. As the plane taxied up to the ramp, he was invited to come to the doorway, where the stewardess told him that a jet was waiting for him nearby to take him back to Durango. He boarded the T-39 Sabreliner that belonged to the Air Force general based in Denver and made it back to Durango. From there a limousine took him to Rockwood, but he had to walk the rest of the perilous way down the rails and over the Animas River on the moonless night back to the resort.[90] In the meantime, Kodak technicians were beginning to understand why Aden had advised against the invar mirrors. It was taking 3,000 hours, rather than the estimated 800 hours, to grind, polish, test, and coat each mirror, and so their team had fallen far behind in their production schedule.[91]

The next meeting on GAMBIT-3 that Aden attended was at CIA Headquarters. Kodak had given up on the invar mirrors, which were replaced by lightweight fused silica mirrors developed by Corning. Afterward, DCI McCone invited Aden, Land, and McMillan to his suite on the top floor of the headquarters at Langley for lunchtime discussions about the new mirror material. Aden assured McCone that what he had witnessed in the optical industry could overcome any challenges related to size. McMillan pulled Aden aside and asked if he would meet with the USAF Special Projects Team in El Segundo, California, for further discussions about GAMBIT-3, not realizing that by being McMillan's consultant his relationship with the Agency would be compromised, given the ongoing competition the Air Force and the Agency for control of aerial surveillance.[92]

The first launch of GAMBIT-3 was on July 29, 1966, lasting five days and successfully reading out several targets. Although the number of targets and resolution achieved is still classified, the intelligence acquired was the best of any reconnaissance satellite to that point. Indeed, the amount of data gathered from

GAMBIT-3 up to June 1967, along with concurrent data from CORONA and GAMBIT-1, swamped photointerpreters.[93]

Among those data were reports of strange aircraft in Baku, Azerbaijan, on the Caspian Sea in the Soviet Union. Beginning in the early 1960s, the Soviets developed *ekranoplans*, which were wing-in-ground-effect (WIG) aircraft designed to fly just above flat surfaces (usually water, but also snow-covered land, ice fields, etc.); lift was provided by the ground effect of compression of the ram air stream between the wings and water. Such craft have a higher lift-drag ratio and are therefore more fuel-efficient and faster than on water. At such low altitudes they are also invisible to radar. The Central Hydrofoil Bureau, led by designer Rostislav Alexeyev, produced several WIGs for the Soviet military to be used for such missions as troop transport, anti-submarine warfare, and anti-shipping strikes. One of those craft was the KM (*korahbl'-maket*, or "prototype ship"), built at the Volga River shipyard near Gorky (now Nizhny Novgorod) and then transported secretly downriver to the Caspian Sea for testing. To that point it was the largest aircraft in the world, with a fuselage 330 feet long, wingspan of 130 feet, and gross weight of 948,000 pounds; on one of its flights it weighed an astounding 1,299,300 pounds. Its first flight was in mid-October 1966, powered by 10 Dobrynin VD-7 turbojets, eight in two groups of four on the forward fuselage and two flanking the fin leading edge. Optimal flight altitude was a mere 13–46 feet. The KM flew transport missions for 15 years before it crashed in 1980 because of pilot error and sank with no casualties.[94]

It was likely GAMBIT-3 that revealed the existence of the KM to US intelligence in 1967. Aden was present at a top-secret Pentagon briefing where he saw a photo of the KM along a dock in Baku on the Caspian Sea and heard the nickname given to it: Caspian sea monster. Looking at the surveillance photos, he noticed that jet smoke had darkened the top surface of the wings. The Pentagon briefer asserted that the jets impinged up on the bottom of the wings to provide lift. Aden disagreed: "See the smoke stain on the top of the wing? The plane uses the Bernoulli effect. It is that effect that provides aircraft lift by moving the air faster over the upper surface than the bottom. The jet over the wing provides lift." The speaker argued that his briefing charts ("the bible to a briefer") contradicted Aden's scientific explanation, and the aircraft engineers present toed the official line. At another briefing, photos showed three smaller "monsters" beside a larger one, and the group joked that the monster had had babies. Those were likely the *Orlyonok* (Eaglet), a troop-carrier assault WIG aircraft with a fuselage 190 feet long and a wingspan of 103 feet.[95]

Another revelation via GAMBIT-3 was an installation in the mountains above Dushanbe, the capital of Tajikistan. Satellite photos showed 10 holes in the ground, six in a hexagon and four in a skewed square. Next to them, a large building was being constructed in the shadow of the mountain, along with a power line. It was unclear whether the unit was an anti-satellite weapon, an interferometer, a high-resolution monitor for US satellites passing overhead, or something else.[96]

The missions of GAMBIT-1 were eventually canceled in June 1967. The last GAMBIT-3 flight was in April 1984; by then, the HEXAGON satellite program (KH-9) had been operational for 13 years and would continue until 1986, providing even more intelligence on Soviet weapons development and deployment than any program to that point.[97] There is nothing declassified or in the Meinel autobiography to indicate that Aden contributed substantially to HEXAGON, nor had Phil Pressel (the Perkins-Elmer engineer chiefly responsible for HEXAGON optical systems) ever heard of Aden (personal communication with AP), likely because Aden's involvement, if any, was as a member of a review committee for the Agency.[98] However, Aden was deeply involved in Project FULCRUM and caught in the middle of the growing struggle between the NRO and the CIA for control of orbiting reconnaissance, which ultimately led to HEXAGON.

FULCRUM and HEXAGON

In May 1963, DCI John McCone directed a panel under Edwin Purcell of Harvard University to make recommendations for the future of satellite surveillance. The panel recommended improving CORONA instead of embarking on a new satellite program, but it was not what McCone wanted to hear. Land disagreed with the other members of the Purcell Committee, wanting a system with the broad coverage of CORONA but the high resolution of GAMBIT. So, later that autumn, McCone instructed DDST Bud Wheelon to explore replacing CORONA altogether with a system that could provide the resolution of 2–4 feet needed by the photointerpreters. Wheelon proposed a program code-named FULCRUM consisting of (1) two 60-inch focal length stereo cameras with ground resolution of 2–4 feet for a strip 360 miles wide, (2) 68,000 feet of film seven inches wide that could cover 11 million square miles on each mission, and (3) a reentry vehicle. In contrast, McMillan thought that CORONA's resolution could be improved with a longer focal-length lens and that a new satellite program was unnecessary. In June 1964, McCone directed McMillan to make FULCRUM an NRO project but to assign all responsibility for research, development, and operation to the CIA. McCone also asked Land to convene a panel to consider this new system. On the Land Panel besides Land were Allen Donovan, Sidney Drell, Richard Garvin, Spurgeon Keeny, Jr., Donald Ling, Lundahl, and Aden. The committee met on June 26, 1964, and recommended the need for a new system to Wheelon and McCone, a conclusion supported also by reports from Itek and Space Technology Laboratories of TRW. Deputy Director of Defense Research and Engineering, Eugene Fubini, was more circumspect and thought it necessary to compare old CORONA results, new CORONA results, the Land Panel's recommendations, and GAMBIT-1 results before committing to FULCRUM.[99]

Wheelon monitored Itek's progress on FULCRUM and asked Aden to make a site visit on his behalf. Aden, who was at that time director of the Steward Observatory, reported back to Wheelon on June 8, 1964: "My brief exposure to the proposed Itek system impressed me that the group working on this job appear to have all the potential problem areas that I expressed to you quite well in hand."[100]

On July 2, 1964, Wheelon presented a plan for starting FULCRUM, encompassing not only Itek design studies but expanding the overall role of the CIA in space systems, essentially bypassing the input of the NRO and McMillan. At Wheelon's request, Itek and Space Technology Laboratories had proposed a payload with two counter-rotating Itek cameras in a three-axis stabilized satellite and a recovery vehicle, all launched on a Titan II booster. The Matsukov cameras would have f/3.0 lenses with a 60-inch focal length, resolving 2.7–4.0 feet from an altitude of 100 miles. McMillan was blindsided by the proposal, which had been commissioned without consulting the NRO. He took the briefing book and accelerated an NRO program called S-2 already in the planning stages by the Secretary of the Air Force/Special Projects Office to compete with FULCRUM. It was not the first time that Wheelon and McMillan had crossed swords. The antagonism between Wheelon and McMillan began long before either arrived in Washington. Wheelon had submitted a paper on radio waves to *Physical Review*, published by the American Physical Society. McMillan served as one of the peer reviewers and rejected the paper outright. Wheelon felt that McMillan's review was uninformed and complained to the editor, who agreed to publish the paper after all. Now the two, with McCone supporting Wheelon and Secretary of Defense Robert McNamara supporting McMillan, were representing different departments battling for control over spy satellites, and not even President Kennedy was able to resolve their personal and professional differences. Aden was caught in the middle, and in a letter to him on CIA letterhead, Wheelon referred to the "Perils of Pauline" in the reconnaissance field, which made it difficult for the Agency to recruit staff members. Wheelon also mentioned that he had passed along Aden's proposal to Eugene Fubini in support of his [Aden's] "new organization," which probably referred to the College of Optical Sciences.[101]

In the fall of 1964, Itek had completed preliminary designs for both S-2 and FULCRUM; the system judged the better of the two would help resolve the power struggle. Eastman Kodak and Fairchild Camera were also working on S-2, a simpler design with a ground resolution of 3–4 feet. Disagreements over the stability of the proposed 120-degree scan angle of FULCRUM between the Agency and Itek in late 1964 and early 1965 caused Itek to drop out of the FULCRUM program precipitously, and so the Agency transferred Itek's papers and working prototype to Perkin-Elmer. McMillan quickly transferred the S-2 program from Eastman Kodak to Itek, leaving Kodak to work on the Air Force's Manned Orbiting Laboratory (KH-10). In April 1965, Lyndon Johnson replaced McCone as DCI with Vice Admiral William Raborn, Jr. Another panel led by Land reviewed the status of FULCRUM and S-2 and recommended that all the studies continue to be funded. A new agreement worked out by Raborn and Secretary of Defense Cyrus Vance assigned responsibility for the optical sensor subsystem of FULCRUM to the CIA and the engineering of the spacecraft, recovery vehicle, and booster to the NRO/Air Force, yet McMillan continued to invest NRO funds in S-2, which among other issues led to his resignation on September 30, 1965. On April 22, 1966, the United States Intelligence Board (USIB) signed off with Perkin-Elmer on a new search system similar to FULCRUM that would avoid the unstable 120-degree scan, code-named HEXAGON (KH-9). The HEXAGON ("Big Bird") contract was awarded to Perkin-Elmer in October 1966.[102]

Three mechanical innovations from the engineers at Perkin-Elmer made HEXAGON one of the most effective reconnaissance satellites to that point: (1) the *twister*, which allowed a short section of film to be rotated during photography while still moving fast in a linear direction; (2) the *air bar*, which assured film flatness and optimal tracking, and (3) the *looper*, which was an intermediate film storage apparatus that could allow for all film speeds and positions in preparation for the following photographic frame. The upshot was that film was not wasted as it rotated synchronously with the optical bar as it moved linearly. In other words, the entire optical system rotated, not just the cameras. According to Phil Pressel, film could be advanced at an astounding 200 inches per second, double the speed mentioned by Aden in *Echoes from a Simpler Time*. Land visited the Perkin-Elmer plant and was "blown away" by a demonstration of the twister and film speed.[103]

As there is no declassified evidence that Aden was involved in the surveillance and mapping functions of HEXAGON, it lies beyond the scope of this study, but in its launch history from June 1971 to April 1984 the recovery vehicles returned 29,000 photographs covering 38 million square miles. On each mission it could provide stereoscopic photography of up to 90 percent of Sino-Soviet territory. The intelligence acquired was of paramount importance for the United States in negotiating the Strategic Arms Limitation Talks Agreements with the Soviet Union from a position of strength.[104]

MOL/DORIAN (KH-10) and KENNEN (KH-11)

In 1957 the Air Force proposed a manned spaceplane named Dyna-Soar (a predecessor of the Space Shuttle), to be built by Boeing, launched by a Titan IIIC booster, and intended for military surveillance, servicing satellites, scientific tests, etc. Concurrent with development of Dyna-Soar, NASA had been commissioned in 1961 by President Kennedy to send men to the Moon and was developing a two-man version of the successful Mercury program named Gemini. The next year the Air Force supported the plan for the Military Orbital Development System with five Gemini B missions in 1965. But NASA was planning its own space station, called the Manned Orbital Research Laboratory, using Apollo mission hardware. In December 1963, following a comparative cost review of Dyna-Soar and Gemini B missions and probably taking into account the mounting costs of waging the Vietnam War, Secretary of Defense Robert McNamara announced the cancellation of Dyna-Soar, replaced by a proposal for a Manned Military Orbiting Laboratory (President Johnson later removed "Military" from the name).[105]

In a press conference on August 25, 1965, President Johnson approved development of the Manned Orbiting Laboratory (MOL) by the Department of Defense at a cost of $1.5 billion. Douglas Aircraft Company was selected to design and build the spacecraft to house the astronauts. General Electric was contracted to plan the space experiments. It would be launched by a Titan IIIC booster. Once in orbit, two astronauts in a NASA Gemini capsule would dock and then enter to perform experiments; the Gemini capsule would be the return vehicle for the astronauts. The first launch with a two-man crew was scheduled for 1968, with four more manned flights later that year.[106] The publicized objectives of the MOL program were (1) to learn what humans can do in space and its use for military purposes; (2) to develop technology and equipment to advance manned and unmanned space flight; and (3) to conduct experiments with that technology and equipment.[107]

Even before Johnson's announcement, the Agency and the NRO saw the potential of the MOL for photo-reconnaissance purposes; the covert studies fell under the code name DORIAN (KH-10). The primary mirror had a diameter of just under 72 inches, and it and the tracking mirrors were of an eggcrate design. The photographic system was capable of delivering a theoretical resolution of 4 inches and, given atmospheric turbulence, an actual resolution of 9 inches. However, Land of the PSAC Reconnaissance Panel argued that not enough effort had been made to explore "alternatives to the use of man" and recommended an automated camera payload that could be operated with or without an astronaut aboard. The Air Force continued to insist that a ground resolution of less than one foot required a manned system. The disagreements between the PSAC and Air Force continued until Robert McNamara made the decision to proceed with the manned system in order to achieve better than one-foot resolution. Three years later, however, the Agency was arguing that this resolution was optimistic and not worth the burgeoning costs and scheduling delays, especially because

HEXAGON, which was being developed in parallel, could provide equivalent or better resolution at a fraction of the cost of DORIAN. Moreover, the Agency felt that if the Soviets could take down a U-2, they could use an anti-ballistic missile to attack the MOL and jeopardize the safety of the astronauts. Fear of human losses had become more acute with the launchpad deaths of Virgil "Gus" Grissom, Edward White, and Roger Chaffee during an Apollo 1 rehearsal on January 27, 1967. Swelling outlays for the Vietnam War had already raised the hackles of fiscal conservatives and bipartisan antiwar activists in Congress. Consequently, MOL/DORIAN was canceled in early June 1969 with the approval of Congress and President Nixon.[108] In July of that year, Aden was again secretly contracted to serve as consultant to the Air Force "on optical systems and materials, advanced technology and exploratory research beyond the state-of-the-art related to satellite reconnaissance" for 12 months with a fee of $10,000 (not declassified until 2020). The Personal Services Contract, signed by new NRO director John McLucas, extended Aden's classified work into the first Nixon administration.[109]

While having a lunch of sandwiches at the Pentagon, Aden learned of the cancellation of DORIAN, which suddenly made seven 72-inch mirror blanks surplus. Over time he persuaded security personnel not to destroy them and proposed using them for a multiple mirror telescope, Project COLT, so named because the mirrors would be arranged like the bullets in the cylinder of a Colt six-shooter. According to a redacted NRO memo declassified in July 2015, Aden prepared a Statement of Work discussing the design and performance of the optical system and detectors. As there was no overlap with ongoing military projects, USAF Captain Lawrence Pence recommended to Undersecretary of the Air Force John McLucas that the Air Force contract with "Dr. Meinel's group" in a memo dated January 19, 1971. Not long after that, a Lockheed C-5 cargo jet landed just after midnight at Davis-Monthan Air Force Base in Tucson. With the engines still running, the crated mirrors were quickly unloaded without an exchange of paperwork and were moved by a University of Arizona vehicle to a secure warehouse in the city. After parting handshakes, the C-5 departed as stealthily as it had arrived. When astronomers later asked about the source of the mirrors, Aden simply and truthfully replied that they were experimental and that the government had no further use for them. Chapter 9 discusses how the mirror blanks were used in the Multiple Mirror Telescope.[110]

In the late 1960s, Aden often traveled from Tucson to meet with the Air Force Special Projects Team in El Segundo, California—after his friend Brockway McMillan retired in 1965—to "spitball" new concepts for large space cameras. One of his ideas was to deploy the mirror outside the spacecraft and use cross-axis relay mirrors to send images to several cameras. While there, he met his future boss and advocate at the Jet Propulsion Laboratory, (then) Lt. Colonel Lew Allen (1925–2010). After receiving his MS and PhD in physics from the University of Illinois, Allen served in Special Projects from 1965 to 1968 as Deputy Director for Advanced Plans before transferring to the Pentagon as deputy director and then director of the NRO, then back to El Segundo as the director of the Office of Special Projects and also deputy commander for Satellite Systems of the Air Force Space and Missile Systems Organization. With increasing promotions in rank, he served as director of the NRO Program A, deputy director of the CIA, and director of the NSA. General Allen was Vice Chief of Staff and Chief of Staff of the Air Force in 1978. After retiring from active duty in 1982, he was director of JPL and headed up NASA's Deep Space Exploration Program. The other Air Force officer whom Aden met at Special Projects in El Segundo was (then) Lt. Commander Robert Geiger (1923–2013). Geiger graduated from the US Naval Academy and received his MS in aeronautical engineering from MIT. He was Allen's Assistant Deputy Director for Advanced Plans and later became director of the NRO Program C and as Rear Admiral served as the Chief of Naval Research until his retirement from active military service in 1978.[111]

Both men teamed with Aden on a project in the late 1960s to determine the optimum ground resolution from the atmosphere by flying a 36-inch aperture camera on a balloon from about 80,000 feet, similar to the resolution that a 300-inch aperture camera could achieve from orbit. Itek assembled the camera, launched from Luke Air Force Base west of Phoenix. Aden's role was to monitor atmospheric conditions for the two weeks before and after each of two flights. Volunteer observers measured the strength of the shadow bands (caused by air turbulence) observed through telescopes on the ground; they learned that conditions on the flight days were similar to those of average conditions. The balloon was aloft for two days, recording targets such as 4×8 plywood sheets with Air Force standard resolution targets on them. The average resolution was even less than 2 inches, the figure that Aden had calculated a few years earlier at the secret meeting at the RAND Corporation (see earlier discussion) and later confirmed by laser interferometry experiments from a U-2 plane in over-flight. The message was clear: it was possible to construct large space cameras producing images that would not be affected by atmospheric turbulence.[112] The HEXAGON program was one of the benefactors of this new information, as was KENNEN (CRYSTAL, KH-11).

One of the potential problems—and implicit fears—with the bucket system beginning with CORONA became reality in May 1964. CORONA Flight 78 (Mission 1005) in the KH-4A series was launched on April 27, 1964. Vandenberg AFB sent a recovery command to the satellite on its 47th orbit, but pyrotechnic charges used to separate the capsule had shorted out. Subsequent commands were also received, but the capsule remained intact. Even the backup system named LIFEBOAT was unresponsive. So the satellite continued in orbit for six more days until drag brought it lower and lower. On May 26, observers in Maracaibo, Venezuela, saw five burning objects in the night sky. In early July two farm workers came upon a golden kettle cradling a roll of charred film near La Fría in Táchira State, bordering Colombia. They took the bucket to their employer, who tried unsuccessfully to sell it to villagers and even smugglers and so resorted to hawking the contents as household utensils and toys. Word of the unusual object spread, and a commercial photographer sent an image of it to the US Embassy in Caracas. CORONA team members posing as Air Force officials flew down to Caracas, bought the bucket from the Venezuelan government, and explained it away as a NASA experiment that failed.[113]

Worries about loss of recovery vehicles evaporated with the advent of electro-optical imaging in the 1970s. Both the Air Force and the Agency competed to perfect the system. Aden consulted for the Air Force design that used a tape-storage camera developed by CBS Laboratories and produced high-resolution images transmitted to ground stations, harking back to the original concept of nonrecoverable satellites that issued from the RAND Corporation in the 1950s but replacing the need for television camera systems and radio links. In the first missions, the mirror had a diameter of 7.6 feet but was later enlarged to 7.9 feet, the same as the Hubble Space Telescope. It was 64 feet long, 10 feet in diameter, and eventually weighed 43,220 pounds. In many respects, it was the prototype of the Hubble Space Telescope, although their purposes and equipment were different. Given the code name KENNEN (KH-11), the first launch was in December 1976. Successive launches, however, incorporated a new CCD system adopted and promoted by the Agency. William Boyle and George Smith had published a paper in April 1970 that described the design and uses of the CCD. Now CCD image sensors could capture photons and other radiated energy particles in pixels and transmit them to a ground station in real time via Satellite Data Systems spacecraft in an orbit that would keep it 8–9 hours over Soviet territory. The orbit was higher than previous Keyhole satellites: 150 × 250 miles.[114]

KENNEN was the blackest of the satellite programs to that point, and knowledge of its existence was highly restricted within the Agency and Air Force. Over the span of 37 years, 16 KH-11 missions were launched. It was also one of the most strategically important programs in the history of satellite surveillance. For example, it was used to locate the American hostages inside the US Embassy in Tehran in April 1980, to assess loss of heat-shield tiles on the space shuttle *Columbia* in 1981, and to show the damage to the nuclear reactor at Chernobyl in April 1986. In 1992, KH-11 was given several enhancements to the optics and rate of data transmission and was renamed CRYSTAL.[115]

OXCART

We had developed orbital cameras [in CORONA] that were doing a great job but were too far away in orbit and too small to yield the detailed resolution of important objects on the ground. So one day Bud [Wheelon] convened a group of us "advisors" and unveiled his plan to build the fastest plane ever developed. It would be fast enough that even a ground-based missile couldn't reach it before it was "long gone." The Lockheed team then disclosed to us a fantastic airplane— more a missile than an airplane.

—Aden Meinel[116]

Origins

As soon as the first U-2 overflight in 1956, the Agency's Deputy Director for Plans, Richard Bissell, realized that the Soviets were tracking the flights and so tried to reduce the U-2's radar cross section with Project RAINBOW, as mentioned earlier. It only introduced a new set of problems, such as added weight which limited altitude, leading to the program's cancellation in May 1958. Meanwhile the Soviets were spending billions of rubles in developing missile systems to down the planes. Like Bissell, Kelly Johnson of Lockheed foresaw the limited life span of the U-2; both agreed that a successor to the U-2 should be operational before satellites made aircraft reconnaissance obsolete. In April 1958, Lockheed began to study the feasibility of an aircraft with a reduced radar cross section that could travel 4,000 nautical miles at an altitude of 90,000 feet and at a speed of Mach 3.0—four times faster than the U-2 and five miles higher—much like Rich and Janos's comparison between an Indy 500 race car and a covered wagon, respectively.[117]

It was in early 1958 that the president's scientific advisor, James Killian, recommended feasibility studies for a new reconnaissance aircraft. After Eisenhower signed off on it, Richard Bissell convened an advisory panel chaired by Land and including Purcell, Donovan, H. Guyford Stever, Eugene Kiefer, and Courtland Perkins, with participation by Charyk (Air Force Assistant Director for R&D), Garrison Norton (Assistant Secretary of the Navy), and Bissell himself. Proposals from two firms, Lockheed and Convair, were given serious consideration. The Panel recommended the one from Convair for a ramjet-powered Mach 4.0 aircraft (called the FISH) that would be launched on the B-58 Hustler. It minimized radar returns (which was Eisenhower's primary goal), whereas the Lockheed design gave more priority to aerodynamic performance. Eisenhower rarely acted precipitously, which frustrated vendors and government personnel alike and earned him the nickname

"Speedy Gonzales" at Lockheed's Skunk Works. He recommended additional research and development, so both Convair and Lockheed were sent back to the drawing board for the next several months before submitting new proposals.[118]

Lockheed experimented with ramjet engines and a combination of ramjets and turbojets. Just as the U-2 was called "Angel" because of its high-altitude operations, the new configurations were designated Archangel, or A for short, followed by consecutive numbers. The A-1 was a single-seat, two-engine airplane with wings mounted more than halfway back on the bullet-shaped fuselage, clearly built for speed. But the problem with supersonic speed was twofold: how to mitigate the high temperatures generated and how to maintain a low radar cross section (RCS) despite the heat. In April 1959, Johnson proposed the A-11, which added downward-sloping chines and lowered the RCS by 90 percent. That was still not enough, and the Land committee immediately rejected it. By April 1959, Johnson had flown back and forth a dozen times to Washington and Agency headquarters in Langley, Virginia, to promote the A-12, a ground-launched, single-pilot jet with two J58 P&W engines. Cesium was added to the fuel to reduce detection of the after-burner plume and thereby lower the RCS. It could reach Mach 3.2 at 97,500 feet with a range of 4,600 miles. To withstand the heat at such a speed, Johnson chose not to build the A-12 out of steel or aluminum but a titanium alloy. It ultimately weighed 96,000 pounds unfueled and stretched to 108 feet long with a double delta wing mid-fuselage, two P&W J58 engines mounted mid-wing, and canted rudders.[119]

Meanwhile, Convair also used P&W J58 engines on its KINGFISH, but instead mounted them inside the fuselage to reduce the RCS. Smaller than the A-12, it had a stainless steel honeycomb "skin" and flattened wings that stretched from just behind the cockpit all the way aft with leading edges made of Pyroceram, a glass-ceramic material manufactured by Corning Glass and also used in telescope mirrors (see Chapter 7, section "The 80-inch telescope," subsection "The mirror"). Both proposals were submitted on August 20, 1959, to a panel comprising representatives of the Department of Defense, Air Force, and CIA. Although Agency members of the panel were swayed by the low RCS of the KINGFISH, others were concerned about past cost overruns and production delays by Convair. In contrast, Lockheed had a proven track record. It had produced the U-2 on time, under budget, and in the secure premises of Skunk Works, so it was given a four-year contract until January 1, 1960, to reduce the RCS of the A-12. Subsequent research was conducted under the code name OXCART in August 1959, replacing the former code name GUSTO. The new code name has been reported as computer-generated and/or perhaps selected by OXCART Agency project manager John Parangosky or Bud Wheelon. Having been to India and seen how slow oxcarts are, Aden commented to Wheelon on the irony of the name given to the fastest aircraft at that time.[120]

From its conception, OXCART was one of the most classified secrets every held by the US government, as high or higher than the Manhattan Project. Production companies such as Skunk Works were invisible to the outside world. Compartmentalization was extreme, and work in the different areas was strictly on a need-to-know basis. Support contractors had pseudonyms, such as C&J Engineering (among many) for Lockheed, and United Aircraft Company for Perkin-Elmer/Pratt & Whitney. The Agency performed its own security clearance investigations rather trust them to the FBI. T. D. Barnes[121] estimated that less than 5 percent of those involved with OXCART knew that the CIA was involved. Only 1 percent knew of a clearance requirement other than (but not above) the top-secret level, which was Sensitive Compartmentalized Information (SCI). Working with Bud Wheelon's CIA team from 1962 to 1968 and then for the Air Force at the Pentagon from 1963 to 1971, Aden had one of the highest clearances available and was read into most projects, many still classified. The Agency did not acknowledge the existence of the A-12 until 1991 and did not begin declassifying OXCART until 2007.[122]

Testing

A mockup of the A-12 was constructed at Skunk Works and transported for RCS testing to Area 51 by EG&G beginning in November 1959 and lasting a full 18 months until a satisfactory RCS was achieved. Dick McEwen, who was one of 20 men working in Special Projects to reduce the RCS, recalled that the model of the A-12 was made of wood and aluminum and mounted upside down on a pylon (pole) 2.5 feet in diameter and 60 feet high. The pole was raised and lowered by a hydraulic lift with both low-frequency and high-frequency antennas. In a pit beside the pole were a sump pump, controls to operate the rotator, and a hydraulic pump and fluid. The head on the pylon could tilt fore and aft and rotate 360 degrees. Later a US Army NIKE radar tracking system with a transponder was installed to track the airplane. Engineers were in the control room at all times to watch the radar returns and decide what changes needed to be made. On February 11, 1960, the Agency signed a contract to build 12 aircraft for $96.6 million, although the engineering problems associated with the searing heat resulting from flying faster than a speeding bullet at high altitudes caused the price tag to skyrocket over the next few years.[123]

Production

Flying at Mach 3.2 (more than half a mile a second) at up to 97,600 feet would heat the A-12's skin to 900°F, the nose to 800°F, the engine cowlings to 1,200°F, and the cockpit windshield to 620°F. That eliminated construction out of aluminum, which would lose its strength at 300°F, leaving stainless steel as the obvious alternative. But the weight of steel meant more internal support, more fuel, less range, and less altitude. Lockheed engineer Henry Combs had recommended the use of titanium for the exhaust nozzles on the afterburner of the F-104 Starfighter a decade earlier and now pondered whether it was possible to construct the entire A-12 out of a titanium alloy, the first time it had ever been done. Titanium was as strong as stainless steel but half the weight and could resist the high temperatures and pressures. Johnson was happy to entertain any material that would lower the weight, even if it meant coping with new sets of problems every day, not the least of which was the supply of titanium. The earliest deliveries of titanium came from the Titanium Metals Corporation, but most of those were of poor quality. Unfortunately, the world's largest exporter of the metal was the Soviet Union, so the Agency purchased large quantities of it from the Soviets themselves, using third parties and dummy corporations, without raising their suspicions.[124]

Once the problem of supply was solved, another and another arose to take its place. The hardness of titanium required new machining tools. The heat in the cockpit required the driver to wear a pressurized suit with cooling and oxygen supplies. Air was redirected from the engine compressor into a fuel air cooler and expansion turbine into the cockpit, lowering the ambient temperature from 200°F to a pleasant 70°–80°F. New hydraulics, pumps, lubricants, fuels, sealants, electronics, windshields, fittings, landing gears, tires, etc., had to be found or engineered to withstand the searing temperatures. B. F. Goodrich invented a special rubber with aluminum particles that gave the tires radiant cooling; filling them with nitrogen instead of air made them safer. Shell Oil developed a safe, high-flash-point fuel, and nitrogen was added to the fuel tanks to prevent vapor ignition. Tank depots to store this special fuel were established at Edwards AFB and Beale AFB in California and also at US bases in Eiselson, Alaska; Thule, Greenland; Kadena, Okinawa; and Incirlik, Turkey. A smart valve supplied the hottest fuel to the engines, while the cooler fuel supplied the landing gear and electronics. Penn State developed a special oil, something like 10–400 weight, but at a high price. To reach speeds of Mach 3.2, Lockheed engineered movable inlet cones for the P&W J58 engines to regulate the airflow into the inlet, resulting in an astonishing 84 percent propulsion efficiency at Mach 3.[125]

The cameras for OXCART were among the most critical elements of successful missions, and those also were engineered to overcome problems. Perkin-Elmer, Eastman Kodak, and Hycon produced three different photographic systems. There were two pairs of Perkin-Elmer Type-I cameras developed by Rod Scott, Aden, and the CIA team, one pair aimed forward and to each side, and the other pair aimed toward the rear for stereo photography. The cameras had f/4.0 18-inch lenses and 5,000 feet of 6.6-inch film. Stereo pairs of photographs could cover an area 71 miles wide with a ground resolution of 12 inches. Several new features were used: (1) a reflecting cube instead of a prism for a scanner; (2) a film supply designed to minimize weight shift; (3) constant-velocity film transport so that stereo images could be placed next to one another on the same piece of film; and (4) airbars for the film systems. Eastman Kodak offered the Type-II camera, another stereo system with a 21-inch lens and 8-inch film that could cover 60 miles and a ground resolution of 17 inches. Hycon's camera, designed by James Baker, the Type-IV camera, was an updated version of the B camera for the U-2. It could carry 12,000 feet of film and had a ground resolution of only 8 inches. All three cameras were purchased for OXCART, but all were only as good as the windows protecting the cameras. The exterior of the window could reach 550°F, but the interior would be 150°F. Such heat differentials would cause optical distortion and render even the best cameras useless. Corning Glass Works ingeniously solved the problem with a quartz glass window fused to the metal frame with high-frequency sound waves. Later, in 1964, Texas Instruments developed an infrared stereo camera for U-2s in Project TACKLE to find out if the People's Republic of China was producing weapons-grade nuclear material. The FFD-4, as it was known, was adapted for the A-12.[126]

In November 1960, with the election of Kennedy as president, Kelly Johnson began to fear that the whole project would lose funding for three reasons. One was the already mounting costs, 80 percent of them material costs. The original contract was delivery of 12 A-12s for $96.6 million, but by October 1961 only 10 aircraft could be delivered at a cost of $161.2 million. However, the Air Force contracted with Lockheed to produce three aircraft modified after the A-12 to carry a second crewman and three air-to-air missiles—Project KEDLOCK. The AF-12 (A for "article" and F for "fighter"), later called the YF-12A, was intended to intercept enemy bombers before reaching the United States and is still today the fastest and heaviest interceptor ever built. Like the A-12, it could travel at Mach 3.2 and 90,000 feet. KEDLOCK was canceled in early 1968 for budgetary reasons. The second factor that threatened further development of OXCART was the discovery in the early 1960s that the Soviets had a new radar system spanning nine districts from Leningrad to the Far East named TALL KING. With a frequency range of 169–175 megacycles, a pulse repetition of 185–202 pulses per second, and an estimated range of 300 miles, its antennas expanded the range of detection, which added more pressure to reduce OXCART's RCS even more. Finally, there were cost overruns at P&W on the J58, but Richard Bissell was able to secure $38 million from the US Navy to make up the shortfall.[127]

Nonetheless, in February 1962, the first A-12 produced, named Article 121, was shipped (with the wings removed) from Skunk Works to Area 51, arriving three days later. The second, Article 122, left on June 26, 1962, and the third, Article 123, in August of that year. Each convoy of a dozen or so support and security vehicles escorted the wide-load trailers through the Mojave Desert, stopping only at preapproved motels and restaurants along the way. There were 18 convoys in all. Four more A-12s, including the two-seat trainer A-12T called the Titanium Goose, were secretly shipped to Area 51 from Skunk Works by the end of 1962. The rear cockpit of the Goose filled the space that would have been occupied by the camera in the single-seat air-craft. On one of those treks through the desert, the wide-load trailer clipped an oncoming Greyhound bus. To preserve secrecy, project managers arranged to pay Greyhound $4,890 for damages under the table rather than let the inci-dent be publicized in court.[128]

At Area 51, the P&W J58 engines were not yet ready, so J75 engines were mounted instead. As soon as the fuel was loaded, though, 68 leaks in the fuel tanks appeared. The A-12 had been designed to accommodate expansion in the skin during the heat in flight, and so when the titanium was cold, there were gaps in the expansion joints that were filled by sealants. However, the fuel soft-ened those sealants, causing leaks. It was a problem that was never solved. To minimize fuel loss and reduce takeoff weight, just enough fuel was added on the ground to get the A-12s airborne, about 50,000 pounds. After takeoff, the fuel tanks would be topped off in flight by tankers. As soon as the A-12 reached alti-tude and speed, the leaks stopped.[129]

Driver selection was as rigorous as it was for the U-2. Candidates had to be fighter pilots with reserve Air Force commissions and married, preferably with two children. If the candidates had domineering, demanding wives, they would be disqualified. They had to be under 6 feet tall and 175 pounds to fit in the cockpit and between 25 and 40 years old. They underwent exhaustive physical and psychological exams, as well as background and security checks. Out of 16 candidates on the short list, 11 made the cut. These, too, were sheep-dipped: they had to resign their Air Force commissions and become temporary civilian employees of the Agency. Payment arrangements were similar to those for U-2 pilots. When their contract ended, the Air Force rescinded the resignations and awarded promotions.[130]

The first unofficial flight out of Area 51 was on April 25, 1962, with Louis Schalk in the driver's seat. He took it about a mile and a half at altitude of 20 feet. The following day was Article 121's first official flight, again piloted by Schalk. Both flights encountered minor problems that were easily corrected. When the J58 engines finally arrived from P&W in June 1962, the J75 engines were replaced with some difficulty, and the first J58-equipped flight took off on January 15, 1963. The chines had to be rebuilt to further improve the RCS to make the air-craft more invisible to TALL KING radar. Engine difficulties caused by the shock waves at supersonic speeds and also by foreign objects sucked from the runway kept engineers busy solving these new complications.[131]

Before the year was out, two more versions of OXCART were ordered by the Agency and the Air Force, in addition to the A-12 and YF-12A. One was known as Project TAGBOARD, in which motherships (M-21) carried ramjet-powered drones (D-21) "piggyback" just forward of the tail section. The drones could reach Mach 3.3 over a range of 3,000 miles. The Agency passed the project off to the Air Force, which later substituted B-52 bombers to carry the drones. TAGBOARD was canceled in July 1971 after a launch accident killed a crewman and no successful missions. The second was another reconnaissance aircraft with a two-seat cockpit and an array of photographic and electronic equipment that would become the SR-71 Blackbird.[132]

Surfacing

As early as late 1962, the Department of Defense considered publicizing ("surfacing") the YF-12A interceptor as a cover for the top-secret A-12 to explain the exorbitant defense spending. Some journalists and those in the aerospace industry had also heard of the program, and the Kennedy administration feared it would soon appear in the media. Defense Secretary Robert McNamara in particular wanted at least part of OXCART to be surfaced, but DCI John McCone successfully opposed it, along with Killian and Land, on the grounds that any public disclosure of the plane would allow the Soviets to create a radar complex to counter those features and therefore the value of the plane for reconnaissance.[133] But another argument for surfacing was later presented to Kennedy: Lockheed had been shown favoritism in funding for supersonic technology among contractors to the tune of $700 million already received from the government. That same technology, if shared, would benefit civilian supersonic transport, which was then under consideration by Congress. McCone conveyed all this to the president on November 12, 1963. Kennedy advised the Agency and Department of Defense to prepare a proposal for surfacing for his consideration when he returned to the White House from a trip to Dallas, Texas, in a little over a week. He never did.[134]

A week after Kennedy's assassination, McCone met with President Johnson, Secretary of State Dean Rusk, National Security Advisor McGeorge Bundy, and Secretary McNamara to brief the president on the OXCART program and why the question of surfacing had been raised: a crash that might be made public, visibility in flight, and unfair competition to contractors other than Lockheed and P&W. McCone again recommended waiting, rather than expose an intelligence resource, but Rusk and McNamara urged the president to make the program public as soon as practicable for the reasons listed earlier. McNamara noted that if the president presented the YF-12A to Congress he could eliminate criticism, save $75 million (the cost of a B-70 supersonic bomber then under consideration), and spread supersonic technology among engineers and scientists. Another argument in favor of surfacing, not mentioned in the memo of the meeting, was that TALL KING now made it almost impossible for OXCART to overfly the Soviet Union undetected. Johnson ordered preparation of a surfacing paper but deferred further action until the following spring.[135]

Although the members of the PFIAB and Lt. General Marshall Carter were still opposed to surfacing OXCART, McCone ultimately believed that TALL KING would eliminate the possibility of any overflights of the Soviet Union.[136] On February 2, 1964, he sent out a memo to US Intelligence Board members advising them of the release of a "very limited" portion of the OXCART-KEDLOCK program. They were reminded that security restrictions remained in effect and that they were never to discuss the program or any portion thereof. Even those no longer involved, such as Dulles and Bissell, were notified, as was Aden, who received a call from the Agency. Aden interpreted the impending press conference as fulfilling Johnson's need to show that he was actively monitoring what the Soviet Union was doing, presumably to raise his approval rating, which was 79 percent by the end of the month. It might also have been staged to demonstrate that the United States was not falling behind the Soviet Union in military technology, as Johnson's presidential opponent that year, Barry Goldwater, was claiming. In either case, the motivation was probably political, and Johnson was—if nothing else—a master politician throughout his career in Congress and the White House, as emphasized in the comprehensive biographical series about Johnson by Robert Caro.[137] The CIA did not trust Johnson, in part because he could not uphold the secrecy of the projects at Area 51. The mistrust was probably mutual after the Bay of Pigs invasion of 1961 failed miserably.[138]

At a National Security Council meeting on February 28, 1964, McCone, Bundy, Rusk, and McNamara discussed the previously noted reasons for surfacing the A-11, *not* the A-12, a cover for the latter. All present agreed with the decision. The next day President Johnson announced the development of the A-11, which he said had been tested in flights over 2,000 miles an hour at altitudes greater than 70,000 feet. The purpose of announcing it, he said, was to "allow the orderly exploitation of this advance technology in our military and commercial planes." He cited the "air-to-air missile system" of the A-11, which in fact referred to the canceled YF-12A interceptor. Kelly Johnson was the one who wrote the president's press release, intending to preserve the secrecy of the A-12, so this was not an error on Johnson's part. Johnson was so anxious to boost his image that he decided to surface the planned Blackbird successor to the A-12, originally called the RS-71 (for Reconnaissance-Strike), on July 25. At the last minute, Air Force General Curtis LeMay changed the designation to SR for Strategic Reconnaissance, but the press release still read RS-71. This gave rise to the oft-repeated, mistaken interpretation that Johnson had transposed the letters in his speech.[139]

Downtime

Successive convoys left the Skunk Works in 1963 and 1964 with A-12s, YF-12s, and the M-21s of TAGBOARD bound for Area 51. By mid-1965 the entire A-12 fleet had been delivered.[140] Tests of the A-12 prior to long-range flights revealed several unrelated problems. On May 24, 1963, pilot Ken Collins had to eject when water froze in the pitot tube, which led to an incorrect reading of the air speed. He was unhurt when the plane stalled, but Article 3 was lost. On July 4, 1964, Article 133 piloted by Bill Park crashed during approach. He ejected sideways and landed safely on the first swing of his parachute. A crash during takeoff of Article 126 occurred on December 28, 1965, because of mis-wired pitch and yaw gyros on board; pilot Mele Vojvodich, Jr., ejected safely. Finally, on November 20, 1965, the first successful long-range flight lasted over six hours, reaching Mach 3.29 and an altitude of 90,000 feet. OXCART was now ready for its first missions.[141]

Ben Rich, one of Lockheed's brilliant engineers, convinced Kelly Johnson to paint the A-12s black to radiate heat away during friction and thereby lower the wing temperature by 35 degrees. Johnson was at first reluctant to add more weight but acquiesced the next day. By late 1964, all had been painted, and the so-called Blackbirds were born (Figure 11.4; consult the color insert for this figure). An alternative origin for the Blackbirds is one told by T. D. Barnes, who worked for years at Area 51 on the A-12 and other aircraft. Barnes wrote that Bissell had come up with a blue-black paint during Project RAINBOW with the U-2s and insisted that Lockheed use it on the A-12s to make it blend better at high flight; a metallic additive to the paint raised its price to $400 a gallon.[142]

Overflights of the Soviet Union were ruled out, but A-12s were on standby, even before refitting with P&W J58 engines, for surveillance of Cuba in November 1964 as Project SKYLARK. Five pilots and five aircraft were readied but never flew, leaving the missions to the U-2. For the better part of the next three years, deployment of OXCART was a political football while its performance and equipment were being perfected. At issue was the perception in some quarters that with the development of the SR-71 the A-12 had become "the world's most advanced anachronism."[143] In November 1965 the Bureau of the Budget raised objections about the costs of both OXCART and SR-71 programs and suggested phasing out OXCART by September 1966. A group including Deputy Director of the CIA Carl Duckett, Carl Fischer of the Bureau of the Budget, and Herbert Bennington of the Defense Department visited the Skunk Works to examine the

OXCART program in relation to the SR-71. The ensuing Fischer-Bennington Report recommended phasing out OXCART by September 1966, making way for the SR-71 to assume manned reconnaissance missions by the end of 1967. Both the Department of Defense and the Agency opposed the recommendation for different reasons. McNamara said that the SR-71 would not be operational by September 1966, and DCI Richard Helms argued that the A-12's Type I camera, which Aden had helped design, covered twice as much area as the SR-71 cameras and yielded better resolution. Nonetheless, President Johnson recommended mothballing the OXCART fleet by January 1968. In the meantime, OXCART remained on call for overflights of Cuba and Southeast Asia, although there was growing reluctance to send the A-12 to Cuba so as not to disturb the relatively peaceful status quo.[144]

In contrast to recommendations of the Fischer-Bennington Report, the PFIAB (chaired by Clark Clifford from 1963 to 1968) identified gaps in clandestine intelligence coverage of North Vietnam and South China, particularly development of nuclear weapons and military buildups and movements, but also China's ability to influence and support political and military activity in Southeast Asia and mainly North Vietnam.[145] In three years the Agency had lost four U-2s over China, and the Air Force also never recovered several drones from overflights there. In July 1966 the Agency recommended deployment of A-12 aircraft to provide early warning of a Chinese military buildup or major movement into North Vietnam and that the aircraft be held in readiness for use over North Vietnam if necessary, despite continuing opposition from the Departments of State and Defense.[146]

While waiting for mission assignments, A-12 drivers continued to train out of Area 51, setting speed and distance records for all aircraft to that time. On December 21, 1966, Lockheed test pilot Bill Park flew over 10,198 statute miles in 6 hours at an average speed of 1,659 miles per hour (including refueling in flight at 602 miles per hour). The first OXCART fatality occurred on January 5, 1967, when Article 125 ran out of fuel 70 miles short of Groom Lake. Agency pilot Walter Ray, married only three months, ejected but could not separate from the seat and was killed upon impact with the ground. As a cover for OXCART, the Air Force told the press that an SR-71 was lost in Nevada.[147]

BLACK SHIELD

The Agency alerted President Johnson to the possibility that surface-to-surface missiles were being transported to North Vietnam, and so Johnson requested a proposal. On May 15, 1967, DCI Richard Helms submitted a plan named Project BLACK SHIELD to use OXCART to overfly North Vietnam out of Kadena Air

Base on Okinawa. Although CORONA (KH-4) could provide low-resolution photos of railyards, for instance, only one mission a year was sufficiently cloudless. Better resolution was also needed to identify ballistic missiles and related equipment as well as to detect camouflaged sites. Since the beginning of 1967, 37 U-2 missions covered portions of North Vietnam, but again cloud cover obscured targets. The U-2s were also vulnerable to launches of SA-2 surface-to-air missiles, and so coverage was limited to the northwest of the country where there were no SAM installations. Low-level drones had narrow coverage, and high-level ones were likewise subject to missile attacks. OXCART aircraft installed with operational electronic counter measures (ECM) would be virtually invulnerable to SA-2 missiles at high altitudes and speed and could provide ground resolutions of 1.0–3.5 feet. The proposal reached Johnson's desk the next day, when he signed off on it. On May 31, 1967, three A-12s equipped with Type-1 cameras designed by Rod Scott and Aden flew to Kadena; two arrived in about six hours with two mid-air refuelings, but the third had navigation problems and stopped at Wake Island overnight. Mele Vojvodich, who had been forced to eject during takeoff in December 1965, flew the first mission on May 31, 1967, at Mach 3.1 and 80,000 feet. He was able to fly undetected in clear weather over 10 priority targets and 70 of the 190 known SAM sites. The cameras photographed those sites, as well as ports, airfields, and army barracks in Haiphong and Hanoi. Three and a half hours after departure, he was back in Kadena. Film was couriered to Eastman Kodak and received for interpretation at NPIC within 24 hours of the mission.[148]

There were reports that the Viet Cong were transporting weapons and ammunition into Cambodia and storing them in caches, although the French, who had invested heavily in that country, disputed the reports. Two BLACK SHIELD missions (BX6737 and BX6738) over Cambodia confirmed the existence of several arms caches along two rivers near the South Vietnamese border.[149]

In subsequent missions the rest of the year, only four radar signals from A-12s were detected on the ground. Given the A-12's speed, a single-pass mission kept it over Vietnam only 12.5 minutes and a two-pass mission 21.5 minutes. As it had a minimum 86-mile turning radius (sometimes up to 125 miles), it sometimes crossed over the Chinese border and indeed was able to photograph Chinese airfields, military facilities, and ports. Although SA-2 missiles were fired at A-12s on October 28 and October 30, they detonated behind the aircraft. From the first flight until the last over North Vietnam on March 8, 1968, OXCART provided high-resolution photographs of all of North Vietnam's significant airfields, the port of Haiphong where two or three Soviet freighters were docked, MRBM sites, railroads, bridges, the Demilitarized Zone, and post-strike bomb damage to power plants, airfields, etc. However, it found no evidence of surface-to-surface missile deployment to North Vietnam.[150]

OXCART'S exemplary performance and photo-intelligence over North Vietnam led the White House and Congress to re-evaluate the program's termination. The Agency claimed that OXCART flew higher and faster than the Air Force's SR-71 and had better cameras, while the Air Force argued that its spy plane had better sensors, three sets of cameras, infrared detectors, side-looking radar, and electronic intelligence collection equipment. There was no conclusive way to settle the debate than to have a "flyoff" between the two. So on October 20 and October 25, 1967, they flew test missions over the United States, followed by another on identical flight paths, separated by an hour. The data collected were compared by both organizations. Results showed that OXCART's Type-1 camera was superior to the SR-71's Operational Objective camera but that the infrared, radar, and electronic gear of the SR-71 supplied additional valuable intelligence. This was enough information to give OXCART another temporary reprieve when President Johnson decided to maintain both programs, even as military costs in Southeast Asia kept rising.[151]

On January 23, 1968, the USS *Pueblo* was captured by North Korean forces, who tortured the 83 crew members for 11 months thereafter. President Johnson considered using airpower to force their release, but DCI Richard Helms recommended OXCART to locate the ship and men, arguing that it could photograph all of North Korea in less than 10 minutes undetected. On January 26, driver Jack Weeks made three passes over the southern part of North Korea and DMZ. He photographed the ship in an inlet in Wonsan Bay on the east coast, surrounded by two patrol boats and two Komar-class missile boats, but no signs of a military buildup. He later flew out of Kadena on June 4, 1968, in Article 129 on a routine test flight but disappeared without a trace east of the Philippines. A second BLACK SHIELD mission over North Korea was requested on January 29, 1968, to photograph seven major airfields, naval bases, and ground-force activity there. Although Dean Rusk was reticent to approve it, fearing that it would be shot down, the mission was given the green light and was flown by Frank Murray on February 19, 1968; Murray had been an F-101 chase pilot "sheep-dipped" from the Air Force to the CIA to replace the late Walter Ray. Jack Layton piloted the third flight of BLACK SHIELD— and the final flight of OXCART—on May 8. Altogether there were 29 BLACK SHIELD missions: 24 over North Vietnam, 2 over Cambodia, and 3 over North Korea.[152]

Termination

Over the objections of National Security Advisor Walter Rostow, the PFIAB, some congressmen, the President's Scientific Advisory Committee, and DCI

Richard Helms (who called the OXCART "the most sophisticated operationally proved, manned reconnaissance system ever developed"), a new study to determine the fate of the remaining A-12s was ordered. After considering four proposed alternatives, including the continuation of OXCART under Agency management, Defense Secretary Clark Clifford supported President Johnson's original decision to phase out the program, and Johnson signed off on it on May 21, 1968. The SR-71 assumed all future overflights of North Vietnam and would become the most valuable reconnaissance vehicle of the Vietnam War. By mid-June the remaining A-12s were mothballed back in Palmdale, California. Other than the J58 engines, they could not be cannibalized for parts because the A-12 was smaller than the SR-71, and the Type-1 Perkin-Elmer cameras were too large for the compartment in the SR-71.[153] One of the A-12s now rests high on poles outside the entrance to the San Diego Air & Space Museum at Balboa Park as a memorial to the genius of Kelly Johnson and Lockheed engineers and to the A-12's contributions during the Cold War as the first stealth plane. Others are currently on display in Palmdale, New York City, Los Angeles, Langley (Virginia), and in Alabama at Huntsville, Birmingham, and Mobile.[154]

Significance

The 1950s and 1960s were a crucible for innovative achievements in aerodynamics and aerial surveillance by spy planes and satellites. From their origins in 1957 through phase-out in 1968, Lockheed produced 15 A-12s, 3 YF-12As, and 31 SR-71s. Altogether they logged more than 7,300 flights and 17,000 hours of air time. More than 2,400 hours were spent at more than Mach 3.0. Given the delays in the A-12's deployment for political and budgetary reasons and the speed in developing the SR-71, it is understandable that it was surpassed in some ways by the SR-71 even before its first missions over North Vietnam. Although the A-12 was originally intended to overfly the Soviet Union undetected, Soviet advances in radar made that impossible, obviating the aircraft's original raison d'etre. The U-2 had a faster response time and did not require planning for in-flight refueling and emergency landing fields. Also, the A-12 was high-maintenance in terms of costs, time, and required personnel, which is how the Department of Defense repeatedly justified terminating the CIA program in favor of Air Force projects. However, unlike the U-2, the A-12 and SR-71 were never shot down by virtue of their maximum speed and altitude. Apart from the valuable military intelligence gleaned by BLACK SHIELD, the years of OXCART's engineering firsts in developing innovative designs, engines, fuels, the Type-1 camera, RCS reduction, manufacture with titanium, etc., pioneered technologies for stealth aircraft in the skies today.[155]

For his service to the US Air Force and national security, sometime in the late 1960s (exact date unknown) Aden was presented with a certificate of appreciation signed by Air Force Chief of Staff General John P. McConnell and also by Secretary of the Air Force Harold Brown, who went on to serve as president of Caltech from 1969 to 1977 before becoming Secretary of Defense under President Jimmy Carter and receiving the Presidential Medal of Freedom in 1981 (Figure 11.5; consult the color insert for this figure).

Aden and the shifting role of science in the Cold War

Among the posts and accomplishments on his résumé, Aden listed his appointments as scientific consultant to the CIA from 1962 through 1968 and to the Secretary of the US Air Force from 1963 through 1971. His high-level security clearance during the administrations of several US presidents is evident in his and Marjorie's autobiography and correspondence, and we may never know the full extent of his involvement in matters of aerial surveillance during the Cold War. We do know that he was often summoned—at government expense—for high-level meetings in Washington, DC, Boston, California, and elsewhere while director of the Steward Observatory, the Optical Sciences Center, and College of Optical Sciences; his frequent absences for undisclosed reasons were noted by many alumni and emeriti faculty of the College interviewed for this book. We know that he contributed to the satellite program in varying degrees from CORONA (KH-4) through KENNEN (KH-11) and to OXCART, as recounted earlier. Aden was a trusted associate of fellow scientists Albert "Bud" Wheelon in the Agency, General Lew Allen, Jr., of the Joint Chiefs of Staff, Director Brockway McMillan of the National Reconnaissance Office, and Din Land of Polaroid. He was caught in the middle of the ongoing feud between the Agency and the Air Force and ultimately chose to consult only for the Air Force because he grew tired of flying to Washington when California's aircraft facilities and its Air Force bases were so much closer to his Arizona home.[156] Over the decades he witnessed firsthand how differently presidential administrations valued input from scientists into policies promoting national security.

Even before World War II ended, Vannevar Bush was considering how to fund military technology for national defense afterward. Should the OSRD continue as before, or would a new agency to support scientists be required in peacetime? Bush felt that the duplication of military research projects was wasting resources and also that it was necessary to fund pure scientific research, not just industrial and military applications, to ensure continuing innovation by new generations of scientists and thereby remain steps ahead of rival nations. He sent a letter to

President Roosevelt asking for his thoughts. On November 17, 1944, Roosevelt replied, asking four questions still relevant today: (1) While safeguarding military security, what can be done to publicize contributions to science during the war in order to stimulate new enterprises and provide jobs for returning servicemen and others? (2) How do we organize a program to continue medical research in the war against disease? (3) How can government abet research by public and private organizations? (4) How might government support young scientists, such as those whose educations had been interrupted by the war? Bush formed separate committees of experts to tackle each of the questions. Land of Polaroid, who would be so instrumental in defining and conceptualizing needs for aerial surveillance during the Cold War, served on the committee to advise on question 3. They recommended the establishment of a National Science Foundation—which Bush readily adopted—to distribute monies to educational and nonprofit research institutions for scientific research, create scholarships and fellowships in the natural sciences, enable dissemination of scientific and technical data, support international cooperation in science, and come up with ways to facilitate the transition between pure research and industrial applications. With the input of that committee and three others, Bush drafted a long response to Roosevelt, published as *Science: The Endless Frontier*. In it, Bush advocated ample funding for both pure and applied science, as well as the need for scientific consultation in military research and governmental decision-making.[157]

Determined not to relinquish control to a centralized civilian body like a National Science Foundation and responding to Bush's threat to shut down the OSRD when the war ended, both the Army Air Force (not yet separate entities at that point) and the Navy organized their own cadre of scientific consultants. Army Air Chief Henry "Hap" Arnold invited Theodore von Kármán of Caltech to chair a Scientific Advisory Group commissioned to envision an Air Force dependent on scientists; it would lead to Project RAND. Edward Bowles, consultant to Arnold, successfully proposed that a scientific panel contract with Douglas Aircraft Company to develop new weapons. Bowles simultaneously pursued Army policymakers and called for West Point to establish an academic major or minor in technology designed to solve military problems.[158]

The Office of Naval Research (ONR), founded in August 1946 by Admiral Harold Bowen to promote nuclear propulsion for naval vessels and directed by Bush, helped to shape postwar science policies until the Korean War. The ONR supported academic disciplines, individuals, and institutions and actively campaigned for financial support and training of scientists and administrators to make the United States the leader in global science to fill the gap left by the wartime devastation of European scientific resources. University Presidents Killian (MIT) and DuBridge (Caltech) praised its efforts to restore and expand basic research in all fields of science after the war, either by contracts or grants (the latter not until 1959 after Sputnik). The ONR funded not only research costs but also conferences and travel to them. It provided access to naval test ranges, ships, and aircraft for experiments and gave wide latitude to university administrators managing research finances. Chief scientist for the ONR was Alan Waterman, a Yale physicist who would later become first director of the National Science Foundation (NSF) in 1951 and serve until 1963. Succeeding him was Emanuel Piore, Director of Research at IBM and a member of the President's Science Advisory Committee under Eisenhower, along with Land (Polaroid), Waterman (NSF), Killian (MIT), Lauritsen (Caltech), Rabi (Columbia), and others.[159]

In July 1947, Congress passed a bill creating a National Science Foundation along the lines of Bush's plan, but Truman summarily vetoed it on constitutional grounds that the foundation's director and policies would be decided by a board appointed by the president, not by the president himself. He was leery of giving scientists a dominant role, "the last word," as the administration's science advisor Don Price put it. He finally signed a new bill in the spring of 1950 that established the NSF with Waterman as its first director (appointed by the president) and a charter much like that of Bush's *Endless Frontier* but without military and medical research components.[160]

With the postwar influence and success of the ONR and NSF, it should not be surprising that Eisenhower should call upon scientists to help calm a wary public after the launch of Sputnik and affirm American superiority in both science and defense. In December 1957 he appointed Killian as Special Assistant to the President for Science and Technology (later followed by Kistiakowsky) and included the President's Scientific Advisory Committee (PSAC) in December 1957 in his inner circle. He balanced the tremendous influence of the Department of Defense with the common sense of his PSAC to reach unbiased decisions about military programs and expenditures. Killian and Land were also on the President's Board of Consultants on Foreign Intelligence Activities, and the two of them were in virtual control of intelligence-gathering activities for years, mainly in the CIA and National Security Agency. They were also at the forefront of convincing Eisenhower to expand science education in America, which materialized as the National Defense Education Act of 1958 that stressed training in mathematics, science, and modern foreign languages and doubled the number of students attending college by 1970. Land and his wife personally donated $12.5 million anonymously to build the Science Center for undergraduate science and mathematics at Harvard University.[161]

Robert McNamara, formerly president of the Ford Motor Company, introduced increased financial accountability for government funding when he became Secretary of Defense in 1961 under President Kennedy. He held the position throughout the Kennedy and Johnson administrations and vigorously fought for Defense Department control of aerial surveillance projects and cost containment. Suddenly relevance, pragmatism, and bean-counting entered into awarding and administering military contracts. Long gone were the free-wheeling days of ONR contracts and grants.

President Kennedy, who had campaigned in part on the "missile gap" while Eisenhower needed to remain tight-lipped about the classified data emerging from U-2 and CORONA photography that clearly showed otherwise, learned the truth when he took office and suddenly developed a keen interest in surveillance and photointerpretation. Arthur Lundahl, director of the NPIC, recalled that Kennedy wanted explanations of technical data in terms that the average person could understand. For example, in describing how the U-2's cameras could image an area 100 nautical miles long and 200 miles wide on 10,000 feet of film, Lundahl made the analogy of photointerpreters using a magnifying glass to scan a roll of film that stretched from the White House to the Capitol and back. Kennedy was so impressed that he called on Lundahl to repeat the analogy for U-2 briefings at the White House. Killian and Land said that Kennedy simply had difficulty understanding technical information and did not rely on members of the PFIAB as much as Eisenhower did. Kennedy kept Eisenhower in the loop on U-2 photography and on the Cuban Missile Crisis, during which Kennedy actively sought Ike's counsel.[162] Although there were some who believed that it was dangerous to entrust important decisions to a few unelected and unaccountable scientists, most of Kennedy's inner circle welcomed them because as experts they could be trusted to be objective and independent.[163]

In contrast, Lyndon Johnson had little time for basic science unless it benefited society directly. He did not understand complex technology and ignored scientists in general. On one occasion while being briefed in the simplest terms the scientists knew, he was at the other end of the table talking on the telephone. One reason for his contempt might be that he associated university professors and students with the relentless demonstrations and attacks on his policies in Vietnam and did not receive the support from them that he wanted for his Great Society program.[164]

When Richard Nixon was elected in 1968, he reorganized the White House science and technology section and soon eliminated his science advisor and President's Advisory Committee, in part because some committee members opposed his policies. The influence of science in the White House had reached the lowest point in modern American history. When Gerald Ford became president upon Nixon's resignation, he proposed in June 1975 that there be established an Office of Science and Technology Policy with a director who would serve as the president's science advisor. It also established a Federal Science and Technology Survey Committee to assess the entire science and technology effort. Ford signed H.R. 10230 into law as the National Science and Technology Policy, Organization, and Priorities Act on May 11, 1976. The following year, newly elected President Jimmy Carter abolished the PFIAB, only to have Ronald Reagan reinstitute it in 1981 shortly after his landslide defeat of Carter in 1980.[165]

The launch of Sputnik I in 1957 had sparked a renaissance of science, technology, and science education on the perception that America lagged behind the Soviets in those fields and so made it vulnerable to another surprise attack like Pearl Harbor, only much worse. Not only did it accelerate plans at the RAND Corporation for satellite surveillance and at Lockheed for OXCART in which Aden was deeply involved, it also precipitated an increase of 19.8 percent in funding of NSF, amounting to $30,750,000 funded by Congress that would be used to help develop Kitt Peak National Observatory under Aden's directorship (see Chapter 7).[166]

Notes

1. Burrows, W. E. 1986. *Deep black: Space espionage and national security*. Random House, New York, 92; Yenne, B. 2014. *Area 51 black jets*. Zenith Press, Minneapolis, Minnesota, 29; Peebles, C. 1997. *The Corona project: America's first spy satellites*. Naval Institute Press, Annapolis, Maryland, 17; Taubman, P. 2003. *Secret empire: Eisenhower, the CIA, and the hidden story of America's space espionage*. Simon & Schuster, New York, 23; Green, C. M., and Lomask, M. 1970. *Vanguard: A history*. National Aeronautics and Space Administration, SP-4202, Washington, DC, 1.

2. Hitchcock, W. I. 2018. *The age of Eisenhower: American and world in the 1950s*. Simon & Schuster, New York, 380.

3. Hitchcock, *The age of Eisenhower*, 380; Sambaluk, N. M. 2015. *The other space race: Eisenhower and the quest for aerospace security*. Naval Institute Press, Annapolis, Maryland, 63, 67, 117.

4. Burrows, *Deep black*, 93.

5. Public Papers of the Presidents of the United States, Dwight D. Eisenhower: 1957, https://quod.lib.umich.edu/p/ppotpus/4728417.1957.001?rgn=main;view=fulltext, accessed November 6, 2019.

6. Draft by J. F. Dulles and White House Statement, DDE Presidential Papers 1953–1961, Ann Whitman File, Administration Series, Box 37, Folder: U.S. Satellites, Dwight D. Eisenhower Presidential Library; Mieczkowski, Y. 2013. *Eisenhower's Sputnik moment: The race for space and world prestige*. Cornell University Press, Ithaca, New York, 65.

7. Hitchcock, *The age of Eisenhower*, 379–80; Green and Tomask, *Vanguard*, 196.

8. Green and Tomask, *Vanguard*, 43–51, 57.

9. Hitchcock, *The age of Eisenhower*, 388; Green and Tomask, *Vanguard*, 210, 217.

10. *Washington Post*, July 30, 1958; Thomas, E. 2012. *Ike's bluff: President Eisenhower's secret battle to save the world*. Back Bay Books, New York, 313–14; Sambaluk, *The other space race*, 129.

11. Burrows, *Deep black*, 95; Dienesch, R. M. 2016. *Eyeing the red storm*. University of Nebraska Press, Lincoln, 139.

12. Letter from Riggs Monfort, Information and Privacy Coordinator, Central Intelligence Agency, to Alec Pridgeon, June 27, 2019.

13. Neufeld, M. J. 2007. *Von Braun: Dreamer of space, engineer of war*. Random House, New York, 199–200.

14. Taubman, *Secret Empire*, 57–8.

15. Baker, D. 2016. *US spy satellites 1959 onwards (all missions, all models)*. Haynes, Sparkford, UK, 21–3; Peebles, *The Corona project*, 4–6.

16. Peebles, *The Corona project*, 6–9.

17. D. Baker, *US spy satellites*, 28; Peebles, *The Corona project*, 9–12; Pedlow, G. W., and Welzenbach, D. E. 2016. *The Central Intelligence Agency and overhead reconnaissance: The U-2 and OXCART Programs, 1954–1974*. Skyhorse Publishing, New York, 5–6, 21–3; McElheny, V. K. 1998. *Insisting on the impossible: The life of Edwin Land*. Perseus Books, Cambridge, Massachusetts, 284–7.

18. McElheny, *Insisting on the impossible*, 288–9; Pedlow and Welzenbach, *The Central Intelligence Agency*, 25.

19. Pedlow and Welzenbach, *The Central Intelligence Agency*, 24–5.

20. McElheny, *Insisting on the impossible*, 290–1; D. Baker, *US spy satellites*, 29.

21. Peebles, *The Corona project*, 13; McElheny, *Insisting on the impossible*, 289; Richelson, J. T. 2002b. *The wizards of Langley: Inside the CIA's Directorate of Science and Technology*. Westview Press, Boulder, Colorado, 29; Brugioni, D. A., and McCort, R. F. 1988. Personality: Arthur C. Lundahl. The art of aerial photography. *Programmatic Engineering & Remote Sensing* 54: 271, https://www.cia.gov/library/readingroom/docs/CIA-RDP86B01053R000100090067-3.pdf; accessed November 18, 2019.

22. McElheny, *Insisting on the impossible*, 14.

23. McElheny, *Insisting on the impossible*, 197.

24. Brugioni and McCort, Personality: Arthur C. Lundahl, 271–2.

25. Pedlow and Welzenbach, *The Central Intelligence Agency*, 32–5; Peebles, *The Corona project*, 18–19; Yenne, *Area 51 black jets*, 29; Richelson, *The wizards of Langley*, 11–12.

26. Rich, B. R., and Janos, L. 1994. *Skunk Works: A personal memoir of my years at Lockheed*. Little, Brown, and Company, New York, 107–8.

27. Rich and Janos, *Skunk Works*, 7, 111; Barnes, T. D. 2018a. *The Angels. Book One: The CIA Area 51 chronicles*. Self-published. 71.

28. Yenne, *Area 51 black jets*, 30–1; Pedlow and Welzenbach, *The Central Intelligence Agency*, 11–13, 16–18, 35–9, 41, 44, 46; Richelson, *The wizards of Langley*, 13; Memo, A Unique Opportunity for Comprehensive Intelligence, to DCI from Edwin Land, November 5, 1954; Richelson, J. T. (ed.) 2002a. The U-2, OXCART, and the SR-71: U.S. aerial espionage in the Cold War and beyond. National Security Archive Electronic Briefing Book No. 74. https://nsarchive2.gwu.edu/NSAEBB/NSAEBB74/; accessed September 10, 2019.

29. National Reconnaissance Office, US Military, and Department of Defense. 2017c. *20th century spy in the sky satellites: Secrets of the National Reconnaissance Office*. Vol.

5. *Leaders, founders, pioneers.* Progressive Management Publications, Washington, DC, 56–7, 201–2.

30. Pedlow and Welzenbach, *The Central Intelligence Agency,* 53, 57–60; Richelson, *The wizards of Langley,* 14; Barnes, T. D. 2017. *The secret genesis of Area 51.* The History Press, Charleston, South Carolina, 53; Pace, S. 2016. *The projects of Skunk Works.* Voyageur Press. Minneapolis, Minnesota, 86; Taubman, *Secret Empire,* 151–2; Shaw, B. 1989. Origins of the U-2: Interview with Richard M. Bissell, Jr. *Air Power History* 36: 15–24, 19.

31. Meinel, A. B., and Meinel, M. P. 2002a. *Echoes from a simpler time.* Unpublished, 134.

32. https://geographicalimaginations.com/tag/atomic-bomb/;   accessed   November 11, 2019.

33. Rich and Janos, *Skunk Works,* 121–2.

34. Peebles, *The Corona project,* 23–4; Pedlow and Welzenbach, *The Central Intelligence Agency,* 103.

35. Pedlow and Welzenbach, *The Central Intelligence Agency,* 66–7; Haines, G. K. 1998. The National Reconnaissance Office: Its origins, creation, and early years. In D. A. Day, J. M. Logsdon, and B. Latell (eds.), *Eye in the sky: The story of the Corona spy satellites,* 143–56. Smithsonian Institution Press, Washington, DC, 150–3; Organization and Delineation of Responsibilities: Project Oilstone, August 2, 1955, National Security Archive, Document 4, https://nsarchive2.gwu.edu/NSAEBB/NSAEBB74/U2-04.pdf; accessed October 16, 2019.

36. Barnes, *The Angels,* 182, 187–90; Powers, F. G., and Gentry, C. 2004. *Operation Overflight: A memoir of the U-2 incident.* Potomac Books, University of Nebraska Press, Lincoln, xiv; Powers, F. G., Jr., and Dunnavant, K. 2019. *Spy pilot: Francis Gary Powers, the U-2 incident, and a controversial Cold War legacy.* Prometheus Books, Amherst, New York, 42; Taubman, *Secret Empire,* 159.

37. Barnes, *Secret genesis,* 75, 96–7, 10.

38. Taubman, *Secret Empire,* 177.

39. Richelson, *The wizards of Langley,* 15.

40. Pedlow and Welzenbach, *The Central Intelligence Agency,* 113, 117, 119.

41. Pedlow and Welzenbach, *The Central Intelligence Agency,* 119–20, 139–41, 144, 146; Barnes, *The Angels,* 271–2; Taubman, *Secret Empire,* 187.

42. https://www.history.com/this-day-in-history/nikita-khrushchev-challenges-united-states-to-a-missile-shooting-match; accessed November 12, 2019; Pedlow and Welzenbach, *The Central Intelligence Agency,* 137–8.

43. Killian, J. R., Jr. 1977. *Sputnik scientists and Eisenhower: A memoir of the first Special Assistant to the President for Science and Technology.* MIT Press, Cambridge, Massachusetts, 13–14, 310–14.

44. Taubman, *Secret Empire,* 87; Killian, *Sputnik scientists,* xv; Ambrose, S. E. 1990. *Eisenhower: Soldier and president.* Simon & Schuster, New York, 452; McElheny, *Insisting on the impossible,* 310–16.

45. Killian, *Sputnik scientists,* 35–6.

46. Killian, *Sputnik scientists,* 219, 239, 241.

47. Memo of Conference with the President by Brigadier General A. J. Goodpaster, February 10, 1958, White House Office, Office of the Staff Secretary, Records, Subject Series, Alphabetical Subseries, Box 14, Folder: Intelligence Matters (4), Dwight D. Eisenhower Presidential Library.

48. https://www.rand.org/content/dam/rand/pubs/reports/2015/R262z1.pdf; accessed November 14, 2019.

49. Burrows, *Deep black*, 84; Davies, M. E., and Harris, W. R. 1988. *RAND's role in the evolution of balloon and satellite observation systems and related U.S. space technology.* RAND Corporation, Santa Monica, California, 69–70; Hall, R. C. 1998. Postwar strategic reconnaissance and the genesis of Corona. In D. A. Day, J. M. Logsdon, and B. Latell (eds.), *Eye in the sky: The story of the Corona spy satellites*, 86–118. Smithsonian Institution Press, Washington, DC, 42; Peebles, *The Corona project*, 25; Dienesch, *Eyeing the red storm*, 140.

50. Davies and Harris, *RAND's role*, 70, 86–7; Dienesch, *Eyeing the red storm*, 143, 150.

51. Dienesch, *Eyeing the red storm*, 155–7; Peebles, *The Corona project*, 43–4, 51; Hall, Postwar strategic reconnaissance, 111; National Reconnaissance Office, US Military, and Department of Defense. 2017d. *20th century spy in the sky satellites: Secrets of the National Reconnaissance Office.* Vol. 6. *Corona: America's first satellite program.* Progressive Management Publications, Washington, DC, 15–18; Burrows, *Deep black*, 104.

52. Peebles, *The Corona project*, 4; Taubman, *Secret Empire*, 238; Day, D. A., Logsdon, J. M., and Latell, B. 1998. Introduction. In D. A. Day, J. M. Logsdon, and B. Latell (eds.), *Eye in the sky: The story of the Corona spy satellites*, 1–18. Smithsonian Institution Press, Washington, DC, 273.

53. Memorandum titled Project CORONA—Security in 1960, March 17, 1959, White House Office, Office of the Staff Secretary, Records, Subject Series, Alphabetical Subseries, Box 15, Folder: Intelligence matters (10). Dwight D. Eisenhower Presidential Library; National Reconnaissance Office, Vol. 6. *Corona*, 28.

54. Memorandum for record from A. J. Goodpaster, April 21, 1958, White House Office, Office of the Staff Secretary, Records, Subject Series, Alphabetical Subseries, Box 14, Folder: Intelligence Matters (6), Dwight D. Eisenhower Presidential Library.

55. Hall, Postwar strategic reconnaissance, 112–13, 117; Peebles, *The Corona project*, 45–8; Lewis, J. E. 2002. *Spy capitalism: Itek and the CIA.* Yale University Press, New Haven, Connecticut, 90; National Reconnaissance Office, Vol. 6. *Corona*, 21–2; Smith, F. D. 1997. The design and engineering of Corona's optics. In *Corona between the sun and the earth: The first NRO reconnaissance eye in space* (ed. R. A. McDonald), 111–20. The American Society for Photogrammetry and Remote Sensing, Bethesda, Maryland, 115.

56. Wheelon, A. D. 1998. Corona: A triumph of American technology. In D. A. Day, J. M. Logsdon, and B. Latell (eds.), *Eye in the sky: The story of the Corona spy satellites*, 29–47. Smithsonian Institution Press, Washington, D.C., 34–7.

57. Peebles, *The Corona project*, 272–5; National Reconnaissance Office, Vol. 6. *Corona*, 38, 43–4.

58. Memorandum from Eisenhower to agency heads, August 26, 1960, White House Office, Office of the Staff Secretary, Records, Subject Series, Alphabetical Subseries, Box 15, Folder: Intelligence Matters (18), Dwight D. Eisenhower Presidential Library.

59. Peebles, *The Corona project*, 276–96; Day, D. A. 1998. The development and improvement of the Corona satellite. In D. A. Day, J. M. Logsdon, and B. Latell (eds.), *Eye in the sky: The story of the Corona spy satellites*, 48–85. Smithsonian Institution Press, Washington, DC, 63–5.

60. Day, The development and improvement of the Corona satellite, 65; National Reconnaissance Office, Vol. 6. *Corona*, 134–5; Peebles, *The Corona project*, 121; Smith, Corona's optics, 116–17.

61. Day, The development and improvement of the Corona satellite, 67, 69, 75, 77, 80, 82; Peebles, *The Corona project*, 244, 282–312; National Reconnaissance Office, Vol. 6. *Corona*, 135–6, 138–9; Lewis, *Spy capitalism*, 226; Meinel and Meinel, *Echoes*, 134.

62. D. Baker, *US spy satellites*, 91–2; Peebles, *The Corona project*, 132–7; Richelson, *The wizards of Langley*, 56–7.

63. Peebles, *The Corona project*, 139–44, 264–6; National Reconnaissance Office, Vol. 6. *Corona*, 141–2; May, E. R. 1998. Strategic intelligence and U.S. security. In D. A. Day, J. M. Logsdon, and B. Latell (eds.), *Eye in the sky: the story of the Corona spy satellites*, 21–8. Smithsonian Institution Press, Washington, DC; Brugioni, D. A. 2010. *Eyes in the sky: Eisenhower, the CIA and Cold War aerial espionage.* Naval Institute Press, Annapolis, Maryland, 392–3.

64. Letter from A. B. Meinel to Ed Meinel, September 22, 2009.

65. Interview with Helmut Abt by Alec Pridgeon, April 14, 2016, Tucson, Arizona.

66. Meinel and Meinel, *Echoes*, 132, 342; Reed, S. G., Van Atta, R. H., and Deitchman, S. J. 1990. DARPA technical accomplishments: An historical review of selected DARPA projects. Vol. 1. Institute for Defense Analyses, Alexandria, Virginia, Chapter 10: 1, 3.

67. Meinel and Meinel, *Echoes*, 132.

68. Meinel and Meinel, *Echoes*, 132; A. B. Meinel, unpublished correspondence.

69. Meinel and Meinel, *Echoes*, 132, 343–4.

70. A. B. Meinel, unpublished correspondence; this is not to be confused with the "Phoenix Lights" UFO phenomenon in March 1997.

71. Meinel and Meinel, *Echoes*, 132–4; email from A. B. Meinel to W. Patrick McCray, October 12, 1998; Richelson, J. T. 2003. Civilians, spies and blue suits: The bureaucratic war for control of overhead reconnaissance 1961–1965. National Security Archive Monographs. https://nsarchive2.gwu.edu/monograph/nro/; accessed September 10, 2019, 19.

72. Memorandum to the Special Assistant to the President for National Security Affairs from John McCone, September 10, 1963, Papers of President Kennedy, National Security Files, Departments and Agencies, Box 271, John F. Kennedy Presidential Library and Museum.

73. Richelson, *The wizards of Langley*, 68–72; Richelson, Civilians, spies and blue suits, 29–38, 40–3.

74. Email from A. B. Meinel to Mark Drach-Meinel, undated; personal communications, Ed Meinel and David Drach-Meinel.

75. Stern, S. M. 2005. *The week the world stood still: Inside the secret Cuban Missile Crisis.* Stanford University Press, 18–19.

76. Pedlow and Welzenbach, *The Central Intelligence Agency*, 210–11, 222–3; Burrows, *Deep black*, 117; Stern, *The week the world stood still*, 54.

77. Email from A. B. Meinel to Mark Drach-Meinel, undated; personal communications, Ed Meinel and David Drach-Meinel.

78. Stern, *The week the world stood still*, 157–8.

79. Stern, *The week the world stood still*, 174; Brugioni, D. A. 1990. *Eyeball to eyeball: The inside story of the Cuban Missile Crisis.* Random House, New York, 462; Powers, Jr., and Dunnavant, *Spy pilot*, 67.

80. Stern, *The week the world stood still*, 207.

81. National Reconnaissance Office, US Military, and Department of Defense. 2017a. *20th century spy in the sky satellites: Secrets of the National Reconnaissance Office.* Vol. 1. *Gambit photoreconnaissance satellite 1963–1984.* Progressive Management Publications, Washington, DC, 140–2; D. Baker, *US spy satellites*, 123–35; Haines, G. K. 1997. Critical to US security: The development of the GAMBIT and HEXAGON satellite reconnaissance systems. https://www.nro.gov/Portals/65/documents/hist ory/csnr/gambhex/Docs/Critical%20to%20US%20Security.pdf, 2, 13–14.

82. National Reconnaissance Office, Vol. 1. *Gambit photoreconnaissance satellite 1963–1984*, 141–2; D. Baker, *US spy satellites*, 123–4.

83. Haines, Critical to US security, 19–20.

84. Haines, Critical to US security, 21.

85. Memorandum to Secretary of Defense and Director of Central Intelligence from McGeorge Bundy, January 24, 1962, National Security Files, Bromley K. Smith Series, Box 470, Foreign Intelligence Advisory Board, 1/61–4/62, John F. Kennedy Presidential Library and Museum.

86. Meinel and Meinel, *Echoes*, 135.

87. Memo for the Special Assistant to the President for National Security Affairs from John McCone, December 23, 1963, National Security File—Intelligence File, Folder: Foreign Intelligence Advisory Board, Document 23b, Lyndon Johnson Library and Museum.

88. Haines, Critical to US security, 32–43; National Reconnaissance Office, Vol. 1. *Gambit photoreconnaissance satellite 1963–1984*, 141–2, 189, 210; D. Baker, *US spy satellites*, 130–1.

89. Haines, Critical to US security, D. Baker, *US spy satellites*, National Reconnaissance Office, Vol. 1. *Gambit photoreconnaissance satellite 1963–1984*, 216–23.

90. Meinel and Meinel, *Echoes*, 136–8.

91. National Reconnaissance Office, Vol. 1. *Gambit photoreconnaissance satellite 1963–1984*, 222; Haines, Critical to US security, 47; letter from A. B. Meinel to Brockway McMillan, August 29, 1964, Harvill Papers, Special Collections at the University of Arizona Libraries.

92. Meinel and Meinel, *Echoes*, 137–8; D. Baker, *US spy satellites*, 133; Haines, Critical to US security, 47.
93. D. Baker, *US spy satellites*, 133.
94. Komissarov, *Russia's ekranoplans*, 3–4, 17.
95. Meinel and Meinel, *Echoes*, 141–2; Komissarov, S. 2002. *Russia's ekranoplans: The Caspian Sea monster and other WIG craft*. Midland, Hinckley Publishing, UK, 6, 17, 19.
96. Meinel and Meinel, *Echoes*, 142.
97. National Reconnaissance Office, Vol. 1. *Gambit photoreconnaissance satellite 1963–1984*, 222, 241; Haines, Critical to US security, 49–52.
98. Meinel and Meinel, *Echoes*, 140.
99. Haines, Critical to US security, 60–3; National Reconnaissance Office, US Military, and Department of Defense. 2017b. *20th century spy in the sky satellites: Secrets of the National Reconnaissance Office*. Vol. 2. *Hexagon photoreconnaissance satellite 1971–1986*. Progressive Management Publications, Washington, DC, 25–7; Richelson, *The wizards of Langley*, 122–3.
100. Lewis, *Spy capitalism*, 234–5.
101. Haines, Critical to US security, 63–71; National Reconnaissance Office, Vol. 2. *Hexagon photoreconnaissance satellite 1971–1986*, 25–31; Richelson, *The wizards of Langley*, 112, 122–30; Lewis, *Spy capitalism*, 224–61; D. Baker, *US spy satellites*, 138; Taubman, *Secret Empire*, 345; letter to A. B. Meinel from A. D. Wheelon, August 17, 1964.
102. National Reconnaissance Office, Vol. 2. *Hexagon photoreconnaissance satellite 1971–1986*, 31–42; Lewis, *Spy capitalism*, 234–61; Haines, Critical to US security, 70–1; Taubman, *Secret Empire*, 347; Richelson, *The wizards of Langley*, 126–9; Richelson, Civilians, spies and blue suits, 58; D. Baker, *US spy satellites*, 138–9; Center for the Study of National Reconnaissance. 2012. *A history of the Hexagon program*. Chantilly, Virginia, 41; Pressel, P. 2013. *Meeting the challenge: The Hexagon KH-9 reconnaissance satellite* (editor-in-chief N. Allen). American Institute of Aeronautics and Astronautics, Reston, Virginia.
103. Meinel and Meinel, *Echoes*, 140; Pressel, *Meeting the challenge*, 8–9, 83, 118–21; email from Phil Pressel to A. Pridgeon, January 2, 2020.
104. D. Baker, *US spy satellites*, 149–50; Haines, Critical to US security, 100.
105. D. Baker, *US spy satellites*, 156–8; Outzen, J. D. (ed.). 2015. *The DORIAN files revealed: A compendium of the NRO's Manned Orbiting Laboratory documents*. National Reconnaissance Office, Center for the Study of National Reconnaissance, Chantilly, Virginia, 8, 25–30; Sambaluk, *The other space race*, 224.
106. Press release/conference, August 25, 1965, National Security Files-Intelligence, Document no. 4, Lyndon B. Johnson Library and Museum; Burrows, *Deep black*, 236.
107. News release from the Office of the Assistant Secretary of Defense (Public Affairs), August 25, 1965, National Security File—Charles E. Johnson, Document no. 22, Lyndon B. Johnson Library and Museum.

108. Outzen (ed.), *The DORIAN files revealed*, 40, 93, 98–9, 155, 163; Burrows, *Deep black*, 236; D. Baker, *US spy satellites*, 159, 161–3; McElheny, *Insisting on the impossible*, 339.

109. https://www.nro.gov/Portals/65/documents/foia/declass/Archive/NARP/1969%20NARPs/SC-2018-00033_C05111728.pdf.

110. Personal correspondence of A. B. Meinel; letter from A. B. Meinel to W. P. McCray, November 9, 1998; interview with Robert Shannon by James Breckinridge, February 27, 2019; https://www.nro.gov/Portals/65/documents/foia/declass/mol/816.pdf.

111. Meinel and Meinel, *Echoes*, 139; National Reconnaissance Office, Vol. 5. *Leaders, founders, pioneers*, 44, 103–4.

112. Meinel and Meinel, *Echoes*, 139–40.

113. National Reconnaissance Office, Vol. 6. *Corona*, 137–8; Peebles, *The Corona project*, 159–60, 294.

114. Burrows, *Deep black*, 243–9; D. Baker, *US spy satellites*, 165–71; Meinel and Meinel, *Echoes*, 140–1; Richelson, *The wizards of Langley*, 198–201; Boyle, W. S., and Smith, G. E. 1970. Charge coupled semiconductor devices. *Bell System Technical Journal* 49: 587–93.

115. Burrows, *Deep black*, 249; Richelson, *The wizards of Langley*, 20; D. Baker, *US spy satellites*, 167–8.

116. Email from A. B. Meinel to Mark Drach-Meinel, undated.

117. Pedlow and Welzenbach, *The Central Intelligence Agency*, 268–9; Rich and Janos, *Skunk Works*, 193, 195; Johnson, C. L. 1968. History of the OXCART program. Report SP-1362. Lockheed Aircraft Corporation, Burbank, California. Approved for release July 2007, 1.

118. Central Intelligence Agency (CIA), Directorate of Science & Technology (DST). 2016. History of the Office of Special Activities (OSA). Chapter 20. From Inception to 1969. Washington, DC. Released on appeal by Interagency Security Classification Appeals Panel (ISCAP) (final release), 2016, http://www.governmentattic.org/20docs/CIAhistOSAincep-1969u.pdf, 4, 9; Pedlow and Welzenbach, *The Central Intelligence Agency*, 270, 273; Memo to James Killian from Land Panel, November 15, 1958, White House Office of the Special Assistant for Science and Technology, Box 2, Folder: Intelligence (June–Nov. 1958), Dwight D. Eisenhower Presidential Library; Robarge, D. 2012. *Archangel: CIA's supersonic A-12 reconnaissance aircraft*. 2nd ed. Central Intelligence Agency, Washington, DC, 5–6; Johnson, History of the OXCART program, 3; Rich and Janos, *Skunk Works*, 195.

119. Johnson, History of the OXCART program, 2–4; Pedlow and Welzenbach, *The Central Intelligence Agency*, 272, 282; Rich and Janos, *Skunk Works*, 197–200; Robarge, *Archangel*, 5–7.

120. Pedlow and Welzenbach, *The Central Intelligence Agency*, 281–2, 284–5; Robarge, *Archangel*, 7; email from A. B. Meinel to Mark Drach-Meinel, undated.

121. Barnes, T. D. 2015. *Soaring with the eagles: Autobiography of Area 51 Veteran TD Barnes*. Self-published.

122. Meinel and Meinel, *Echoes*, 135; Barnes, *Soaring with the eagles*, 167–8, 181, 197, 248; Barnes, T. D. 2018c. *CIA Project OXCART. Area 51 Nevada*. Self-published, 15.

123. Pedlow and Welzenbach, *The Central Intelligence Agency*, 285, 288; Johnson, History of the OXCART program, 9; Robarge, *Archangel*, 11; Rich and Janos, *Skunk Works*, 200; interview with Dick McEwen by AP on January 26, 2019; Barnes, T. D. 2018b. *The Archangels. Book Two: The CIA Area 51 chronicles*. Self-published, 55, 76; Brugioni, *Eyes in the sky*, 216.

124. Pedlow and Welzenbach, *The Central Intelligence Agency*, 288; Rich and Janos, *Skunk Works*, 201–3.

125. McIninch, T. P. 1971. The Oxcart story. Studies in Intelligence 15: 1–25, https://www.cia.gov/library/center-for-the-study-of-intelligence/kent-csi/vol15no1/pdf/v15i1a01p.pdf; Pedlow and Welzenbach, *The Central Intelligence Agency*, 289; Robarge, *Archangel*, 25; Merlin, P. W. 2011. *Images of aviation: Area 51*. Arcadia Publishing, Charleston, South Carolina, 75; Brugioni, *Eyes in the sky*, 217.

126. Pedlow and Welzenbach, *The Central Intelligence Agency*, 290–2; McIninch, The Oxcart Story, 5, https://www.cia.gov/library/center-for-the-study-of-intelligence/kent-csi/vol15no1/pdf/v15i1a01p.pdf; letter from A. B. Meinel to Mark Drach-Meinel, undated; https://www.cia.gov/library/readingroom/docs/DOC_0000645397.pdf, 23–4; Robarge, *Archangel*, 36–8.

127. https://www.cia.gov/library/readingroom/docs/DOC_0000645397.pdf, 23–4; Rich and Janos, *Skunk Works*, 215; Johnson, History of the OXCART program, 11; Pace, *The projects of Skunk Works*, 110, 113; Pedlow and Welzenbach, *The Central Intelligence Agency*, 29, https://www.cia.gov/library/readingroom/docs/CIA-RDP78T04759A002100010003-1.pdf; Barnes, *The Archangels. Book Two*, 162.

128. Barnes, *CIA Project OXCART*, 30; Murray, *Once upon a time at Area 51*, 49; Pedlow and Welzenbach, *The Central Intelligence Agency*, 299; Yenne, *Area 51 black jets*, 81.

129. Pedlow and Welzenbach, *The Central Intelligence Agency*, 296–7; Johnson, History of the OXCART program, 12; Robarge, *Archangel*, 30; Yenne, *Area 51 black jets*, 79–80; https://www.cia.gov/library/readingroom/docs/DOC_0000645397.pdf, 28; http://www.governmentattic.org/20docs/CIAhistOSAincep-1969u.pdf; Murray, F. 2014. *Once upon a time at Area 51*. Self-published, 45; Barnes, *CIA Project OXCART*, 28.

130. Barnes, *The Archangels. Book Two*, 86–9; Barnes, *CIA Project OXCART*, 61; http://www.governmentattic.org/20docs/CIAhistOSAincep-1969u.pdf.

131. Pedlow and Welzenbach, *The Central Intelligence Agency*, 298–300; https://www.cia.gov/library/readingroom/docs/DOC_0000645397.pdf, 30–32; Johnson, History of the OXCART program, 12–13; Robarge, *Archangel*, 40–1; Pace, *The projects of Skunk Works*, 109.

132. Pedlow and Welzenbach, *The Central Intelligence Agency*, 301–2; https://www.cia.gov/library/readingroom/docs/DOC_0000645397.pdf, 33–4; http://www.governmentattic.org/20docs/CIAhistOSAincep-1969u.pdf; memo from J. Patrick Coyne to George McBundy, May 8, 1963, National Security Files, Bromley K. Smith Series, Box 470, Foreign Intelligence Advisory Board, John F. Kennedy Presidential Library and Museum; Barnes, *The Archangels. Book Two*, 163.

133. Memo for the record, meeting held in office of James Killian, Boston, January 12, 1963, National Security File - Intelligence, Box 4, Folder: National Reconnaissance Program [1 of 2], Document 14, Lyndon B. Johnson Library and Museum; memo for McGeorge Bundy from J. Patrick Coyne, January 20, 1963, National Security File—Intelligence, Box 4, Folder: National Reconnaissance Program [1 of 2], Document 8a, Lyndon B. Johnson Library and Museum; http://www.governmentattic.org/20d ocs/CIAhistOSAincep-1969u.pdf.

134. Pedlow and Welzenbach, *The Central Intelligence Agency*, 301–3; https://www.cia. gov/library/readingroom/docs/DOC_0000645397.pdf, 34–5.

135. https://www.cia.gov/library/readingroom/docs/DOC_0000645397.pdf, 35–36; Pedlow and Welzenbach, *The Central Intelligence Agency*, 303–4; memo from John McCone regarding surfacing OXCART, November 29, 1963, meetings with President, 11/23/63–12/27/63, Lyndon B. Johnson Library and Museum.

136. Memo to Robert McNamara from Lt. Gen. Marshall Carter, February 1, 1964, National Security File—Intelligence, Box 9, Folder: Aircraft Contingency (2), Document 65a, Lyndon B. Johnson Library and Museum.

137. Caro, R. A. 1983. *The years of Lyndon Johnson: The path to power*. Alfred A. Knopf, New York; Caro, R. A. 1990. *The years of Lyndon Johnson: Means of ascent*. Alfred A. Knopf, New York; Caro, R. A. 2002. *The years of Lyndon Johnson: Master of the Senate*. Alfred A. Knopf, New York. Caro, R. A. 2012. *The years of Lyndon Johnson: The passage of power*. Alfred A. Knopf, New York.

138. https://www.cia.gov/library/readingroom/docs/DOC_0000645397.pdf, 36–37; Pedlow and Welzenbach, *The Central Intelligence Agency*, 304–5; memo to USIB principals from John McCone, February 28, 1964, NRO Program, Sanitized Volume 2, Document S23, Lyndon B. Johnson Library and Museum; Meinel and Meinel, *Echoes*, 135; https://news.gallup.com/poll/116677/presidential-approval-ratings-gallup-historical-statistics-trends.aspx; Yenne, *Area 51 black jets*, 86; Johnson, History of the OXCART program, 113.

139. Summary record of the National Security Council Meeting prepared by Bromley Smith, February 28, 1964, National Security File—Intelligence, Box 9, Folder: Aircraft Contingency (2), Document 53, Lyndon B. Johnson Library and Museum; Press conference No. 6 of the President of the United States, National Security File - Intelligence, Box 9, Folder: Aircraft Contingency (2), Document 19, Lyndon B. Johnson Library and Museum; Pedlow and Welzenbach, *The Central Intelligence Agency*, 304–5; Yenne, *Area 51 black jets*, 84, 86; https://www.cia.gov/ library/readingroom/docs/DOC_0000645397.pdf, 36–7; Johnson, History of the OXCART program, 15; Polmar, N., Bessette., J. F., Bryan, H., Carey, A. C., Gorn, M., Graff, C., and Veronico, N. A. 2016. *Spyplanes: The illustrated guide to manned reconnaissance and surveillance aircraft from World War I to today*. Voyageur Press. [Place of publication not given], 222; http://www.governmentattic.org/20docs/CIAhistO SAincep-1969u.pdf; Barnes, *Soaring with the eagles*, 115; Caro, *The years of Lyndon Johnson* (1983), (1990), (2002), (2012).

140. Murray, *Once upon a time at Area 51*, 14.

141. Pedlow and Welzenbach, *The Central Intelligence Agency*, 306–8; Johnson, History of the OXCART program, 14, 17–18; McIninch, The Oxcart Story, https://www.cia.gov/library/center-for-the-study-of-intelligence/kent-csi/vol15no1/pdf/v15i1a01p.pdf; https://www.cia.gov/library/readingroom/docs/DOC_0000645397.pdf, 37–39; Barnes, *Soaring with the eagles*, 191–2; Barnes, *CIA Project OXCART*, 71–6.

142. Rich and Janos, *Skunk Works*, 203, 205–9; personal communication from T. D. Barnes to AP, October 5, 2019.

143. Richelson, *The wizards of Langley*, 138.

144. Pedlow and Welzenbach, *The Central Intelligence Agency*, 312–13; McIninch, The Oxcart Story, https://www.cia.gov/library/center-for-the-study-of-intelligence/kent-csi/vol15no1/pdf/v15i1a01p.pdf; Richelson, *The wizards of Langley*, 138–40; Johnson, History of the OXCART program, 18; https://www.cia.gov/library/readingroom/docs/DOC_0000645397.pdf, 42–3; memo to President Johnson from Walter Rostow, September 21, 1966, National Security File—Intelligence, Box 8, Folder: OXCART (2 of 3), Document 18, Lyndon B. Johnson Library and Museum; http://www.governmentattic.org/20docs/CIAhistOSAincep-1969u.pdf; Barnes, *CIA Project OXCART*, 87.

145. Memo to Walter Rostow from J. Patrick Coyne, June 8, 1966, FIAB, Volume 2, Document 36a, Lyndon B. Johnson Library and Museum.

146. Richelson, *The wizards of Langley*, 140; memo to Walter Rostow from Richard Helms, August 5, 1966, National Security File—Intelligence, Folder: OXCART (2 of 3), Document 24a, Lyndon B. Johnson Library and Museum; memo to the President from Walter Rostow on deployment of OXCART, July 5, 1966, National Security File—Intelligence, Folder: OXCART (3 of 3), Document 28a, Lyndon B. Johnson Library and Museum; OXCART deployment proposal, 1966, National Security File—Intelligence, Folder: OXCART (3 of 3), Document 28h, Lyndon B. Johnson Library and Museum.

147. https://www.cia.gov/library/readingroom/docs/DOC_0000645397.pdf, 44; Pedlow and Welzenbach, *The Central Intelligence Agency*, 314; Barnes, *CIA Project OXCART*, 79, 81.

148. Richelson, *The wizards of Langley*, 142–3; https://www.cia.gov/library/readingroom/docs/DOC_0000645397.pdf, 46–8; Pedlow and Welzenbach, *The Central Intelligence Agency*, 317–18; memo on BLACK SHIELD from Richard Helms to the 303 Committee, May 15, 1967, National Security File—Intelligence, Box 8, Folder: OXCART (1 of 3), Document 4a, Lyndon B. Johnson Library and Museum; Johnson, History of the OXCART program, 18; email from A. B. Meinel to Mark Drach-Meinel, undated; http://www.governmentattic.org/20docs/CIAhistOSAincep-1969u.pdf; Barnes, *CIA Project OXCART*, 90.

149. Brugioni, *Eyes in the sky*, 21.9; http://www.governmentattic.org/20docs/CIAhistOSAincep-1969u.pdf.

150. Richelson, *The wizards of Langley*, 143; Pedlow and Welzenbach, *The Central Intelligence Agency*, 317–19; https://www.cia.gov/library/readingroom/docs/DOC_0000645397.pdf, 48; memo on intelligence assessment of BLACK SHIELD returns

to date, June 28, 1967, National Security File—Intelligence, Box 8, Folder: OXCART (1 of 3), Document 1, Lyndon B. Johnson Library and Museum; memo on preliminary assessment of BLACK SHIELD photography of October 28–30, National Security File, Folder: Country File Vietnam, Box 88, Document 1a, Lyndon B. Johnson Library and Museum; Barnes, *Soaring with the eagles*, 147.

151. Richelson, *The wizards of Langley*, 144; https://www.cia.gov/library/readingroom/docs/DOC_0000645397.pdf, 54–5; Pedlow and Welzenbach, *The Central Intelligence Agency*, 323–4; Brugioni, *Eyes in the sky*, 220.

152. https://www.cia.gov/library/readingroom/docs/DOC_0000645397.pdf, 51; Pedlow and Welzenbach, *The Central Intelligence Agency*, 320; Robarge, *Archangel*, 83–7; Rich and Janos, *Skunk Works*, 245; memo from the Directorate of Intelligence concerning preliminary assessment of BLACK SHIELD mission 6847 over North Korea, January 29, 1968, National Security File—Country File Asia & the Pacific, Folder: Korea Pueblo Incident, Misc. Volume 1, Document 13, Lyndon B. Johnson Library and Museum; memo for the Chairman, Joint Chiefs of Staff from Lt. General Joseph Carroll, January 29, 1968, National Security File—Intelligence, Box 9, Folder: NRO, Document 1b, Lyndon B. Johnson Library and Museum; Murray, *Once upon a time at Area 51*, 59; Barnes, *Soaring with the eagles*, 192; Barnes, *The Archangels. Book Two*, 241–2; Barnes, *CIA Project OXCART*, 91; http://www.governmentattic.org/20docs/CIAhistOSAincep-1969u.pdf.

153. Richelson, *The wizards of Langley*, 145–6; Pedlow and Welzenbach, *The Central Intelligence Agency*, 324–5; https://www.cia.gov/library/readingroom/document/cia-rdp75b00159r000100070014-9.

154. http://roadrunnersinternationale.com/articlestoday.html; http://www.governmentattic.org/20docs/CIAhistOSAincep-1969u.pdf; Brugioni, *Eyes in the sky*, 220.

155. https://www.cia.gov/library/readingroom/docs/DOC_0000645397.pdf, 50; Pedlow and Welzenbach, *The Central Intelligence Agency*, 326–7; Robarge, *Archangel*, 97–99; McIninch, The Oxcart Story, https://www.cia.gov/library/center-for-the-study-of-intelligence/kent-csi/vol15no1/pdf/v15i1a01p.pdf.

156. Meinel and Meinel, *Echoes*, 135–6.

157. Zachary, G. P. 1997. *Endless frontier: Vannevar Bush, engineer of the American century*. The Free Press, New York, 218–23; Bush, V. 1945. *Science: The endless frontier. A report to the President on a program for postwar scientific research.* Reprinted July 1960 by the National Science Foundation, Washington, DC, 3–4, 44, 75.

158. Zachary, *Endless frontier*, 227–30, 315.

159. Sapolsky, H. M. 1990. *Science and the Navy: The history of the Office of Naval Research.* Princeton University Press, 9–10, 38–9, 41, 42, 45, 49; Killian, *Sputnik scientists*, 246–7; Zachary, *Endless frontier*, 328–9; U.S. President's Science Advisory Committee Records, 1957–1961, Box 5, Folder 1, Dwight D. Eisenhower Presidential Library; White House Office, Office of the Special Assistant for National Security Affairs, Records 1952–1961, NSC Series, Subject Subseries, Box 16, Dwight D. Eisenhower Presidential Library.

160. Zachary, *Endless frontier*, 332, 369.

161. Killian, *Sputnik scientists*, 243–4, 247; McElheney, *Insisting on the impossible*, 311, 316–17.

162. Brugioni, *Eyes in the sky*, 398–400, 402.

163. Sapolsky, *Science and the Navy*, 100.

164. Sapolsky, *Science and the Navy*, 57, 100; Brugioni, *Eyes in the sky*, 402; Killian, *Sputnik scientists*, 254–5.

165. Killian, *Sputnik scientists*, 255–6; Sapolsky, *Science and the Navy*, 100; McElheny, *Insisting on the impossible*, 339; Brugioni, *Eyes in the sky*, 405.

166. Edmondson, F. K. 2005. *AURA and its US National Observatories*. Cambridge University Press, 80.

12

In Space at Last

By 1982, at age 60, Aden had been promoted to Regents Professor at UA. He had created a national observatory for all astronomers across the United States and had established the College of Optical Sciences, a national center for interdisciplinary basic and applied research in optical science and engineering. In addition, Aden and Marjorie had placed the quest to create affordable power from the Sun on a firm engineering and science foundation by writing two textbooks, teaching several courses, and lecturing widely around the globe on the subject. But by then, Aden complained that he had no students, no teaching responsibilities, and felt disconnected from the Optical Sciences Center he had created. He was looking for new challenges to channel his energies.

His earliest passion was aerospace engineering and rockets, with optics and astronomy a close second. In 1983 he and Marjorie could not resist a chance to combine the two by joining Breckinridge's group at NASA/Jet Propulsion Laboratory (JPL) to study advanced optics for astronomical space missions.

From 1984 until they officially retired from JPL at the age of 73, Aden and Marjorie worked tirelessly, collaborating with younger astronomers, optical scientists, and engineers to develop the next generation of visionary missions, propulsion systems, and space telescopes with unique capabilities. The Meinels' early efforts can be seen today in certain aspects of the Spitzer Space Telescope mission, repair of the Hubble Space Telescope, optical systems to characterize exoplanets, deployable space telescopes, membrane optics, and the soon-to-be-launched (2021) 6.5-m James Webb Space Telescope (JWST), along with several smaller missions. Aden was well known at DOD and NASA in Washington, DC, and enjoyed sharing with senior government leadership the excitement of science and advanced telescopes/instruments and their importance to national security and astrophysics.[1]

Background—Space science

Scientific investigations that require measurements and observations from platforms in space are called *space science*. The space science era began with the launch of the NASA/JPL Explorer 1, on February 1, 1958. This was 120 days after the Soviets first demonstrated that humans could create an artificial satellite

With Stars in Their Eyes. James B. Breckinridge and Alec M. Pridgeon, Oxford University Press. © Oxford University Press 2022. DOI: 10.1093/oso/9780190915674.003.0012

by placing Sputnik 1 into stable orbit. Explorer 1 was equipped with scientific instruments and a transmitter and made the scientific discovery of the Van Allen radiation belts that surround and protect life on Earth from the intense radiation and particles that blow through space. In the beginning, commercial cameras, laboratory sensors, and airplane radio transmitters were hardened for the space environment. Engineering and inventive technologies were devoted to the expensive propulsion, structural and aerodynamic control, and challenges of just rising from Earth's surface and into stable orbit. As time passed, access to space became routine. Instruments grew in sophistication to expand our understanding of Earth as a planet and into the depths of space to discover our place in the universe.

Although not part of James Van Allen's 1958 science team, 10 years earlier (1948) Aden had laid the groundwork for Van Allen by discovering that Earth's magnetic field was creating a current that focused solar protons to the polar regions and create auroras. This was work he continued at Yerkes Observatory as a young astronomer using the Schmidt camera/spectrometers he built for his 1948 dissertation.[2]

By the time access to space became routine in the early 1960s, much of the "easy" science had been completed, and more complicated instruments were required. At first, cameras on spacecraft were used exclusively by engineers to evaluate the behavior of structures at launch and in space. In fact, the early missions to the planets had no cameras on the payload. But the public-relations value of space images soon resulted in every mission using cameras or images to interpret the data for the public and provide context for the scientific research. Serendipitous discoveries became too expensive. A process to identify specific science measurement objectives, define new instrument capabilities, and create new sensor technologies was needed to justify expensive flight programs to sponsoring agencies. It became extremely important to maximize the science return with minimum cost risk and demonstrate this clearly to funding agencies and the American taxpayer.

With his experience in optical system design, the engineering fabrication of large ground-based telescopes and instruments, practical background in rocket technology, and an in-depth understanding of astronomical research priorities, Aden was uniquely qualified to make significant contributions to space science during his short but highly successful career at NASA/JPL during the 1990s. As a JPL distinguished senior scientist who was given ready access to the highest levels of management in the organization, he was available to help younger engineers succeed and reveled in their youthful enthusiasm. His charismatic personality and strong intellect enabled him to move freely across organizational barriers and focus on solving hard technical problems. Most engineers and scientists did not feel threatened by his presence and his engaging manner.

Space telescopes versus ground-based telescopes

The universe radiates energy to reveal itself across the entire electromagnetic spectrum from gamma rays associated with picometer (10^{-12} m) wavelengths to extremely low-frequency, 3Hz (10^5 km) wavelengths. No single telescope or instrument system is sensitive across such a broad spectral range. Telescopes that cover the wavelength interval between 100 nm (far UV) and about 100 micron (submillimeter) wavelength are conventionally called "optical" because their engineering performance is conveniently described by waves and photons. Ground-based optical telescopes look at the universe from beneath an absorbing, scattering. and turbulent atmosphere, which limits angular resolution to typically no better than 0.5 arcseconds. Interstellar space, however, is mostly empty and contains few dust and solid particles to scatter light. If there is something that limits angular resolution in space for the UV/optical/IR region of the spectrum, it is unknown. The limit is our ability to build space-based optical instruments capable of extremely high angular resolution. In space, many more diverse physical processes are observable to extend our understanding of the universe. In a similar manner, there is no known barrier to imaging fainter and fainter objects other than our ability to build space telescopes with larger light-collecting surface areas.

Building a telescope for operation in space presents many challenges. The mass must be low to lift the telescope into orbit, but the mechanical stiffness of the telescope must be high to focus light. Without Earth's atmosphere to blur images, the mechanical stiffness of the space telescope became more important for high performance than for a telescope on the ground. Once launched, access to the telescope for repairs and new instruments is either expensive or unavailable. Therefore, the complicated system must have extremely high reliability. Pointing and tracking of the telescope to point at an object continuously is fully automatic, with no opportunity for interaction from an observer. Space-to-ground communication is needed to command the telescope to point to a new object and then to downlink the data. All space telescopes operate in an extreme thermal radiation environment: bright hot sunlight illuminating one side and cold black sky on the other, which cause the telescope to bend, snap, and pop. This is particularly problematic when the telescope is repointed, and the sunlight moves to illuminate a different portion of the structure and change the temperature on sections of the telescope by hundreds of degrees.

Between 1966 and 1972, NASA launched four Orbiting Astronomical Observatories (OAO). Two of these telescopes were successful: OAO-2 and OAO-3. The former contained a 16-inch telescope feeding four instruments which used the 105–500 nm bandwidth region, and OAO-3 contained a 32-inch aperture spectrophotometer covering 100–300 nm bandwidth and had an X-ray instrument. During this time Aden was deeply involved with DOD surveillance telescopes in space and the founding of the Optical Sciences Center, leaving no time to participate in the OAO program.

The success of those two OAO missions that achieved operational status led to many proposals from several groups of astronomers for their own space telescopes to operate in spectral bands across the electromagnetic spectrum. NASA Headquarters quickly realized that finding funds for a swarm of expensive space telescopes was unlikely. In 1979 a committee of the National Academy of Sciences, chaired by Peter Meyer and Harlan Smith, suggested that the astronomers group their efforts into four select regions of the spectrum and create a Great Observatory program.

Astronomers record data across the entire electromagnetic spectrum, from gamma rays to long-wave radio. Earth's atmosphere absorbs radiation in three important spectral windows and transmits distorted images in the visible region of the spectrum. Astronomers must use telescopes in space to observe in those four spectral windows that are absorbed or distorted severely by Earth's atmosphere. NASA engineers would build each telescope, and the academic science community would compete to provide instruments and reduce and interpret the data. NASA would integrate the instruments onto the telescope and then launch and operate it for the benefit of the community of astronomers. Observing time on the telescope/instrument would be awarded competitively based on priorities established by peer scientists without conflict of interest. This approach was agreed to by all and has been successful over the past 50 years. Aden's experience with DOD surveillance telescopes in space and his doctorate in astronomy fit perfectly into working on the Great Observatory program.

Although not directly involved with the Hubble Space Telescope (HST), Aden had made significant contributions to its construction through his consulting activities over many years with the major aerospace contractors who built the optics and mechanical structures for it and with DOD in telescope development. However, he had no role—nor did NASA invite him to participate—in optical testing and system acceptance. The HST could not have been built without significant technology contributions from the large space optics defense contractors: Perkin-Elmer, Itek, and Eastman Kodak. Aden was funded by the Air Force to support the optics programs at those defense contractors. His work with the Air Force was classified, and NASA managers and technologists were unaware of his contributions to space telescopes for surveillance applications.

NASA managers rely on input from the scientific community to establish science mission priorities prior to mission-formulation studies, which estimate mission cost, schedule, and performance as part of the request to Congress for multi-year funding. Space science administrators and managers in the Science Mission Directorate (SMD) look to the National Academy of Sciences as a source of information to prioritize space science measurements. The astronomy division of SMD collaborated with the astronomy division of NSF in 1980 to provide financial support for a survey of astronomy and astrophysics perspectives for the 1980s by the National Research Council of the National Academy of Sciences. Academy members elected George B. Field of the Harvard Smithsonian Center to chair a 20-member astrophysics committee to identify astronomical science priorities for the 1980s. This committee was chartered to advise government funding agencies on astronomical research priorities: (1) status of previous recommendations, (2) development of ground and space telescopes and instruments, and (3) summary of knowledge in the field and societal impact.

This committee advised NASA to establish four major new programs which they considered to be critically important for the rapid and effective progress of astronomical research in the 1980s. Members of the committee were unanimous in their recommendations with the following order of priority: (1) an Advanced X-ray Astrophysics Facility (AXAF); (2) a very long baseline (VLB) array of radio telescopes to produce radio images with an angular resolution of 0.3 milliarcseconds; (3) a new technology telescope (NTT) of the 15-m class operating from the ground at wavelengths of 0.3–20.0 microns (µ) wavelength; (4) a 10-m-aperture-class Large Deployable Reflector (LDR) in space to carry out spectroscopic and imaging operations in the far infrared and submillimeter wavelength regions of the spectrum that are inaccessible for study from the ground. The AXAF was funded by NASA and named for the world-famous astronomer, and Aden's mentor at Yerkes, Subrahmanyan Chandrasekhar, now referred to as the Chandra X-ray Observatory, listed in Table 12.1. The other two ground-based observatories were the responsibility of the NSF astronomy

Table 12.1. List of the Great Observatory Space Telescope Missions Developed and Operated by NASA for the Benefit of the Astronomy Community

Name	Approval Date	Launch Date	Wavelength Band	Comments
Hubble Space Telescope	1978	1990	0.2–1.1 microns	Currently operational
Compton Gamma Ray Observatory	1977	1991	20 keV–30 GeV (40 pm-60 am)	De-orbited 2000
Chandra X-ray Observatory	1976	1999	0.12–12.0 nm	Currently operational
Spitzer Space Telescope	1983	2003	3.6–160.0 microns	Decommissioned 2020

division. The fourth mission recommended, the LDR, became the subject of intense study in the 1980s. Some of the resulting technology was integrated into the JWST, and part of the planned science was absorbed into the Spitzer Space Telescope. Table 12.1 shows the four Great Observatory missions and their current (2020) status. Since each mission covered its own unique spectral region, there was little common technology, and the optimum telescope and instrument architectural design required for each is different.

In the early 1980s, the one remaining unexplored region of the electromagnetic spectrum blocked by the Earth's atmosphere and unobservable from the ground was the submm and mm (approximately 10μ to 1-mm bandwidth) where spectral lines of important molecular structures and significant spatial structure in the universe lie. The Jet Propulsion Laboratory had designed and built the Infrared Astronomical Satellite (IRAS), which was launched in 1983 to map the entire sky at 12, 25, 60, and 100μ wavelength. This JPL survey mission made many new discoveries and demonstrated the scientific value of a more comprehensive, detailed far IR to submm mission. The LDR, an IR and submm mission, was a natural for mission concept development at JPL. It was conceived as having a 20- to 50-m diameter aperture and diffraction-limited at 50μ wavelength. The precise wavelength bandpass was to be determined years later, based on then-current science measurement priorities and the status of instrument technology.

The NASA Astrophysics Division was busy during the 1980s with building, launching, and preparing to operate the HST. It was integrated into the Space Shuttle bay, waiting for the next launch opportunity, when the Space Shuttle *Challenger* disaster occurred in 1986. As a result, the HST was not launched into orbit until April 24, 1990.

Compelling missions

In addition to the National Academy of Sciences study led by George Field in the 1980s, NASA issued a request to its science centers for ideas for future missions that would be both scientifically compelling and technologically interesting to the aerospace and defense industries. These studies continued throughout the decade, and Aden contributed significantly to several of them.

NASA uses their mission development process to obtain the information needed to identify the scope, risk, and cost of a mission in order to obtain science advocacy, identify the engineering reality, and justify financial support for the mission. Today science missions are developed in response to priorities negotiated among members of specific scientific disciplines, such as astronomy, atmospheric science, planetary science, etc. These science priorities are mapped onto science measurement objectives, which in turn are mapped onto design concepts for instrument and spacecraft hardware and science analysis software. In most cases, this process requires several years of negotiations among scientists and engineers before a cost-effective and scientifically productive mission is conceived and a system proposal that covers cost, schedule, and performance is ready to be prepared and reviewed. These space science mission development teams work on different sets of science measurement objectives to define a cost-effective mission to recommend to the broader science community and move forward to Congress for funding. Each of these design teams is staffed by experts in the required engineering subsystems, such as space structures, thermal management, orbital mechanics, optical engineering, detectors, controls and pointing, and digital electronics data processing. Experts from the science team explain the science measurement objectives and calibration accuracy to ensure that the instrument being developed will indeed perform the measurements needed for the highest quality measurement possible.

A point design concept for the flight spacecraft/instrument system is created using input from each engineering discipline. In some cases, inventions are needed to meet subsystem requirements, and these are documented in a technology development plan or road map. Major cost impacts occur when projects need to stop work while unplanned inventions are needed to solve an unanticipated problem. Years may be required to create all of the individual inventions needed to reduce the cost and schedule risk needed to preserve science measurement capability.

The system engineering design concept is used to reduce the risk of one subsystem not performing and delaying the hardware development. Clearly, when a project enters its building phase where millions of dollars are flowing per month and, say, 250 engineers are working, the project cannot stop, and engineers are paid for remaining idle while individual inventions are being constructed to create a critical subsystem. It is extremely important that all the new technology be ready when a project enters its flight build phase.

At the end of 1981, NASA directors were charged to identify candidate missions for future consideration. The average citizen was losing interest in the space program. The United States had won the space race with several manned missions to the Moon. *Voyager* was on its way to Jupiter and Saturn, *Viking 1* and *Viking 2* had landed on Mars, *Mariner 10* flew by Mercury and Venus, and *Mariner 2* and *Mariner 5* probed the atmosphere of Venus. The Soviet Union had sent several missions to Venus, including the first successful lander. It seemed our solar system was adequately explored. What was next?

The LDR mission study gave NASA an opportunity to implement a unique management approach to define the mission objectives rapidly and to document clearly the technologies needed for LDR. Rather than go through the usual years-long process of creating a point design and issuing a call for proposals to study the technology needs for the LDR mission, NASA convened a workshop of over 120 technical and scientific experts representing 10 different scientific/technical fields from industry, government, and academia during June 21–25, 1982, at the Asilomar Conference Grounds in Pacific Grove, California, on the Monterey Peninsula. A day was devoted to the definition of both the problem and a preliminary mission concept in the form of presentations from NASA engineers and scientists. Participants were divided into 10 groups representing their areas of expertise. During the following day and a half, the engineering groups discussed and summarized their perception of the technical challenges, while the scientists imagined the profound discoveries that would be enabled by such a mission. On Wednesday, everyone convened in the large conference room to listen to presentations from each panel and to discuss technology and science priorities interactively. On Thursday and Friday morning this process was repeated again in small groups and group presentations. Friday afternoon and Saturday were devoted to report-writing for publication. By Sunday all had returned home.

Breckinridge led the optical design and engineering panel at this Asilomar meeting and invited Aden and Marjorie, who were at UA, to participate. Aden clearly understood the science measurement objectives and their trade-offs, as well as the complexity of space flight hardware and the constraints of operating a telescope and instruments in space.[3] He moved from room to room during the week, providing on-the-spot presentations in real time and input on the science measurement value and space optical system engineering challenges to the systems and simulations, structures and materials, controls and pointing, science instruments, detectors and thermal management panels. Preliminary mission plans and technology development plans were created, and Aden and Marjorie continued their work on LDR through 1984 and 1985.

Bruce Murray, then director of JPL, was nearing the end of his five-year term and resigned from the Laboratory on June 30, 1982, a week after the Asilomar meeting. He returned to the Caltech campus in his role as professor of planetary sciences to continue his academic career in teaching and research. About six months prior to his departure, he commissioned several small teams of JPL scientists and engineers to meet and discuss new spacecraft, instruments, and the future of space science exploration. One of these new concepts was missions to explore planets around stars other than our own Sun. As the lead NASA center for the exploration of our solar system, it seemed logical that JPL should investigate doing the same for distant solar systems.

The deputy director assumed leadership of JPL after Murray left until a new director could be found. In the spring of 1982, Lew Allen was retiring as Chief of Staff of the Air Force after a 39-year distinguished career, and he expressed interest in Caltech/JPL to his aerospace industry colleagues, some of whom were Caltech trustees. The nationwide search ended when Allen was appointed director in July 1982. Allen had earned his PhD in nuclear physics at the University of Illinois in 1954 with an experimental dissertation on high-energy photonuclear reactions. He held the distinction of being the first scientist/engineer with a PhD to achieve the rank of general. Allen led JPL from 1982 until 1990, during the period that included the launches of the *Galileo* mission to study Jupiter and its moons, *Magellan* to map the surface of Venus, and the IRAS to survey the night sky at IR wavelengths, as well as flybys of Uranus and Neptune by Voyager 2.

Aden and Marjorie move to JPL

After participating in the Asilomar conference in 1982, Aden and Marjorie expressed interest in retiring from their 20-year career at UA and joining the JPL optics section. In 1984, Breckinridge, who had been Aden's student in 1970, managed a section of 75 optical engineers at JPL responsible for the design and construction of space-flight optical systems. He also represented optics technology at NASA mission development workshops. He prepared the paperwork to advertise job openings for both a senior scientist and his wife. The Meinels and their work were not known to then-current senior-level managers at either JPL or NASA, so Breckinridge prepared himself for a major political battle with management to hire them. Lew Allen, Aden's good friend from his consulting work for the Air Force and now director of JPL and vice president of Caltech, smoothed the way. Aden was hired by Caltech/JPL as a distinguished visiting scientist and Marjorie as his research associate, both reporting to Breckinridge. It had been 40 years since they worked for Caltech, and they were considered Institute rehires when they signed up in 1984. They sold the home they built 25 years earlier in Tucson and moved to a house in Pasadena, about halfway between the houses where Marjorie and Aden grew up.

Their eight-year career at JPL included working with younger engineers and scientists in support of their careers and developing sound visions for several deep-space missions and technology to support those missions. Rather than provide a chronology of events focused on specific engineering or science challenges, their accomplishments fell into six technical areas:

1. The LDR system study to create technology fundamental to the JWST, to be launched 40 years later.
2. The Thousand Astronomical Units (TAU) mission to make high-precision, fundamental measurements of the distance from Earth to most objects in our galaxy. This mission concept, based on an interstellar spacecraft sent to explore the nearby universe, has yet to mature.

3. Space optical systems for the detection and characterization of planets in solar systems external to ours. Some of our most compelling space-science questions back then are active areas of research today: Are we alone? Where did we come from? How did we get here?

4. Interferometry to recover images across the surface of stars, measure precision stellar motions, and reveal details of planetary formation and evolution.

5. Repair of the Hubble Space Telescope (HST). Aden and Marjorie stimulated many ideas, created candidate solutions, and participated in the science/engineering teams that defined the repair of the optical failure.

6. The ground-based Advanced Electro Optical System (AEOS) telescope of the Air Force. Aden consulted with colleagues at the Air Force Research Laboratory to design and build a ground-based, 3.67-m aperture satellite tracking camera now deployed and in regular use on Haleakalā, Maui, Hawaii.

Exoplanets

In 1982, before Aden and Marjorie Meinel arrived at JPL, Breckinridge completed a study showing that planets outside our solar system could be directly imaged combining the coronagraph technology developed and used by solar astronomers with the technology of space-based stellar telescopes.[4] This technique enables direct measurement of the surface and atmosphere spectra from exoplanets. It caught the attention of Aden, and he started to investigate and quickly came to the same conclusion. Today, NASA's exoplanet program has grown as the nation has become enthralled with the possibility of discovering terrestrial exoplanets that might support life as we know it. As a contractor to NASA/JPL but employed at UA, Aden participated in some of these teams to create visionary missions for NASA. Aden and Marjorie made several visits to NASA Headquarters in Washington, DC, to discuss the viability of the program. Today NASA supports a significant program in exoplanet science and discovery.

Large Deployable Reflector

The one remaining unexplored region of the electromagnetic spectrum obscured by Earth's atmosphere and unobservable from the ground is the far IR, submm and mm (approximately 10μ–1-mm) regions where spectral lines of important molecular structures and interesting cosmic structure in the universe lie. The

LDR's challenge was to create one or more point design concepts for a telescope/instrument system with a 20-m diameter aperture and diffraction-limited at 50μ wavelength, which was capable of spectroscopy, wide-field imaging, and polarimetry. The specific science measurement objectives, precise wavelength band passes, specific fields of view, and angular resolution were to be determined years later based on the then-current science measurement priorities and the status of instrument and detector technology.

The development of new space telescopes requires 20 or 30 years of planning and technology development. Under the leadership of Charles Pellerin, NASA director for astrophysics, funding was made available to study an LDR. Products from the study included engineering details of the hardware concepts and a technology development plan that documented the technical sequence of events to be followed to provide the basis for the engineering development of such a telescope. The overall aim was to stimulate scientific investigations across this unknown region of the spectrum and design a technology program, both of which would take years to complete before a mission to explore this new spectral window could be defined. The science measurement objective of this new space system was to record the cold gas and dust in the universe and understand the formation of molecules, stellar systems, evolution of galaxies, and cosmology, among many other topics.

The LDR required a new telescope/instrument system and mission architecture. In the case of HST, the telescope architecture is that of the classic ground-based telescope in use for hundreds of years. The 2.4-m size of the primary mirror approximates that of the 100-inch telescope that was dedicated in 1917 on Mt. Wilson Observatory (MWO). Classic ground-based optical astronomy telescope configurations like that used for HST have a large amount of open volume contained within the telescope tube. In HST, light reflects from a primary mirror to a distant secondary mirror and then directly back through a hole in the center of the primary mirror. The distance between the primary mirror and the secondary mirror is about three times the diameter. In the classic telescope this volume remains open to let light pass. This empty volume is therefore unused for spacecraft components or telescope and instruments. The HST, with its 2.4-m aperture and five science instruments, filled the entire 4.5-m × 18.0-m bay of the Space Shuttle.

It became quickly obvious that the 20-m LDR telescope needed a new structural paradigm. The space technology to assemble a telescope in space from parts that were launched by separate spacecraft was many years away. Astronomers knew that they could afford only one launch. Could they unfold or deploy a large telescope after a single launch? If that were possible, a larger-aperture telescope could be placed into orbit by using that open volume to store structural parts of the telescope, which would then be deployed or folded out of the way of the light path after reaching orbit. A larger-aperture telescope with more light-gathering power could be placed on orbit.

A mission vision was established and technology-development road maps created during two major meetings held June 21–25, 1982 (already mentioned), and March 17–22, 1985, at the Asilomar Conference Grounds in Pacific Grove, California, attended by astronomers with a wide range of interests and engineers from industry, government, and academia. This work included several industry contracts for systems studies.[5]

The objective of these LDR studies was to identify a conceptual point design that satisfies three conditions: (1) enables high-priority scientific measurements, (2) does not violate laws of physics, and (3) appears reasonable from an engineering perspective to build after the agency makes investments in select technologies. This conceptual point design is carried forward in detail sufficient to define a technology-development program focused on achieving the measurement capability of the space telescope system. The wavelength range of scientific interest bridges the gap between radio and optical.

Another question arose. Is the system to have an architecture similar to that for a radio telescope or an optical telescope? No one had built a telescope for this region of the electromagnetic spectrum before. Which parts of each architecture do we take and merge together to make our submm imaging telescope? Radio telescopes are point radiometers, that is, they measure the amount of electromagnetic power at a point on the sky. Optical telescopes record power at multiple points across a region of the sky (field of view) to form an image.

The original guidelines required that the telescope be attached to the Shuttle and assembled by astronauts. These studies were pursued for a while until the cost estimates became too high, and it was moved to be a deployable free-flyer. To stimulate discussion, JPL radio engineers created a design that appeared to be a radio antenna (Figure 12.1).[6] The f/ratio on the primary mirror was far too small to have a desirable field of view for astronomical imaging. Telescopes that use this architecture are optimum for point detectors or radiometers and are used in radio astronomy.

Figure 12.2 shows the Meinel 1985 design concept for the LDR.[7] Note the astronaut on the right side of the light baffle. Aden visualized a two-stage optics system where the first stage, with an optically imperfect aperture characteristic of a large mirror, would gather radiation and focus it into a second smaller optical stage. The second stage would clean up the image, much like eyeglasses do, and focus a high-acuity image onto a detector. In Figure 12.2, light comes in from the top, reflects from the segmented primary, passes to reflect from the secondary, and converges to the first focal region. This image is not sharp and sits within a hole in the center of the quaternary mirror shown. The light expands to fill the tertiary (3rd reflection) mirror, reflects back to the quaternary (4th reflection),

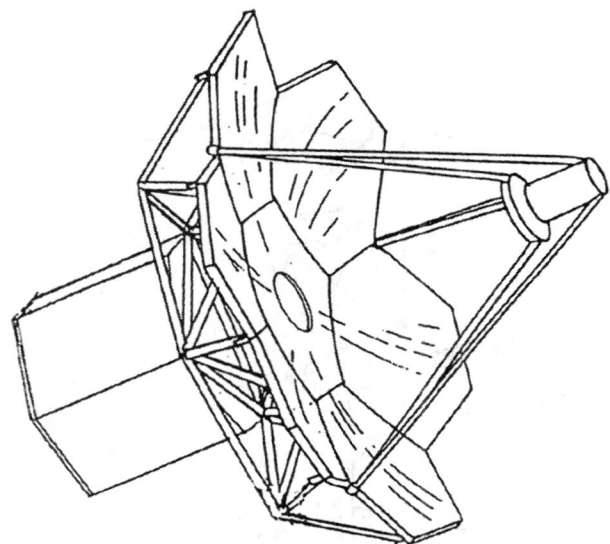

Figure 12.1. Large Deployable Telescope concept based on a radio telescope architecture.

Drawing by Aden Meinel.

and then converges to the final focus. The optical power (curvature) on the secondary and the tertiary are such that an in-focus image of the segmented primary mirror falls onto the quaternary. In Aden's vision the quaternary is segmented to match 1:1 the segments on the primary. The tip-tilt errors of the segments on the primary are compensated for by tip-tilts in the opposite sense of the segments that comprise the quaternary.

The results of these early 1980s workshops created a mission vision and initiated technology development for the next-generation space telescope which became the JWST. The LDR was NASA's first telescope to deploy the primary mirror in segments to the precision needed to form a diffraction-limited, 6.5-m clear aperture space telescope.

One early (1983) innovative concept by Aden and Marjorie showed how a 20 × 8-m aperture telescope could be configured to fit into the Space Shuttle bay, deployed with relatively simple operations by an astronaut, and provide high-value science.[8] They developed a sound engineering rationale and analysis of signal-to-noise ratio for the concept and discussed the relationship between the sunshield design and the orbital inclination, concluding the paper by proposing use of the LDR as a basic module to permit the construction of supergiant space telescopes and interferometers, both for IR/submm studies.

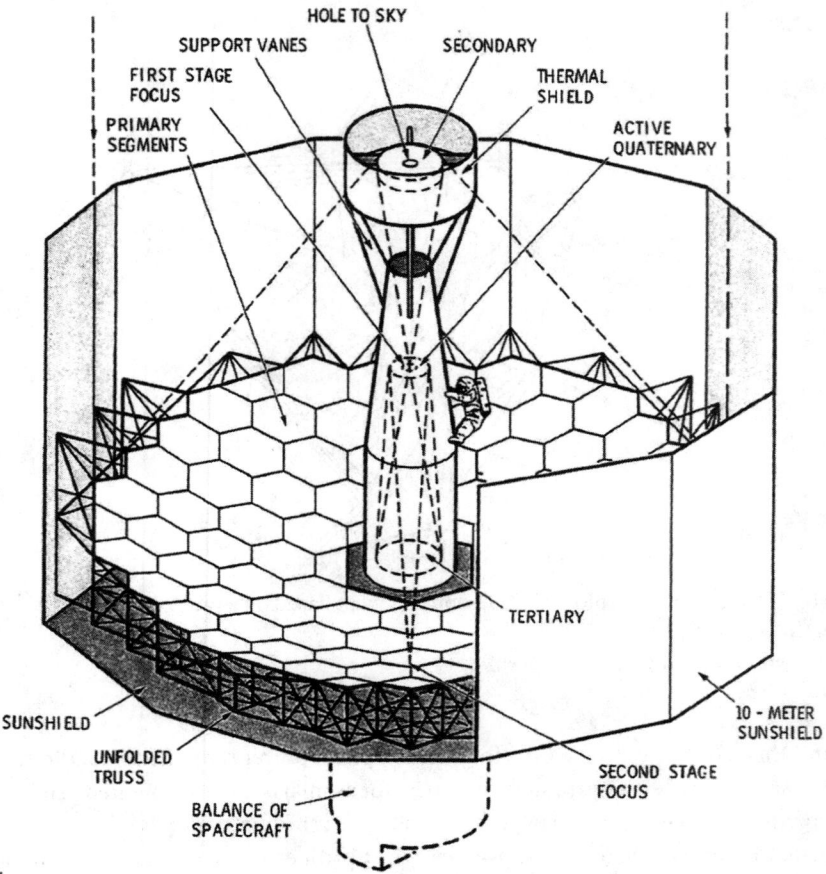

Figure 12.2. Meinel design concept for the Large Deployable Reflector.
Drawing by Aden Meinel.

Aden and Marjorie pursued different technologies for the optical fabrication and test of mirror segments and mirror structures suitable for cost-effective deployment in space. One innovative approach was to build a thin optical mirror that could be rolled up for launch, then unrolled in space with surface-correction using adaptive optics.[9] Segmented space telescopes require innovative techniques to test the off-axis parabolic surface. Aden and Marjorie also developed several ways to convert theory into practice for the LDR project.[10] They completed a detailed study of the optical system design for a highly corrected two-stage optical system and provided concepts for its construction and test and commented on scientific applications.[11]

Although the LDR was never funded to a complete mission, it was highly successful in bringing together engineers and scientists from industry, academia, and government and focusing their interests, attention, and creative energy on ideas for a mission vision and the definition of a series of technology development plans that would support several future, compelling NASA space-science missions. The Meinels showed astronomers how to be proactive in the design of their telescopes and not just sit back and take what the telescope industry delivered to them, as they had for ground-based telescopes prior to Aden's innovative telescopes for KPNO during the late 1950s. This broad creative new community of scientists and engineers created the architecture for a cost-effective JWST over the past 20 years.

Thousand Astronomical Units (TAU) mission

One of the innovative missions discussed in 1982 was sending a probe to a nearby star/planet system: Alpha Centauri, 4.367 light years away. To arrive at Alpha Centauri in the minimum amount of time for the purpose of placing an instrumented science satellite on orbit about the star would require spending the first half of the mission accelerating and the second half decelerating. Even traveling to the nearest star would require over 100 years before science data could be returned. One scientist suggested a stellar penetrator mission, which would send an instrumented spacecraft like a high-speed javelin to penetrate a star. At high speed the instrument would spend only picoseconds inside the star, not enough time to obtain an instrument reading. These studies of sending probes to other star systems were soon dropped in favor of building optical systems that use the photons radiating from stars, exoplanets, and surrounding gas clouds.

Aden led a study of the opto-mechanical and propulsion system to a send a 1.5-m-aperture imaging system 1,000 astronomical units from our solar system to record images from a distant point and compare those images with images recorded with a similar telescope in Earth orbit to obtain absolute parallaxes out to a range of 500 kiloparsecs (1 parsec = 3.62 light years) from Earth. Comparing these two images would give astronomers absolute measurements of distances across our galaxy.[12] This would enable precision calibration of the absolute luminosity of Cepheid variables and consequently an improved distance scale for the universe. Aden and Marjorie named the mission TAU, for thousand astronomical units.

Launch date for a TAU mission was to be well into the first decade of the 21st century, but it will likely not happen in our lifetimes, if ever. Study of TAU has focused on the technologies required to carry out this ambitious mission and the identification of preliminary scientific rationale for such a deep-space flight.

A 1-MW nuclear-powered electric propulsion (NEP) system forms the baseline method for achieving the high velocities required. A solar system escape velocity of 106 km/s is needed to propel the TAU vehicle to 1,000 AU in 50 years. The NEP system must accelerate the vehicle for about 10 years before this velocity is attained because of the extremely low thrust nature of the xenon-fueled ion engines. At the end of the thrusting phase, the NEP system is jettisoned to allow the TAU spacecraft and science experiments to coast out to 1,000 AU. Another important technology for TAU is advanced optical communication systems for transmitting science data to Earth. Aden calculated that a 1-m optical telescope combined with a 10-W laser transponder can transmit 20 kbps to a 10-m tele-scope in Earth orbit from 1,000 AU.

The potential of NEP was investigated as the enabling technology for achieving a mission to a thousand astronomical units within 50 years. By means of a 1,000 AU baseline, the primary objective was to make measurements of distances to the stars in our own galaxy and beyond. In addition, several unique deep-space studies in astronomy, astrophysics, cosmology, and space plasma physics can be carried out. Technology requirements were compared with current plans for both nuclear space power and ion propulsion research.

Missions into near interstellar space have been under study for at least the past 15–20 years. These early mission studies focused on the scientific object-ives related to the understanding of the distant solar and interstellar environ-ments, including magnetic fields, energetic particles, and dust. Studies at JPL in the mid-1970s by Len Jaffe were the first to discuss the possibility of making si-multaneous stellar observations from Earth and from a spacecraft situated 500–1,000 AU from Earth in order to obtain trigonometric parallaxes which could be used to obtain the distances to stars throughout our galaxy. In addition, other scientific objectives studied were to characterize the heliopause, study the inter-stellar medium, observe low-energy cosmic rays, measure interplanetary gas and dust, and determine the mass of the solar system. This study was also the first to baseline an NEP system as the most promising propulsion technology. Such deep-space missions received renewed interest at JPL at the instigation of Aden and Allen because of the recognition that true unmanned interstellar missions to our nearest neighbors are at present not feasible and because of the increased interest in understanding the cosmic distance scale and its implications on cos-mology. The primary scientific objective of the TAU mission was to establish an accurate cosmic distance scale throughout our galaxy and perhaps beyond. Given the present 2 AU baseline (diameter of Earth's orbit), trigonometric par-allax allows astronomers to measure distances accurately only out to about 100 parsecs (pc). Given the same accuracy of measurement, TAU will be able to ex-pand the measurement of distances out to about 50 kpc. This range will bring the stars in the galaxy within our distant measurement capability and perhaps allow us to make accurate measurements of distances out to the Magellanic Clouds. Such measurements will reduce our uncertainty in the expansion rate of the universe, help to determine the age of the galaxy, assist in understanding galactic structure, and provide important distance information which go into the studies of stellar evolution. Depending on the exact location of the solar helio-pause and the progress of the Voyager spacecraft, TAU may permit the first in situ measurements of the plasma environment across the heliosphere into the tenuous interstellar medium.

Once into interstellar space, TAU will sample galactic magnetic fields, ener-getic particles, and gas, dust, and plasma environments. In addition, outside the heliosphere TAU will measure low-frequency galactic and extragalactic radio emissions which cannot be observed within the heliosphere because of the noise background caused by solar wind and planetary magnetospheres. By providing TAU with an accurate laser transponder, the spacecraft and Earth can act as end masses in an electromagnetically tracked, free-mass gravitational wave detector. The gravitational waves of interest are those that were created during the early cosmological processes of the Big Bang, which may be detectable by TAU with several orders of magnitude higher accuracy than currently available. Thus, TAU will have a major impact on the fields of astronomy, astrophysics, space physics, and cosmology.

Hubble Space Telescope

Aden and Marjorie had been at JPL a few years before HST was launched on mission STS-31 of Space Shuttle *Discovery* on April 24, 1990. By June, word was received that something appeared to be wrong with the optical system.

The 100-inch Hooker and the 200-inch Hale telescopes use primary and sec-ondary mirrors that are parabolic surfaces of rotation. In contrast, the HST has a Ritchey-Chrétien (RC) optical system with two hyperbolic mirrors providing high-quality images over a wider field of view. The light reflected from the con-cave primary mirror converges to a convex secondary mirror and then reflects back toward the primary and passes through a hole in the primary into the in-strument package. The formulas of the surfaces, the surface-figuring techniques to shape the mirrors, and the optical test methodologies were developed by George Ritchey at MWO in the early 1910s. They were independently developed by Henri Chrétien in France, and hence the system is called Ritchey-Chrétien. This two-surface design is free of third-order coma and spherical aberration, but does suffer from fifth-order coma, astigmatism, or relatively severe field curva-ture. To achieve this exceptional performance, it is necessary to put aberrations on the primary that the secondary then removes. Ideally, the most accurate test requires that the telescope be assembled and illuminated with collimated light from a point source at infinity. The cost of such a test for a telescope as large as HST was too expensive to consider. The surface on the primary mirror was fabricated and tested separately from the fabrication and test of the secondary mirror, and the two were first tested together at the spacecraft integration facility in Sunnyvale, California, where an inexpensive and incomplete test was paid for by NASA. This was the state of technology in the 1980s. Today computer-generated holograms (CGH) enable straightforward testing of the primary and secondary separately. However, this technology was not available in the 1980s when the HST mirrors were being fabricated.

In 1960, almost 20 years before the HST mirror fabrication started, Aden selected the RC optical configuration for the f/2.3, 2.1-m (84-inch) ground-based telescope on Kitt Peak. Ten years before NASA started to figure the 2.4-m HST primary/secondary mirror system, Aden and Don Loomis successfully figured the 2.1-m RC telescope for the NSF at Kitt Peak. Four years later, they successfully figured the 2.3-m (90-inch) RC Bok telescope for Steward Observatory. The key to building a successful RC two-mirror telescope system is optical testing. They developed techniques for successfully figuring and testing large-aperture RC optical systems under difficult conditions. In the case of the 84-inch, there was no optical test tower available, and the first time the primary and secondary were used together was in the telescope on the mountain. The first time the telescope was assembled at Kitt Peak, the images were perfect. In the mid-1960s, many years before the HST 2.4-m mirror system was built, Aden designed, fabricated, and tested the successful 2.1-m KPNO telescope that saw first light on September 15, 1964.

Aden devoted time to the study of the cost of telescopes and included this subject in the class he taught at OSC from 1967 to 1971. Small f/ratios reduce the cost of domes but require a deeply curved primary mirror, which is difficult to shape optically (i.e., figure). He achieved this successfully for the 2.1-m telescope by raising the temperature of the Pyrex glass until it was slightly viscous and slumping the disk over a mandrel. However, the 90-inch blank for the Bok Telescope is fused silica, a high-temperature material with a viscosity that could not be controlled as well as that for Pyrex. It was necessary to grind this concave. Aden had one of his students complete a thesis studying grinding rates, hardness of abrasives, and time to complete a required optical figure.[13] Much of the expense in figuring an optical surface is the cost of labor. Understanding the rates of grinding and polishing mirrors of different sizes and materials is used to estimate project development cost. These comprehensive studies are published and available to all manufacturers of large optics. This knowledge should have provided a clue to the technical managers that something was wrong with the processing of the HST mirror when it was necessary to add a second, unplanned grinding shift to remove more glass to shift the focus of the marginal rays farther from the mirror vertex and thereby overcorrect spherical aberration.

Aden's role on HST

Aden ended his consultancy with Perkin-Elmer in 1972, when he stepped down as director of OSC. It is a mystery why, during the design and construction of the optical segment of the telescope and the optical instruments between 1975 and 1985, NASA never engaged Aden or any of the optical telescope experts at the College of Optical Sciences, or members of the optical engineering communities in Tucson, at Lick Observatory, or at Mt. Palomar, in an advisory capacity. It could have been that Aden did not respond to the NASA call for technical advisors at the start of the project because he was too busy with his work on the MMT and solar energy, or that he applied to be an advisor and was not selected by NASA managers for some unknown reason. Nancy Roman, the NASA astronomy manager for HST in 1975, had worked with Aden at Yerkes as a research associate. In 1953 Aden was looking for a young astronomer to collaborate with him on his funded stellar and auroral spectroscopy work. Roman's Yerkes research contract came up for renewal, but Aden hired Helmut Abt instead on the recommendation of astronomer Jesse Greenstein at Caltech,[14] and Roman moved on to the Naval Research Laboratories (NRL). Whatever the reason, HST development proceeded without the experience of ground-based telescope-builders.

Many members of the astronomy community had been waiting years for an opportunity to use the HST. NASA had launched an aggressive publicity campaign to incite global interest and to keep funding sources in Congress enthusiastic. When NASA let the contract for the construction of the telescope in 1978, it forecast the launch for 1983. University professors told their students to plan to receive data from the telescope by 1984. But, of course, there were many schedule delays. Students needed to select different research topics because the data they needed were not going to be available in time for them to finish their doctorates.

Technology development for the Wide Field and Planetary Camera (WF/PC) detectors had been frozen in time before 1980 in order to build the flight detectors on schedule for the launch of the telescope. After the Space Shuttle *Challenger* accident on January 28, 1986, a new schedule for the integration of the HST into the shuttle bay was created, along with a new forecast date for the launch. About 18 months before launch in 1990, NASA Headquarters provided JPL with funds to start building the WF/PC-2 to replace the first camera. Detector technology had evolved quickly, and new ones were available to increase the sensitivity of the system significantly. Fortunately, recognizing that detector technology was evolving rapidly, NASA management directed JPL about eight months before the launch to begin ordering parts for a near-duplicate of WF/PC that became known as WF/PC-2. This new camera was planned for the first servicing mission, nominally scheduled for several months after the initial launch of HST.

The HST was finally launched on April 24, 1990, aboard Space Shuttle *Discovery* and was placed into orbit on the next day. Two months later, on Monday, May 21, 1990, the *Los Angeles Times* published the first HST image of two stars (isolated white-dots) against a black sky. Everyone at NASA/JPL was expecting an image of a recognizable scene, such as a planetary nebula or perhaps Jupiter in stunning detail. It was immediately obvious to several optical engineers and astronomers at JPL that something was wrong. Aden suspected that the radius of curvature on one or both of the two mirrors of the RC system was wrong. On June 21, 1990, NASA announced publicly that there was indeed a serious problem with its orbiting telescope's ability to focus on distant stars and galaxies.

On Thursday, June 28, Jim Breckinridge arranged to meet on July 2 with JPL director Allen on the subject of HST, along with Aden and Marjorie, astronomer Arthur Vaughan (formerly of MWO), and John Trauger. Sitting at the round table in Allen's office, each person provided his or her explanation of how the error could have occurred and suggested how it could be fixed. They all agreed that if the error were only on the primary mirror, then a fix using the JPL/Caltech back-up camera, WF/PC-2, would be straightforward—not simple, but possible. It was the primary science imaging camera on the telescope, and without it the science mission would be severely compromised. In a few days, Aden was to discover another issue that would complicate the engineering changes required for the new camera.

For a few days the broad science community believed they should ask NASA to bring the telescope back to Earth for repair and relaunch. At $500M per shuttle launch (two would be needed) and another predicted $200M+ to disassemble, refigure, and replace the mirror, the cost was deemed impossible and too risky. Finding hundreds of millions of dollars in the federal budget was impossible. Most believed that once down, it would never be launched again. In addition, engineers reported that the Shuttle landing loads would be too large, and crew safety waivers might not be approved to return the 12,000-kg telescope to Earth.

Thoughts quickly moved to see if the wavefront error might be fixed in the instruments in orbit. Fortunately, the HST was built with latches and grab rods to enable on-orbit replacement of instruments. But the focal plane of the telescope was shared with five instruments. The telescope wavefront error affected all five, and each of the five would be rendered scientifically useless because of the telescope error. A decision was made to sacrifice one of the instruments to insert a mechanism that would deploy optical correctors for three of the instruments and to ensure that the modified WF/PC would contain its own optical correction capability. The deployable optical system was called Corrective Optics Space Telescope Axial Replacement (COSTAR) and was designed to correct HST's primary mirror aberration for light focused on the Faint Object Camera (FOC), Goddard High Resolution Spectrograph (GHRS), and the Faint Object Spectrograph (FOS). The COSTAR wavefront correction system was designed and built by Ball Aerospace and was successfully deployed into the telescope on the first servicing mission.[15] It was in use until 2009, when the needed wavefront correction was built into each instrument. This instrument and its development did not involve Aden and will not be discussed further here. This narrative will concentrate on the development of the WP/PC-2.

Observing programs had been set up in the operations pipeline to create the highest priority, high angular-resolution science. These plans were quickly shelved while scientists came to grips with the degraded telescope and scurried to develop observing plans for productive scientific measurements that could be made with the flawed telescope until it was corrected. On July 4, 1990, Breckinridge and Allen flew to Washington, DC, from Los Angeles for the first meeting of the Allen Committee to investigate the cause of the problem and recommend actions to prevent future occurrences. Breckinridge remembers riding in a taxi with Allen, creeping through heavy traffic near the Washington Mall while 4th of July fireworks were bursting overhead. His thought was how beautiful and how innocent it was in the presence of the new adventure on which they were embarking.

The next day they walked from their hotel to a conference room on the 4th floor of a nondescript federal building near the railroad tracks, not far from NASA Headquarters. There they were met by Len Fisk, NASA administrator for science; John Mangus, optics expert from the Goddard Space Flight Center; Charles Pellerin, administrator for the Astronomy and Physics Division; George Rodney, associate administrator for safety and mission quality; and George E. Reese, associate general counsel. The HST Optical Systems Failure Board of Investigation was formed under the leadership of Allen, and its participants were sworn in by a federal judge and signed conflict-of-interest paperwork.

In response to science community's worries that their concerns were not being heard, NASA Headquarters formed the Hubble Independent Optics Review Panel (HIORP), composed of many astronomers and optical scientists. Aden and Marjorie, Art Vaughan, Duncan Moore (director of the Institute of Optics at the University of Rochester), and a group from the Space Telescope Science Institute (STScI) led this series of meetings over the next few months and developed written reports for the astronomy community and the official NASA Failure Review Board.

As active members of the HIORP, Aden and Marjorie participated in endless meetings across the country, explaining different hardware and software solutions being worked on at JPL and Ball Aerospace Corporation. Aden's first-hand knowledge of rocket flight hardware, his in-depth understanding of science measurement objectives, and his extensive reputation as a member of the broad astronomy community served the Panel well during this period. He listened to every proposal, no matter how wild, thought through its practicality, and responded patiently. His conversations would begin "Did you look at . . . ?" or "Did you know that . . . ?" He used his excellent lecture skills to help younger optical scientists and astronomers understand the actual technical issues facing JPL engineers and STScI scientists. In this way, he squelched many rumors before they circulated among management and the press, and relieved technical teams from many unnecessary distractions to their work.

The HST was unique in that its instruments can be replaced while the telescope remains on orbit. Several months before the launch of the telescope, NASA asked JPL to begin the process of building a second WF/PC to same specifications as the first, but with much more sensitive next-generation detectors. NASA project managers estimated it would take 36 months to build, test, and calibrate a second camera (WF/PC-2). Fortunately, JPL engineers had ordered several long-lead items to be manufactured. Some were already in house, and other parts were on the way. This shortened the schedule and enabled the relatively fast response to the crises. When Aden and Marjorie were home at JPL they participated actively in the design teams that were building this replacement camera with its optical elements to correct the HST wavefront error.

An example of Aden's generosity toward younger engineers is shown in his work on alignment issues between the WF/PC and the HST. During a meeting of the engineering team, he calculated on the back of an envelope that the wavefront error in the telescope implied that, as a result of the manufacturing error on the primary mirror, the alignment of the WF/PC-2 needed to be adjusted to better than 0.003 inch, far better than would be required were the error not present. No one else knew this. He went to the blackboard and described his calculations, and immediately others discussed it and how to solve the problem. Some team members proposed using a tip-tilt adaptive optics mirror in the optical path. (Thus, in 1993 NASA flew the first successful adaptive optics sensor and mechanism in space.) Aden backed off and let the younger team members run with ideas and take credit for it. During such conversations, Aden was always good at convincing the engineers that they themselves had thought of the idea first.

What is the nature of the error?

The surface on the 2.4-m primary mirror was ground and polished to the wrong shape because of an error in the optical test null corrector used to provide the reference curvature for the three-dimensional, doubly curved mirror surface. The invention of the null corrector by Abe Offner increased the accuracy and reduced the cost and time required to test large aperture astronomical telescopes optically.[16] Unfortunately, the construction of the null corrector used by the contractor to fabricate the HST primary was flawed. This led to the wrong curvature on the finished 2.4-m mirror, an error that was not discovered until after the telescope was inserted into orbit at significant expense. The error is on only one surface in the optical system.

The nature of the error is shown in Figure 12.3. The top figure shows the way the mirror should have been built. The incoming rays (shown coming from the left) were planned to reflect from the primary mirror in the original design and intersect the telescope axis at the focus as shown. The bottom figure shows the way the incoming rays (shown from the left) were misdirected by the error in curvature in the fabricated mirror. Note that the design required that all of the rays that strike the mirror come to one focus, whereas the fabricated mirror surface shows the marginal ray coming to a focus (crossing the axis) at a different point than the paraxial ray. Since the focus of the marginal rays is to the left of the

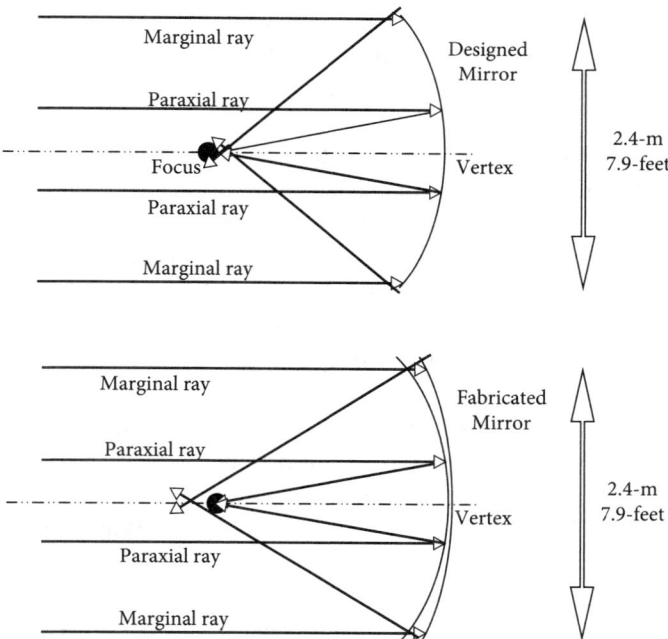

Figure 12.3. Hubble Space Telescope primary mirror error described without reference to the secondary mirror. See text for details.
Drawing by J. Breckinridge.

paraxial focus, the optical system is said to exhibit overcorrected spherical aberration. The magnitude of the error is shown schematically in the lower figure by looking at the difference in the curvature of the solid line (representing the fabricated mirror) and the dashed line (representing the correct surface). At the edge of the mirror, the correct surface is displaced from the manufactured surface by about 1 micron.

The manufacturing and test process resulted in a mirror surface that was too flat. Too much material was removed from the surface away from the center, and although the mirror was fabricated to be smooth to Lambda over 100, the base curvature of the mirror was wrong. The telescope was aberrated, and simple refocusing or translation of the image plane along the axis would not correct the aberration. The aberration is called spherical because the base spherical surface is not correct. Additional mirrors or lenses needed to be inserted into the optical system.

In the interest of completeness, details on how this happened and when the error might have been discovered are summarized in "The Hubble Space Telescope Optical Systems Failure Report," written by Allen and the team members.[17]

Optical layout of the flawed telescope with the WF/PC

Figure 12.4 shows an optical schematic of the HST with the WF/PC. The drawing is not to scale. The telescope mirror diameter is 2.4 m (94 inches). The dotted box shown to the left of mid-center surrounds the telescope assembly. The dotted box in the lower portion of the drawing identifies the volume of the WF/PC optical instrument. The telescope system was designed to disconnect the WF/PC structure from the telescope and replace it during a reservicing mission. This enables camera upgrades during the mission to replace detectors and aging optics.

Light enters the system from the top center-left, passing down to reflect from the primary mirror (wavefront error) to the secondary, then the beam converges back down along the telescope axis to reflect from the Pick-off flat mirror. At this point, the beam reflects to the right in the drawing to form an image of the sky at the HST telescope focus located at a 4-sided (-faced) pyramid mirror of the WF/PC instrument. At the pyramid mirror the field of view is segmented into four sections and reflected to four fold mirrors. (Note: only two of the four fold mirrors are shown in this two-dimensional drawing because the other two are rotated 90 degrees into and out of the plane of the paper.) Each of the four image segments of the full field of view are passed through their own individual small Cassegrain finite conjugate relay telescope. Ground-based digital image

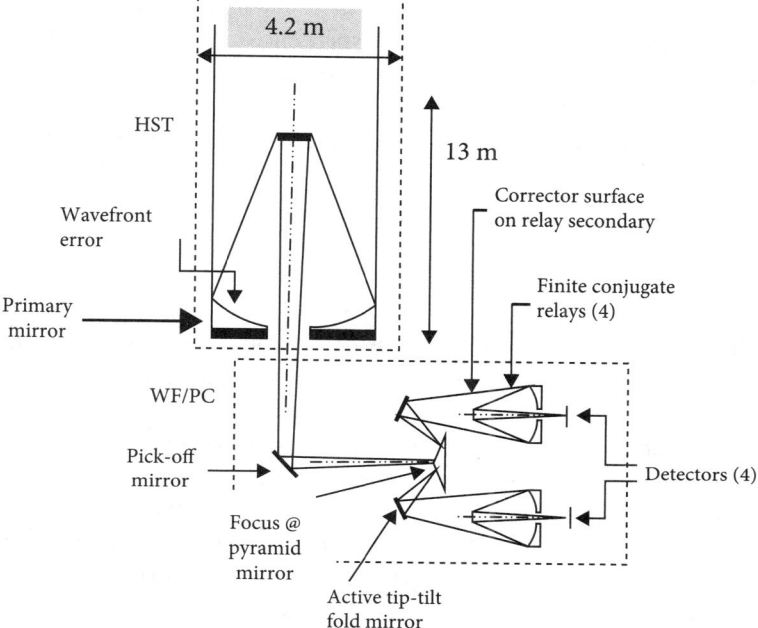

Figure 12.4. Cross-section view of the Hubble Space Telescope with the WF-PC mounted at the axial focus. Image not to scale. The telescope is a cylinder 13.2 m (43.5 ft.) long and 4.2 m (14.0 ft.) in diameter. The 4.2-m telescope tube diameter was necessary to accommodate the 2.4-m clear aperture mirror. The WF/PC, shown as the dotted box lower right of the telescope, is the size and approximate shape of a baby grand piano. The entire telescope system weighs 12,246 kg (about 27,000 lb.). The telescope system was designed for an on-orbit replacement of the WF/PC weighing 270 kg.

Drawing by J. Breckinridge.

processing is applied to fuse the four images into one for scientific interpretation, publication, and public consumption.

The original hardware had four more relay telescopes, clocked at 45 degrees relative to these four. The original WF/PC had eight small, finite, conjugate relay telescopes. Four of the relay telescopes had focal lengths to give wide field of view, and the other four gave a narrow field for the planetary camera, hence the name Wide Field and Planetary Camera.

Light from each of these small relay telescopes then falls onto its own 800 × 800-pixel CCD detectors. After detection, digital image processing is used to reassemble the images into a 1,600 × 1,600-pixel image of the celestial field of view for scientific analysis. In an RC system, the surfaces of both the primary and the secondary are hyperbolas of rotation with axes superposed and form the axis of the two-mirror telescope system. The hyperbolic surface shapes are in addition to the base spherical curvatures of the mirrors. These spherical surfaces provide the optical power to define the location of the image plane. The hyperbolic contour or optical figure determines the image quality at all points in the field of view.

Where to place the correction?

The 12-ton telescope and instrument assembly could not be returned safely back to Earth for repair of the shape of the primary mirror. Given the optical layout shown earlier in Figure 12.3, optical scientists were left with the problem of where to place the optical correction (eyeglasses), what curvatures are needed on the lens or mirrors, and how to access the on-orbit telescope to make the repair in the minimum amount of time with the least cost to the taxpayer. Several candidate locations were considered for the corrector and were proposed to the review panel, including Aden. Two of the ideas suggested were to replace the telescope secondary mirror or to place a 2.4-m diameter lens over the surface. Both were impractical. The HST was designed to have its five instruments replaced, including the WF/PC.

NASA scientists wanted excellent correction over as wide a field of view as possible. This limited the location in the optical path to a place where a high-quality image of the telescope system exit pupil could be positioned. The constraint to minimize cost led to making the most use of already manufactured camera parts. An image of the exit pupil (the primary mirror) fell near the secondary optical element in the four telescope relay optical systems.

The axis of the 2.4-m telescope now needed to be aligned relative to that of the WF/PC to within 0.003 inches. The WF/PC relay optics contain aspheric primary and secondary mirrors. Each of the eight primary mirrors has its own axis, as do each of the secondaries. For the required diffraction limited performance, the axes of the primary and secondary mirrors needed to be superposed, the axes of each of the eight relay telescopes needed to be superposed, and that combined axis needed to be superposed on the axis of the HST.

An obvious plan was to change the shape of the small secondary mirror on each of the four relays and thereby place the phase conjugate or opposite wavefront error to that error on the primary. When this suggestion came up, Aden leapt to his feet, ran to the blackboard, and proceeded to lecture the panel on spherical aberration and optical system lateral magnification. The basic problem was that the error is on the 2.4-m mirror, and the plan was to correct that with a mirror about 2 cm in diameter. That is a minification of 110 times. The minification changes the image of the mirror diameter but not the magnitude of the wavefront error that needs to be corrected. Therefore, the slopes of the needed correction increase by a factor of 110, and the base radius of curvature for the correction needed to decrease by a factor of 110. This size difference causes the system to be much more sensitive to alignment errors. Aden quickly calculated on the blackboard that the alignment tolerance between the optical axis of the HST system and the optical axis of the WF/PC would be 0.003 inches if they were to place the correction on each of the secondaries of the camera.

Recall that the WF/PC and HST were designed to be separated by the astronauts during a space walk to swap out the cameras. But the design of the telescope latching system, that is, the latches that clamp a new WF/PC that was carried to orbit by the Shuttle to the on-orbit HST, were no better than 0.1 inches, which was sufficient were there no primary mirror error. During his off-the-top-of-his-head blackboard calculation, Aden showed that a new WF/PC could not correct the error. Misalignment would introduce an aberration that looked like on-axis coma, and images would appear worse. What to do?

He asked if there were some way to use adaptive optics to adjust from the ground images recorded on-orbit. James Fanson of the JPL team said yes, and he moved to the blackboard to show everyone how a new form of adaptive optics would be developed to enable the telescope/instrument system to be corrected.

In a short two hours, the optical system architecture was defined, a new adaptive optics was found, and several months of development work were planned that eventually fixed the primary scientific imaging system for the HST during the first servicing mission.

On-orbit prescription

But what should the new aspheric optical prescription for the secondaries be? What would be the new on-orbit optical prescription of the telescope? How accurately could it be determined from the surface of Earth?

No one had ever measured the optical prescription of a telescope to nanometer precision while it was in space, thousands of kilometers from scientists and engineers, much less do it with the whole world looking on. If the HST were to be corrected, this precise measurement needed to be made quickly to support both the redesign of WF/PC and all future instruments.

NASA established the Hubble Aberration Recovery Program (HARP) to collect CCD digital images of stars at different field points across the focal plane and through HST focus as viewed through the WF/PC. Any small wavefront errors inside the WF/PC were added to much larger error on the primary mirror. These star images were examined in detail by 15 different teams to determine the on-orbit optical prescription of the telescope. The collections were done in two stages.

The error was quickly localized to the 2.4-m primary mirror and attributed to an error in the shape of the doubly curved surface. This type of error is described in terms of a conic constant, κ, where all points on a curved surface are described by the equation:

$$z = \frac{C(x^2 + y^2)}{1 + \sqrt{1 - (1 + \kappa)C^2(x^2 + y^2)}}$$

where $C = 1/R$. C is called curvature of the surface, and R is the radius of curvature of the reference sphere. Table 12.2 shows values for the conic constant κ for conic sections rotated about the system optical axis. The primary mirror for the RC optical system is optically figured to be a hyperboloid with a conic constant around 1.014 ± 0.0005. Table 12.3 summarizes measurements of the HST primary mirror conic constant.

Aden obtained his value for the primary mirror conic constant using the analog tools he and Loomis developed during the optical testing of the KPNO 84-inch (2.1-m) and the 90-inch (2.3-m) Bok Telescope primary mirrors 25 years earlier. The method Aden used is described in detail in the technical literature.[18] Several other groups, using far more modern sophisticated numerical analysis tools, arrived at nearly the same value.

Aden's quiet and calm demeanor and the fact he had 30 years more experience than anyone else in the room led to a better understanding on the part of all. During the meetings, when astronomers would propose preposterous solutions, Aden quietly explained engineering limitations. Often he was the most respected

Table 12.2. Values of the Conic Constant in Terms of the Common Names Used to Describe Surfaces of Revolution

Value of the Conic Constant	Figure of Revolution
$\kappa = 0$	Sphere
$\kappa = -1$	Paraboloid
$-1 < \kappa < 0$	Ellipsoid (prolate spheroid)
$\kappa < -1$	Hyperboloid
$\kappa > 0$	Oblate spheroid

Table 12.3. Summary of Measurements Made of the HST Primary Mirror Conic Constant Based on 18 Independent Measurements Using Data Recorded at Different Field Points in the Aberrated Image and Through Focus

Conic Constant	Rms WFE	Number of Independent Measures in This Range	Comments
−1.0127	0.24		
		4	
−1.0132	0.25		
		5	
−1.0136	0.26		
		1	WF/PC-2 used this
−1.0140	0.27		Meinels' measure
		3	COSTAR used this
−1.0145	0.28		
		3	
−1.0149	0.29		
		2	
−1.0154	0.30		

Note: Column 1 shows the range of the conic, column 2 gives the equivalent rms wavefront error, and column 3 gives the number of independent measurements within the range values given in column 1 and 2. The comments column identifies the values used by WF/PC-2 and COSTAR error bars provided by the investigator and converts the conic constant estimates into wavefront errors in units of microns at 632.8 nm.

person in the room, spoke freely with authority, and became the arbiter of arguments.

Once the optical prescription of the telescope/instruments had been determined, the values were turned over to another team who built, tested, and calibrated the corrective hardware for flight.

Finally, on December 2, 1993, the WF/PC-2 was launched aboard Space Shuttle *Endeavour* with a crew of seven, whose tasks were to dock with the Hubble and repair or replace three gyroscopes, two electronic control units, eight fuses, and two solar panels. They also installed two new magnetometers and COSTAR (Corrective Optics Space Telescope Axial Replacement) with 10 mirrors to redirect light from Hubble's secondary mirror to three instruments. On the third day of the mission, astronauts Jeffrey Hoffman and Story Musgrave removed WF/PC-1 and installed WF/PC-2 in its place (Figure 12.5a, b; consult the color insert for this figure). After some telescope alignments, sharp images of stars, nebulas, and galaxies dazzled the world, beginning on December 18, 1993.[19]

Imaging interferometry

The angular resolution of a telescope is determined by the physical distance between the two extremes of the light-collecting regions across the reflecting surface of the primary mirror or lens. The brightness of the image is determined by the total surface area of the mirror or lens. For 300 years, astronomers used monolithic, continuous-surface mirrors and lenses for their astronomical telescopes and imaging systems.

Albert Michelson, who was awarded the 1907 Nobel Prize in Physics, invented a new instrument called an optical interferometer, which increased the angular resolution of an optical telescope without building a primary mirror that was single, large, expensive, and difficult to handle. In 1891, Michelson used two small apertures in front of the 36-cm refractor at Lick Observatory to measure the diameters of four of Jupiter's moons, obtaining results consistent with measurements made using a visual micrometer at the 92-cm refractor at Lick. On November 21, 1919, Albert Michelson proposed to George Ellery Hale that an interferometer be added to Mt. Wilson's 100-inch telescope to make direct measurements of the diameters of distant stars.[20] Between 1920 and 1930, Michelson and Francis Pease measured the diameters of several stars. These successful measurements had a profound impact on modern astrophysics, and a larger (15-m) interferometer was built on Mt. Wilson to measure the diameters of many more distant stars. Modern-day interferometry is a technique astronomers use to create high angular-resolution images of stars, gas, dust, and galaxies.[21]

Ten years later, when Aden was observing at MWO, he worked beside this large instrument. He became familiar with the details of Michelson's work through his father-in-law, Edison Pettit, and learned even more about optical interferometry from Fred Pearson, Michelson's optician, when they were both at Yerkes Observatory about 1950. In 1991, while Aden was at JPL, the Keck Observatory announced plans to build a second 10-m telescope in proximity to the 10-m telescope then under construction. He started to develop conceptual optical system designs to understand how these two large telescopes, sitting in proximity to each other, could be combined to create an even larger aperture with a sensitivity and optical angular resolution of about 2 milliarcseconds, exceeding any telescope in the world then being planned.

By August 1992, Aden and Marjorie had completed a comprehensive study of the opto-mechanical configuration of a Keck Interferometric Imaging Array (KIIA). This 133-page concept definition study included designs for outrigger telescopes, optical phase delay lines, beam compressors, and beam injectors. They also showed how the optical subsystems would be packaged or nested around the two 10-m Keck Telescopes and provided dimensioned, assembly-view mechanical drawings. Aden proposed an array of portable 1.5-m telescopes to supplement the optical signals provided by the 10-m apertures. He proposed an optical array of six telescopes, two 10-m aperture and four 1.5–2.0-m aperture. This detailed information was sufficient to provide funding agencies cost estimates and the astronomical sciences community confidence that the optical interferometer would provide an invaluable research tool. Figure 12.6 shows one of the many drawings from their final report. Aden and Marjorie visited the site weekly to monthly during construction to review the latest progress made by the core design team, which included Gerard van Belle, Mark Colavita, and Gautam Vasisht. Marjorie would also provide technical feedback.[22]

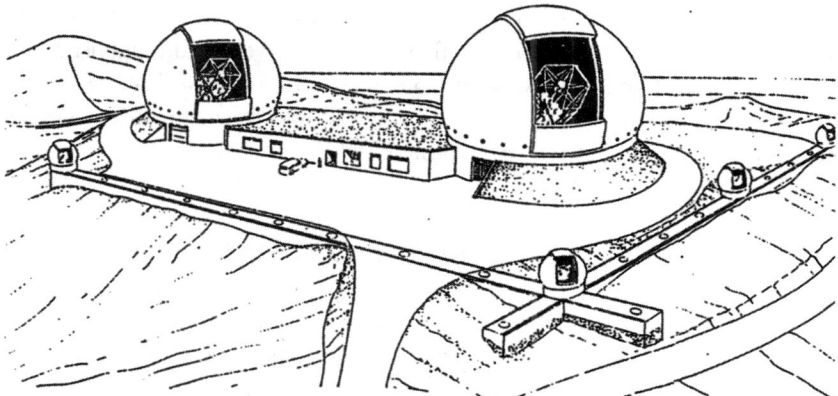

Figure 12.6. Concept for the Keck Interferometric Imaging Array (KIIA). Portable 1.5-m "outrigger" telescopes are shown nested around the two 10-m Keck Telescopes. The large volume underneath and between the two Keck Telescopes was used by the laboratory to combine the white-light fringes coherently to reconstruct diffraction-limited images with high angular resolution.
Drawing by Aden Meinel.

Retirement from JPL/Caltech

By 1994, Lew Allen had retired and the new director, Ed Stone, reorganized the observational systems division along the lines of instrument manufacturing rather than technical disciplines. The optics section was abolished and groups partitioned out into other sections with hardware instrument construction and not research priorities. Breckinridge transferred to the JPL defense program office and eventually to Washington, DC, to manage technology-development programs for NASA and NSF. Aden and Marjorie retired from JPL in 1995 at age 73 and moved their office north from Pasadena along the California coast to their home on Shoreline Drive in Santa Barbara, which had a view straight to the ocean from their living room. For the following 10 years they continued to work, first as temporary on-call, off-site employees for JPL, and then just to keep busy.

At this time Marjorie was a full partner in all of Aden's work. She collaborated on his mathematical analyses, performed calculations, completed publication drawings, and wrote and edited text for journal articles and reports. A close friend, Nasrat Raouf, would drive up from Pasadena to help with their hardware and software IT problems so that they could sustain their productivity. Aden enjoyed deep-sea fishing and would go out on a charter boat about once a month. They were active members in the local Lutheran church where Marjorie was the substitute organist.

They converted a bedroom in their home into an office where they could work quietly together without interruption from the usual JPL meetings and telephone calls. Between 1998 and 2004 they researched, wrote, and published 11 scientific papers in four diverse areas, an enviable record for someone 50 years younger: membrane optics, telescope cost models, opto-mechanical deployment architectures, and geometrical optics design.

Membrane aperture space telescopes

Could one unfurl a lightweight membrane only a few mm thick by either a lightweight mechanical system or inflation, and then compensate the aberrated wavefront using adaptive optics on a small optical device to create an extremely large (20+ m) aperture in space to gather data for astronomical investigations and new discoveries? It was this line of thinking that led Aden to investigate large membrane space optics. Bob Freeland, one of Aden's collaborators at JPL, was the principal investigator for a free-flyer experiment for the IN-STEP Inflatable Antenna Experiment (IAE) on Shuttle Flight STS-77.[23] The IAE successfully inflation-deployed a 14-meter-diameter membrane reflective structure on STS-77 in 1996 and partially deployed its secondary support structure. Breckinridge, in collaboration with Aden and Marjorie, investigated the feasibility of upgrading this technology to be useful for optical/IR astronomical space telescopes.[24]

Large telescope primary apertures are needed to collect light sufficient to measure faint astronomical details near the threshold of detection. In addition, if these apertures are of high precision, they enable high-resolution imaging. Much of the cost of a space telescope is in the mass and volume needed for a classic large primary aperture. Finding new technology to reduce this mass and volume would revolutionize astronomical discovery. One step to reduce the volume per unit aperture light-collecting area is to segment the primary mirror and fold the segments, packaging them into a smaller volume to fit into the nose-cone of the launch vehicle payload. A deployment mechanism and metrology alignment system are then needed to unfold or deploy the segments into the full aperture. This adds mass back into the system, but modern launch vehicles such as the Ariane are capable of high-lift, as we see for the new James Webb Space Telescope.

Membrane lightweight large-aperture space optics were studied for a new DARPA program.[25] Originally conceived for a contribution to the NASA terrestrial planet finder (TPF) program, Aden and Marjorie proposed the use of space-inflatables to deploy large membrane apertures.[26] Aden and Marjorie's objective was to discover whether, if the structural issues could be solved, the image quality could be made sufficient to provide value to astronomy. If yes, then what would the technology development road map look like to achieve reality for the system? This road map is extremely important to guide and inform NASA technology managers where investments are needed to support these next-generation space telescopes with superior capabilities.

A membrane large-telescope primary aperture might solve both the mass and volume problem for space telescope applications.[27] Perhaps a 20-m aperture could be launched in current vehicles. Diffractive optical elements perform well for single wavelength sources, such as lasers in optical LIDAR systems, which do not require achromatic performance. Astronomical applications require broadband white light and thus are required to be achromatic to concentrate optical power to record images. Aden and Marjorie investigated technologies to make achromatic space-optics membrane systems and suggested solutions.[28]

Cost models for planning large-aperture ground and space telescopes

Work in collaboration with Gerard van Belle summarizes all of the work Aden did over 50 years on analyzing the cost of ground- and space-based astronomical telescopes.[29] As director of Kitt Peak National Observatory, Aden had become sensitive to the cost of astronomical instruments/telescopes. How to get the most discoveries per dollar was behind many of his concepts and designs.

Aden and Marjorie's analysis concluded that for telescope apertures built prior to 1980, costs scaled as aperture size to the 2.8 power. After 1980, "traditional" monolithic mirror telescope costs were scaled as aperture to the 2.5 power. Large, segmented mirror telescopes (Keck, etc.) appear to deviate from this relationship with significant cost savings, although it is unclear what power law such structures follow, but the exponent is less than 2.5. The value of this work is to provide telescope system architects with clues to high-cost subsystems and where architectural decisions affect cost. Published in 2004, this work provides the most comprehensive analysis based on cost for existing telescope systems today.

A new telescope without instruments is not of much value. Each new telescope requires new instruments. These new instruments are often the size and cost of the previous-generation telescopes. Developing a diverse set of instruments often costs almost as much as the new telescope into which they will be integrated.

Deployment architectures for large sparse-aperture space telescope systems

Early in the development of radio telescopes, astronomers discovered that they could accurately record the time radio signals that arrived at separate radio antennas through significant computational effort and reconstruct an image of the source that created those signals.[30] This technology, called interferometry or aperture synthesis, was refined over the years. The NSF radio telescope facilities near Socorro, New Mexico, and the Atacama Large Millimeter/submillimeter Array (ALMA) in Chile regularly record radio signals, which are then converted to digital images for scientific analysis.

Like many astronomers before them, and motivated by the success of the MMT, Aden and Marjorie sought to apply the principles of radio interferometry to create high-fidelity images in the UV/optical/IR region of the spectrum. Two technologies establish sparse aperture image quality: (1) detectors, and (2) the physical position and size of the mirrors across the aperture being synthesized using digital image processing. Aden and Marjorie investigated several innovative ways to deploy the mirrors or elements of a sparse aperture optical systems in space. The impact of this work has been minimal. After the optics section was disbanded by JPL management and the Meinels retired to Santa Barbara, they had become disconnected from their peers in Pasadena. They were unaware that optical science and technology had moved on to show that sparse apertures could not provide the high-quality images that astronomers wanted and needed.[31]

Spherical primary telescope with aspheric correction at a small pupil

During the summer of 1954, Aden and the family were at Yerkes Observatory. With encouragement from Strömgren and observatory funding, Aden packed up the family and drove first to Lick Observatory and then to Pasadena to discuss his new idea for a 500-inch (12-m) ground-based optical telescope. He wanted to know the reaction of C. D. Shane of Lick Observatory and Ira Bowen of Palomar Observatory to his concept. Bowen asked McMath to join the meeting with Aden, which was held in Bowen's office.

Aden's highly innovative idea is to use a fixed 500-inch diameter doubly curved spherical primary mirror, composed of many smaller spherical mirrors tiled together into a continuous surface with narrow gaps. This spherical primary would point up to the zenith or possibly be tilted to point along the meridian. At this concept review meeting, Bowen and McMath expressed doubts about the proposed configuration, in particular the temporal stability of the proposed fixed bowl of spherical mirrors and the fabrication and test of a needed full-aperture aspheric corrector. Bowen pointed out the advantages of making a fixed spherical mirror pointing down to gather the light that reflects from a large siderostat composed of an array of flat (zero-power) mirror elements for a siderostat. This was a concept he was thinking about for the future. One advantage is that the 10-m siderostat would turn only half of the zenith distance angle, as would a conventional telescope.

Aden discussed his work in which he placed the aspheric correction surface near the focus of the sphere and examined the field aberrations. He identified oblique spherical aberration as the limiting factor to a wide field-of-view system.[32] This project was set aside when Aden left Yerkes to lead the Kitt Peak National Observatory site exploration and construction.

Years later, when Aden was working on large telescope concepts with his colleagues in China, this issue came up again. He was able to continue his work studying these aberrations and found solutions to correct them over a reasonable field of view for the Large Sky Area Multi-Object Fiber Spectroscopic Telescope (LAMOST). It is a siderostat telescope that uses a segmented fixed spherical primary mirror telescope with an adaptive segmented aspheric siderostat mirror. The aspheric mirror is active so that the aspheric shape of the corrector is adjusted as the tilt of the aspheric siderostat mirror is changed.[33] The telescope is located at Xinglong Station, Hebei province, China.

The field of view of a spherical mirror has no field-dependent aberrations; every point in the field is identical because a spherical mirror has no unique optical axis. Each partitioned field requires correction only for spherical aberration, independent of where the partitioned optical subsystem is located in the telescope field of view. It is therefore possible to place identical small aspheric correctors in the pupil formed inside each partitioned subsystem.[34]

Henderson, Nevada

Family members reported that by 2003 Marjorie was having serious short-term memory problems.[35] Aden found the situation difficult. He wanted her to be able to learn to remember things and became frustrated when he was unable to help her. He had lost his intellectual partner who for 60 years was the first person to hear and understand his concepts and ideas. She had helped to write, organize thoughts, edit text, and add her own creative contributions. By 2005, Marjorie had about a one-minute memory and had fallen twice, requiring visits to the emergency room. Their youngest son, David Drach-Meinel, a practicing Lutheran minister, would travel from Henderson, Nevada, to Santa Barbara on a moment's notice. Caring for her caused Aden to collapse every few months and be hospitalized himself. David would stay with Marjorie at the Santa Barbara house until Aden recovered and could return home.

By September 2006, their failing health required them to be closer to family. Aden's small machine shop was abandoned. Their files, drawings, library, and papers were discarded, along with much of their furniture, to move to Henderson and a new house on Eclipsing Stars Drive, not far from David and his family. They joined Christ the Servant Lutheran Church where both David and his wife Diane were (and still are) pastors. There Aden and Marjorie met Ursula and Ralph "Jerry" Licon, who fell in love in Garmisch, Bavaria, in 1955 when she worked at a bed-and-breakfast and he was stationed there in the original US Army Special Forces (Green Berets); Jerry served two tours in Vietnam and also on the space capsule recovery team for astronauts Alan Shepard (1961) and John Glenn (1962). They all shared wartime experiences and remained close until Jerry's death in 2010 and Aden's death the following year.[36]

While in Henderson, Aden wrote a popular article for *Physics Today* on galaxy collisions and ordinary dark matter.[37] His final scientific research paper was presented by their daughter at the IAU symposium held in Tenerife, Spain, November 3–7, 2008.[38] Aden was unable to participate because of his poor health. Marjorie passed away of natural causes on Tuesday, June 24, 2008, at age 86.

Aden's last few months were spent at an assisted-living facility only a block from David's home. David would take him out to a four-hour lunch every day. They always paused in the church parking lot to look at the solar panels that covered the church roof, to which Aden would always remark: "I never thought I'd live to see the day." Aden passed away in his sleep on Sunday, October 2, 2011, at the age of 88. His ashes, along with Marjorie's, were scattered in scenic Red Rock Canyon National Conservation Area, located 17 miles west of downtown Las Vegas.

Awards

Singly and jointly, Aden and Marjorie were honored throughout their careers, beginning as early as 1952, when Aden received the Optical Society of America's Adolph Lomb Medal, which was established in 1940 in the name of the first treasurer of the OSA and awarded to a person who made a contribution to optics early in his or her career. Two years later, he won the American Astronomical Society Helen B. Warner Prize for a significant contribution to observational or theoretical astronomy during the five years preceding the award. The only stipulation is that the astronomer must be younger than 36 years of age in the year of the award or within eight years of receipt of a doctorate. In 1980 Aden was presented with the OSA's highest award, the Frederick Ives Medal established in 1928, "for his contributions to thermal solar energy, analysis of the principles of coherently combined, independent telescopes, and the leadership he has given to several major optical and astronomical research centers."[39] Four years later, he won the SPIE (Society of Photo-Optical Instrumentation Engineers) George W. Goddard Award in Space and Airborne Optics "for the invention and development of a new process or technique, technology, instrumentation, or system."[40] In 1990 he was awarded the American Astronomical Society's George Van Biesbroeck Prize, which biennially honors a person for "long-term extraordinary or unselfish service to astronomy, often beyond the requirements of his or her paid position,"[41] and also an Honorary Doctor of Science degree from the University of Arizona. His final individual honor, in 1993, was NASA's Exceptional Scientific Achievement Medal "for exceptional scientific contributions (specific, concrete scientific achievements) toward achievement of the NASA mission."[42]

Jointly, in both 1992 and 2000, Aden and Marjorie received the SPIE Rudolf and Hilda Kingslake Award in Optical Design "in recognition of significant achievement in the field of optical design, including the theoretical or experimental aspects of optical engineering."[43] Both were also presented with the SPIE Gold Medal, the Society's highest honor, in 1997 for "outstanding engineering or scientific accomplishments in optics, photonics, electrooptics, or imaging technologies or applications."[44]

Tom Gehrels (1925–2011), whose brother Cornelius died at the Mittelbau-Dora camp visited by Aden in the spring of 1945 (see Chapter 4), was a doctoral student of Gerard Kuiper at the University of Chicago/Yerkes Observatory before he became professor of planetary sciences at UA and a close friend of

Aden. Gehrels made sky surveys using the 48-inch Schmidt telescope at Palomar Observatory for years, looking mainly for asteroids and comets. He would send the photographic plates to Ingrid and Cornelius (Kees) van Houten at the University of Leiden in the Netherlands for analysis. The team named thousands of asteroids and several comets. Two asteroids were discovered in the main asteroid belt between the orbits of Mars and Jupiter in this way on September 24, 1960. One of them, 4 km in diameter and magnitude 14.1, was named 4065 Meinel (2820 P-L) in Aden's honor. The other one, 6 km in diameter and magnitude 13.5, was named 4064 Marjorie (2126 P-L).[45]

In more than one sense, then, Aden and Marjorie were in space at last.

Notes

1. Letter from B. I. Edelson, NASA Associate Administrator for Space Science and Applications, dated July 11, 1985, welcoming Aden and Marjorie Meinel to JPL and the NASA community.
2. Meinel, A. B. 1948. The near infrared spectrum of the airglow and aurora. *Publications of the Astronomical Society of the Pacific* 60: 373.
3. Meinel, A. B. 1959. Astronomical observations from space vehicles. *Publications of the Astronomical Society of the Pacific* 71: 369–80 (Reprinted as Kitt Peak National Observatory, Contributions no. 1, 1959); Meinel, A. B. 1961. New frontiers of astronomical technology. *Science* 134: 1165–71 (Reprinted as *New frontiers in astronomical instrumentation*. Kitt Peak National Observatory, Contributions no. 11, 1961e).
4. Breckinridge, J. B., Kuper, T. G., and Shack, R. V. 1982. Space telescope low scattered light camera—a model. *Proceedings SPIE* 331: 395–403.
5. Agnew, D. L., and Jones, P. A. 1986. Large Deployable Reflector (LDR) system concept and technology definition study, May 30, 1986. Ames Research Center Contract NAS2-11861 NASA contractor report CR 177413 performed by Eastman Kodak Company, Government Systems Division, Rochester, New York.
6. Swanson, P. N., and Null, G. W. 1982. Large Deployable Reflector-Pathfinder Study. JPL Report # D-195 December 1982, NASA Jet Propulsion Laboratory, Pasadena, California.
7. Meinel, A. B., Meinel, M. P., Stacy, J. E., Saito, T. T., and Paterson, S. R. 1986. Wave front correctors by diamond turning. *Applied Optics* 25: 824–5.
8. Meinel, A. B., Meinel, M. P., and Woolf, N. J. 1983. Deployable reflector configurations. *Proceedings SPIE* 0383. Deployable Optical Systems (December 1, 1983), https://doi.org/10.1117/12.934917.
9. Romeo, R., Meinel, A. B., Meinel, M. P., and Chen, P. C. 2000. Ultralightweight and hyperthin rollable primary mirror for space telescopes. *Proceedings SPIE* 4013: 634–9. https://doi.org/10.1117/12.393998.
10. Meinel, A. B., and Meinel, M. P. 1989. Optical testing of off-axis parabolic segments without auxiliary optical test elements. *Optical Engineering* 28: 71, https://doi.org/10.1117/12.7976904.

11. Meinel, A. B., and Meinel, M. P. 1992. Two-stage optics: High acuity performance from low acuity optical systems. *Optical Engineering* 31: 2271–81, https://doi.org/10.1117/12.59946; Meinel, A. B., Meinel, M. P., and Breckinridge, J. B. 1994. Deployable space telescopes for planetary and astronomical missions. *Proceedings SPIE* 2214, 250–5, Space Instrumentation and Dual-use Technologies (June 8, 1994), https://doi.org/10.1117/12.177664.

12. Nock, K. T., Breckinridge, J. B., Buratti, B. J., Lesh, J., Meinel, A. B., Meinel, M. P., Ramirez, M., and Sercel, J. C. 1986. Thousand Astronomical Unit voyage: A deep space mission. *Bulletin of the American Astronomical Society* 18: 1012.

13. Weinwig, S. A. 1974. *Comparative grinding rates of optical glasses.* Master's thesis, University of Arizona.

14. Personal communication from Helmut Abt to J. Breckinridge, 2019.

15. Crocker, J. H. 1993. Engineering the COSTAR. *Optics and Photonics News* 4: 22–6.

16. Offner, A. 1963. A null corrector for paraboloidal mirrors. *Applied Optics* 2: 154–5.

17. National Aeronautics and Space Administration. 1990. The Hubble Space Telescope Optical Systems Failure Report. NASA-TM-103443.

18. Meinel, A. B., Meinel, M. P., and Schulte, D. H. 1993. Determination of the Hubble Space Telescope effective conic constant error from direct image measurements. *Applied Optics* 32: 1715–19, https://doi.org/10.1364/AO.32.001715.

19. Zimmerman, R. 2010. *The universe in a mirror: The saga of the Hubble Telescope and the visionaries who built it.* Princeton University Press, 152–4, 161–2.

20. Michelson, A. A. 1920. On the application of interference methods to astronomical measurements. *Astrophysical Journal* 51: 257–62.

21. Buscher, D. F. 2015. *Practical optical interferometry: Imaging at visible and infrared wavelengths.* Cambridge University Press; Labeyrie, A., Lipson, S. G., Nisenson, P. 2006. *An introduction to optical stellar interferometry.* Cambridge University Press.

22. Interview of Gerard van Belle by James Breckinridge in conjunction with 2019 AAS Meeting, Seattle, Washington, on January 10, 2019.

23. Freeland, R., Bilyeu, G. D., Veal, G. R., Steiner, M. D., and Carson D. E. 1998. Large inflatable antenna flight experiment results. *Acta Astronautica* 41: 267–77.

24. Breckinridge, J. B., Meinel, A. B., and Meinel, M. P. 1998. Inflation-deployed camera and hyper-thin mirrors. *Proceedings SPIE* 3356: 780–7.

25. Meinel, A. B., and Meinel, M. P. 2002b. Large membrane space optics: Imagery and aberrations of diffractive and holographic achromatized optical elements of high diffraction order. *Optical Engineering* 41: 1995–2007.

26. Meinel, A. B., and Meinel, M. P. 2002c. Parametric dependencies of high-diffraction-order achromatized aplanatic configurations that employ circular or crossed-linear diffractive optical elements. *Applied Optics* 41: 7155–66; Meinel, A. B., and Meinel, M. P. 2000. Inflatable membrane mirrors for optical passband imagery. *Optical Engineering* 39: 541–50; Breckinridge, J. B., Meinel, A. B., and Meinel, M. P. 1998. Inflation-deployed camera and hyper-thin mirrors. *Proceedings SPIE* 3356: 780–7.

27. Hyde, R. A. 1999. Eyeglass. 1. Very large aperture diffractive telescopes. *Applied Optics* 38: 4198–212; Barton, I. M., Britten, J. A., Dixit, S. N., Summers, L. J., Thomas, I. M., Rushford, M. C., Lu, K., Hyde, R. A., and Perry, M. D. 2001. Fabrication of large-aperture lightweight diffractive lenses for use in space. *Applied Optics* 40: 447–51.

28. Meinel, A. B., and Meinel, M. P. 2002b. Large membrane space optics.

29. Van Belle, G. T., Meinel, A. B., and Meinel, M. P. 2004. The scaling relationship between telescope cost and aperture size for very large telescopes. *Proceedings SPIE* 5489, Ground-based Telescopes (September 28, 2004), https://doi.org/10.1117/12.552181.

30. See Breckinridge, J. B., Bryant, N., and Lorre. J. 2008. Innovative pupil topographies for sparse aperture telescopes and SNR. *Proceedings SPIE* 7013, https://doi.org/10.1117/12.787011.

31. Meinel, A. B., and Meinel, M. P. 2000. Spherical primary telescope with aspheric correction at a small internal pupil. *Applied Optics* 39: 1–4.

32. Meinel, A. B. 1953. Aspheric field correctors for large telescopes. *Astrophysical Journal* 118: 335–44.

33. Wang, S.-G., Su, D.-Q., and Chu, Y.-Q. 1996. Special configuration of a very large Schmidt telescope for extensive astronomical spectroscopic observation. *Applied Optics* 35: 5155–64.

34. Ramsey, L. W., and Weedman, D. W. 1984. The Penn State spectroscopic survey telescope. In: Very large telescopes, their instrumentation and programs. *Proceedings of the International Astronomical Union Colloquium* 79: 851–60.

35. Personal communication to J. Breckinridge from David Drach-Meinel in December 2020.

36. Interview of Ursula Licon by A. Pridgeon on May 1, 2018, in Henderson, Nevada.

37. Meinel, A. B. 2007. Galaxy collisions and ordinary dark matter. *Physics Today* 60: 14.

38. Meinel, A. B., and Meinel, B. 2009. Evidence of a magnetic sheath around a jet from NGC 6543. In *IAU Symposium Proceedings Series* (eds. K. G. Strassmeier, A. G. Kosovichev, and J. E. Beckman), 133–4. Cambridge University Press..

39. https://www.osa.org/en-us/get_involved/awards_and_honors/awards/award_descriptions/ivesquinn/.

40. https://spie.org/about-spie/awards-programs/awards-listing/spie-george-w-goddard-award-in-space-and-airborne-optics.

41. https://aas.org/grants-and-prizes/george-van-biesbroeck-prize.

42. https://nasapeople.nasa.gov/awards/nasamedals.htm.

43. https://spie.org/about-spie/awards-programs/awards-listing/spie-rudolf-and-hilda-kingslake-award-in-optical-design-.

44. https://spie.org/about-spie/awards-programs/awards-listing/spie-gold-medal-award.

45. https://ssd.jpl.nasa.gov/sbdb.cgi.

13

Slowly Fades the Supernova

Every person born leaves a legacy for those around them and those who follow for generations. Each one of us is the sum total of life's experiences, based on interaction with others and opportunities taken. Aden and Marjorie Meinel's legacy will stretch far into the future. They lived lives of discovery, driven by curiosity about the world around them and the insatiable drive to help others unselfishly. Aden's personality was not that of an empire builder. He knew how to marshal the efforts of others and gave charismatic leadership to important national initiatives in scientific research, engineering, and education.

Just as important, he knew when to step aside for young persons to flourish. Robert (Bob) Breault was interviewed for this volume on February 21, 2019. Bob was a Top Gun instructor at Nellis Air Force Base before leaving the Air Force to enroll at UA, where he obtained his MS in 1972 and PhD in 1979. He is chairman and founder of the Breault Research Organization (BRO), which sells optical engineering software to over 40 countries. He helped redesign the Hubble and Spitzer Space Telescopes, XMM-Newton X-ray Space Observatory for the European Space Agency, Laser Interferometer Gravitational Wave Observatory (LIGO), and the Cassini-Huygens mission. He recalled the day that he knocked on Aden's office door after the end of Aden's course that he had taken.

> Being a former fighter pilot and very brash and bold, I went up to his office. "Sir, you've given us these projects, and I'm making a Cassegrain. I'm making a flat. I'm making a prism. I'd like to finish the project. The course is over. I've gotten my grade. I enjoy pushing glass." He stared at me for a few seconds. "Bob, I like that. Come with me." And so we marched out of his office across the hall to the elevator and went down. He led me into the back room. He was looking around. "This area I declare the student optical fabrication shop. These guys who have been taking my course can work in here until they finish their projects." Two years later I finished pushing glass.[1]

Another student named John Lytle, who received his MS in 1969 with Roland Shack as advisor, was interviewed on February 28, 2017. He, too, clearly recalled the day when he met with Aden.

With Stars in Their Eyes. James B. Breckinridge and Alec M. Pridgeon, Oxford University Press. © Oxford University Press 2022. DOI: 10.1093/oso/9780190915674.003.0013

He had an idea for null testing Ritchey-Chrétien telescope assemblies at finite conjugates. He put a sketch on my desk and said, "See if this works." And so I modeled it in the computer. Actually, I didn't even need to do that. I just did some basic first-order calculations and regretfully told Aden, "This isn't going to work. There's a problem with baffling, and there's a problem with obscuration." He said, "Well, see if you can find a way to *make* it work." So I fiddled around with the idea a bit, and I came up with a way to turn concept inside out and actually managed to come up with a construction that had a viable obscuration ratio, testing the telescope assembly at finite conjugates. The setup was autostigmatic in this form, including some sub-diameter optics to achieve the necessary correction. He glanced at it and said, "Write it up." I said, "Okay, I'll do my best." I wasn't a great writer at that point in my life, but I wrapped some words and drawings around the idea and gave it to him for his approval. He said, "Change this, fix that, and publish it." I replied, "What do I do to publish it?" I'd never published anything. He calmly replied, "Send it to *Applied Optics*." So I spent some considerable time fine-tuning the manuscript to make it read as clearly as possible. Then I put his name at the top and my name below it and showed it to him again. He looked at it, and I saw him cross out one item. He handed it back to me and said, "Go for it." I looked at it, and he had crossed his name out. I said, "Hang on, it was your idea." He said, "But it was your execution. I've got enough thunder. Don't worry about stealing any of mine." I just thought that was really noble of him.

Lytle's later visit to Aden's office illustrated how devoted Aden was to giving students the same sort of opportunities that had been offered to him at Pasadena Junior College and Caltech.

I walked into his office one day, and I said, "You know, Dr. Meinel, I think I'd rather do optical design than build telescope mirrors." He said, "That's fine." So he started sending me around the country to sample the various lens design programs in the industry, which would include the Itek program . . . and . . . program at Perkin-Elmer, to name a couple. Ultimately, Dr. Meinel dispatched me to get familiar with most of these programs, to learn a little bit about how each operated. I realize now that he didn't do this to support the Optical Sciences Center but to give me some experience, which was something I later learned he was famous for. He really provided me some interesting opportunities.[2]

Aden repeatedly said that once a discovery is made, it cannot be made again. The path that scientific inquiry takes is paved by discoveries of new scientific facts about the universe. Many of these discoveries are dead ends; others open up new windows into the universe, expand our knowledge and understanding of the world about us, and provide opportunities for intellectual stimulation in the years ahead. Aden and Marjorie left us with a world of new facts, new facilities, revolutionary technology for sustainable energy, and improved national defense to influence and protect future generations.

His youthful experiences as an accomplished classical pianist, machinist, woodworker, optician, draftsman, and mathematician prepared him well for his later career. His inquiring mind propelled him through high school in Pasadena to Mt. Wilson Observatory and Caltech. In 1944, at the age of 22, he was one of the first to design multistage rockets, not for military purposes but to collect solar spectrographic data at high altitudes. With his experience in rockets and rocket launchers at Caltech, he was selected as a Navy Ensign to retrieve rocket and optics technology from Nazi Germany on the front lines as the war in Europe was winding down. During Operation Paperclip, Aden assisted the evacuation of 231 optical scientists and their families from Jena to Heidenheim in 1945, among them Lou Bruckner from Zeiss. Some went to the States and worked at Wright-Patterson Air Force Base, like Bruckner, who later helped Aden draft the curriculum for the Optical Sciences Center (OSC).

In 1946 he entered the University of California at Berkeley as a sophomore and within three years had completed his PhD in astronomy by opening a new window in our understanding of auroras and the relationship between Earth and the Sun. Aden designed, built, and used the instrument he invented to record data unknown before and interpreted the astrophysics correctly: protons from the Sun caused Earth's auroras. During this time, he also married his high school sweetheart and fathered three children before his 27th birthday on November 22, 1949.

What was there left to do? They could have lived out the next 62 years quietly in research and teaching at almost any university of their choice, but they opted to leave a broad legacy, driven by discovery in several areas: astronomy, optical fabrication and testing, engineering and education in the optical sciences, the foundations of solar thermal energy independence, national defense, and major education facilities. They influenced the lives of perhaps millions of people around the globe.

Astronomical sciences and instrumentation

Discovery is the engine of scientific investigations in astronomy. Discoveries are made, new theories are developed to explain these observations, predictions of new observations are made, new apparatus built to make new measurements to either verify or dispute theory. And the cycle continues. Aden contributed to all phases in this full cycle. In the field of astronomy, he designed and built new instruments and used them to discover new facts about the universe.

In high school he built Schmidt cameras and telescopes at the Mt. Wilson Optical shop. He used this knowledge after the war to build his own instrument, record spectra of the night sky auroras, and interpret correctly the science behind his measurements.

The Meinel spectrograph on a 36-inch telescope was just as sensitive as an ordinary spectrograph on an 80-inch telescope and increased access to sky for many more young astronomers. Special small f/ratio optics and modern high-efficiency diffraction gratings revolutionized spectroscopy in the 1960s. He laid the groundwork for modern large-telescope spectrographs and with Helmut Abt created a new atlas of spectra. Aden took the risk and built the cost-effective 84-inch KPNO telescope at f/2 by being the first person to heat-slump a large light-weighted glass mirror blank and design and apply new optical test methods. He was the first astronomer to use a near-zero expansion fused silica, lightweight mirror for a prudent telescope (Bok 90-inch).

He designed and led the construction of the world's first segmented large-aperture astronomical telescope (the MMT) and demonstrated the utility of sparse large-aperture telescopes at optical wavelengths. Aden developed concepts for deployable space telescopes using both mirrors and lightweight rolled membranes for primary mirrors. Together Aden and Marjorie traveled the world, lecturing on astronomical telescopes and instrumentation in Europe and Asia. Their work enabled Hyderabad University to build the largest (at that time) astronomical telescope in India and stimulated productive research programs by several Indian students. Aden and Marjorie's many trips to China contributed to several telescopes, including the revolutionary LAMOST telescope located at Xinglong Station in Hebei province in China. Several generations of observational astronomers across the world built their careers on scientific data recorded with the many astronomical telescopes and instruments that incorporated Aden's technical innovations.

Aden selected Kitt Peak mountain, along with the city of Tucson, Arizona, and the UA campus, for the nation's astronomical observatory. Since its founding in 1957, this research facility of the National Science Foundation has supported hundreds of permanent staff and has trained hundreds of young astronomers in observational astrophysics. As director of KPNO, Aden recruited and hired outstanding astronomers and engineers and plotted the facilities layout on the mountain 53 miles southwest of Tucson. In addition, he participated in the design and engineering implementation of the mountaintop telescopes and instruments that revolutionized modern observational astrophysics. Generations of astronomers graduated and moved forward to universities, government, and industry across the United States and around the world to advance our understanding of the universe.

A few years later, Aden moved to UA, where he led the expansion of the Steward Observatory's astronomical academic and research programs. He recruited Gerard P. Kuiper, one of his mentors from Yerkes Observatory, to join UA. Kuiper went on to establish a center of excellence in lunar and planetary science by forming the Lunar and Planetary Laboratory (LPL) with a PhD program and its own astronomical observatory in the Catalina Mountains north of Tucson. Over the past 60 years, the LPL facilities and graduates have supported lunar and planetary exploration from the first Moon landings by Apollo to today's plans for a Mars manned lander.

In 2009, 70 years after starting work at Mt. Wilson observatory, Aden published his final scientific paper, which remains controversial. In the *International Astronomical Union Symposium Proceedings Series*, Aden and Marjorie announced a compelling discovery. They suggested that a jet of cosmic rays accompanied by an incredibly strong magnetic field had traveled 220 parsecs from the planetary nebula NGC 6543 (the "Cat's Eye Nebula") and irradiated Earth and caused Pleistocene extinctions. The only evidence was from ice cores in Greenland and Antarctica. The paper was never published in mainstream journals because neither the geophysicist nor biologists would accept the premise in the absence of proof.[3]

Optical system engineering, fabrication, and testing

Optical engineers have a saying: "If we can test it, we can build it." Aden's optical system engineering, mirror fabrication, and optical test skills that he developed for ground-based astronomical instruments were also applicable to space optical systems and commercial industry for imaging systems. Along with his optician, Don Loomis, and Stanley Bashkin in the Department of Physics at UA, Aden invented ion-beam polishing, which is used in industry today to place complicated but smooth optical figures on dielectric surfaces for industrial applications. The 84-inch and 90-inch telescope primary mirrors were fabricated by Loomis in a laboratory with a low ceiling and no access to either the center of curvature or to the focus of the mirrors. Aden invented the optical test procedure that enabled the mirrors to be fabricated and deliver near diffraction-limited performance.

Solar power

Aden and Marjorie were the first persons to recognize that the most efficient use of solar thermal energy is to collect sunlight over large areas and convert the thermal energy into heat and then use that heat to run steam-powered electric generators to distribute power over a grid system. Prior to their work, it was believed that the most efficient use of solar heat was for each dwelling to have its own thermal hot water unit. But cost-effectiveness was not possible on that scale at that time, leading the National Academy of Sciences in 1970 to see no economic future in the high-flying solar dreams of the 1950s. Aden and Marjorie, however, disagreed in an article in the *Tucson Daily Citizen* that same year, saying that thousand-gigawatt solar stations ("farms") in the western United States could supply much of the nation's energy needs and even provide power for desalination plants. They later moderated their initial concept to megawatt solar farms. Members of Congress, industry, and even actor Marlon Brando endorsed their plans enthusiastically and sought answers to a slew of questions, but the "old guard"—establishment solar scientists—demurred on the basis that the large-scale concepts were still unworkable, especially at reasonable costs. When research funding for solar farms slowly diminished, Aden and Marjorie reversed course and abandoned their efforts to promote solar energy, turning instead to endorse nuclear- and coal-generated power as more viable and sustainable, at least for the foreseeable future.

Not until the 1980s would interest in (and funding for) solar farms begin to emerge at state and municipal levels as technology and lower PV prices caught up with the Meinel concept. By the middle of that decade, there were huge solar power plants constructed in southern California. By the beginning of 2020, there were over 82 GW of installed photovoltaics in the United States, 40 percent of that rooftop and agricultural photovoltaics and 60 percent utility scale. There are 900 MW of Meinel-type solar energy-generating stations in the deserts of California and Arizona, including the 354 MW Solar Energy Generating Systems (SEGS) facility in San Bernardino County, California (Figure 13.1), and also 400 MW of power tower stations in California.

Education

Since Aden established the academic program in optical sciences, over 3,000 students have earned either a BS, MS, or PhD degree in optical sciences from UA. Today, the Wyant College of Optical Sciences, which evolved from the OSC founded by Aden in 1967, houses the largest and most comprehensive academic program in optical sciences and engineering in the Western world.

The College training and research program spans four areas of emphasis: optical engineering and design, image science, photonics, and optical physics. The first, optical engineering and design, covers those areas of opto-mechanical, structural, thermal, dynamics, coatings, materials, radiation propagation, and detectors that affect the image quality and brightness of the optical system as it

Figure 13.1. Part of the 354 MW SEGS solar complex in northern San Bernardino County, California.
Courtesy of the US Department of the Interior, Bureau of Land Management.

will be used. Image science is dedicated to extracting information from images and understanding how man interprets those images. Photonics is the science and engineering of integrating optics into systems and devices for the purpose of reducing cost and improving the function of electronic systems or replacing electronics completely. It includes fiber optics and on-chip miniature optical systems. Optical physics is the study of the creation of light and its interaction with matter. These four areas cover the broad interdisciplinary scope of optics: the generation, propagation, and detection of light, and image interpretation.

Graduates become professors, scientists, engineers, and entrepreneurs working in academia, industry, government, and business around the globe. The College's annual budget for sponsored research exceeds $20 million. By 2020 there were over 3,000 graduates with BS, MS, or PhD degrees in optics. Currently there are over 100 faculty, teaching more than 75 undergraduate and graduate classes to hundreds of students.

National defense

From the moment at a RAND meeting when Aden showed that the average resolution of images taken from cameras aimed downward from space without atmospheric turbulence was 2 inches or less, the defense establishment was stunned. The race was on to test the theory, and confirmation came later with laser interferometry from a U-2 in overflight. His ideas and input into camera and mirror design were incorporated in OXCART and every aerial surveillance satellite from KH-4 through KH-6 (CORONA) to KH-11 (KENNEN/CRYSTAL). Aden's knowledge of rockets and missiles helped Bud Wheelon and the CIA with photointerpretation of Soviet emplacements in Cuba that helped to end the missile crisis in October 1962. Satellite surveillance beginning with CORONA enabled verification of compliance with nuclear arms agreements so as to dissect facts from rhetoric. Satellites yielded reliable data to estimate missile inventories and in that way were critical to ending the Cold War. Their ability to identify breaches of arms control agreements was the foundation in negotiating the Nuclear Non-Proliferation Treaty (1968), SALT I (1972), and SALT II (1979).

University of Arizona Campus

Figure 13.2 (consult the color insert for this figure) shows a portion of the UA campus dedicated to the optical, planetary, and astronomical sciences. Scientific progress thrives on the proximity of scientists and their laboratories/staff to one another. Aden arranged for these facilities to be in proximity. Five buildings are identified in the figure: (1) the Richard Caris Telescope Mirror Laboratory; (2) the Meinel Building with the James C. Wyant College of Optical Sciences; (3) the Kuiper Building with the Lunar and Planetary Laboratory; (4) the National Optical and Infrared Astronomical Observatory (KPNO headquarters); and (5) Steward Observatory and Department of Astronomy. Were it not for the vision and energy of Aden, these facilities would be quite different, or entirely missing from the campus landscape.

University of Arizona president Richard Harvill and Aden negotiated with the state for funds and eminent domain rights to acquire the city block across the street from the Steward Observatory for the national observatory laboratory and offices (Figure 13.2, #4), thus concentrating the physical facilities for astronomical research and education into one area on campus. Aden recruited Gerard Kuiper to join UA, and Kuiper established the LPL facilities and recruited faculty under NASA sponsorship. Aden and Harvill arranged to locate the building (Figure 13.2, #3) adjacent to KPNO headquarters and across the street from Steward Observatory.

Aden made the decision to locate KPNO headquarters near the Department of Astronomy on campus, and the university responded with plans to expand its academic program in the astronomical sciences. When he was director of Steward Observatory, Aden implemented a building addition (Figure 13.2, #5) to accommodate the department expansion. Later additions brought the facility to the size we see today.

The 80,000-square-foot core building (center) of the OSC was designed by Aden and paid for by the US Air Force. Today, after two additional expansions, the entire 157,000-square-foot building (Figure 13.2, #2), which contains laboratories, offices, and classrooms, is his namesake.

As director of Steward Observatory, Aden recruited Roger Angel and provided laboratory space for the initial development of large-aperture, spin-cast, lightweight telescope mirrors now used in state-of-the-art, ground-based telescopes and the Giant Segmented Mirror Telescope (GSMT). The Richard F. Caris Mirror Lab (Figure 13.2, #1), under the stands on the north side of the football stadium, houses the optical fabrication and test facilities for these giant 8-m mirrors.

Aden's comprehensive understanding of observational astronomy and atmospheric physics led to his selection of Kitt Peak mountain for the US National Observatory. Without Aden, none of these facilities would be there. After his departure from Kitt Peak in March 1960, the mountaintop continued to develop, until today a consortium of universities and institutes operates 22 optical telescopes and two radio telescopes, enabling research and education for hundreds of faculty and students who qualify for merit-based research founded on compelling science proposals, competitively selected from institutions and academic departments around the world (Figure 13.3; consult the color insert for this figure).

International influence

Over their lifetime, Aden and Marjorie spent months teaching international scientists and engineers around the world practical solar thermal energy, astronomy, and how to build telescopes and instruments for observational astrophysics. Three principal locations were the Summer School in Trieste, Italy, where they taught solar energy and astrophysics; Osmania University in Hyderabad, India, where they lectured on telescopes, instruments, and solar energy; and several institutions in Beijing, Nanjing, and Shanghai, China.

Industry

Optics is a science and technology fundamental to video communication, medical and satellite imaging, surgical lasers, fiber optics, observational sciences, vision, remote-sensing agriculture and climate, video games, intelligence and national defense, and many others. Optics pervades all of society today. The Optical Society of America initiatives in the early 1960s were led by senior members of industry and government who recognized the serious need to create a highly skilled workforce in the optical sciences. By 2018 the global optics and photonics industry had gross revenues of $282 billion, more than 1,000 times what they were in 1965.[4] During his interview referenced earlier, Bob Breault shared his own experience after graduation from UA to illustrate this well:

> I was the first local formal business to start from the Optical Sciences Center, and for 20 years the other companies were "competitors." When I graduated in 1979, I was offered a $77,000 job in El Segundo. I already had $88,000 in contracts. . . . We got funded up until 1996 by state money and sometimes the city or county money. At the time we were making $233 million a year as gross revenue. That's excluding Raytheon. We collected about $10,000 a year in dues. In 2006 we had 2,300 employees. In 1996 we were making $2.3 billion, and we'd gone up to 23,000 employees in the state. If you wanted to put in $100,000 over 10 years and get $2 billion net gain out of it, you're very happy. . . . In December 1992 *Business Week* put us on the cover as Optics Valley. The Optical Sciences Center was the heart and brain of the entity.

Aden Meinel's vision 55 years ago and the support he received from President Richard Harvill and many others propelled UA to become the leading center in optics graduate education.

The late Stephen Jacobs (1928–2019), one of Aden's first photonics hires at OSC as a specialist in laser technology, helped to pioneer quantum electronics with Marlan Scully and taught it at two-week summer schools.[4] Jacobs used laser interferometry to identify optical materials that were dimensionally stable over wide ranges of temperature and time; his results were so precise that his laboratory was effectively a National Bureau of Standards for dimensional stability. He collaborated on projects such as determining quartz homogeneity for the Gravity Probe B relativity experiment, identifying the most homogeneous glass for spin casting, and developing ultra-stable Invar 36 for the Cassini spacecraft.[5] Photonics now dominates our daily lives, from consumer electronics, solar applications, manufacturing, and medicine to defense and entertainment.[6]

One of Aden's many gifts that was so useful in establishing OSC was in recognizing talents in academia and industry and then filling faculty niches with them—Bob Noble, Roland Shack, Bob Shannon, Stephen Jacobs, Nick Stavroudis, Phil Slater, Bill Wolfe, James Eyer, Clarence Babcock, Arthur Turner, Bernhard Seraphin, Gerard Kuiper, and others. His hands-off management style with faculty and students alike increased their self-confidence and self-sufficiency, which in turn snowballed into more curiosity, more students, more courses offered, more discoveries and inventions, more papers, more awards. Their accomplishments are in one sense his also, just as students owe much to their teachers, although he would never presume to take credit. With Marjorie constantly by his side, he transformed the lives of those who would go on to make their marks in science and industry and train others. Together they formed one massive star, and with their deaths the shock waves of the supernova are still producing new stars in astronomy, optics, and solar engineering.

Notes

1. Interview with Robert Breault by J. Breckinridge on February 21, 2019, in Tucson, Arizona.
2. Interview with John Lytle by J. Breckinridge on February 28, 2017, in Tucson, Arizona.
3. Interview with Helmut Abt by A. Pridgeon on April, 14, 2016, in Tucson, Arizona.
4. Interview of Stephen Jacobs by J. Breckinridge on September 29, 2016, in Tucson, Arizona.
5. https://www.optics.arizona.edu/news-events/news/quicknews/reflections-stephen-f-jacobs.
6. SPIE. 2020. Optics and photonics industry report: An in-depth assessment of the global optics and photonics industry, highlighting industry trends and profiling key companies involved. SPIE, Bellingham, Washington, 4.

Acronyms and Abbreviations

AAS	American Astronomical Society
ABMA	Army Ballistic Missile Agency
AEC	Atomic Energy Commission
AEPG	Army Electronics Proving Ground
AF-CRL	Air Force-Cambridge Research Laboratories
AMOS	ARPA Midcourse Optical Station
AR	Aircraft rocket
ARPA	Advanced Research Projects Agency
ASR	Antisubmarine rocket
AURA	Association of Universities for Research in Astronomy
AXAF	Advanced X-ray Astrophysics Facility
BIA	Bureau of Indian Affairs
BuOrd	Bureau of Ordnance
CAFT	Combined Advance Field Teams
CCD	Charge-coupled device
CIA	Central Intelligence Agency (the Agency)
CIC	Counterintelligence Corps
CIT	California Institute of Technology (Caltech)
COSTAR	Corrective Optics Space Telescope Axial Replacement
DARPA	Defense Advanced Research Projects Agency
DCI	Director of Central Intelligence (CIA)
DDP	Deputy Director for Plans (CIA)
DDR	Deputy Director for Research (CIA)
DDS&T	Deputy Director of Science and Technology
DISIC	Dual improved stellar index camera
DNRO	Director of the National Reconnaissance Office
DOD	Department of Defense
DST	Directorate of Science & Technology (CIA)
ECM	Electronic counter measures
EG&G	Edgerton, Germeshausen & Grier, Inc.
ELT	Extremely large telescope
FFAR	Forward-firing aircraft rocket
FOV	Field of view
GALCIT	Guggenheim Aeronautical Laboratory, California Institute of Technology
HIORP	Hubble Independent Optics Review Panel
HST	Hubble Space Telescope
HVAR	High-velocity aircraft rocket

IAS	Institute of Aeronautical Sciences
IAU	International Astronomical Union
ICBM	Intercontinental ballistic missile
IDA	Institute for Defense Analysis
IGY	International Geophysical Year
IR	Infrared
IRAS	Infrared astronomical satellite
IRBM	Intermediate-range ballistic missile
JPL	Jet Propulsion Laboratory
JWST	James Webb Space Telescope
KPNO	Kitt Peak National Observatory
LDR	Large deployable reflector
LPL	Lunar and Planetary Laboratory (UA)
LST	Landing ship, tanks; large space telescope
MAD	Magnetic anomaly detector
MIT	Massachusetts Institute of Technology
MMT	Multiple Mirror Telescope
MOL	Manned orbiting laboratory
MRBM	Medium-range ballistic missile
MWO	Mt. Wilson Observatory
NAO	National Astronomical Observatory
NARA	National Archives and Records Administration
NAS	National Academy of Sciences
NASA	National Aeronautics and Space Administration
NavTechMisEu	Naval Technical Mission in Europe
NDRC	National Defense Research Committee
NEP	Nuclear-powered electric propulsion
NIMO	National Institute of Modern Optics
NOAO	National Optical Astronomy Observatory
NOTS	Naval Ordnance Test Station
NPIC	National Photographic Interpretation Center
NRO	National Reconnaissance Office
NSF	National Science Foundation
OAO	Orbiting Astronomical Observatory
OCI	Office of the Coordination of Information
ODM	Office of Defense Mobilization
ONI	Office of Naval Intelligence
ONR	Office of Naval Research
ORR	Office of Research and Reports
OSA	Optical Society of America; Office of Special Activities
OSC	Optical Sciences Center
OSI	Office of Scientific Intelligence
OSRD	Office of Scientific Research and Development
OSS	Office of Strategic Services

PBY	Patrol Boat-Y
P-E	Perkin-Elmer
PFIAB	President's Foreign Intelligence Advisory Board
PID	Photographic Intelligence Division
PJC	Pasadena Junior College
PSAC	President's Scientific Advisory Committee
PV	Photovoltaics
P&W	Pratt & Whitney
RAND	Research and development
RCS	Radar cross section
RIT	Rochester Institute of Technology
SALT	Southern African Large Telescope; Strategic Arms Limitation Talks
SAM	Surface-to-air missile
SAMOS	Satellite and Missile Observation System
SAMSO	Space and Missile Systems Organization
SCI	Sensitive compartmentalized information
SMD	Science Mission Directorate
SPIE	Society of Photo-Optical Instrumentation Engineers
STScI	Space Telescope Science Institute
TAD	Temporary additional duty
TAU	Thousand astronomical units
TCP	Technological Capabilities Panel
UA	University of Arizona
USAF	United States Air Force
USAFFE	US Army Forces in the Far East
USIB	United States Intelligence Board
WF/PC	Wide-field planetary camera
WIG	Wing-in-ground (effect)

Literature by Aden Meinel and Marjorie Pettit Meinel

(arranged first chronologically and then alphabetically by author and title)

1940s

Pettit, M. S. 1944. The long-period variable star RT Cygni. *Astronomical Society of the Pacific* 56: 107–11.

Pettit, M. S. 1944. *A study of the long-period variable star RT Cygni.* Master's thesis. Claremont Colleges, Claremont, California.

Meinel, A. B. 1948. The near infrared spectrum of the airglow and aurora. *Publications of the Astronomical Society of the Pacific* 60: 373.

Meinel, A. B. 1949. *A spectrographic study of the night sky and aurora in the near infrared.* PhD dissertation, University of California at Berkeley.

1950s

Meinel, A. B. 1950. Evidence for the entry into the upper atmosphere of high-speed protons during auroral activity. *Science* 112: 590.

Meinel, A. B. 1950. Hydride emission bands in the spectrum of the night sky. *Astrophysical Journal* 111: 207–8.

Meinel, A. B. 1950. Identification of the 6560Å emission in the spectrum of the night sky. *Astrophysical Journal* 111: 433.

Meinel, A. B. 1950. A new band system of N^+_2 in the auroral spectrum. *Comptes rendus de l'Académie des Sciences* 231: 1049.

Meinel, A. B. 1950. A new band system of N^+_2 in the infrared auroral spectrum. *Astrophysical Journal* 112: 562–3.

Meinel, A. B. 1950. O_2 emission bands in the infrared spectrum of the night sky. *Astrophysical Journal* 112: 464–8.

Meinel, A. B. 1950. OH emission bands in the spectrum of the night sky. I. *Astrophysical Journal* 111: 555–64.

Meinel, A. B. 1950. OH emission bands in the spectrum of the night sky. II. *Astrophysical Journal* 112: 120–30.

Meinel, A. B. 1950. On the entry into the earth's atmosphere of 57-keV protons during auroral activity. *Physical Review* 80: 1096–7.

Meinel, A. B. 1950f. On the entry into the Earth's atmosphere of high speed protons during auroral activity. *Science* 112 (2916): 590.

Meinel, A. B. 1950. Strong permitted OI and NI lines in the infrared auroral spectrum. *Transactions of the American Geophysical Union* 31: 21–4.

Meinel, A. B. 1951. The analysis of auroral emission bands from the A^2II state of N^+_2. *Astrophysical Journal* 114: 431–7.

Meinel, A. B. 1951. The auroral spectrum from 6200 to 8900 A. *Astrophysical Journal* 113: 583–8.

Meinel, A. B. 1951. Book review of *Atlas der Restlinen*, vol. III: *Spektren seltener Metalle und einiger Metalloide* (by A. Gatterer and J. Junkes). *Astrophysical Journal* 114: 371.

Meinel, A. B. 1951. Doppler-shifted auroral hydrogen emission. *Astrophysical Journal* 113: 50–4.

Meinel, A. B. 1951. Science takes a look at the Northern Lights. *Naval Research Reviews*, Office of Naval Research.

Meinel, A. B. 1951. The spectrum of the airglow and the aurora. *Reports on Progress in Physics* 14: 121.

Meinel, A. B. 1952. Book review of *The aurorae* (by Liev Harang). *Astrophysical Journal* 115: 336.

Meinel, A. B. 1952. Book review of *Grating spectrum of iron* (by A. Gatterer). *Astrophysical Journal* 115: 335–6.

Meinel, A. B. 1952. Excitation mechanisms in the aurora. L'Étude optique de L'Atmosphère Terrestre: Communications présentées qu colloque international tenu à l'Institut d'Astrophysique de l'Université de Liège les 3 et 4 Septembre 1951. *Mémoires de la Société royale des sciences de Liège*, Sér. 4, tome 12: 203–14.

Meinel, A. B., and C. Y. Fan. 1952. Laboratory reproduction of auroral emissions by proton bombardment. *Astrophysical Journal* 115: 330–1.

Meinel, A. B., and Swings, P. 1952. The spectra of the night sky and aurora. In *The atmospheres of the Earth and planets* (ed. G. P. Kuiper), 2nd ed., pp. 159–210. University of Chicago Press.

Fan, C. Y., and Meinel, A. B. 1953. Laboratory ionic-impact emission spectra. *Astrophysical Journal* 118: 205–13.

Meinel, A. B. 1953. Aspheric field correctors for astronomical telescopes (abstract). *Journal of the Optical Society of America* 43: 811.

Meinel, A. B. 1953. Aspheric field correctors for large telescopes. *Astrophysical Journal* 118: 335–44.

Meinel, A. B. 1953. Book review of *Prism and lens making* (by F. Twyman). *Astrophysical Journal* 117: 240.

Meinel, A. B. 1953. Magnetic fields of galactic dimensions. *Publications of the Astronomical Society of the Pacific* 65: 289.

Meinel, A. B. 1953. Origin of the continuum in the night-sky spectrum. *Astrophysical Journal* 118: 200–4.

Meinel, A. B., and Schulte, D. H. 1953. A note on auroral motions. *Astrophysical Journal* 117: 454–5.

Chamberlain, J. W., Fan, C. Y., and Meinel, A. B. 1954. A new O_2 band in the infrared auroral spectrum. *Astrophysical Journal* 120: 560–2.

Meinel, A. B. 1954. The morphology of the aurora. *Proceedings of the National Academy of Sciences of the United States of America* 40: 943–50.

Meinel, A. B. 1954. The near infrared spectrum of the aurora. *Proceedings of the Conference on Auroral Physics*, 1954, p. 75.

Meinel, A. B. 1954. Protons and the aurora. *Proceedings of the Conference on Auroral Physics*, 1954, p. 41.

Meinel, A. B. and Chamberlain, J. W. 1954. Emission spectrum of twilight, night sky and aurora. In *The solar system*, vol. II. *The Earth as a planet* (ed. G. P. Kuiper), pp. 514–75. University of Chicago Press.

Meinel, A. B., Negaard, B. J., and Chamberlain, J. W. 1954. A statistical analysis of low-latitude aurorae. *Journal of Geophysical Research* 59: 407–13.

Morgan, W. W., Meinel, A. B., and Johnson, H. M. 1954. Spectral classification with exceedingly low dispersion. *Astrophysical Journal* 120: 506–8.

Meinel, A. B. 1955. Optical instrumentation for upper atmospheric physics. Invited paper. *Journal of the Optical Society of America* 45: 899–900.

Meinel, A. B. 1955. Systematic auroral motions. *Astrophysical Journal* 122: 206–7.

Roach, F. E., and Meinel, A. B. 1955. The height of the nightglow by the van Rhijn method. *Astrophysical Journal* 122: 530–53.

Roach, F. E., and Meinel, A. B. 1955. Nightglow heights: A reinterpretation of old data. *Astrophysical Journal* 122: 554–8.

Meinel, A. B. 1956. Airglow and aurora. *Smithsonian Contributions to Astrophysics* 1: 95.

Meinel, A. B. 1956. An F/2 Cassegrain camera. *Astrophysical Journal* 124: 652–7.

Meinel, A. B. 1956. Galactic research with small Schmidt cameras. *Mitteilungen der Astronomischen Gesellschaft* 7: 15.

Meinel, A. B. 1957. Book review of *Astronomical optics and related subjects* (ed. Zdenek Kopal). *Publications of the Astronomical Society of the Pacific* 69: 369.

Meinel, A. B. 1957. Book review of *The sun* (by Giorgio Abetti). *Journal of the Optical Society of America* 47: 1137.

Meinel, A. B. 1957. Extremely innovative design for a f/1.5 minimum mass telescope. KPNO Technical Report #12. Office of the Executive Secretary. August 15, 1957.

Meinel, A. B. 1957. A report on the National Astronomical Observatory facility (abstract). *Astronomical Journal* 62: 27.

Meinel, A. B. 1958. Final report on the site selection survey for the National Astronomical Observatory. AURA, Kitt Peak National Observatory (Reprinted as Kitt Peak National Observatory, Contributions no. 45, October 1963).

Meinel, A. B. 1958. The National Observatory at Kitt Peak. *Sky & Telescope* 17: 492–9.

Meinel, A. B. 1958. Report on the Kitt Peak Observatory. *Astronomical Journal* 63: 308.

Meinel, A. B. 1959. Astronomical observations from space vehicles. *Publications of the Astronomical Society of the Pacific* 71: 369–80 (Reprinted as Kitt Peak National Observatory, Contributions no. 1, 1959).

Meinel, A. B. 1959. Ratio spectra for G8 stars (abstract). *Astronomical Journal* 64: 341–2.

Meinel, A. B., and Golson, J. C. 1959. Spectral classification from ratio spectra. *Publications of the Astronomical Society of the Pacific* 71: 445–50 (Reprinted as Kitt Peak National Observatory, Contributions No. 2, 1959).

1960s

Meinel, A. B. 1960. Book review of *Vistas in astronomy*, vol. 2 (ed. M. Alperin and H. F. Gregory). *Publications of the Astronomical Society of the Pacific* 72: 429.

Meinel, A. B. 1960. Preliminary spectrophotometric observations of Nova Herculis, 1960 (abstract). *Astronomical Journal* 64: 494.

Meinel, A. B. 1961. Astronomical seeing and observatory site selection. In *Stars and stellar systems: telescopes* (ed. G. P. Kuiper), pp. 154–75. University of Chicago Press.

Meinel, A. B. 1961. Design considerations for a large aperture orbital telescope: report. *Mémoires de la Société royale des sciences de Liège*, Sér. 5, tome 4: 49–59 (Reprinted as Kitt Peak National Observatory, Contributions no. 7, 1961).

Meinel, A. B. 1961. Design of reflecting telescopes. In *Telescopes, stars and stellar systems* (ed. G. P. Kuiper and B. M. Middlehurst), p. 25. University of Chicago Press.

Meinel, A. B. 1961. New frontiers of astronomical technology. *Science* 134: 1165–71 (Reprinted as *New frontiers in astronomical instrumentation*. Kitt Peak National Observatory, Contributions no. 11, 1961).

Meinel, A. B. 1961. New frontiers of optics (abstract). *Journal of the Optical Society of America* 51: 471.

Meinel, A. B. 1962. Advances in astronomical technology. *The Indian & Eastern Engineer*. (104th anniversary number; reprinted in *Annual Report, Smithsonian Institution*, 1963, 293).

Meinel, A. B. 1962. Book review of *Space astrophysics* (ed. W. Liller). *Applied Optics* 1: 164.

Meinel, A. B. 1962. Continuous H2 emission in B-star atmospheres. *Astronomical Journal* 67: 581.

Meinel, A. B. 1962. Detection of molecular hydrogen in stellar atmospheres. *Publications of the Astronomical Society of the Pacific* 74: 523–4.

Meinel, A. B. 1962. High resolution optical space telescopes. Space Age Astronomy, Proceedings of an International Symposium held August 7–9, 1961, at the California Institute of Technology in conjunction with the 11th General Assembly of the International Astronomical Union (eds. A. J. Deutsch and W. B. Klemperer), p. 236. Academic Press, New York.

Meinel, A. B. 1962. Infrared astronomy (abstract). *Astronomical Journal* 67: 118.

Meinel, A. B., and Hoxie, D. T. 1962. On the spectrum of lightning in the Venus atmosphere. *Publications of the Astronomical Society of the Pacific* 74: 329–30.

Johnson, H. L., and Meinel, A. B. 1963. Infrared color index of M31 at 2.2 microns. *Astrophysical Journal* 138: 1317–19.

Meinel, A. B. 1963. Astronomy. (Report of American Astronomical Society meeting, Tucson, Arizona, April 17–20, 1963). *Science* 141: 178–81.

Meinel, A. B. 1963. On the ultraviolet continuous spectrum of B stars. *Astrophysical Journal* 137: 321–6.

Meinel, A. B. 1963. Spectrophotometry of Nova Herculis 1960. *Astrophysical Journal* 137: 834.

Meinel, A. B. 1963. Steward Observatory folded-camera grating spectrograph. *Astronomical Journal* 68: 285.

Meinel, A. B., and Abt, H. A. 1963. *Final report on the site selection survey for the National Astronomical Observatory*. Contributions of the Kitt Peak National Observatory no. 45.

Meinel, M. P., and Meinel, A. B. 1963. Late twilight glow of the ash stratum from the eruption of Agung volcano. *Science* 142: 582–3.

Meinel, A. B., Middlehurst, B., and Whitaker, E. 1963. Low-latitude noctilucent cloud of June 15, 1963. *Science* 141: 1176–8.

Bashkin, S., and Meinel, A. B. 1964. Laboratory excitation of the emission spectrum of a nova. *Astrophysical Journal* 139: 413–16.

Bashkin, S., Meinel, A. B., Malmberg, P. R., and Tilford, S. G. 1964. Optical atomic spectroscopy with a van de Graaff accelerator. *Physics Letters* 10: 63–64.

Meinel, A. B. 1964. Some aspects of graduate study in the optical sciences (abstract). *Journal of the Optical Society of America* 54: 1385.

Meinel, A. B. 1964. Symposium on instrumental astronomy: Introduction. *Astronomical Journal* 69: 317–18.

Meinel, A. B., and Bashkin, S. 1964. Optical spectra from fast ion beams (abstract). *Astronomical Journal* 69: 552–3.

Meinel, A. B., and Meinel, C. P. 1964. Low–latitude noctilucent cloud of 2 November 1963. *Science* 143: 38–9.

Meinel, A. B., and Salanave, L. E. 1964. N^+_2 emission in lightning. *Journal of Atmospheric Sciences* 21: 157–60.

Meinel, A. B. 1965. A flat-field relay optical system. Optical Sciences Center Technical Report 5.

Meinel, A. B. 1965. General specifications for an 84- or 100-inch telescope for erection at the Kitt Peak observing station, Steward Observatory of the University of Arizona. Optical Sciences Center Technical Report 4.

Meinel, A. B. 1965. Introduction to the design of astronomical telescopes. Optical Sciences Center Technical Report 1.

Meinel, A. B. 1965. A stationary table optical polisher. Optical Sciences Center Technical Report 2.

Meinel, A. B., Bashkin, S., and Loomis, D. A. 1965. Controlled figuring of optical surfaces by energetic ionic beams (abstract). *Applied Optics* 4: 1674.

Meinel, A. B., and Canright, R. B. 1965. Observation of the Elf lunar flare (abstract). *Publications of the Astronomical Society of the Pacific* 77: 136.

Meinel, A. B., Schulte, D. H., and Wynne, C. G. 1965. Automatic optical designing for astronomy (abstract). *Publications of the Astronomical Society of the Pacific* 77: 136–7.

Noble, R. H., Statham, R. B., and Meinel, A. B. 1965. Effect of surface conductance on ionic polishing (abstract). *Journal of the Optical Society of America* 55: 1580.

Bashkin, S., Fink, D., Malmberg, P. R., Meinel, A. B., and Tilford, S. G. 1966. Collisional excitation atomic spectra in accelerated beams of light elements. *Journal of the Optical Society of America* 56: 1064–75.

Meinel, A. B. 1966. Mirror materials: other possibilities. In *The construction of large telescopes*. International Astronomical Union Symposium No. 27 (ed. D. L. Crawford), pp. B5–B9. Academic Press, New York.

Meinel, A. B. 1966. New approaches to very large telescopes. In *The construction of large telescopes*. International Astronomical Union Symposium No. 27 (ed. D. L. Crawford), pp. G11–G16. Academic Press, New York.

Meinel, A. B. 1966. Other possibilities. The Construction of Large Telescopes, Proceedings from Symposium no. 27 held in Tucson, Arizona, Pasadena and Mount Hamilton, California, U.S.A., April 5–12, 1965 (ed. D. L. Crawford). International Astronomical Union Symposium no. 27, p. 54. Academic Press, London.

Meinel, A. B., Akasofu, S. I., and Chapman, S. 1966. The aurora. In *Handbuch der Physik*, vol. 49/1 (ed. A. A. Winkelmann), pp. 1–158. Springer-Verlag, Berlin.

Meinel, A. B., Baranne, A., Wynne, C. G., and Schulte, D. H. 1966. Optical design. In *The construction of large telescopes*. International Astronomical Union Symposium No. 27 (ed. D. L. Crawford), pp. A9–A33. Academic Press, New York.

Meinel, A. B., and Meinel, M. P. 1966. Book review of *Twilight: A study in atmospheric optics* (by G. V. Rozenberg). *Science* 154: 1160.

Meinel, A. B., and Shack, R. V. 1966. A new design class of wide-field cameras for wide pass-band uses. Optical Sciences Center Technical Report 9.

Meinel, A. B., and Shack, R. V. 1966. A wide-angle all-mirror ultraviolet camera. Optical Sciences Center Technical Report 6.

Shack, R. V., and Meinel, A. B. 1966. A three-mirror flat-field photographic objective (abstract). *Journal of the Optical Society of America* 56: 545.

Meinel, A. B. 1967. Experience with lens design programs at the University of Arizona. *Journal of the Society of Motion Picture Television Engineers* 76: 201.

Meinel, A. B., Bok, B. J., Fitch, W. S., Hilliard, R. L., Taylor, D. J., and White, R. E. 1967. Specifications for the Cassegrain instruments including the Cassegrain observing platform, Steward Observatory 90-inch telescope. Optical Sciences Center Technical Report 16.

Meinel, A. B., and Meinel, M. P. 1967. Volcanic sunset-glow stratum: Origin. *Science* 155: 189.

Crawford, D. L., Meinel, A. B., and Stockton, M. W. 1968. (ed.). *A symposium on support and testing of large astronomical mirrors.* Optical Sciences Center Technical Report 30.

Meinel, A. B. 1968. Twenty years of physics: optics. *Physics Today* 21: 40–3.

Meinel, A. B., Aveni, A. F., and Stockton, M. W. 1968. *Catalog of emission lines in astrophysical objects.* Optical Sciences Center Technical Report 27.

Meinel, A. B., and Wilkerson, G. W. 1968. An f/2 focal reducer for the 60-inch U.S. Naval Observatory telescope. Optical Sciences Center Technical Report 7.

Meinel, A. B. 1969. Astronomical telescopes. In *Applied optics and optical engineering*, vol. 5 (ed. R. Kingslake), pp. 133–82. Academic Press, New York.

1970s

Meinel, A. B. 1970. Aperture synthesis using independent telescopes. *Applied Optics* 9: 2501–4.

Meinel, A. B. 1971. The Optical Sciences Center of the University of Arizona. Part 1. *Applied Optics* 9: 2413.

Meinel, A. B. 1971. A 1.8-m lightweight doubly asymmetric equatorial telescope design. *Applied Optics* 10: 249–56.

Meinel, A. B., Eyer, J. A., Noble, R. H., and Slater, P. N. 1971. The Optical Sciences Center: Its history, organization, and relation to government and industry. *Applied Optics* 10: 243–8.

Meinel, A. B., and Meinel, M. P. 1971. Is it time for a new look at solar energy? *Bulletin of the Atomic Scientists* 27: 32–7.

Meinel, A. B., and Meinel, M. P. 1971. Large scale conversion of solar energy (abstract). *Transactions—American Geophysical Union* 52: 814.

Meinel, A. B., and Meinel, M. P. 1971. *Power for the people.* Parker, Tucson.

Meinel, A. B., and Meinel, M. P. 1972. A harvest of solar energy. *Optical Sciences Newsletter* 6: 68–75.

Meinel, A. B., and Meinel, M. P. 1972. Physics looks at solar energy. *Physics Today* 25: 44–50.

Meinel, A. B., and Meinel, M. P. 1972. Solar economics. *Physics Today* 25: 13.

Meinel, A. B., and Meinel, M. P. 1972. Solar energy—the possible dream? *Aware* No. 17, 3–5.

Meinel, A. B., and Meinel, M. P. 1972. Thermal performance of a linear solar collector. American Society of Mechanical Engineers. Paper 72-WA/SOL-7.

Meinel, A. B., Shannon, R. R., Whipple, F. L., and Low, F. J. 1972. A large multiple mirror telescope (MMT) project. *Optical Engineering* 11: 33–7.

Meinel, A. B. 1974. Optical interfaces in solar energy utilization. *Proceedings SPIE, Effective Systems Integration and Optical Design* (March 1, 1974). https://doi.org/10.1117/12.954217.

Meinel, A. B., and Meinel, M. P. 1974. *Solar energy conversion, lecture notes, solar energy conversion* (eds. A. N. Mancini and I. F. Quercia), pp. 1–285. International College of Applied Physics, Italy.

Meinel, A. B., and Meinel, M. P. 1974. Solar power: option or illusion? Teller Center for Physics and Society, University of Colorado.

Meinel, A. B., and Meinel, M. P. 1974. The village energy center: A new option for solar energy utilization by Sahel communities? UNESCO, Paris.

Meinel, A. B. 1975. Optical interfaces in solar energy utilization. In *Effective systems integration and optical design*; Proceedings of the Seminar, San Diego, California, August 21–23, 1974 (A75-37330 17-35), pp. 12–16. Palos Verdes Estates, California, Society of Photo-Optical Instrumentation Engineers, 1975.

Meinel, A. B. 1975. Solar energy conversion, 2nd course. In *Lecture notes, solar energy conversion* (eds. A. N. Mancini and I. F. Quercia), pp. 19–46. International College of Applied Physics, Italy.

Meinel, A. B., and Meinel, M. P. 1975. Solar energy conversion. In *Lecture notes, solar energy conversion* (eds. A. N. Mancini and A. Silvestrini), pp. 1–356. University of Catania Press, International College of Applied Physics, Italy.

Meinel, A. B., and Meinel, M. P. 1975. Stratospheric dust-aerosol event of November 1974. *Science* 188: 477–8.

Meinel, A. B. 1976. Concentrating collectors. In *Solar energy engineering* (ed. A. A. M. Sayigh), pp. 183–216. Academic Press, New York.

Meinel, A. B. 1976. Cost effectiveness considerations for selective surfaces in solar collectors. Invited paper, UNESCO/WMO Symposium, Geneva, Switzerland, September 1976.

Meinel, A. B. The MMT. 1976. Invited paper, Tel Aviv University, Israel, November 11, 1976.

Meinel, A. B. 1976. The MMT and solar energy. Invited papers, Niels Bohr Institute, Copenhagen, Denmark, November 1976.

Meinel, A. B. 1976. Progress in development and application of selective surfaces. In *Heliotechnique and development*; Proceedings of the International Conference, Dhahran, Saudi Arabia, November 2–6, 1975. Volume 1 (A77-19043 06-44), pp. 166–79. Cambridge, Massachusetts, Development Analysis Associates.

Meinel, A. B., Baranne, A., Baum, W. A., Dolfus, A., Duchesne, M., Godoli, G., Hunter, A., de Jager, C., Livingston, W. C., Mikhel'Son, N., Sedmak, G., Sinvhal, S. D., and Walker, M. 1976. Instruments and techniques (Instruments et techniques). Meinel, A. B., Baranne, A., Baum, W. A., Dolfus, A., Duchesne, M., Godoli,. G., Hunter, A., de Jager, C., Livingston, W. C., Mikhel'Son, N., Sedmak, G., Sinvhal, S. D., and Walker,

M. 1976. Instruments and techniques (Instruments et techniques). *Transactions of the International Astronomical Union* 16A: 19–30.

Meinel, A. B., McKenney, D. B., and Meinel, M. P. 1976. A modular fixed-mirror Brayton-cycle solar power system. In International Conference on Solar Electricity, Toulouse, France, March 1–5, 1976, Reports (A77-21776 08-44), pp. 867–76. Toulouse, Centre National d'Etudes Spatiales.

Meinel, A. B., and Meinel, M. P. 1976. *Applied solar energy: An introduction.* Addison-Wesley, Reading, Massachusetts.

Meinel, A. B., and Meinel, M. P. 1976. Energy transfer in a large-scale thermal solar power farm. *Solar Energy* 18: 177–81.

Meinel, A. B., and Meinel, M. P. 1976. The exotic energy options: Separating reality from illusion. In *Proceedings of the International Conference on Physics in Industry,* Dublin, Ireland, pp. 19–30. Pergamon Press, Oxford.

Meinel, A. B., and Meinel, M. P. 1976. Solar energy: The options. Invited paper, Symposium on Utilization of Sun Energy, Stockholm, Sweden, November 2, 1976.

Meinel, A. B., and Meinel, M. P. 1976. Solar energy in the United States. Invited paper, Solar Energy Symposium, Ben Gurion University, Beersheba, Israel, November 1976.

Meinel, A. B., and Meinel, M. P. 1976. Solar photothermal power generation. *Environmental Conservation* 3: 15–21.

Meinel, A. B., Meinel, M. P., and Shaw, G. E. 1976. Trajectory of Mt. St. Augustine 1976 eruption ash cloud. *Science* 193: 420–2.

Osborn, D. E., Meinel, A. B., and Beauchamp, W. T. 1976. A solar collector modeling technique for grain drying application. In *Sharing the sun: Solar technologies in the seventies,* vol. 7, 50–63. International Solar Energy Society, Winnipeg, Manitoba, Canada.

Meinel, A. B. 1977. An overview of the technological possibilities of future telescopes. Conference on Optical Telescopes of the Future, ESO/CERN, Geneva, December 12, 1977.

Meinel, A. B. 1977. *Ten-meter monolithic option* (ed. M. P. Meinel). University of California, Santa Cruz.

Meinel, A. B. 1977. Very large telescopes using siderostats. Conference on Optical Telescopes of the Future, ESO/CERN, Geneva, December 12, 1977.

Meinel, A. B., and Meinel, M. P. 1977. Economic aspects of solar energy. First International Scientific Forum on an Acceptable World Energy Future, November 7, 1977.

Meinel, A. B., and Meinel, M. P. 1977. Hard realities of the soft-path to a solar future. First International Scientific Forum on an Acceptable World Energy Future, November 7, 1977.

Meinel, A. B., and Meinel, M. P. 1977. National energy/gross national product trajectories. First International Scientific Forum on an Acceptable World Energy Future, November 7, 1977.

Meinel, A. B., and Meinel, M. P. 1977. Nuclear power costs (part 2). Hearings before a Subcommittee of the Committee on Government Operations, September 22, 1977, pp. 1403–30. US Government Printing Office, Washington, DC.

Meinel, A. B., and Meinel, M. P. 1977. Photosynthetic and water efficiency of Salsola pestifer. In International Solar Energy Society, Annual Meeting, Orlando, Florida, June 6–10, 1977, Proceedings. Sections 14–25. (A78-11212 01-44), pp. 25.11–25.13. Cape Canaveral, Florida, International Solar Energy Society.

Meinel, A. B., and Meinel, M. P. 1977. "Soft" energy paths: Reality and illusion. *Electric Perspectives* 77: 24–7.

Meinel, A. B., and Meinel, M. P. 1977. Solar energy. First International Scientific Forum on an Acceptable World Energy Future, November 7, 1977.

Meinel, A. B., and Meinel, M. P. 1977. Solar energy options for arid lands. Conference on Alternate Strategies for Desert Development and Management, UNITAR, June 6, 1977.

Meinel, A. B., Meinel, M. P., and McGowan, J. G. 1977. Applied solar energy: An introduction. *American Journal of Physics* 45: 499. https://doi.org/10.1119/1.11017.

Meinel, A. B. 1978. An overview of the technological possibilities of future Telescopes. In *ESO Conference on Optical Telescopes of the Future* (eds. E. F. Pacini, W. Richter, and R. N. Wilson), pp. 13–26. ESO, Geneva.

Meinel, A. B., and Meinel, M. P. 1978. Distributed solar power systems: An option. *Energy* 3: 23–5.

Meinel, A. B., and Meinel, M. P. 1978. Energy/GNP trajectory study. *Electric Perspectives* 78.

Meinel, A. B., and Meinel, M. P. 1978. Energy, you and the future. In *Nuclear energy and alternatives* (eds. Kadiroglu, O. Kemal, A. Perlmutter, and L. Scott), pp. 37–48. Ballinger Publishing Co., Pensacola, Florida.

Meinel, A. B., and Meinel, M. P. 1978. Hearings before the Subcommittee on Advanced Energy Technologies and Energy Conservation, Research, Development and Demonstration, February 7, 1978, pp. 1–46. US Government Printing Office, Washington, DC.

Meinel, A. B., and Meinel, M. P. 1978. Large optical telescopes of the 21st century. Symposium on Important Advances in 20th Century Astronomy, May 20, 1978.

Meinel, A. B., and Meinel, M. P. 1978. A reconcentrator facility for materials testing at the STTF. Golden, Colorado, April 1978.

Meinel, A. B., and Meinel, M. P. 1978. Solar energy and the village energy center. Workshop on Solar Energy for Rural Development, Solar Pond, and Desalination. National Solar Energy Convention of India, December 18, 1978.

Meinel, A. B., and Meinel, M. P. 1978. Solar energy, fuel of the future? Viewpoints, Christian Perspectives on Social Concerns, Augsburg Publishing House.

Meinel, A. B., and Meinel, M. P. 1978. Solar energy: Promises and problems. Seminar on the hard and soft energy path: Do we have a choice? South Dakota School of Mines and Technology, August 30, 1978.

Meinel, A. B., and Meinel, M. P. 1978. The solar path: How soon? Public forum debate with Amory Lovings, April 27, 1978.

Meinel, A. B., and Meinel, M. P. 1978. An update on economic aspects of solar energy. In *Nuclear energy and alternatives* (eds. Kadiroglu, O. Kemal, A. Perlmutter, and L. Scott), pp. 543–72. Ballinger Publishing Co., Pensacola, Florida.

Meinel, A. B., Meinel, M. P., and Henson, C. M. 1978. Energy, income inequalities, and human welfare. Second International Scientific Forum on an Acceptable World Energy Future, November 27, 1978.

Eckhardt, S. K., and Meinel, A. B. 1979. Comparison of performance of several fixed-mirror concentrators. First International Symposium on Non-Conventional Energy, Trieste, Italy, August–September 1979.

Meinel, A. B. 1979. An autocollimating f/1 camera for use with eschelle spectrographs using image converters. *Proceedings SPIE* 0172, Instrumentation in Astronomy III (May 3, 1979). https://doi.org/10.1117/12.957114.

Meinel, A. B. 1979. Cost-scaling laws applicable to very large optical telescopes. *Optical Engineering* 18: 645–7.

Meinel, A. B. 1979. Multiple mirror telescopes of the future. In *The MMT and the future of ground-based astronomy. Proceedings of a symposium held to mark the dedication of the Multiple Mirror Telescope at the Mount Hopkins Observatory, Arizona, on May 9, 1979* (ed. T. C. Weekes), pp. 7–21. SAO Special Report #385.

Meinel, A. B., Meinel, E. S., and Meinel, W. B. 1979. Fresnel mirror and lens concentrators. First International Symposium on Non-conventional Energy, Trieste, Italy, August–September 1979.

Meinel, A. B., and Meinel, M. P. 1979. Cloudy days ahead for solar energy: Are advocates of the soft path giving us a hard sell? *The Sciences* 19: 18–21.

Meinel, A. B., and Meinel, M. P. 1979. Congressional Briefing, House Subcommittee on Energy Development, February 28, 1979.

Meinel, A. B., and Meinel, M. P. 1979. Cost scaling laws for solar concentrators. First International Symposium on Non-Conventional Energy, International Center for Theoretical Physics, Trieste, Italy, August–September 1979.

Meinel, A. B., and Meinel, M. P. 1979. Energy/GNP trajectories: The relationship between economic growth and energy consumption. In *Ethics and energy. Decisionmakers bookshelf* (ed. F. J. Abbate), vol. 5, pp. 1–14. Edison Electric Institute, Washington, DC.

Meinel, A. B., and Meinel, M. P. 1979. Photovoltaic concentrator modules. First International Symposium on Non-conventional Energy, Trieste, Italy, August–September 1979.

Meinel, A. B., and Meinel, M. P. 1979. Report of the working group on solar cold storage in tropical climates. First International Symposium on Non-conventional Energy, Trieste, Italy, August–September 1979.

Meinel, M. P., and Meinel, A. B. 1979. Energy for the future: The solar contribution. First International Symposium on Non-Conventional Energy, International Center for Theoretical Physics, Trieste, Italy, August–September 1979.

Meinel, M. P., Meinel, A. B., Thompson, T., and Friederichs, D. 1979. Biomass conversion using weed species. First International Symposium on Non-conventional Energy, Trieste, Italy, August–September 1979.

1980s

Meinel, A. B., and Meinel, M. P. 1980. Aden and Marjorie Meinel's China trip October–November 1979. *Applied Optics* 19: 2666–9.

Meinel, A. B., and Meinel, M. P. 1980. Design considerations for a 10-meter optical table telescope. *Applied Optics* 19: 2683–7.

Meinel, A. B., and Meinel, M. P. 1980. Design considerations for a 10-meter optical table telescope. Optical and infrared telescopes for the 1990s. Kitt Peak National Observatory, vol. 1, 95.

Meinel, A. B., and Meinel, M. P. 1981. Optics in solar energy. International Commission on Optics XII, Graz, Austria, September 1981. Taylor & Francis, London.

Meinel, A. B., and Meinel, M. P. 1980. Some comments on scaling law information in relation to very large telescope cost goals. Optical and infrared telescopes for the 1990s, Kitt Peak National Observatory, vol. 2, p. 1027.

Meinel, A. B., Meinel, M. P., Hu, N., Hu, Q., and Pan, C. 1980. Minimum-cost 4-meter telescope developed at 4 October 1979 Nanjing study of telescope design and construction. *Applied Optics* 19: 2670–9.

Meinel, A. B., Meinel, M. P., Hu, N., Hu, Q., and Pan, C. 1980. Optical and infrared telescopes for the 1990s. Kitt Peak National Observatory, vol. 1, 61.

Meinel, A. B., and Meinel, M. P. 1981. Optimum solution for spherical primary mirror with two and three aspheric corrector plates located near focus. *Applied Optics* 20: 3627–9.

Meinel, A. B., and Meinel, M. P. 1981. Options for next generation telescopes. In *Current trends in optics* (eds. F. T. Arecchi and F. R. Aussenegg), pp. 40–54. International Commission on Optics-12, Graz, Austria, 1981. Taylor & Francis, London.

Meinel, A. B., and Meinel, M. P. 1981. Prospects of solar energy. In *Current trends in optics* (eds. F. T. Arecchi and F. R. Aussenegg), pp. 9–14. International Commission on Optics-12, Graz, Austria, 1981. Taylor & Francis, London.

Meinel, A. B. 1982. Cost relationships for nonconventional telescope structural configurations. *Journal of the Optical Society of America* 72: 14–20.

Meinel, A. B., and Meinel, M. P. 1982. Optical figuring process, Dupont 2.6-meter telescope. *Applied Optics* 21: 4198–200.

Meinel, A. B., and Meinel, M. P. 1982. Optical system design study for the University of Texas 300-inch telescope. *Proceedings SPIE* 0332, Advanced Technology Optical Telescopes I (November 4, 1982). https://doi.org/10.1117/12.933518.

Meinel, A. B., and Meinel, M. P. 1982. Optimum solution for spherical primary mirror with two and three aspheric corrector plates located near focus. Part 2. *Applied Optics* 21: 1323–5.

Breckinridge, J. B., Swanson, P. N., Meinel, A. B., and Meinel, M. P. 1983. Perception for a large deployable reflector telescope. *Proceedings SPIE* 0444, Advanced Technology Optical Telescopes II (November 3, 1983). https://doi.org/10.1117/12.937956.

Meinel, A. B. and Meinel, M. P. 1983. Energy for the future: The world view. *Annals of Nuclear Energy* 10: 209–19. Pergamon Press, New York.

Meinel, A. B., and Meinel, M. P. 1983. Self-null corrector test for telescope hyperbolic secondaries. *Applied Optics* 22: 520–1.

Meinel, A. B., and Meinel, M. P. 1983. Self-null corrector test for telescope hyperbolic secondaries: Comments. *Applied Optics* 22: 2405.

Meinel, A. B., and Meinel, M. P. 1983. *Sunsets, twilights, and evening skies*. Cambridge University Press.

Meinel, A. B., Meinel, M. P., and Woolf, N. J. 1983. Deployable reflector configurations. *Proceedings SPIE* 0383. Deployable Optical Systems (December 1, 1983). https://doi.org/10.1117/12.934917.

Meinel, A. B., Meinel, M. P., and Woolf, N. J. 1983. Multiple aperture telescope diffraction images. *Applied Optics and Optical Engineering*, vol. IX (eds. R. R. Shannon and J. C. Wyant), pp. 150–201. Academic Press, New York.

Meinel, A. B. 1984. Book review of *Treatise on solar energy*, vol. 1: *Fundamentals of solar energy* (by H. P. Garg). *Journal of Solar Energy Engineering* 106: 376–7.

Meinel, A. B., and Meinel, M. P. 1984. Aberrations of an infrared chopping secondary. *Applied Optics* 23: 2675–6.

Meinel, A. B., and Meinel, M. P. 1984. Book review of *Rare halos, mirages, anomalous rainbows, and related electromagnetic phenomena: A catalog of geophysical anomalies* (by W. R. Corliss). *ICARUS (International Journal of Solar System Studies)* 57: 719–20. Cornell University, Ithaca, New York.

Meinel, A. B., and Meinel, M. P. 1984. *Optical telescope*. McGraw-Hill Yearbook of Science and Technology 1983, McGraw-Hill, New York.

Meinel, A. B., and Meinel, M. P. 1984. Zero-coma condition for decentered and tilted secondary mirror in Cassegrain/Naysmith configuration. *Optical Engineering* 23: 801. https://doi.org/10.1117/12.7973386.

Meinel, A. B., Meinel, M. P., Shannon, R. R., and Parks, R. E. 1984. Mirrors: Fabrication, processing, testing and tolerances. In *Very large telescopes, their instrumentation and programs*; Proceedings of the Colloquium, Garching, West Germany, April 9–12, 1984 (A85-36926 17-89), pp. 109–17; discussion, p. 117. Garching, West Germany, European Southern Observatory.

Meinel, A. B., Meinel, M. P., and Su, D. Q. 1984. Hyperbolic primary telescope configuration. *Applied Optics* 23: 2469–71.

Meinel, A. B., Meinel, M. P., Su, D. Q., and Wang, Y. N. 1984. Four-mirror spherical-primary submillimeter telescope design. *Applied Optics* 23: 3020–3.

Meinel, A. B., Meinel, M. P., and Tull, R. G. 1984. Optical engineering aspects of the University of Texas 7.6-meter telescope (abstract). Symposium on Large-aperture Optics Fabrication. 1984 Annual Meeting Optical Society of America.

Meinel, A. B., and Meinel, M. P. 1985. Aberrations of an IR chopping secondary: Authors' reply to comment 1. *Applied Optics* 24: 943–4.

Meinel, A. B., and Meinel, M. P. 1985. Aberrations of an IR chopping secondary: Authors' reply to comment 2 and continuation of authors' reply to comment 1. *Applied Optics* 24: 944.

Meinel, A. B., and Meinel, M. P. 1985. An overview of optical telescope technology. *Proceedings SPIE* 0542, Optical Fabrication & Testing Workshop (October 22, 1985). https://doi.org/10.1117/12.948156.

Meinel, A. B., and Meinel, M. P. 1985. Very large optics of the future. *Journal of the Optical Society of America* 2: 83.

Meinel, A. B., and Meinel, M. P. 1985. A view in the mirror—or through the looking glass. *Proceedings SPIE* 0571, Large Optics Technology (ed. G. M. Sanger), pp. 2–8. Society for Photo-Optical Instrumentation Engineers, Bellingham, Washington.

Meinel, A. B., Meinel, M. P., and Wang, Y. N. 1985. Compact Nasmyth focal reducer camera. *Applied Optics* 24: 2751–2.

Stacy, J. E., Meinel, A. B., and Meinel, M. P. 1985. Upgrading telescopes by active pupil wavefront correction. *Proceedings SPIE* 0554, 1985 International Lens Design Conference (February 14, 1986). https://doi.org/10.1117/12.949222.

Swanson, P. N., Breckinridge, J. B., Diner, A., Freeland, R. E., Irace, W. R., McElroy, P. M., Meinel, A. B., and Tolivar, A. 1985. A system concept for a moderate cost large deployable reflector (LDR). *Optical Engineering* 25: 1045–54.

Meinel, A. B., and Meinel, M. P. 1986. Close star encounters. *Bulletin of the American Astronomical Society* 18: 711.

Meinel, A. B., and Meinel, M. P. 1986. Limits to size for terrestrial telescopes: Two-stage optics. *Proceedings SPIE* 0628, Advanced Technology Optical Telescopes III (August 20, 1986). https://doi.org/10.1117/12.963567.

Meinel, A. B., and Meinel, M. P. 1986. Optical and infrared space optics challenges. 33rd Annual Meeting, American Astronautical Society, AAS-86–270.

Meinel, A. B., and Meinel, M. P. 1986. Thousand astronomical unit voyage: A deep space parallax opportunity. *Bulletin of the American Astronomical Society* 18: 1012.

Meinel, A. B., and Meinel, M. P. 1986. Very large optics of the future. *Optics News* 12: 9–14.

Meinel, A. B., and Meinel, M. P. 1986. Wind deflection compensated, zero-coma telescope truss geometries. *Proceedings SPIE* 0628, Advanced Technology Optical Telescopes III (August 20, 1986). https://doi.org/10.1117/12.963557.

Meinel, A. B., Meinel, M. P., and Stacy, J. E. 1986. Compensation for primary reflector wavefront error. Patent application, NASA, Pasadena, California.

Meinel, A. B., Meinel, M. P., Stacy, J. E., Saito, T. T., and Paterson, S. R. 1986. Wave front correctors by diamond turning. *Applied Optics* 25: 824–5.

Nock, K. T., Breckinridge, J. B., Buratti, B. J., Lesh, J., Meinel, A. B., Meinel, M. P., Ramirez, M., and Sercel, J. C. 1986. Thousand Astronomical Unit voyage: A deep space mission. *Bulletin of the American Astronopmical Society* 18: 1012.

Patterson, S. R., Saito, T. T., Meinel, A. B., Meinel, M. P., and Stacy, J. E. 1986. Diamond turning wavefront correctors: Opening new optical design flexibilities. *Proceedings SPIE* 0676. Ultraprecision Machining and Automated Fabrication of Optics, p. 10 (February 13, 1987). https://doi.org/10.1117/12.939509.

Meinel, A. B., and Meinel, M. P. 1987. At sunset (A). *Journal of the Optical Society of America* 4: 108.

Meinel, A. B., and Meinel, M. P. 1987. Optical and infrared space optics challenges. In *Aerospace century XXI: Space sciences, applications, and commercial developments*; Proceedings of the Thirty-third Annual AAS International Conference, Boulder, Colorado, October 26–29, 1986. San Diego, California, Univelt (Advances in the Astronautical Sciences. Vol. 64, Pt. 3, pp. 1137–47).

Meinel, A. B., and Meinel, M. P. 1987. Telescope structures: An evolutionary overview. In *Structural mechanics of optical systems II*; Proceedings of the Meeting, Los Angeles, California, January 13–15, 1987 (A88-34486 13-74), pp. 2–5. Bellingham, Washington, Society of Photo-Optical Instrumentation Engineers.

Meinel, A. B., and Meinel, M. P. 1988. At sunset. *Optics News* 14: 6–13.

Meinel, A. B., and Meinel, M. P. 1988. Evolution of ground and space based optical systems during the past 30 years. Space Optics for Astrophysics and Earth and Planetary Remote Sensing, 1988 technical digest series, vol. 10, p. 130. Summaries of papers presented at the Space Optics for Astrophysics and Earth and Planetary Remote Sensing Topical Meeting, September 27–29, North Falmouth, Massachusetts.

Meinel, A. B., Meinel, M. P., and Breckinridge, J. B. 1988. Wavefront correction in very large monolithic or segmented mirror telescopes. *Very Large Telescopes and their Instrumentation*, ESO Conference and Workshop Proceedings, Proceedings of a ESO Conference on Very Large Telescopes and Their Instrumentation, held in Garching, March 21–24, 1988, Garching, European Southern Observatory (ESO), 1988 (ed. M.-H. Ulrich), p. 479.

Meinel, A. B., Meinel, M. P., and Synnott, S. P. 1988. Astrophysical object models for modeling of interferometer configurations. ESO Conference Workshop Proceedings, no. 29, part 2, pp. 1075–8.

Meinel, A. B., and Meinel, M. P. 1989. Active wavefront control challenges of the NASA large deployable reflector (LDR). *Proceedings SPIE* 1114, Active Telescope Systems (September 20, 1989). https://doi.org/10.1117/12.960844.

Meinel, A. B., and Meinel, M. P. 1989. Attaching precise to lightweight supports. *Applied Optics* 28: 3779.

Meinel, A. B., and Meinel, M. P. 1989. Error-compensated telescope. *Applied Optics* 28: 3546.

Meinel, A. B., and Meinel, M. P. 1989. The lunar configurable array telescope (LCAT) (abstract). NASA Technical Reports 19990009648.

Meinel, A. B., and Meinel, M. P. 1989. Optical testing of off-axis parabolic segments without auxiliary optical test elements. *Optical Engineering* 28: 71. https://doi.org/10.1117/12.7976904.

1990s

Breckinridge, J. B., Meinel, A. B., and Meinel, M. P. 1998. Inflation-deployed camera and hyper-thin mirrors. *Proceedings SPIE* 3356: 780–7.

Meinel, A. B., and Meinel, M. P. 1990. Is the Space Telescope a mistake? No! *Sky & Telescope* 81: 356.

Meinel, A. B., and Meinel, M. P. 1990. The Lunar Configurable Array Telescope (LCAT). *The Next Generation Space Telescope*, Proceedings of a workshop held at Baltimore, Maryland, September 13–15, 1989 (eds. P-Y Bely, C. J. Burrows, and G. D. Illingworth), p. 177. NASA, Washington, DC.

Meinel, A. B., Meinel, M. P., and Breckinridge, J. B. 1990. Wavefront control of large optical systems. *Proceedings SPIE* 1271, Adaptive Optics and Optical Structures (August 1, 1990). https://doi.org/10.1117/12.20406.

Meinel, A. 1991. Venus in stereo. *Sky & Telescope* 82: 118.

Meinel, A. B., and Meinel, M. P. 1991. Infrared and the search for extrasolar planets. *Proceedings SPIE* 1540, Infrared Technology XVII (December 1, 1991). https://doi.org/10.1117/12.48730.

Meinel, A. B. 1992. Letter. *Optics and Photonics News* 3: 2.

Meinel, A. B., and Meinel, M. P. 1992. Two-stage optics: High acuity performance from low acuity optical systems. *Optical Engineering* 31: 2271–81. https://doi.org/10.1117/12.59946.

Meinel, A. B., Meinel, M. P., and Schulte, D. H. 1993. Determination of the Hubble Space Telescope effective conic constant error from direct image measurements. *Applied Optics* 32: 1715–19. https://doi.org/10.1364/AO.32.001715.

Meinel, A. B., Meinel, M. P., and Breckinridge, J. B. 1994. Deployable space telescopes for planetary and astronomical missions. *Proceedings SPIE* 2214, 250–5, Space Instrumentation and Dual-use Technologies (June 8, 1994). https://doi.org/10.1117/12.177664.

Meinel, A. B., Meinel, M. P., and Breckinridge, J. B. 1994. FAST: A folded astronomical space telescope. *Proceedings SPIE* 2199, 603–8, Advanced Technology Optical Telescopes V (June 1, 1994). https://doi.org/10.1117/12.176228.

Meinel, A. B., Meinel, M. P., Gilman, J. J., Reppert, P., and Stone, J. L. 1994. Photovoltaic's promise still less than sunny? *Physics Today* 47: 15. https://doi.org/10.1063/1.2808624.

2000s

Meinel, A. B., and Meinel, M. P. 2000. Inflatable membrane mirrors for optical passband imagery. *Optical Engineering* 39: 541–50.

Meinel, A. B., and Meinel, M. P. 2000. Spherical primary telescope with aspheric correction at a small internal pupil. *Applied Optics* 39: 1–4.

Romeo, R., Meinel, A. B., Meinel, M. P., and Chen, P. C. 2000. Ultralightweight and hyperthin rollable primary mirror for space telescopes. *Proceedings SPIE* 4013: 634–9. https://doi.org/10.1117/12.393998.

Meinel, A. B., and Meinel, M. P. 2001. Comparison of lens and Fresnel null correctors. *Applied Optics* 40: 3688–97.

Meinel, A. B., and Meinel, M. P. 2002. Large membrane space optics: Imagery and aberrations of diffractive and holographic achromatized optical elements of high diffraction order. *Optical Engineering* 41: 1995–2007.

Meinel, A. B., and Meinel, M. P. 2002. Large sparse-aperture space optical systems. *Optical Engineering* 41: 1983–94.

Meinel, A. B., and Meinel, M. P. 2002. Large membrane space optics: imagery and aberrations of diffractive and holographic achromatized optical elements of high diffraction order. *Optical Engineering* 41: 1995-2007.

Meinel, A. B., and Meinel, M. P. 2002. Parametric dependencies of high-diffraction-order achromatized aplanatic configurations that employ circular or crossed-linear diffractive optical elements. *Applied Optics* 41: 7155–66.

Meinel, A. B., and Meinel, M. P. 2003. Extremely large sparse aperture telescopes. *Optics and Photonics News* 14: 26–9.

Meinel, A. B., and Meinel, M. P. 2004. Optical phased array configuration for an extremely large telescope. *Applied Optics* 43: 601–7.

Van Belle, G. T., Meinel, A. B., and Meinel, M. P. 2004. The scaling relationship between telescope cost and aperture size for very large telescopes. *Proceedings SPIE* 5489, Ground-based Telescopes (September 28, 2004). https://doi.org/10.1117/12.552181.

Meinel, A., Meinel, M. Drach-Meinel, D., and Meinel, B. 2006. Observational evidence of cosmic rays from the Cat's Eye planetary nebula 6543. 36th COSPAR Scientific Assembly. Held July 16–23, 2006, in Beijing, China. Meeting abstract from the CD-ROM, #2765.

Meinel, A. B. 2007. Galaxy collisions and ordinary dark matter. *Physics Today* 60: 14.

Meinel, A. B., and Meinel, M. 2007. Planetary nebula NGC 6543 as the source of the spikes of cosmic rays recorded in the Greenland and Antarctic ice cores. Asymmetrical Planetary Nebulae IV, held in La Palma, June 18–22, 2007. Published online at http://www.iac.es/proyect/apn4, article 25.

Meinel, A. B. 2009a. Reminiscences: Aden goes to war. *Optics and Photonics News* 20: 16–17.

Meinel, A. B. 2009b. Reminiscences: Aden returns from the war. *Optics and Photonics News* 20: 20–1.

Meinel, A. B., and Meinel, B. 2009. Evidence of a magnetic sheath around a jet from NGC 6543. In *IAU Symposium Proceedings Series* (eds. K. G. Strassmeier, A. G. Kosovichev, and J. E. Beckman), pp. 133–4. Cambridge University Press.

2010s

Meinel, A. B. 2014. Personal reflections of the first decade. Appendix 1, James C. Wyant Optical Sciences Center/College of Optical Sciences. 50 years of excellence, *Proceedings SPIE* 9186, p. 25.

Meinel, M. S. 2014. Pettit, Edison. In *Biographical encyclopedia of astronomers* (ed. T. Hockey et al.). Springer, New York. https://doi.org/10.1007/978-1-4419-9917-7_1079.

APPENDIX 3

Select Publications of Edison Pettit, Hannah Steele Pettit, and Helen Pettit Knaflich

Edison Pettit

McMath, R. R., and Pettit, E. 1937. Prominences of the active and sun-spot types compared. *Astrophysical Journal* 85: 279–303 + plates.

McMath, R. R., and Pettit, E. 1938. Motions in the loops of prominences of the sunspot type, Class IIIb. *Publications of the Astronomical Society of the Pacific* 50: 56–7.

McMath, R. R., and Pettit, E. 1938. Prominence studies. *Astrophysical Journal* 88: 244–77 + plates.

McMath, R. R., and Pettit, E. 1939. The Doppler effect in an eruptive prominence. *Publications of the Astronomical Society of the Pacific* 51: 7–10.

Pettit, E. 1925. The forms and motions of the solar prominences. *Publications of the Yerkes Observatory*. Vol. 3, Part 4: iii–vi, 205–40 + plates.

Pettit, E. 1928. The optical properties of fused and crystal quartz. *Publications of the Astronomical Society of the Pacific* 40: 200–1.

Pettit, E. 1929. The dimensions of thermocouple receivers. *Publications of the Astronomical Society of the Pacific* 41: 272–4.

Pettit, E. 1932. Characteristic features of solar prominences. *Astrophysical Journal* 76: 9–43.

Pettit, E. 1932. Measurements of ultra-violet solar radiation. *Astrophysical Journal* 75: 185–221.

Pettit, E. 1934. The reflecting properties of aluminum-surfaced mirrors. *Astronomical Society of the Pacific* 46: 27–31.

Pettit, E. 1938. The highest eruptive prominence. *Publications of the Astronomical Society of the Pacific* 50: 13–15.

Pettit, E. 1939. Limb-darkening of the sun. *Publications of the Astronomical Society of the Pacific* 51: 321–7.

Pettit, E. 1940. Radiation measurements on an eclipsed moon. *Astrophysical Journal* 91: 408–21.

Pettit, E. 1940. Spectral energy-curve of the Sun in the ultraviolet. *Astrophysical Journal* 91: 159–85.

Pettit, E. 1941. The interference polarizing monochromator. *Publications of the Astronomical Society of the Pacific* 53: 171–81.

Pettit, E. 1941. The rotation of a tornado prominence. *Publications of the Astronomical Society of the Pacific* 53: 289–90.

Pettit, E. 1943. The properties of solar prominences as related to type. *Astrophysical Journal* 98: 6–19.

Pettit, E. 1943. Visual magnitudes of Nova Puppis 1942. *Publications of the Astronomical Society of the Pacific* 55: 14–20.

Pettit, E. 1944. Frequency of eruptive prominences. *Publications of the Astronomical Society of the Pacific* 56: 124–5.

Pettit, E. 1946. An eruptive prominence of record height, June 4, 1946. *Publications of the Astronomical Society of the Pacific* 58: 310–14.

Pettit, E. 1946. The light-curves of T Coronae Borealis. *Publications of the Astronomical Society of the Pacific* 58: 153–6.

Pettit, E. 1947. The canals of Mars. *Publications of the Astronomical Society of the Pacific* 59: 5–11.

Pettit, E. 1951. The motions of eruptive prominences. *Publications of the Astronomical Society of the Pacific* 63: 237–44.

Pettit, E. 1954. Magnitudes and color indices of extragalactic nebulae determined photo-electrically. *Astrophysical Journal* 120: 413–37.

Pettit, E. 1958. The visual magnitudes of double stars. *Astronomical Journal* 63: 324–8.

Pettit, E., and Nicholson, S. B. 1922. The application of vacuum thermocouples to problems in astrophysics. *Astrophysical Journal* 56: 295–317.

Pettit, E., and Nicholson, S. B. 1924. Radiation measures on the planet Mars. *Publications of the Astronomical Society of the Pacific* 36: 269–72.

Pettit, E., and Nicholson, S. B. 1927. Radiometric measures of certain faint red stars. *Publications of the Astronomical Society of the Pacific* 39: 241–2.

Pettit, E., and Nicholson, S. B. 1928. Stellar radiation measurements. *Astrophysical Journal* 68: 279–308.

Pettit, E., and Nicholson, S. B. 1930. Lunar radiation and temperatures. *Astrophysical Journal* 71: 102–35.

Pettit, E., and Nicholson, S. B. 1930. Spectral energy-curve of sun-spots. *Astrophysical Journal* 71: 153–62.

Pettit, E., and Nicholson, S. B. 1933. Measurements of the radiation from variable stars. *Astrophysical Journal* 78: 320–53.

Pettit, E., and Nicholson, S. B. 1936. Radiation from the planet Mercury. *Astrophysical Journal* 83: 84–102.

Pettit, E., and Nicholson, S. B. 1955. Temperatures on the bright and dark sides of Venus. *Astrophysical Journal* 67: 293–303.

Pettit, E., and Steele, H. B. 1918. The application of Schaeberle's method in the photography of the corona at Matheson, Colorado, June 8. *Popular Astronomy* 26: 466–80.

Hannah Steele Pettit

Miller, J. A., Barton, S. G., Steele, H. B., and Pitman, J. H. 1914. *Determination of the parallaxes of fifty stars, and description of the instruments and of the methods employed.* Sproul Observatory Publications (4).

Miller, J. A., Pitman, J. H., and Steele, H. B. 1920. *The parallaxes of fifty stars determined at the Srpoul Observatory.* Sproul Observatory Publications (5).

Pettit, H. S. 1919. The proper motions and parallaxes of 359 stars in the Cluster h Persei. *Popular Astronomy* 27: 671–2.

Pettit, H. S. 1919. *The proper motions and parallaxes of 359 stars in the Cluster h Persei, derived from the photographs made with the 40-inch refractor of the Yerkes Observatory.* PhD dissertation. University of Chicago.

Pettit, E., and Steele, H. B. 1918. The application of Schaeberle's method in the photography of the corona at Matheson, Colorado, June 8. *Popular Astronomy* 26: 466–80.

Van Biesbroeck, G., and Pettit, H. S. 1919. Stellar parallaxes, derived from photographs made with the 40-inch refractor of the Yerkes Observatory. *Astronomical Journal* 32: 63–4.

Van Biesbroeck, G., and Pettit, H. S. 1920. *Parallaxes of fifty-two stars*. Publications of the Yerkes Observatory. Vol. 4, Part 3. University of Chicago Press.

Helen Pettit Knaflich

Barbier, D., and Pettit, H. 1952. Photometric observations of the airglow and of the aurora borealis at College, Alaska. *Annales de Geophysique* 8: 232.

Bedinger, J. F., and Knaflich, H. B. 1966. Observed characteristics of iononspheric winds. *Radio Science* 1: 156–68.

Bedinger, J. F., Knaflich, H., Manring, E., and Layzer, D. 1968. Upper-atmosphere winds and their interpretation—1. Evidence of strong nonlinearity of the horizontal flow above 80km. *Planetary and Space Science* 16: 159–93.

Kenney, J. F., and Knaflich, H. B. 1967. A systematic study of structured micropulsations. *Journal of Geophysical Research* 72: 2857–69.

Kenney, J. F., Knaflich, H. B., and Liemohn, H. B. 1968. Magnetospheric parameters determined from structured micropulsations. *Journal of Geophysical Research* 73: 6737–49.

Knaflich, H. B., and Kenney, J. F. 1967. IPDP events and their generation in the magnetosphere. *Earth and Planetary Science Letters* 2: 453–9.

Knaflich, H. B., Kenney, J. F., and Hessler, V. P. 1968. Longitudinal spread of pc 1 micropulsations in the magnetosphere. *Nature* 217: 1134–6.

Liemohn, H. B., Kenney, J. F., and Knaflich, H. B. 1967. Proton densities in the magnetosphere from pearl dispersion measurements. *Earth and Planetary Science Letters* 2: 360–6.

Manring, E., St. Amand, P., Pettit, H. B., Roach, F. E., Williams, D. R., and Weldon, R. G. 1954. Simultaneous observations of nightglow 5 577 at two stations. *Annales d'Astrophysique* 17: 186–96.

Pettit, H. B. 1944. The trajectories of eruptive prominences. *Publications of the Astronomical Society of the Pacific* 56: 21–6.

Pettit, H. B., Roach, F. E., St. Amand, P., and Williams, D. R. 1954. A comprehensive study of atomic emissions in the nightglow. *Annales de Geophysique* 10: 326.

Roach, F. E., and Pettit, H. B. 1951. On the diurnal variation of [OI] in the nightglow. *Journal of Geophysical Research* 56: 325–53.

Roach, F. E., Pettit, H. B., Tandberg-Hanssen, E., and Davis, D. N. 1954. Observations of the zodiacal light. *Astrophysical Journal* 119: 253–73.

Roach, F. E., Pettit, H. B., and Williams, D. R. 1950. The height of the atmospheric OH emission. *Journal of Geophysical Research* 55: 183–90.

Roach, F. E., Pettit, H. B., Williams, D. R., St. Amand, P., and Davis, D. N. 1953. A four-year study of [OI] 5577 Å in the nightglow. *Annales d'Astrophysique* 16: 185–205.

Roach, F. E., Williams, D. R., and Pettit, H. 1953. Diurnal variation of [O I] 5577 in the nightglow. *Astrophysical Journal* 117: 456–9.

St. Amand, P., Pettit, H. B., Roach, F. E., and Williams, D. R. 1955. On a new method of determining the height of the nightglow. *Journal of Atmospheric and Terrestrial Physics* 6: 189–97.

References

Abt, H. A. 2009. The Kitt Peak 2.1-meter telescope: An unusually innovative telescope (abstract). *Bulletin of the American Astronomical Society* 41: 187.

Abt, H. A., Morgan, W. W., and Strömgren, B. 1957. A description of certain galactic nebulosities. II. *Astrophysical Journal* 126: 322 + 23 figures.

Abt, H. A. 2020. *A stellar life*. Palmetto Publishing, Charleston, South Carolina.

Adams, W. S. 1938. George Ellery Hale (1868–1938). *Astrophysical Journal* 87: 369–88.

Adams, W. S. 1947. Early days at Mount Wilson. *Publications of the Astronomical Society of the Pacific* 59: 213–304.

Adams, W. S., and Dunham, T., Jr. 1934. The B band of oxygen in the spectrum of Mars. *Astrophysical Journal* 79: 308–16.

After Action Report, Third U.S. Army, 1 August 1944–9 May 1945. Volume II. Parts 1–9. Staff section reports. U.S. Army Military History Institute, Washington, DC.

Agnew, D. L., and Jones, P. A. 1986. Large Deployable Reflector (LDR) system concept and technology definition study, May 30, 1986. Ames Research Center Contract NAS2-11861 NASA contractor report CR 177413 performed by Eastman Kodak Company, Government Systems Division, Rochester, New York.

Ahlberg, K. L. 2008. *Transplanting the great society: Lyndon Johnson and food for peace*. University of Missouri Press, Columbia.

Ambrose, S. E. 1990. *Eisenhower: Soldier and president*. Simon & Schuster, New York.

Anonymous. 1951. Report of Committee. *Astronomical Journal* 56: 147–8.

Anonymous. 1955. Astronomical photoelectric conference, sponsored by the National Science Foundation and Lowell Observatory at Flagstaff, Arizona, August 31 to September 1, 1953. *Astrophysical Journal* 60: 17–18.

Baade, W. 1956. The period-luminosity relation of the Cepheids. *Publications of the Astronomical Society of the Pacific* 68: 5–16.

Babcock, H. D. 1938a. Address of the retiring president of the Society in announcing the award of the Bruce Gold Medal to Dr. Edwin Hubble. *Publications of the Astronomical Society of the Pacific* 50: 87–96.

Babcock, H. D. 1938b. George Ellery Hale. *Publications of the Astronomical Society of the Pacific* 50(295): 156–65.

Baker, D. 2016. *US spy satellites 1959 onwards (all missions, all models)*. Haynes, Sparkford, UK.

Ballard, S. 1962. Optical activities in the universities. *Applied Optics* 1: 96.

Barnes, T. D. 2015. *Soaring with the eagles: Autobiography of Area 51 Veteran TD Barnes*. Self-published.

Barnes, T. D. 2017. *The secret genesis of Area 51*. The History Press, Charleston, South Carolina.

Barnes, T. D. 2018a. *The Angels. Book One: The CIA Area 51 chronicles*. Self-published.

Barnes, T. D. 2018b. *The Archangels. Book Two: The CIA Area 51 chronicles*. Self-published.

Barnes, T. D. 2018c. *CIA Project OXCART: Area 51 Nevada*. Self-published.

Barton, I. M., Britten, J. A., Dixit, S. N., Summers, L. J., Thomas, I. M., Rushford, M. C., Lu, K., Hyde, R. A., and Perry, M. D. 2001. Fabrication of large-aperture lightweight diffractive lenses for use in space. *Applied Optics* 40: 447–51.

Bashkin, S. 1968. Beam foil spectroscopy. *Applied Optics* 7: 2341–50.

Bashkin, S., Fink, D., Malmberg, P. R., Meinel, A. B., and Tilford, S. G. 1966. Collisional excitation atomic spectra in accelerated beams of light elements. *Journal of the Optical Society of America* 56: 1064–75.

Bell, T. E. 2007. Roger Hayward: Forgotten artist of optics. *Sky & Telescope* 114: 31–7.

Benedict, Jr., R., Breckinridge, J. B., and Fried, D. L. 1994. Introduction: Atmospheric compensation technology. *Journal of the Optical Society of America* A11: 257–62.

Berg, A. S. 1998. *Lindbergh*. Berkley Books, New York.

Berry, W. 1830. *County genealogies: Pedigrees of the families in the county of Kent*. Sherwood, Gilbert and Piper, London.

Biberman, L. M., and Zandt Williams, V. 1964. *Optics: An Action Program*. Task III: Basic long range research programs in optics. OSA Member Survey. *Applied Optics* 4: 205–7.

Bowman, N. J. 1957. *The handbook of rockets and guided missiles*. Perastadion Press, Whiting, Indiana.

Boyle, W. S., and Smith, G. E. 1970. Charge coupled semiconductor devices. *Bell System Technical Journal* 49: 587–93.

Breckinridge, J. B. 1976. Interference in astronomical speckle patterns. *Journal of the Optical Society of America* 66: 1240–2.

Breckinridge, J. B. 2012. *Basic optics for the astronomical sciences*. SPIE Press, Bellingham, Washington.

Breckinridge, J., Bryant, N., and Lorre, J. 2008. Innovative pupil topographies for sparse aperture telescopes and SNR. *Proceedings SPIE* 7013-3E.

Breckinridge, J. B., Kron, G. E., and Pappiashvili, I. 1964. Transfer efficiency and storage capacity of electronographic image tubes. *Astronomical Journal* 69: 534–5.

Breckinridge, J. B., Kuper, T. G., and Shack, R. V. 1982. Space telescope low scattered light camera: A model. *Proceedings SPIE* 331: 395–403.

Breckinridge, J. B., Meinel, A. B., and Meinel, M. P. 1998. Inflation-deployed camera and hyper-thin mirrors. *Proceedings SPIE* 3356: 780–7.

Breckinridge, J. B., Page, N. A., Shannon, R. R., and Rodgers, J. M. 1983. Reflecting Schmidt imaging spectrometers. *Applied Optics* 22: 1175–80.

Brokaw, T. 1998. *The greatest generation*. Random House, New York.

Brugioni, D. A. 1990. *Eyeball to eyeball: The inside story of the Cuban Missile Crisis*. Random House, New York.

Brugioni, D. A. 2010. *Eyes in the sky: Eisenhower, the CIA and Cold War aerial espionage*. Naval Institute Press, Annapolis, Maryland.

Brugioni, D. A., and McCort, R. F. 1988. Personality: Arthur C. Lundahl. The art of aerial photography. *Programmatic Engineering & Remote Sensing* 54: 270–2.

Buck, A. L. 1982. *A history of the Energy Research and Development Administration*. US Department of Energy, Washington, D.C.

Buckley, C. 2017. Wreckage of U.S.S. *Indianapolis*, lost for 72 years, is found in the Pacific. https://nyti.ms/2vP87GF. Accessed April 19, 2019.

Bungay, S. 2015. *The most dangerous enemy: A history of the Battle of Britain*. Aurum Press, London.

Burbidge, E. M., Lynds, C. R., and Stockton, A. N. 1968. Further observations of quasi-stellar objects with absorption-line spectra: Ton 1530, PKS 0237-23, and PHL 938*. *Astrophysical Journal* 152: 1077–93.

Burns, J., III, Hiltner, W. A., and Miller, R. 1956. Image converters with thin protective foils. *Astronomical Journal* 61: 172.

Burrows, W. E. 1986. *Deep black: Space espionage and national security*. Random House, New York.

Buscher, D. F. 2015. *Practical optical interferometry: Imaging at visible and infrared wavelengths*. Cambridge University Press.

Bush, V. 1945. *Science: The endless frontier. A report to the president on a program for postwar scientific research*. Reprinted July 1960 by the National Science Foundation, Washington, DC.

Butler, C. 1962. The light of the atom bomb. *Science* 138: 483–9.

Caro, R. A. 1983. *The years of Lyndon Johnson: The path to power*. Alfred A. Knopf, New York.

Caro, R. A. 1990. *The years of Lyndon Johnson: Means of ascent*. Alfred A. Knopf, New York.

Caro, R. A. 2002. *The years of Lyndon Johnson: Master of the Senate*. Alfred A. Knopf, New York.

Caro, R. A. 2012. *The years of Lyndon Johnson: The passage of power*. Alfred A. Knopf, New York.

Center for the Study of National Reconnaissance. 2012. *A history of the Hexagon program*. Center for the Study of National Reconnaissance. Chantilly, Virginia.

Central Intelligence Agency (CIA), Directorate of Science & Technology (DST). 2016. History of the Office of Special Activities (OSA). Chapter 20. From Inception to 1969. Washington, DC. Released on appeal by Interagency Security Classification Appeals Panel (ISCAP) (final release), 2016. http://www.governmentattic.org/20docs/CIAh istOSAincep-1969u.pdf.

Chang, I. 1995. *Thread of the silkworm*. BasicBooks, New York.

Chapin, J. C. 1945. *The Fourth Marine Division in World War II*. History and Museums Division, Headquarters, US Marine Corps, Washington, D.C.

Chernow, R. 2010. *Washington: A life*. Penguin Press, New York.

Chernow, R. 2017. *Grant*. Penguin Press, New York.

Chevillard, J-P., Connes, P., Cuisenier, M., Friteau, J., and Marlot, C. 1977. Near infrared astronomical light collector. *Applied Optics* 16: 1817–33.

Christman, A. B. 1971. *History of the Naval Weapons Center, China Lake, California*. Vol. 1. *Sailors, scientists, and rockets*. Naval History Division, Washington, DC.

Christopher, J. 2013. *The race for Hitler's X-planes: Britain's 1945 mission to capture secret Luftwaffe technology*. The History Press, Stroud, Gloucestershire, UK.

Clayton, D. D. 2014. Fowler, William Alfred. In T. Hockey et al. (eds.), *Biographical encyclopedia of astronomers*. Springer, New York. http://doi.org/10.1007/ 978-1-4419-9917-7.

Code, A. D. 1960. Stellar astronomy from a space vehicle. *Astronomical Journal* 65: 278–84.

Code, A. D., Houck, T. E., McNall, J. F., Bless, R. C., and Lillie, C. F. 1970. Ultraviolet photometry from the Orbiting Astronomical Observatory. I. Instrumentation and operation. *Astrophysical Journal* 161: 377–88.

Conrady, A. E. 1929. *Applied optics and optical design*. Oxford University Press. Reprinted 1957, 1988, 1992 by Dover Publications.

Conrady, A. E., and Kingslake, R. 1960. *Applied optics and optical design*, Part Two. Dover Publications, New York.

Cooke, R. 1924. *The visitations of Kent taken in the years 1574 and 1592*. John Whitehead and Son, London.

Cox, C. 1999. Report of the select committee on U.S. national security and military/commercial concerns with the People's Republic of China. 3 vols. US Government Printing Office, Washington, DC.

Crim, B. E. 2018. *Our Germans: Project paperclip and the national security state.* Johns Hopkins University Press, Baltimore, Maryland.

Crocker, J. H. 1993. Engineering the COSTAR. *Optics and Photonics News* 4: 22–6.

Cui, X.-Q., Zhao, Y.-H., Chu, Y.-Q., Li, G.-P., Li, Q., Zhang, L.-P., Su, H.-J., Yao, Z.-Q., Wang, Y.-N., and Xing, X.-Z. 2012. The Large Sky Area Multi-Object Fiber Spectroscopic Telescope (LAMOST). *Research in Astronomy and Astrophysics* 12: 1197–242.

Davies, M. E., and Harris, W. R. 1988. *RAND's role in the evolution of balloon and satellite observation systems and related U.S. space technology.* RAND, Santa Monica, California.

Day, D. A. 1998. The development and improvement of the Corona satellite. In D. A. Day, J. M. Logsdon, and B. Latell (eds.), *Eye in the sky: The story of the Corona spy satellites,* pp. 48–85. Smithsonian Institution Press, Washington, DC.

Day, D. A., Logsdon, J. M., and Latell, B. 1998. Introduction. In D. A. Day, J. M. Logsdon, and B. Latell (eds.), *Eye in the sky: The story of the Corona spy satellites,* pp. 1–18. Smithsonian Institution Press, Washington, DC.

DeVorkin, D. H. 1992. *Science with a vengeance: How the military created the US space sciences after World War II.* Springer-Verlag, New York.

DeVorkin, D. H. 2018. *Fred Whipple's empire: The Smithsonian Astrophysical Observatory, 1955–1973.* Smithsonian Institution Scholarly Press, Washington, DC.

Dienesch, R. M. 2016. *Eyeing the red storm.* University of Nebraska Press, Lincoln.

Dimbleby, J. 2016. *The Battle of the Atlantic: How the Allies won the war.* Oxford University Press.

DuBridge, L. A., and Epstein, P. A. 1959. Robert Andrews Millikan: March 22, 1868–December 19, 1953. *Biographical Memoirs of the National Academy of Sciences,* 31: 239–82. Washington, DC.

Dundas, J. H. 1902. Nemaha County. Digitized by Google.

Edmondson, F. K. 2005. *AURA and its US National Observatories.* Cambridge University Press.

Epstein, L. 1967. All reflecting Schmidt camera. *Publications of the Astronomical Society of the Pacific* 79: 132–5.

Evans, D. S., and Mulholland, J. D. 1986. *Big and bright: A history of the McDonald Observatory.* University of Texas Press, Austin.

Farley, M. L. 1930. Much early history is found in memories of pioneer Pettit family. *Peru Pointer.*

Fea, A. 1908. Secret chambers and hiding places: Historic, romantic, & legendary stories & traditions about hiding-places, secret chambers, etc. 3rd ed. Project Gutenberg, eBook # 13918, accessed September 23, 2018.

Forty, G. 1976. *Patton's Third Army at war.* Arms and Armour, London.

Fowler, W. A. 1975. *Charles Christian Lauritsen, 1892–1968: A biographical memoir.* National Academy of Sciences, Washington, DC. http://www.nasonline.org/publications/biographical-memoirs/memoir-pdfs/lauritsen-charles.pdf; accessed July 22, 2019.

Fowler, W. A. 1987. Supplemental interview by Carol Bugé. Pasadena, California. October 3, 1986. Oral History Project, California Institute of Technology Archives. http://resolver.caltech.edu/CaltechOH:OH_Fowler_W; accessed July 4, 2019.

Fowler, W. A., and Ajzenberg-Selove, F. 1985. *Thomas Lauritsen, 1915–1973: A biographical memoir.* National Academy of Sciences, Washington, DC. http://www.nasonline.

org/publications/biographical-memoirs/memoir-pdfs/lauritsen-thomas.pdf; accessed July 22, 2019.

Freeland, R., Bilyeu, G. D., Veal, G. R., Steiner, M. D., and Carson D. E. 1998. Large inflatable antenna flight experiment results. *Acta Astronautica* 41: 267–77.

Fuller, R. P. 2004. *Last shots for Patton's Third Army*. NETR Press, Portland, Maine.

Gardiner, J. 2004. *Wartime: Britain 1939–1945*. Headline Book Publishing, London.

Gardner, I. C. 1927. Application of the algebraic aberration equations to optical design. *Scientific Papers for the Bureau of Standards* 22: 3–202. US Government Printing Office.

Gehrels, T. 1988. *On the glassy sea: An astronomer's journey*. American Institute of Physics, New York.

Gerrard-Gough, J. D., and Christman, A. B. 1978. *History of the Naval Weapons Center, China Lake, California*. Vol. 2. *The grand experiment at Inyokern*. Naval History Division, Washington, DC.

Gibson, E. 2020. NSF and postwar US science. *Physics Today* 73: 40–6.

Gimbel, J. 1990. German scientists, United States denazification policy, and the "Paperclip Conspiracy." *International History Review* 12: 441–65.

Goldberg. 1959. Astronomy from satellites and space vehicles. *Journal of Geophysical Research* 64: 1765–8.

Goodstein, J. R. 1991. *Millikan's school: A history of the California Institute of Technology*. W. W. Norton & Company, New York.

Goudsmit, S. A. 1947. *Alsos*. Tomash Publishers, Los Angeles.

Green, C. M., and Lomask, M. 1970. *Vanguard: A history*. National Aeronautics and Space Administration, SP-4202, Washington, DC.

Grotrian, W. 1928. *Graphische Darstellung der Spektren von Atomen und Ionen mit ein, zwei und drei Valenzelektronen: Zweiter Teil. Struktur der Materie in Einzeldarstellungen* 7. Springer-Verlag, Berlin.

Groves, L. M. 1962. *Now it can be told: The story of the Manhattan Project*. Da Capo Press, Boston.

Gum, C. S. 1952. A large H II region at galactic longitude 226°. *The Observatory* 72: 151–4.

Gum, C. S. 1955. A survey of southern H_{II} regions. *Memoirs of the Royal Astronomical Society* 67: 155–77.

Haines, G. K. 1997. Critical to US security: The development of the GAMBIT and HEXAGON satellite reconnaissance systems. https://www.nro.gov/Portals/65/documents/history/csnr/gambhex/Docs/Critical%20to%20US%20Security.pdf.

Haines, G. K. 1998. The National Reconnaissance Office: Its origins, creation, and early years. In D. A. Day, J. M. Logsdon, and B. Latell (eds.), *Eye in the sky: The story of the Corona spy satellites*, pp. 143–56. Smithsonian Institution Press, Washington, DC.

Haining, P. 2002. *The flying bomb war: Contemporary eyewitness accounts of the German V-1 and V-2 raids on Britain*. Robson Books, London.

Halacy, D. S. 1973. *The coming age of solar energy*. Avon Books, New York.

Hale, G. E. 1915. *Ten years' work of a mountain observatory: A brief account of the Mount Wilson Solar Observatory of the Carnegie Institute of Washington*. Reprinted in 2013 by HardPress Publishing, Miami, Florida.

Hales, A. L. 1992. Lloyd Viel Berkner. *Biographical Memoirs of the National Academy of Sciences* 61: 2–24. https://doi.org/10.17226/2037.

Hall, R. C. 1998. Postwar strategic reconnaissance and the genesis of Corona. In D. A. Day, J. M. Logsdon, and B. Latell (eds.), *Eye in the sky: The story of the Corona spy satellites*, pp. 86–118. Smithsonian Institution Press, Washington, DC.

Hall, R. C. 2015. Earth satellites, a first look by the U.S. Navy. https://ntrs.nasa.gov/search. jsp?R=19770026119 2019-02; accessed February 5, 2019.

Harrison, T. 1997. Archibald Philip Bard. *Biographical Memoirs of the National Academy of Sciences* 72: 14–26. National Academy Press, Washington, DC.

Hartmann, P., Jedamzik, R., Reichel, S., and Schreder, B. 2010. Optical glass and glass ceramic historical aspects and recent developments: A Schott view. *Applied Optics* 49: D157–76.

Hebden, J. C., Hege, E. K., and Beckers, J. M. 1986. Use of the coherent MMT for diffraction limited imaging. *Proceedings SPIE* 628. https://doi.org/10.1117/12.963510.

Hege, E. K., Beckers, J. M., Strittmatter, P. A., and McCarthy, D. W. 1985. Multiple mirror telescope as a phased array telescope. *Applied Optics* 24: 2565–76.

Herbig, G. 1974. Obituary: Alfred Harrison Joy. *Quarterly Journal of the Royal Astronomical Society* 15: 526–31.

Herge, H. C. 1996. *Navy V-12.* Turner Publishing, Nashville, Tennessee.

Hirshson, S. P. 2002. *General Patton: A soldier's life.* Harper Perennial, New York.

Hitchcock, W. I. 2018. *The age of Eisenhower: American and world in the 1950s.* Simon & Schuster, New York.

Horine, C. 1998. WWII rocket project at Caltech's Eaton Canyon. Historical files, Y1.15. Caltech Archives.

Howard, J. N. 2006. Early OSA efforts in optics education. *Optics News* 17: 16–19.

Huntington, E. B. 1868. *History of Stamford, Connecticut, from its settlement in 1641, to the present time, including Darien, which was one of its parishes until 1820.* Self-published, Stamford, Connecticut.

Huse, W. W. 1957. The California Institute and rocket production. Historical files, Y1.5. Caltech Archives.

Hyde, R. A. 1999. Eyeglass. 1. Very large aperture diffractive telescopes. *Applied Optics* 38: 4198–212.

Ingalls, A. G. 1996. *Amateur telescope making,* vol. 2, Part E: Schmidt cameras, pp. 421– 517. Willmann-Bell, Richmond, Virginia.

Irwin, J. B. 1952. Optimum locations for a photometric observatory. *Science* 115(2983): 223–6.

Isaacson, W. 2007. *Einstein: His life and universe.* Simon & Schuster, New York.

Jacobs, S. F., Sargent, M., III, and Scully, M. O. (eds.). 1974. *High energy lasers and their applications.* Addison-Wesley Publishing Company, Reading, Massachusetts.

Jacobsen, A. 2014. *Operation Paperclip: The secret intelligence program that brought Nazi scientists to America.* Little, Brown, and Company, New York.

Jardine, H. 1821. *Report relative to the tomb of King Robert the Bruce, and the cathedral church of Dunfermline.* Hay, Gall, and Co., Edinburgh.

Johnson, C. L. 1968. History of the OXCART program. Report SP-1362. Lockheed Aircraft Corporation, Burbank, California. Approved for release July 2007.

Johnson, H. L., and Richards, H. L. 1970. Optimum size of infrared photometric telescopes. *Astrophysical Journal* 160: L111–L16.

Johnson, T. 1973. The sunshine spreaders. *New Scientist* 58: 337–9.

Jones, V. C. 1985. *United States Army in World War II. Special studies: Manhattan: The Army and the atomic bomb.* Center of Military History, United States Army, Washington, DC.

Joy, A. H. 1945. T Tauri variable stars. *Astrophysical Journal* 102: 168–95.

Kappler, D., and Steiner, J. 2009. *Schott 1884–2009: From a glass laboratory to a technology group.* Schott AG, Mainz.

Kármán, T. von, and Tsien, H. S. 1938. Boundary layer in compressible fluids. *Journal of the Aeronautical Sciences* 5: 227–32.

Kiaulehn, W. 1959. *The odyssey of 41 glassmakers*. Jenaer Glaswerk Schott & Genossen, Mainz, Germany.

Killian, J. R., Jr. 1977. *Sputnik scientists and Eisenhower: A memoir of the first Special Assistant to the President for Science and Technology*. MIT Press, Cambridge, Massachusetts.

King, B., and Kutta, T. J. 1998. *Impact: The history of Germany's V-weapons in World War II*. SARPEDON, Rockville Centre, New York.

Kohnen, D. A. 2015. Seizing German naval intelligence from the archives of 1870–1945. *Global War Studies* 12: 133–71.

Komissarov, S. 2002. *Russia's ekranoplans: The Caspian Sea Monster and other WIG craft*. Midland Publishing, Hinckley, UK.

Kron, G. E. 1967. A simple design for liquid filters. *Publications of the Astronomical Society of the Pacific* 79: 76–7.

Kron, G. E., Ables, H. D., and Hewitt, A. V. 1969. A technical description of the construction, function, and application of the US Navy electronic camera. *Advances in Electronics and Electron Physics* 28: 1–17.

Kuiper, G. P., and Middlehurst, B. M. 1960. *Stars and stellar systems*, Vol. 1: *Telescopes*. University of Chicago Press.

Labeyrie, A., Lipson, S. G., and Nisenson, P. 2006. *An introduction to optical stellar interferometry*. Cambridge University Press.

Lallemand, A., Duchesne, M., and Walker, M. F. 1960. The electronic camera, its installation and results obtained with the 120-inch reflector of the Lick Observatory. *Publications of the Astronomical Society of the Pacific* 72: 268–87.

Lasby, C. 1971. *Project Paperclip: German scientists and the Cold War*. New Saucerian Press, New York.

Latimer, J. 2019. Tank killer: 5-inch High Velocity Aircraft Rocket (HVAR) "Holy Moses." *The China Laker* 22: 2–3.

Lewis, J. E. 2002. *Spy capitalism: Itek and the CIA*. Yale University Press, New Haven, Connecticut.

Lewis Publishing Company. 1904. *A biographical and genealogical history of southeastern Nebraska*. Chicago, Illinois.

Li, Y., Zhang, Q, Wang, J., Xu, Q., and Ye, H. 2017. Precision grinding, lapping, polishing and post-processing of optical glass. In *Comprehensive materials finishing*, vol. 1 (ed. S. Hashmi), pp. 154–70. Elsevier Science, Amsterdam.

Loferski, J. J. 1972. Some problems associated with large scale production of electrical power from solar energy via the photovoltaic effect. Paper 72-WA/Sol-4, ASME Winter Annual Meeting, New York, November 1972.

Lovins, A. B. 1976. Energy strategy: The road not taken? *Foreign Affairs* 55: 65–96.

Lund, A. S. 1999. *Historic Pasadena: An illustrated history*. Historical Publishing Network, San Antonio, Texas.

MacDonald, C. B. 2012. *The last offensive: United States Army in World War II, European Theater of Operations*. Whitman Publishing, Atlanta, Georgia.

Manchester, W., and Read, P. 2012. *The last lion: Winston Spencer Churchill, defender of the realm 1940–1965*. Little, Brown, and Co., New York.

Martin, W. T., and Navy Department (eds.). 2013. *Arming the fleet: Providing our warfighters the decisive advantage*. 3rd ed. Department of the Navy, Washington, DC.

May, E. R. 1998. Strategic intelligence and U. S. security. In D. A. Day, J. M. Logsdon, and B. Latell (eds.), *Eye in the sky: The story of the Corona spy satellites*, pp. 21–8. Smithsonian Institution Press, Washington, DC.

May, M. M. (ed.) 1999. The Cox Committee Report: An assessment. https://carnegieen dowment.org/pdf/npp/coxfinal3.pdf; accessed August 18, 2019.

McElheny, V. K. 1998. *Insisting on the impossible: The life of Edwin Land*. Perseus Books, Cambridge, Massachusetts.

McIninch, T. P. 1971. The Oxcart story. *Studies in Intelligence* 15: 1–25. https://www.cia.gov/library/center-for-the-study-of-intelligence/kent-csi/vol15no1/pdf/v15i1a 01p.pdf.

Meinel, A. B. 1944. C.I.T. memorandum to I. S. Bowen: High-altitude spectroscopic sounding rocket. March 20, 1944. Walter S. Adams correspondence, Box 46, Huntington Library and Archives.

Meinel, A. B. 1948. The near infrared spectrum of the airglow and aurora. *Publications of the Astronomical Society of the Pacific* 60: 373.

Meinel, A. B. 1950a. A new band system of N^+_2 in the infrared auroral spectrum. *Astrophysical Journal* 112: 562–3.

Meinel, A. B. 1950b. Hydride emission bands in the spectrum of the night sky. *Astrophysical Journal* 111: 207–8.

Meinel, A. B. 1950c. Identification of the 6560Å emission spectrum of the night sky. *Astrophysical Journal* 111: 433.

Meinel, A. B. 1950d. OH emission bands in the spectrum of the night sky. *Astrophysical Journal* 111: 555–64.

Meinel, A. B. 1950e. On the entry into the Earth's atmosphere of 57-keV protons during auroral activity. *Physical Review* 80: 1096–7.

Meinel, A. B. 1950f. On the entry into the earth's atmosphere of high speed protons during auroral activity. *Science* 112 (2916): 590.

Meinel, A. B. 1953. Aspheric field correctors for large telescopes. *Astrophysical Journal* 118: 335–44.

Meinel, A. B. 1956. An F/2 Cassegrain camera. *Astrophysical Journal* 124: 652–7.

Meinel, A. B. 1957. Extremely innovative design for a f/1.5 minimum mass telescope. KPNO Technical Report #12. Office of the Executive Secretary. August 15, 1957.

Meinel, A. B. 1958. Final report on the site selection survey for the National Astronomical Observatory. AURA, Kitt Peak National Observatory (Reprinted as Kitt Peak National Observatory, Contributions no. 45, October 1963).

Meinel, A. B. 1959a. Astronomical observations from space vehicles. *Publications of the Astronomical Society of the Pacific* 71: 369–80.

Meinel, A. B. 1959b. Astronomical observations from space vehicles. *Publications of the Astronomical Society of the Pacific* 71: 369–80 (Reprinted as Kitt Peak National Observatory, Contributions no. 1, 1959).

Meinel, A. B. 1961a. Astronomical seeing and observatory site selection. In G. P. Kuiper (ed)., *Stars and stellar systems: Telescopes*, pp. 154–75. University of Chicago Press.

Meinel, A. B. 1961b. Design considerations for a large aperture orbital telescope. Report. Les Spectres des Astres dans l'Ultraviolet Lointain; communications presentees au dixieme colloque International d'Astrophysique tenu a Liege les 11, 12, 13 et 14 Juillet 1960. Institute d'Astrophysique cointe-sclessin, Belgique, 49–59.

Meinel, A. B. 1961c. Design considerations for a large aperture orbital telescope: Report. *Mémoires de la Société royale des sciences de Liège*, Sér. 5, tome 4: 49–59 (Reprinted as Kitt Peak National Observatory, Contributions no. 7, 1961).

Meinel, A. B. 1961d. New frontiers of astronomical optics. *Journal of the Optical Society of America* 51: 471.

Meinel, A. B. 1961e. New frontiers of astronomical technology. *Science* 134: 1165–71 (Reprinted as *New frontiers in astronomical instrumentation*. Kitt Peak National Observatory, Contributions no. 11, 1961).

Meinel, A. B. 1962. Infrared astronomy (abstract). *Astronomical Journal* 67: 118.

Meinel, A. B. 1964. Some aspects of graduate study in the optical sciences (abstract). *Journal of the Optical Society of America* 54: 1385.

Meinel, A. B. 1978. An overview of the technological possibilities of future telescopes. In *ESO Conference on Optical Telescopes of the Future* (eds. E. F. Pacini, W. Richter, and R. N. Wilson), pp. 13–26. ESO, Geneva.

Meinel, A. B. 1979. Cost-scaling laws applicable to very large optical telescopes. *Optical Engineering* 18: 645–7.

Meinel, A. B. 1982. Cost relationships for nonconventional telescope structural configurations. *Journal of the Optical Society of America* 72: 14–20.

Meinel, A. B. 1996. Letter to William Patrick McCray, November 9, 1998 ff. Unpublished.

Meinel, A. B. 2007. Galaxy collisions and ordinary dark matter. *Physics Today* 60: 14.

Meinel, A. B. 2009a. Reminiscences: Aden goes to war. *Optics & Photonics News* 20: 16–17.

Meinel, A. B. 2009b. Reminiscences: Aden returns from the war. *Optics & Photonics News* 20: 20–1.

Meinel, A. B. 2014. Personal reflections of the first decade. Appendix 1. James C. Wyant Optical Sciences Center/College of Optical Sciences. 50 years of excellence. *Proceedings SPIE* 9186.

Meinel, A. B., and Abt, H. A. 1963. *Final report on the site selection survey for the National Astronomical Observatory*. Contributions of the Kitt Peak National Observatory no. 45.

Meinel, A. B., and Bashkin, S. 1964. Optical spectra from fast ion beams (abstract). *Astronomical Journal* 69: 552–3.

Meinel, A. B., Bashkin, S., and Loomis, D. A. 1965. Controlled figuring of optical surfaces by energetic ionic beams (abstract). *Applied Optics* 4: 1674.

Meinel, A. B., Eyer, J. A, Noble, R. H., and Slater, P. N. 1970. The Optical Sciences Center: Its history, organization, and relation to government and industry. *Applied Optics* 10: 243–7.

Meinel, A. B., and Hoxie, D. T. 1962. On the spectrum of lightning in the Venus atmosphere. *Publications of the Astronomical Society of the Pacific* 74: 329–30.

Meinel, A. B., and Meinel, B. 2009. Evidence of a magnetic sheath around a jet from NGC 6543. In *IAU Symposium Proceedings Series* (eds. K. G. Strassmeier, A. G. Kosovichev, and J. E. Beckman), pp. 133–4. Cambridge University Press.

Meinel, A. B., and Meinel, M. P. 1971. Is it time for a new look at solar energy? *Bulletin of the Atomic Scientists* 27: 32–7.

Meinel, A. B., and Meinel, M. P. 1972a. A harvest of solar energy. *Optical Sciences Newsletter* 6: 68–75.

Meinel, A. B., and Meinel, M. P. 1972b. Physics looks at solar energy. *Physics Today* 25: 44–50.

Meinel, A. B., and Meinel, M. P. 1976a. *Applied solar energy: An introduction*. Addison-Wesley, Reading, Massachusetts.

Meinel, A. B., and Meinel, M. P. 1976b. Solar photothermal power generation. *Environmental Conservation* 3: 15–21.

Meinel, A. B., and Meinel, M. P. 1977. "Soft" energy paths: Reality and illusion. *Electric Perspectives* 77: 24–7.

Meinel, A. B., and Meinel, M. P. 1980. Aden and Marjorie Meinel's China trip October–November 1979. *Applied Optics* 19: 2666–9.

Meinel, A. B., and Meinel, M. P. 1989. Optical testing of off-axis parabolic segments without auxiliary optical test elements. *Optical Engineering* 28: 71. https://doi.org/10.1117/12.7976904.

Meinel, A. B., and Meinel, M. P. 1992. Two-stage optics: High acuity performance from low acuity optical systems. *Optical Engineering* 31: 2271–81. https://doi.org/10.1117/12.59946.

Meinel, A. B., and Meinel, M. P. 2000a. Inflatable membrane mirrors for optical passband imagery. *Optical Engineering* 39: 541–50.

Meinel, A. B., and Meinel, M. P. 2000b. Spherical primary telescope with aspheric correction at a small internal pupil. *Applied Optics*: 39: 1–4.

Meinel, A. B., and Meinel, M. P. 2002a. *Echoes from a simpler time.* Unpublished.

Meinel, A. B., and Meinel, M. P. 2002b. Large membrane space optics: Imagery and aberrations of diffractive and holographic achromatized optical elements of high diffraction order. *Optical Engineering* 41: 1995–2007.

Meinel, A. B., and Meinel, M. P. 2002c. Parametric dependencies of high-diffraction-order achromatized aplanatic configurations that employ circular or crossed-linear diffractive optical elements. *Applied Optics* 41: 7155–66.

Meinel, A. B., Meinel, M. P., and Breckinridge, J. B. 1994. Deployable space telescopes for planetary and astronomical missions. *Proceedings SPIE* 2214: 250–5, Space Instrumentation and Dual-use Technologies (June 8, 1994). https://doi.org/10.1117/12.177664.

Meinel, A. B., Meinel, M. P., and Jacobs, B. M. 2008. *The golden age of astronomy: In the beginning. . . .* Unpublished. Deposited at Special Collections, University of Arizona Libraries, Tucson, Arizona.

Meinel, A. B., Meinel, M. P., and Meinel, B. 2003. *The solar odyssey: Adventures along the way.* Unpublished.

Meinel, A. B., Meinel, M. P., and Schulte, D. H. 1993. Determination of the Hubble Space Telescope effective conic constant error from direct image measurements. *Applied Optics* 32: 1715–19. https://doi.org/10.1364/AO.32.001715.

Meinel, A. B., Meinel, M. P., Stacy, J. E., Saito, T. T., and Paterson, S. R. 1986. Wave front correctors by diamond turning. *Applied Optics* 25: 824–5.

Meinel, A. B., Meinel, M. P., and Woolf, N. J. 1983. Deployable reflector configurations. *Proceedings SPIE* 0383. Deployable Optical Systems (December 1, 1983). https://doi.org/10.1117/12.934917.

Meinel, A. B., Shannon, R. R., Whipple, F. L., and Low, F. J. 1972. A large multiple mirror telescope (MMT) project. *Optical Engineering* 11: 33–7.

Meinel, M. P. 1984. *Optical Center OSCillations,* no. 318, May 25, 1984.

Meinel, M. S. 2014. Pettit, Edison. In *Biographical encyclopedia of astronomers* (eds. T. Hockey et al.). Springer, New York. https://doi-org.lp.hscl.ufl.edu/10.1007/978-1-4419-9917-7_1079.

Menzel, D. H. 1977. Obituaries: Charles Greeley Abbot. *Quarterly Journal of the Royal Astronomical Society* 18: 136–9. http://articles.adsabs.harvard.edu/cgi-bin/nph-iarticl e_query?1977QJRAS..18..136M&defaultprint=YES&filetype=.pdf.

Merlin, P. W. 2011. *Images of aviation: Area 51.* Arcadia Publishing, Charleston, South Carolina.

Michel, J. 1975. *Dora.* Holt, Rinehart, and Winston, New York.

Michelson, A. A. 1920. On the application of interference methods to astronomical measurements. *Astrophysical Journal* 51: 257–62.

Mieczkowski, Y. 2013. *Eisenhower's Sputnik moment: The race for space and world prestige.* Cornell University Press, Ithaca, New York.

Miller, J. A., Pitman, J. H., and Steele, H. B. 1920. The parallaxes of fifty stars determined at the Sproul Observatory (second list). *Sproul Observatory Publications* (5): 1–65.

Mohler, O., and Dodson-Prince, H. 1978. Robert Raynolds McMath (1891–1962). *Biographical Memoirs of the National Academy of Sciences* 49: 185–202.

Morgan, W. W. 1967. Frank Elmore Ross. *Biographical Memoirs of the National Academy of Sciences* 39: 391–402.

Morgan, W. W., Meinel, A. B., and Johnson, H. M. 1954. Spectral classification with exceedingly low dispersion. *Astrophysical Journal* 120: 506–11.

Morgan, W. W., Strömgren, B., and Johnson, H. 1955. A description of certain galactic nebulosities. *Astrophysical Journal* 121: 611–14.

Morris, R. 1883. *William Morgan: or political anti-Masonry, its rise, growth and decadence.* Robert Macoy, New York.

Mumford, G. S. 2014. McMath, Robert Raynolds. In *Biographical encyclopedia of astronomers* (eds. T. Hockey et al.). Springer, New York.

Murray, F. 2014. *Once upon a time at Area 51.* Self-published.

National Aeronautics and Space Administration. 1990. The Hubble Space Telescope Optical Systems Failure Report. NASA-TM-103443.

National Reconnaissance Office, US Military, and Department of Defense. 2017a. *20th century spy in the sky satellites: Secrets of the National Reconnaissance Office.* Vol. 1. *Gambit photoreconnaissance satellite 1963–1984.* Progressive Management Publications, Washington, DC.

National Reconnaissance Office, US Military, and Department of Defense. 2017b. *20th century spy in the sky satellites: Secrets of the National Reconnaissance Office.* Vol. 2. *Hexagon photoreconnaissance satellite 1971–1986.* Progressive Management Publications, Washington, DC.

National Reconnaissance Office, US Military, and Department of Defense. 2017c. *20th century spy in the sky satellites: Secrets of the National Reconnaissance Office.* Vol. 5. *Leaders, founders, pioneers.* Progressive Management Publications, Washington, DC.

National Reconnaissance Office, US Military, and Department of Defense. 2017d. *20th century spy in the sky satellites: Secrets of the National Reconnaissance Office.* Vol. 6. *Corona: America's first satellite program.* Progressive Management Publications, Washington, DC.

Naval Technical Mission in Europe, Technical Report 282-45, Record Group 38, Accession #72A-5983, National Archives at College Park, Maryland.

Naval Technical Mission in Europe. ETO Ordnance Intelligence Report No 267. Investigation of ordnance installations in vicinity of Nordhausen, Germany. Record Group 165. Identifier 2282808. National Archives at College Park, Maryland.

Naval Technical Mission in Europe. ETO Ordnance Intelligence Report No 270. V-2 assembly plant at Nordhausen, Germany. Record Group 165. Identifier 2282812. National Archives at College Park, Maryland.

Naval Technical Mission in Europe. Letter Report 152-45. Survey of optical mfg. methods at Zeiss Co., Jena, and E. Leitz Co., Wetzlar. Record Group 38. Identifier 4346015. National Archives at College Park, Maryland.

Naval Technical Mission in Europe. Letter Report 155-45. Survey of optical design. Record Group 38. Identifier 4346018. National Archives at College Park, Maryland.

Naval Technical Mission in Europe. Letter Report 41-45. German controlled missiles. Record Group 38. Identifier 4345900. National Archives at College Park, Maryland.

Naval Technical Mission in Europe. Letter Report 81-45. Guided missiles—report interrogation of personnel concerned with. Record Group 38. Identifiers 4345939, 4345940, 4345942. National Archives at College Park, Maryland.

Naval Technical Mission in Europe. Letter Report 86-45. Activities of Zeiss plant in Jena. Record Group 38. Identifier 4345946. National Archives at College Park, Maryland.

Naval Technical Mission in Europe. Letter Report 89-45. Lens formulae, E. Leitz Werke. Record Group 38. Identifier 4345949. National Archives at College Park, Maryland.

Naval Technical Mission in Europe. Letter Report 93-45 (A). Filters & high refractive index Glass. Record Group 38. Identifier 4345943. National Archives at College Park, Maryland.

Naval Technical Mission in Europe. Reports and Office Files. Historical data on U.S. Naval Technical Mission in Europe—First narrative. Record Group 38. Identifier 4345763. National Archives at College Park, Maryland.

Naval Technical Mission in Europe. Technical Report 455-45. Investigation of processing equipment and materials used by the German firms in the manufacture of lenses and prisms for military optical instruments. Record Group 38. Identifier 4346502. National Archives at College Park, Maryland.

Neufeld, M. J. 2007. *Von Braun: Dreamer of space, engineer of war*. Random House, New York.

Nicholson, S. B. 1962. Edison Pettit, 1889–1962. *Publications of the Astronomical Society of the Pacific* 74: 495–98.

Noble, R. H., Statham, R. B., and Meinel, A. B. 1965. Effect of surface conductance on ionic polishing (abstract). *Journal of the Optical Society of America* 55: 1580.

Nock, K. T., Breckinridge, J. B., Buratti, B. J., Lesh, J., Meinel, A. B., Meinel, M. P., Ramirez, M., and Sercel, J. C. 1986. Thousand Astronomical Unit voyage: A deep space mission. *Bulletin of the American Astronomical Society* 18: 1012.

Nordhoff, C. 1873. *California: For health, pleasure, and residence. A book for travellers and settlers*. Harper & Brothers, New York. https://www.loc.gov/item/14022123/.

O'Donnell, P. K. 2014. *Operatives, spies and saboteurs: The unknown story of the men and women of World War II's OSS*. Free Press, New York.

Offner, A. 1963. A null corrector for paraboloidal mirrors. *Applied Optics* 2: 154–5.

Osborn, W. 2012. Frank Elmore Ross and his variable star discoveries. *Journal of the American Association of Variable Star Observers* 40: 133–40.

Osterbrock, D. E. 1997. *Yerkes Observatory, 1892–1950: The birth, near death, and resurrection of a scientific research institution*. University of Chicago Press.

Osterbrock, D. E. 2001. *Walter Baade: A life in astrophysics*. Princeton University Press.

Osterbrock, D. E. 2003. Don Hendrix: Master Mount Wilson and Palomar Observatories optician. *Journal of Astronomical History and Heritage* 6: 1–12.

Outzen, J. D. (ed.). 2015. The DORIAN files revealed: A compendium of the NRO's Manned Orbiting Laboratory documents. National Reconnaissance Office, Center for the Study of National Reconnaissance, Chantilly, Virginia.

Pace, S. 2016. *The projects of Skunk Works*. Voyageur Press. Minneapolis, Minnesota.

Page, H. M. 1964. *Pasadena: Its early years*. Lorrin L. Morrison, Los Angeles, Calfornia.

Parker, D. T. 2013. *Building victory: Aircraft manufacturing in the Los Angeles Area in World War II*. Self-published. Cypress, California. Kindle ed.

Pauling, L. 1958. Arthur Amos Noyes: September 13, 1866–June 3, 1936. *Biographical Memoirs of the National Academy of Sciences*, 30: 320–46. Washington, DC.

Pedlow, G. W., and Welzenbach, D. E. 2016. *The Central Intelligence Agency and overhead reconnaissance: The U-2 and OXCART Programs, 1954–1974*. Skyhorse, New York.

Peebles, C. 1997. *The Corona project: America's first spy satellites*. Naval Institute Press, Annapolis, Maryland.

Penman, M. 2014. *Robert the Bruce: King of the Scots*. Yale University Press, New Haven, Connecticut.

Pettit, E. 1932. Characteristic features of solar prominences. *Astrophysical Journal* 76: 9–43.

Pettit, E., and Steele, H. 1918. The application of Schaeberle's method in the photography of the corona at Matheson, Colorado, June 8. *Popular Astronomy* 26: 466–80.

Pettit, M. S. 1944. *A study of the long-period variable star RT Cygni*. Master's thesis. Claremont Colleges, Claremont, California.

Phillips, C. I., and Pasadena Museum of History. 2008. *Images of America: Early Pasadena*. Arcadia Publishing, Charleston, South Carolina.

Pickering, W. H. 1978. Interview by Mary Terrall. Pasadena, California. November 1978–December 19, 1978. Oral History Project, California Institute of Technology Archives. http://resolver.caltech.edu/Caltech OH: OH_Pickering_1; accessed July 4, 2019.

Pogue, F. C. 1952. Why Eisenhower's forces stopped at the Elbe. *World Politics* 4: 356–68.

Polmar, N., Bessette., J. F., Bryan, H., Carey, A. C., Gorn, M., Graff, C., and Veronico, N. A. 2016. *Spyplanes: The illustrated guide to manned reconnaissance and surveillance aircraft from World War I to today*. Voyageur Press. [Place of publication not given].

Powers, F. G., and Gentry, C. 2004. *Operation Overflight: A memoir of the U-2 incident*. Potomac Books, University of Nebraska Press, Lincoln.

Powers, F. G., Jr., and Dunnavant, K. 2019. *Spy pilot: Francis Gary Powers, the U-2 incident, and a controversial Cold War legacy*. Prometheus Books, Amherst, New York.

Press Reference Library (Western edition). 1915. *Notables of the West. Being the portraits and biographies of the progressive men of the West*. International News Service.

Pressel, P. 2013. *Meeting the challenge: The Hexagon KH-9 reconnaissance satellite* (editor-in-chief N. Allen). American Institute of Aeronautics and Astronautics, Reston, Virginia.

Price, E. W., Horine, C. L., and Snyder, C. W. 1998. Eaton Canyon: A history of rocket motor research and development in the Caltech-NDRC-Navy Rocket Program, 1941–1946. 34th AIAA/ASME/SAE/ASEE Joint Propulsion Conference and Exhibit, July 13–15, 1998, Cleveland, Ohio. American Institute of Aeronautics and Astronautics.

Province, C. M. 1992. *Patton's Third Army: A chronology of the Third Army advance, August, 1944 to May, 1945*. Hippocrene Books, New York.

Ramsey, L. W., and Weedman, D. W. 1984. The Penn State spectroscopic survey telescope. In: *Very large telescopes, their instrumentation and programs. Proceedings of the International Astronomical Union Colloquium* 79: 851–60. https://doi.org/10.1017/S0252921100108917.

Reed, S. G., Van Atta, R. H., and Deitchman, S. J. 1990. *DARPA technical accomplishments: An historical review of selected DARPA projects*. Vol. 1. Institute for Defense Analyses, Alexandria, Virginia.

Rich, B. R., and Janos, L. 1994. *Skunk Works: A personal memoir of my years at Lockheed*. Little, Brown, and Company, New York.

Richelson, J. T. (ed.) 2002a. The U-2, OXCART, and the SR-71: U.S. aerial espionage in the Cold War and beyond. National Security Archive Electronic Briefing Book No. 74. https://nsarchive2.gwu.edu/NSAEBB/NSAEBB74/; accessed September 10, 2019.

Richelson, J. T. 2002b. *The wizards of Langley: Inside the CIA's Directorate of Science and Technology*. Westview Press, Boulder, Colorado.

Richelson, J. T. 2003. Civilians, spies and blue suits: The bureaucratic war for control of overhead reconnaissance 1961–1965. National Security Archive Monographs. https://nsarchive2.gwu.edu/monograph/nro/; accessed September 10, 2019.

Robarge, D. 2012. *Archangel: CIA's supersonic A-12 reconnaissance aircraft*. 2nd ed. Central Intelligence Agency, Washington, DC.

Roberts, G. B. (comp.). 1989. *Ancestors of American presidents*. New England Historic Genealogical Society, Boston.

Romeo, R., Meinel, A. B., Meinel, M. P., and Chen, P. C. 2000. Ultralightweight and hyperthin rollable primary mirror for space telescopes. *Proceedings SPIE* 4013: 634–9. https://doi.org/10.1117/12.393998.

Rudkjøbing, M. 1988. Bengt Georg Daniel Strömgren. *Quarterly Journal of the Royal Astronomical Society* 29: 282–4.

Sambaluk, N. M. 2015. *The other space race: Eisenhower and the quest for aerospace security*. Naval Institute Press, Annapolis, Maryland.

Samuel, W. W. E. 2004. *American raiders: The race to capture the Luftwaffe's secrets*. University of Mississippi Press, Jackson, Mississippi.

Sandage, A. 2012. *Centennial history of the Carnegie Institution of Washington*. Vol. 1. *The Mount Wilson Observatory*. Cambridge University Press.

Sapolsky, H. M. 1990. *Science and the Navy: The history of the Office of Naval Research*. Princeton University Press.

Sargent, M., III, Scully, M. O., and Lamb, W. E., Jr. 1974. *Laser physics*. Addison-Wesley Publishing Co., Reading, Massachusetts.

Scott, R. M. 1966. Van Zandt Williams 1916–1966. *Journal of the Optical Society of America* 56: 1149–51.

Scott, R. M. 1982. *Robert the Bruce: King of Scots*. Carroll & Graf, New York.

Sears, D. W. G. 2019. *Gerard P. Kuiper and the rise of modern planetary science*. University of Arizona Press, Tucson.

Seilhamer, G. O. 1908. *The Bard family: A history and genealogy of the Bards of "Carroll's Delight" together with a chronicle of the Bards and genealogies of the Bard kinship*. Kittochtinny Press, Chambersburg, Pennsylvania.

Sellier, A. 2003. *A history of the Dora camp*. Ivan R. Dee, Chicago.

Shack, R. V. 1974. The use of normalization in the application of simple optical systems. *Proceedings SPIE* 54: 155–62. https://doi.org/10.1117/12.954238.

Shane, C. D. 1964. Lick Observatory: The first 75 years. *Publications of the Astronomical Society of the Pacific* 76: 77–87.

Shaw, B. 1989. Origins of the U-2: Interview with Richard M. Bissell, Jr. *Air Power History* 36: 15–24.

Simmons, M. 2020. Bringing astronomy to an isolated mountaintop. https://www.mtwilson.edu/bringing-astronomy-to-an-isolated-mountaintop/; accessed June 8, 2019.

Simons, G. M. 2016. *Operation Lusty: The race for Hitler's secret technology*. Pen and Sword Books, Barnsley, South Yorkshire, UK.

Skidmore, W., Travouillon, T., and Riddle, R. 2006. Evaluation of sonic anemometers as highly sensitive optical turbulence measuring devices for the Thirty Meter Telescope site testing campaign *Proceedings SPIE* 6267. https://doi.org/10.1117/12.671518.

Smith, F. D. 1997. The design and engineering of Corona's optics. In *Corona between the sun and the earth: The first NRO reconnaissance eye in space* (ed. R. A. McDonald), pp. 111–20. The American Society for Photogrammetry and Remote Sensing, Bethesda, Maryland.

Smith, M. C. 2010. Progress and plans for Chinese surveys. GREAT Plenary, Brussels, June 23, 2010.

Snyder, C. W. 1991. Caltech's *other* rocket project: Personal recollections. *Engineering & Science* 54: 2–13.

Spicer, R. B. 1949. *The desert people: A study of the Papago people.* University of Chicago Press.

SPIE. 2020. Optics and photonics industry report: An in-depth assessment of the global optics and photonics industry, highlighting industry trends and profiling key companies involved. SPIE, Bellingham, Washington.

Spitzer, L., Jr. 1960. Space telescopes and components. *Astrophysical Journal* 65: 242–63.

Spitzer, L., Jr. 1966. *Space research: Directions for the future.* National Academy of Sciences Space Science Board publication no. 1403. https://doi.org/10.17226/12410.

Spitzer, L., Jr. 1990. Report to Project RAND: Astronomical advantages of an extra-terrestrial observatory. *Astronomical Quarterly* 7: 131–42.

Srinivasan, S. R. 1986. Astronomy at Osmania University, India. *Journal of the British Astronomical Association* 96: 339–41.

Stanton, D. 2001. *In harm's way: The sinking of the USS* Indianapolis *and the extraordinary story of its survivors.* Henry Holt and Company, New York.

Stebbins, J. 1911. The discovery of eclipsing variable stars. *Astrophysical Journal* 34: 105–11.

Stebbins, J., and Brown, F. C. 1907. A determination of the moon's light with a selenium photometer. *Astrophysical Journal* 26: 326–40.

Steele, J. D. 1930. *Recollections of my life in England and America. With introduction by his great granddaughter Loraine S. McKinstry.* Self-published.

Stern, S. M. 2005. *The week the world stood still: Inside the secret Cuban Missile Crisis.* Stanford University Press.

Stone, F. D., Jr. 1896. *The descendants of George Steele of Barthomley, Cheshire, England, and Chester County, Pennsylvania.* Self-published, Philadelphia.

Stratton, E. A. 1986. *Plymouth Colony: Its history and people, 1620–1691.* Ancestry Publishing, Salt Lake City, Utah.

Strohbehn, J. W. 1971. III. Optical propagation through the turbulent atmosphere. *Progress in Optics* 9: 73–122.

Strömgren, B. 1932. The opacity of stellar matter and the hydrogen content of the stars. *Zeitschrift für Astrophysik* 4: 118–52.

Strömgren, B. 1938. On the helium and hydrogen content of the interior of the stars. *Astrophysical Journal* 87: 520–34.

Strong, J. 1938. *Procedures in experimental physics.* Prentice-Hall, New York.

Struve, O. 1940. Cooperation in astronomy. *Scientific Monthly* 50: 142–7.

Swanson, P. N., and Null, G. W. 1982. Large Deployable Reflector-Pathfinder Study. JPL Report # D-195 December 1982, NASA Jet Propulsion Laboratory, Pasadena, California.

Taubman, P. 2003. *Secret empire: Eisenhower, the CIA, and the hidden story of America's space espionage*. Simon & Schuster, New York.

Thomas, E. 2012. *Ike's bluff: President Eisenhower's secret battle to save the world*. Back Bay Books, New York.

United States Navy. 1944. *The bluejackets manual*. United States Naval Institute, Annapolis, Maryland.

Van Belle, G. T., Meinel, A. B., and Meinel, M. P. 2004. The scaling relationship between telescope cost and aperture size for very large telescopes. *Proceedings SPIE* 5489, Ground-based telescopes (September 28, 2004). https://doi.org/10.1117/12.552181.

van Biesbroeck, G., and Pettit, H. S. 1920. Parallaxes of fifty-two stars. *Publications of the Yerkes Observatory*. Vol. 4, part 3, 87–122.

Van Wyck, K. L. W. 1936. *Genealogy of Pettit families in America: Descendants of John Pettit 1630–1632 first of that name in America*. Reprinted 2013. Isha Books, New Delhi, India.

Vegard, L. 1930. Die Spektren verfstigten Gase und ihre atomtheoretische Deutung. *Annalen der Physik* 6: 487–544.

Versluis, A. 2006. North American esotericism. In *Introduction to new and alternative religions in America* (eds. E. V. Gallagher and W. M. Ashcraft), vol. 3, pp. 92–148. Greenwood Publishing Group, Westport, Connecticut.

Veysey, V. V. Interview by S. K. Cohen. Pasadena, California, July 14 and 21, 1993, and February 4, 1994. Oral History Project, California Institute of Technology Archives. http://resolver.caltech.edu/CaltechOH:OH_Veysey_V; accessed June 20, 2019.

Vivian, J. L. 1897. *The visitations of Cornwall comprising the heralds' visitations of 1530, 1573, & 1620*. William Pollards, Exeter, England.

Wang, J., Xu, Q., and Ye, H. 2017. Precision grinding, lapping, polishing and post-processing of optical glass. In *Comprehensive materials finishing*, vol. 1 (ed. S. Hashmi), pp. 154–70. Elsevier Science, Amsterdam.

Wang, S.-G., Su, D.-Q., and Chu, Y.-Q. 1996. Special configuration of a very large Schmidt telescope for extensive astronomical spctroscopic observation. *Applied Optics* 35: 5155–64.

Wang, Y., Dai, C., Li, W., Meng, X., Dong, H., and Wang. P. 2016. Polishing an off-axis aspheric mirror by ion beam figuring. *Proceedings SPIE* 9683. https://doi.org/10.1117/12.2242718.

Watson, E. C. 2012. Interview by Larry Shirley. Pasadena, California, January 20, 1969. Oral History Project, California Institute of Technology Archives. http://resolver.caltech.edu/CaltechOH_Watson_E; accessed July 4, 2019.

Weinwig, S. A. 1974. *Comparative grinding rates of optical glasses*. Master's thesis, University of Arizona.

Wheelon, A. D. 1998. Corona: A triumph of American technology. In D. A. Day, J. M. Logsdon, and B. Latell (eds.), *Eye in the sky: The story of the Corona spy satellites*, pp. 29–47. Smithsonian Institution Press, Washington, DC.

Whitaker, E. A. 1985. *The University of Arizona's Lunar and Planetary Laboratory: Its founding and early years*. University of Arizona, Tucson.

Whitford, A. E. 1956. The plan for a new American observatory. *Publications of the Astronomical Society of the Pacific* 68: 115–18.

Yenne, B. 2014. *Area 51 black jets*. Zenith Press, Minneapolis, Minnesota.

Zachary, G. P. 1997. *Endless frontier: Vannevar Bush, engineer of the American century*. The Free Press, New York.

Zimmerman, R. 2010. *The universe in a mirror: The saga of the Hubble Telescope and the visionaries who built it*. Princeton University Press.

Name Index

Subject Index

For the benefit of digital users, indexed terms that span two pages (e.g., 52–53) may, on occasion, appear on only one of those pages.

Tables and figures boxes are indicated by *t* and *f* following the page number